THE OXFORD HANDBOOK OF

INNOVATION

Edited by

JAN FAGERBERG

DAVID C. MOWERY

AND

RICHARD R. NELSON

OXFORD
UNIVERSITY PRESS

OXFORD

UNIVERSITY PRESS

Great Clarendon Street, Oxford OX2 6DP

Oxford University Press is a department of the University of Oxford.
It furthers the University's objective of excellence in research, scholarship,
and education by publishing worldwide in

Oxford New York

Auckland Cape Town Dar es Salaam Hong Kong Karachi
Kuala Lumpur Madrid Melbourne Mexico City Nairobi
New Delhi Shanghai Taipei Toronto

With offices in

Argentina Austria Brazil Chile Czech Republic France Greece
Guatemala Hungary Italy Japan Poland Portugal Singapore
South Korea Switzerland Thailand Turkey Ukraine Vietnam

Oxford is a registered trade mark of Oxford University Press
in the UK and in certain other countries

Published in the United States
by Oxford University Press Inc., New York

British Library Cataloguing in Publication Data
Data available

Library of Congress Cataloging in Publication Data
Data available

Typeset by SPI Publisher Services, Pondicherry, India
Printed in Great Britain
on acid-free paper by
CPI Antony Rowe, Chippenham, Wiltshire

ISBN 978–0–19–926455–1 (Hbk.)
ISBN 978–0–19–928680–5 (Pbk.)

9 10 8

PREFACE

In a famous poem, "The Blind Men and the Elephant," John Godfrey Saxe (1816–87) described what may happen when different observers approach the same phenomenon from rather different starting points. In the poem Saxe lets one of the blind men approach the elephant's side. The man finds it to be "very like a wall." Another fits around its leg and concludes that it resembles a tree. And so on. They end up disputing "loud and long." Saxe drew the following moral:

> So oft in theologic wars,
> The disputants, I ween,
> Rail on in utter ignorance
> Of what each others mean,
> And prate about an Elephant
> Not one of them has seen!

The point is, of course, that each "disputant" has a valid insight, but needs to combine it with the insights of others to reach a holistic understanding. If we substitute "innovation" for the elephant and the "social scientists from different disciplines" for the blind men, we come close to understanding the motives that led to the creation of this handbook. Innovation is a multifaceted phenomenon that cannot be easily squeezed into a particular branch of the social sciences or the humanities. Consequently, the rapidly increasing literature on innovation is characterized by a multitude of perspectives based on—or cutting across—existing disciplines and specializations. There is a danger, however, that scholars studying innovation do it from starting points so different that they become unable to—or not interested in—communicating with each other, preventing the development of a more complete understanding of the phenomenon.

The purpose of this volume is to contribute to a holistic understanding of innovation. The volume includes twenty-one carefully selected and designed contributions, each focusing on a specific aspect of innovation, as well as an introductory essay that sets the stage for the chapters that follow. The authors are leading academic experts on their specific topics, and include economists, geographers, historians, psychologists, and sociologists. Some contributors have engineering degrees in addition to their social science degree. Each chapter can be read separately, but most readers will benefit from reading the introductory essay first. Readers interested in pursuing further study on specific topics will find suggestions for additional reading (marked with asterisks) in the reference list at the end of each chapter.

As with all books there is a history behind it. In fact there are several. There is a long history, related to how innovation studies have evolved over the years. Many of the contributions presented here, Chapter 1 in particular, give elements of that story. The shorter history begins in the mid-1990s with the big impetus to innovation research in Europe provided by the "Framework" programmes of the European Commission. Having participated actively in this research for some time, several of the contributors to this volume became interested in establishing a network that could support discussion and evaluation of its results. For this purpose Jan Fagerberg organized in 1999, with the support of the Norwegian Research Council, an international network for innovation studies that met occasionally to discuss selected topics within innovation research. The meetings of this group led to a proposal for a book reflecting our current knowledge on innovation. Oxford University Press was contacted and welcomed the idea. Economic support from the European Commission and the Norwegian Research Council made it possible for the contributors to meet twice to exchange ideas and comment on each other contributions, greatly enhancing the quality and consistency of the volume.

One of the central participants in the network that led to this volume was Keith Pavitt, Professor at SPRU (University of Sussex) and editor of *Research Policy*, the leading journal in the field. With a background in both engineering and economics, Keith was one of the pioneers in cross-disciplinary research on innovation. Characterized by a "fact-finding" approach and a lack of respect for received "grand theories" not supported by solid evidence, he influenced generations of younger researchers and helped put innovation studies on its current "issue-driven," empirically oriented track. Keith enthusiastically supported this book initiative, very quickly (before anybody else) circulated a full draft of a chapter and participated actively in the discussions during the first workshop in Lisbon in November 2002. He died unexpectedly shortly afterwards. The editors and contributors dedicate this book to his memory.

J.F., D.M., R.N.

Oslo, Berkeley, and New York
January 2004

Acknowledgements

Without financial support from the Norwegian Research Council (projects 131468/510 and 139867/510) and the European Commission (the TEARI project—HPSE-CT-2002-60052) this book would not have been realized. We thank Trygve Lande and Helge Rynning from Norwegian Research Council and Nicholaus Kastrinos from the European Commission for their cooperation. The Centre for Technology, Innovation, and Culture (TIK), University of Oslo gave Jan Fagerberg a leave of absence to start working with this project, which he spent at ISEG, Technical University of Lisbon. He would also like to thank ISEG and the Gulbenkian Foundation for helping to make this possible. Similarly, David Mowery would like to thank the Division of Research of the Harvard Business School for a Bower Fellowship during 2003–4 that aided his work on this volume. Manuel Godinho of ISEG helped organize the first workshop in Lisbon in November 2002, and Bart Verspagen of ECIS (University of Eindhoven) similarly assisted in organizing the second workshop in Roermond in June 2003. In addition to the contributors several people participated in these workshops and contributed to the progress of the work, we would particularly like to mention Fulvio Castellacci, João Caraca, Maureen McKelvey, Sandro Mendonça, Richard Stankiewicz, and Mona Wibe. During the final phases of preparing the manuscript for publication, Mike Hobday, Chris Freeman, Ian Miles, and Susan Lees provided invaluable assistance in editing, proofreading, and preparing Keith Pavitt's chapter for publication. During the final phase Charles McCann provided valuable advice to the non-English/American authors. Ovar Andreas Johansson at TIK was a very efficient and helpful project assistant. At Oxford University Press David Musson and Matthew Derbyshire were inspiring and patient partners.

Contents

PART I INNOVATION IN THE MAKING

PART II THE SYSTEMIC NATURE OF INNOVATION

PART III HOW INNOVATION DIFFERS

PART IV INNOVATION AND PERFORMANCE

List of Figures

LIST OF TABLES

LIST OF BOXES

List of Contributors

Virginia Acha Research Fellow, SPRU, University of Sussex, UK.

Bjørn Asheim Professor, Department of Social and Economic Geography and Centre for Innovation, Research and Competence in the Learning Economy (CIRCLE), University of Lund, Sweden, and Centre for Technology, Innovation and Culture (TIK), University of Oslo, Norway.

Susana Borrás Associate Professor, Department of Social Sciences, Roskilde University, Denmark.

Kristine Bruland Professor, Department of History, University of Oslo, Norway.

John Cantwell Professor, Rutgers University, USA and University of Reading, UK.

Charles Edquist Professor, Division of Innovation, Department of Design, Lund Institute of Technology, Lund University, Sweden and Centre for Innovation, Research and Competence in the Learning Economy (CIRCLE), Lund University, Sweden.

Jan Fagerberg Professor, Centre for Technology, Innovation and Culture (TIK), University of Oslo, Norway.

Meric Gertler Professor, Department of Geography and Munk Centre for International Studies, University of Toronto, Canada, and Centre for Technology, Innovation and Culture (TIK), University of Oslo, Norway.

Manuel M. Godinho Associate Professor, ISEG, Universidade Tecnica de Lisboa, Portugal.

Ove Granstrand Professor, Center for Intellectual Property Studies (CIP), Department of Industrial Management and Economics, School of Technology Management and Economics, Chalmers University of Technology, Sweden.

Stine Grodal Doctoral Candidate in Management Science and Engineering, Stanford University, USA.

Bronwyn Hall Professor, Department of Economics, University of California at Berkeley, USA.

Alice Lam Professor, School of Business and Management, Brunel University, UK.

William Lazonick University Professor, University of Massachusetts Lowell, USA and Distinguished Research Professor, INSEAD, France.

Bengt-Åke Lundvall Professor, Department of Business Studies, Aalborg University, Denmark.

Franco Malerba Professor, CESPRI and Istituto di Economia Politica, Bocconi University, Italy.

Ian Miles Professor, PREST, Institute of Innovation Research, University of Manchester, UK.

David C. Mowery Professor, Haas School of Business, University of California at Berkeley, USA.

Rajneesh Narula Professor, Department of International Economics & Management, Copenhagen Business School, Denmark and Centre for Technology, Innovation and Culture (TIK), University of Oslo, Norway.

Richard R. Nelson Professor, Columbia University, USA.

Mary O'Sullivan Associate Professor, Strategy and Management, INSEAD, France.

Keith Pavitt Professor, SPRU, University of Sussex, UK.

Mario Pianta Professor, Faculty of Economics, University of Urbino, Italy.

Walter W. Powell Professor of Education, Sociology, and Organizational Behavior at Stanford University, USA.

Bhaven N. Sampat Assistant Professor, School of Public Policy, Georgia Institute of Technology, USA.

Keith Smith Professor, Department of Industrial Dynamics, Chalmers University of Technology, Sweden.

Nick von Tunzelmann Professor, SPRU, University of Sussex, UK.

Bart Verspagen Professor, Eindhoven Centre for Innovation Studies (Ecis), Eindhoven University of Technology, the Netherlands, and Centre for Technology, Innovation and Culture (TIK), University of Oslo, Norway.

Antonello Zanfei Professor, Faculty of Economics, University of Urbino, Italy.

CHAPTER 1

INNOVATION

A GUIDE TO THE LITERATURE

JAN FAGERBERG

1.1 INTRODUCTION[1]

INNOVATION is not a new phenomenon. Arguably, it is as old as mankind itself. There seems to be something inherently "human" about the tendency to think about new and better ways of doing things and to try them out in practice. Without it, the world in which we live would look very, very different. Try for a moment to think of a world without airplanes, automobiles, telecommunications, and refrigerators, just to mention a few of the more important innovations from the not-too-distant past. Or—from an even longer perspective—where would we be without such fundamental innovations as agriculture, the wheel, the alphabet, or printing?

In spite of its obvious importance, innovation has not always received the scholarly attention it deserves. For instance, students of long-run economic change used to focus on factors such as capital accumulation or the working of markets, rather than on innovation. This is now changing. Research on the role of innovation in economic and social change has proliferated in recent years, particularly within the social sciences, and with a bent towards cross-disciplinarity. In fact, as illustrated in Figure 1.1, in recent years the number of social-science publications focusing on innovation has increased much faster than the total number of such publications.

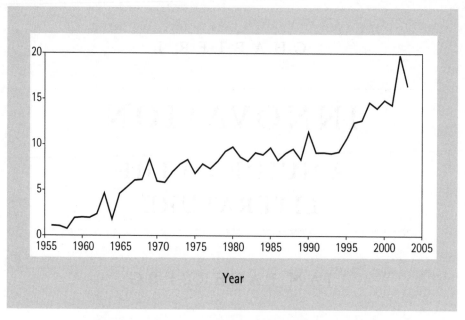

**Fig. 1.1 Scholarly Articles with "Innovation" in the title, 1955–2004
(per 10,000 social science articles)**

Note: The source is the ISI Web of Knowledge, Social Sciences Citation Index (SSCI).

As a result, our knowledge about innovation processes, their determinants and social and economic impact has been greatly enhanced.

When innovation studies started to emerge as a separate field of research in the 1960s, it did so mostly outside the existing disciplines and the most prestigious universities. An important event in this process was the formation in 1965 of the Science Policy Research Unit (SPRU) at the University of Sussex (see Box 1.1). The name of the center illustrates the tendency for innovation studies to develop unde other (at the time more acceptable?) terms, such as, for instance, "science studies" or "science policy studies." But as we shall see in the following, one of the main lessons from the research that came to be carried out is that science is only one among several ingredients in successful innovation. As a consequence of these findings, not only the focus of research in this area but also the notions used to characterize it changed. During the late twentieth/early twenty-first century, a number of new research centers and departments have been founded, focusing on the role of innovation in

---

Box 1.1 SPRU, Freeman, and the spread of innovation studies

SPRU—Science Policy Research Unit—at the University of Sussex, UK was founded in 1965 with Christopher Freeman as its first director. From the beginning, it had a cross-disciplinary research staff consisting of researchers with backgrounds in subjects as diverse as economics, sociology, psychology, and engineering. SPRU developed its own cross-disciplinary Master and Ph.D. programs and carried out externally funded research, much of which came to focus on the role of innovation in economic and social change. It attracted a large number of young scholars from other countries who came to train and work here.

The research initiated at SPRU led to a large number of projects, conferences, and publications. *Research Policy*, which came to be the central academic journal in the field, was established in 1972, with Freeman as the first editor (he was later succeeded by Keith Pavitt, also from SPRU). Freeman's influential book, *The Economics of Industrial Innovation*, was published two years later, in 1974, and has since been revised twice. In 1982, the book, *Unemployment and Technical Innovation*, written by Freeman, Clark, and Soete, appeared, introducing a systems approach to the role of innovation in long-term economic and social change. Freeman later followed this up with an analysis of the national innovation system in Japan (Freeman 1987). He was also instrumental in setting up the large, collaborative IFIAS project which in 1988 resulted in the very influential book, *Technical Change and Economic Theory*, edited by Dosi, Freeman, Nelson, Silverberg, and Soete (both Dosi and Soete were SPRU Ph.D. graduates).

In many ways, SPRU came to serve as a role model for the many centers/institutes within Europe and Asia that were established, mostly from the mid-1980s onwards, combining cross-disciplinary graduate and Ph.D. teaching with extensive externally funded research. Most of these, as SPRU itself, were located in relatively newly formed (so-called "red-brick") universities, which arguably showed a greater receptivity to new social needs, initiatives, and ideas than the more inert, well-established academic "leaders," or at other types of institutions such as business or engineering schools. SPRU graduates were in many cases instrumental in spreading research and teaching on innovation to their own countries, particularly in Europe.

---

economic and social change. Many of these have a cross-disciplinary orientation, illustrating the need for innovation to be studied from different perspectives. Several journals and professional associations have also been founded.

The leaning towards cross-disciplinarity that characterizes much scholarly work in this area reflects the fact that no single discipline deals with all aspects of innovation. Hence, to get a comprehensive overview, it is necessary to combine insights from several disciplines. Traditionally, for instance, economics has dealt primarily with the allocation of resources to innovation (in competition with other ends) and its economic effects, while the innovation process itself has been more or less treated as a "black box." What happens within this "box" has been left to scholars from other disciplines. A lot of what happens obviously has to do with learning, a central topic in cognitive science. Such learning occurs in organized settings (e.g. groups, teams, firms, and networks), the working of which is studied within disciplines such as sociology, organizational science, management, and business studies. Moreover, as economic geographers point out, learning processes tend to be linked to specific contexts or locations. The way innovation is organized and its localization also undergo important changes through time, as underscored by the work within the field of economic history. There is also, as historians of technology have pointed out, a specific technological dimension to this; the way innovation is organized, as well as its economic and social effects, depends critically on the specific nature of the technology in question.

Two decades ago, it was still possible for a hard-working student to get a fairly good overview of the scholarly work on innovation by devoting a few years of intensive study to the subject. Not any more. Today, the literature on innovation is so large and diverse that even keeping up-to-date with one specific field of research is very challenging. The purpose of this volume is to provide the reader with a guide to this rapidly expanding literature. We do this under the following broad headings:

I Innovation in the Making
II The Systemic Nature of Innovation
III How Innovation Differs
IV Innovation and Performance.

Part One focuses on the process through which innovations occur and the actors that take part: individuals, firms, organizations, and networks. As we will discuss in more detail below, innovation is by its very nature a systemic phenomenon, since it results from continuing interaction between different actors and organizations. Part Two outlines the systems perspective on innovation studies and discusses the roles of institutions, organizations, and actors in this process at the national and regional level. Part Three explores the diversity in the manner in which such systems work over time and across different sectors or industries. Finally, Part Four examines the broader social and economic consequences of innovation and the associated policy issues. The remainder of this chapter sets the stage for the discussion that follows by giving a broad overview of some of the central topics in innovation studies (including conceptual issues).

1.2 What is Innovation?

An important distinction is normally made between invention and innovation.[2] Invention is the first occurrence of an idea for a new product or process, while innovation is the first attempt to carry it out into practice. Sometimes, invention and innovation are closely linked, to the extent that it is hard to distinguish one from another (biotechnology for instance). In many cases, however, there is a considerable time lag between the two. In fact, a lag of several decades or more is not uncommon (Rogers 1995). Such lags reflect the different requirements for working out ideas and implementing them. While inventions may be carried out anywhere, for example in universities, innovations occur mostly in firms, though they may also occur in other types of organizations, such as public hospitals. To be able to turn an invention into an innovation, a firm normally needs to combine several different types of knowledge, capabilities, skills, and resources. For instance, the firm may require production knowledge, skills and facilities, market knowledge, a well-functioning distribution system, sufficient financial resources, and so on. It follows that the role of the innovator,[3] i.e. the person or organizational unit responsible for combining the factors necessary (what the innovation theorist Joseph Schumpeter (see Box 1.2) called the "entrepreneur"), may be quite different from that of the inventor. Indeed, history is replete with cases in which the inventor of major technological advances fails to reap the profits from his breakthroughs.

Long lags between invention and innovation may have to do with the fact that, in many cases, some or all of the conditions for commercialization may be lacking. There may not be a sufficient need (yet!) or it may be impossible to produce and/or market because some vital inputs or complementary factors are not (yet!) available. Thus, although Leonardo da Vinci is reported to have had some quite advanced ideas for a flying machine, these were impossible to carry out in practice due to a lack of adequate materials, production skills, and—above all—a power source. In fact, the realization of these ideas had to wait for the invention and subsequent commercialization (and improvement) of the internal combustion engine.[4] Hence, as this example shows, many inventions require complementary inventions and innovations to succeed at the innovation stage.

Another complicating factor is that invention and innovation is a continuous process. For instance, the car, as we know it today, is radically improved compared to the first commercial models, due to the incorporation of a very large number of different inventions/innovations. In fact, the first versions of virtually all significant innovations, from the steam engine to the airplane, were crude, unreliable versions of the devices that eventually diffused widely. Kline and Rosenberg (1986), in an influential paper, point out:

it is a serious mistake to treat an innovation as if it were a well-defined, homogenous thing that could be identified as entering the economy at a precise date—or becoming available at a precise point in time.... The fact is that most important innovations go through drastic changes in their lifetimes—changes that may, and often do, totally transform their economic significance. The subsequent improvements in an invention after its first introduction may be vastly more important, economically, than the initial availability of the invention in its original form. (Kline and Rosenberg 1986: 283)

Thus, what we think of as a single innovation is often the result of a lengthy process involving many interrelated innovations. This is one of the reasons why

Box 1.2 The innovation theorist Joseph Schumpeter

Joseph Schumpeter (1883–1950) was one of the most original social scientists of the twentieth century. He grew up in Vienna around the turn of the century, where he studied law and economics. For most of his life he worked as an academic, but he also tried his luck as politician, serving briefly as finance minister in the first post-World War I (socialist) government, and as a banker (without much success). He became professor at the University of Bonn in 1925 and later at Harvard University in the USA (1932), where he stayed until his death. He published several books and papers in German early on, among these the *Theory of Economic Development*, published in 1911 and in a revised edition in English in 1934. Among his most well-known later works are *Business Cycles* in two volumes (from 1939), *Capitalism, Socialism and Democracy* (1943), and the posthumously published *History of Economic Analysis* (1954).

Very early he developed an original approach, focusing on the role of innovation in economic and social change. It was not sufficient, Schumpeter argued, to study the economy through static lenses, focusing on the distribution of given resources across different ends. Economic development, in his view, had to be seen as a process of qualitative change, driven by innovation, taking place in historical time. As examples of innovation he mentioned new products, new methods of production, new sources of supply, the exploitation of new markets, and new ways to organize business. He defined innovation as "new combinations" of existing resources. This combinatory activity he labeled "the entrepreneurial function" (to be fulfilled by "entrepreneurs"), to which he attached much importance. One main reason for the important role played by entrepreneurs for successful innovation was the prevalence of inertia, or "resistance to new ways" as he phrased it, at all levels of society that entrepreneurs had to fight in order to succeed in their aims. In his early work, which is sometimes called "Schumpeter Mark I," Schumpeter focused mostly on individual entrepreneurs. But in later works he also emphasized the importance of innovation in large firms (so-called "Schumpeter Mark II"), and pointed to historically oriented, qualitative research (case studies) as the way forward for research in this area.

In his analysis of innovation diffusion, Schumpeter emphasized the tendency for innovations to "cluster" in certain industries and time periods (and the derived effects on growth) and the possible contribution of such "clustering" to the formation of business cycles and "long waves" in the world economy (Schumpeter 1939). The latter suggestion has been a constant source of controversy ever since. No less controversial, and perhaps even better known, is his inspired discussion of the institutional changes under capitalism (and its possible endogenous transformation into "socialism") in the book *Capitalism, Socialism and Democracy* (1943).

Sources: Swedberg 1991; Shionoya 1997; Fagerberg 2003.

many students of technology and innovation find it natural to apply a systems perspective rather than to focus exclusively on individual inventions/innovations.

Innovations may also be classified according to "type." Schumpeter (see Box 1.2) distinguished between five different types: new products, new methods of production, new sources of supply, the exploitation of new markets, and new ways to organize business. However, in economics, most of the focus has been on the two first of these. Schmookler (1966), for instance, in his classic work on "Invention and Economic Growth," argued that the distinction between "product technology" and "production technology" was "critical" for our understanding of this phenomenon (ibid. 166). He defined the former type as knowledge about how to *create or improve* products, and the latter as knowledge about how to *produce* them. Similarly, the terms "product innovation" and "process innovation" have been used to character-ize the occurrence of new or improved goods and services, and improvements in the ways to produce these good and services, respectively.[5] The argument for focusing particularly on the distinction between product and process innovation often rests on the assumption that their economic and social impact may differ. For instance, while the introduction of new products is commonly assumed to have a clear, positive effect on growth of income and employment, it has been argued that process innovation, due to its cost-cutting nature, may have a more ambiguous effect (Edquist et al. 2001; Pianta in this volume). However, while clearly distinguishable at the level of the individual firm or industry, such differences tend to become blurred at the level of the overall economy, because the product of one firm (or industry) may end up as being used to produce goods or services in another.[6]

The focus on product and process innovations, while useful for the analysis of some issues, should not lead us ignore other important aspects of innovation. For instance, during the first half of the twentieth century, many of the innovations that made it possible for the United States to "forge ahead" of other capitalist economies were of the organizational kind, involving entirely new ways to organize production and distribution (see Bruland and Mowery in this volume, while Lam provides an overview of organizational innovation). Edquist et al. (2001) have suggested divid-ing the category of process innovation into "technological process innovations" and "organizational process innovations," the former related to new types of machinery, and the latter to new ways to organize work. However, organizational innovations are not limited to new ways to organize the process of production within a given firm. Organizational innovation, in the sense used by Schumpeter,[7] also includes arrangements across firms such as the reorganization of entire industries. Moreover, as exemplified by the case of the USA in the first half of the previous century, many of the most important organizational innovations have occurred in distribution, with great consequences for a whole range of industries (Chandler 1990).

Another approach, also based on Schumpeter's work, has been to classify innov-ations according to how radical they are compared to current technology (Freeman and Soete 1997). From this perspective, continuous improvements of the type referred to above are often characterized as "incremental" or "marginal" innov-ations,[8] as opposed to "radical" innovations (such as the introduction of a totally new type of machinery) or "technological revolutions" (consisting of a cluster of innovations that together may have a very far-reaching impact). Schumpeter focused in particular on the latter two categories, which he believed to be of greater importance. It is a widely held view, however, that the cumulative impact of incremental innovations is just as great (if not greater), and that to ignore these leads to a biased view of long run economic and social change (Lundvall et al. 1992). Moreover, the realization of the economic benefits from "radical" innovations in most cases (including those of the airplane and the automobile, discussed earlier) requires a series of incremental improvements. Arguably, the bulk of economic benefits come from incremental innovations and improvements.

There is also the question of how to take different contexts into account. If A for the first time introduces a particular innovation in one context, while B later introduces the same innovation in another, would we characterize both as innovators? This is a matter of convention. A widely used practice, based on Schumpeter's work, is to reserve the term innovator for A and characterize B as an imitator. But one might argue that, following Schumpeter's own definition, it would be equally consistent to call B an innovator as well, since B is introducing the innovation for the first time in a new context. This is, for instance, the position taken by Hobday (2000) in a discussion of innovation in the so-called "newly industrializing countries" in Asia.[9] One might object, though, that there is a qualitative difference between (*a*) commercializing something for the first time and (*b*) copying it and introducing it in a different context. The latter arguably includes a larger dose of imitative behavior (imitation), or what is sometimes called "technology transfer." This does not exclude the possibility that imitation may lead to new innovation(s). In fact, as pointed out by Kline and Rosenberg (1986, see Box 1.3), many economically significant innovations occur while a product or process is diffusing (see also Hall in this volume). Introducing something in a new context often implies considerable adaptation (and, hence, incremental innovation) and, as history has shown, organizational changes (or innovations) that may significantly increase productivity and competitiveness (see Godinho and Fagerberg in this volume).[10]

Box 1.3 What innovation is not: the linear model

Sometimes it easier to characterize a complex phenomenon by clearly pointing out what it is NOT. Stephen Kline and Nathan Rosenberg did exactly this when they, in an influential paper from 1986, used the concept "the linear model" to characterize a widespread but in their view erroneous interpretation of innovation.

Basically, "the linear model" is based on the assumption that innovation is applied science. It is "linear" because there is a well-defined set of stages that innovations are assumed to go through. Research (science) comes first, then development, and finally production and marketing. Since research comes first, it is easy to think of this as the critical element. Hence, this perspective, which is often associated with Vannevar Bush's programmatic statements on the organization of the US research systems (Bush 1945), is well suited to defend the interests of researchers and scientists and the organizations in which they work.

The problems with this model, Kline and Rosenberg point out, are twofold. First, it generalizes a chain of causation that only holds for a minority of innovations. Although some important innovations stem from scientific breakthroughs, this is not true most of the time. Firms normally innovate because they believe there is a commercial need for it, and they commonly start by reviewing and combining existing knowledge. It is only if this does not work, they argue, that firms consider investing in research (science). In fact, in many settings, the experience of users, not science, is deemed to be the most important source of innovation (von Hippel 1988; Lundvall 1988). Second, "the linear model" ignores the many feedbacks and loops that occur between the different "stages" of the process. Shortcomings and failures that occur at various stages may lead to a reconsideration of earlier steps, and this may eventually lead to totally new innovations.

1.3 Innovation in the Making

Leaving definitions aside, the fundamental question for innovation research is of course to explain how innovations occur. One of the reasons innovation was ignored in mainstream social science for so long was that this was seen as impossible to do. The best one could do, it was commonly assumed, was to look at innovation as a random phenomenon (or "manna from heaven," as some scholars used to phrase it). Schumpeter, in his early works, was one of the first to object to this practice. His own account of these processes emphasized three main aspects. The first was the fundamental uncertainty inherent in all innovation projects; the second was the need to move quickly before somebody else did (and reap the potential economic reward). In practice, Schumpeter argued, these two aspects meant that the standard behavioral rules, e.g., surveying all information, assessing it, and finding the "optimal" choice, would not work. Other, quicker ways had to be found. This in his view involved leadership and vision, two qualities he associated with entrepreneurship. The third aspect of the innovation process was the prevalence of "resistance to new ways"—or inertia—at all levels of society, which threatened to destroy all novel initiatives, and forced entrepreneurs to fight hard to succeed in their projects. Or as he put it: "In the breast of one who wishes to do something new, the forces of habit raise up and bear witness against the embryonic project" (Schumpeter 1934: 86). Such inertia, in Schumpeter's view, was to some extent endogenous, since it reflected the embedded character of existing knowledge and habit, which, though "energy-saving," tended to bias decision-making against new ways of doing things.

Hence, in Schumpeter's early work (sometimes called "Schumpeter Mark I") innovation is the outcome of continuous struggle in historical time between individual *entrepreneurs*, advocating novel solutions to particular problems, and *social inertia*, with the latter seen as (partly) endogenous. This may, to some extent, have been an adequate interpretation of events in Europe around the turn of the nineteenth century. But during the first decades of the twentieth century, it became clear to observers that innovations increasingly involve teamwork and take place within larger organizations (see Bruland and Mowery (Ch. 13), Lam (Ch. 5), and Lazonick (Ch. 2) in this volume). In later work, Schumpeter acknowledged this and emphasized the need for systematic study of "cooperative" entrepreneurship in big firms (so-called "Schumpeter Mark II"). However, he did not analyze the phenomenon in much detail (although he strongly advised others to).[11]

Systematic theoretical and empirical work on innovation-projects in firms (and the management of such projects) was slow to evolve, but during the last decades a quite substantial literature has emerged (see chapters by Pavitt and Lam in this volume). In general, research in this area coincides with Schumpeter's emphasis on uncertainty (Nelson and Winter 1982; Nonaka and Takeuchi 1995; Van de Ven et al. 1999). In particular, for potentially rewarding innovations, it is argued, one may simply not know what are the most relevant sources or the best options to pursue (still less how great the chance is of success).[12] It has also been emphasized that innovative firms need to consider the potential problems that "path dependency" may create (Arthur 1994). For instance, if a firm selects a specific innovation path very early, it may (if it is lucky) enjoy "first mover" advantages. But it also risks being "locked in" to this specific path through various self-reinforcing effects. If in the end it turns out that there actually existed a superior path, which some other firm equipped with more patience (or luck) happened to find, the early mover may be in big trouble because then, it is argued, it may simply be too costly or too late to switch paths. It has been suggested, therefore, that in the early phase of an innovation project, before sufficient knowledge of the alternatives is generated, the best strategy may simply be to avoid being "stuck" to a particular path, and remain open to different (and competing) ideas/solutions. At the level of the firm, this requires a "pluralistic leadership" that allows for a variety of competing perspectives (Van de Ven et al. 1999), in contrast to the homogenous, unitary leader style that, in the management literature, is sometimes considered as the most advantageous.[13]

"Openness" to new ideas and solutions? is considered essential for innovation projects, especially in the early phases. The principal reason for this has to do with a fundamental characteristic of innovation: that every new innovation consists of a new combination of existing ideas, capabilities, skills, resources, etc. It follows logically from this that the greater the variety of these factors within a given system, the greater the scope for them to be combined in different ways, producing new innovations which will be both more complex and more sophisticated. This evolutionary logic has been used to explain why, in ancient times, the inhabitants of the large Eurasian landmass came to be more innovative, and technologically sophisticated, than small, isolated populations elsewhere around the globe (Diamond 1998). Applied mechanically on a population of firms, this logic might perhaps be taken to imply that large firms should be expected to be more innovative than small firms.[14] However, modern firms are not closed systems comparable to isolated populations of ancient times. Firms have learnt, by necessity, to monitor closely each other's steps, and search widely for new ideas, inputs, and sources of inspiration. The more firms on average are able to learn from interacting with external sources, the greater the pressure on others to follow suit. This greatly enhances the innovativeness of both individual firms and the economic systems to which they belong (regions or countries, for instance). Arguably, this is of particular importance for smaller firms, which have to compensate for small internal resources by being good at interacting with the outside world. However, the growing complexity of the knowledge bases necessary for innovation means that even large firms increasingly depend on external sources in their innovative activity (Granstrand, Patel, and Pavitt, 1997; and in this volume: Pavitt; Powell and Grodal; Narula and Zanfei).

Hence, cultivating the capacity for absorbing (outside) knowledge, so-called "absorptive capacity" (Cohen and Levinthal 1990), is a must for innovative firms, large or small. It is, however, something that firms often find very challenging; the "not invented here" syndrome is a well-known feature in firms of all sizes. This arguably reflects the cumulative and embedded character of firm-specific knowledge. In most cases, firms develop their knowledge of how to do things incrementally. Such knowledge, then, consists of "routines" that are reproduced through practice ("organizational memory": Nelson and Winter 1982). Over time, the organizational structure of the firm and its knowledge base typically co-evolve into a set-up that is beneficial for the day-to-day operations of the firm. It has been argued, however, that such a set-up, while facilitating the daily internal communication/interaction of the firm, may in fact constrain the firm's capacity for absorbing new knowledge created elsewhere, especially if the new external knowledge significantly challenges the existing set-up/knowledge of the firm (so-called "competence destroying technical change": Tushman and Anderson 1986). In fact, such problems may occur even for innovations that are created internally. Xerox, for instance, developed both the PC and the mouse, but failed to exploit commercially these innovations, primarily because they did not seem to be of much value to the firm's existing photo-copier business (Rogers 1995).

Thus organizing for innovation is a delicate task. Research in this area has, among other things, pointed to the need for innovative firms to allow groups of people within the organization sufficient freedom in experimenting with new solutions (Van de Ven 1999), and establishing patterns of interaction within the firm that allow it to mobilize its entire knowledge base when confronting new challenges (Nonaka and Takeuchi 1995; Lam, Ch. 5 in this volume). Such organizing does not stop at the gate of the firm, but extends to relations with external partners. Ties to partners with whom communication is frequent are often called "strong ties," while those that are more occasional are denoted as "weak ties" (Granovetter 1973; see Powell and Grodal, Ch. 3 in this volume). Partners linked together with strong ties, either directly, or indirectly via a common partner, may self-organize into (relatively stable) networks. Such networks may be very useful for managing and maintaining openness. But just as firms can display symptoms of path-dependency, the same can happen to established networks, as the participants converge to a common perception of reality (so-called "group-think"). Innovative firms therefore often find it useful to also cultivate so-called "weak ties" in order to maintain a capacity for changing its orientation (should it prove necessary).

1.4 THE SYSTEMIC NATURE OF INNOVATION

As is evident from the preceding discussion, a central finding in the literature is that, in most cases, innovation activities in firms depend heavily on external sources. One recent study sums it up well: "Popular folklore notwithstanding, the innovation journey is a collective achievement that requires key roles from numerous entrepreneurs in both the public and private sectors" (Van de Ven et al. 1999: 149). In that particular study, the term "social system for innovation development" was used to characterize this "collective achievement." However, this is just one among several examples from the last decades of how system concepts are applied to the analysis of the relationship between innovation activities in firms and the wider framework in which these activities are embedded (see Edquist, Ch. 7 in this volume).

One main approach has been to delineate systems on the basis of technological, industrial, or sectoral characteristics (Freeman et al. 1982; Hughes 1983; Carlsson and Stankiewicz 1991; Malerba, Ch. 14 in this volume) but, to a varying degree, to include other relevant factors such as, for instance, institutions (laws, regulations, rules, habits, etc.), the political process, the public research infrastructure (universities, research institutes, support from public sources, etc.), financial institutions, skills (labor force), and so on. To explore the technological dynamics of innovation, its various phases, and how this influences and is influenced by the wider social, institutional, and economic frameworks has been the main focus of this type of analysis. Another important approach in the innovation-systems literature has focused on the spatial level, and used national or regional borders to distinguish between different systems. For example, Lundvall (1992) and Nelson et al. (1993) have used the term "national system of innovation" to characterize the systemic interdependencies within a given country (see Edquist in this volume), while Braczyk et al. (1997) similarly have offered the notion of "regional innovation systems" (see Asheim and Gertler, Ch. 11 in this volume). Since the spatial systems are delineated on the basis of political and administrative borders, such factors naturally tend to play an important role in analyses based on this approach, which has proven to be influential among policy makers in this area, especially in Europe (see Lundvall and Borrás, Ch. 22 in this volume). (Part II of this volume analyzes some of the constituent elements of such systems in more detail.[15])

What are the implications of applying a system perspective to the study of innovation? Systems are—as networks—a set of activities (or actors) that are interlinked, and this leads naturally to a focus on the working of the linkages of the system.[16] Is the potential for communication and interaction through existing linkages sufficiently exploited? Are there potential linkages within the system that might profitably be established? Such questions apply of course to networks as well as systems. However, in the normal usage of the term, a system will typically have more "structure" than a network, and be of a more enduring character. The structure of a system will facilitate certain patterns of interaction and outcomes (and constrain others), and in this sense there is a parallel to the role of "inertia" in firms. A dynamic system also has feedbacks, which may serve to reinforce—or weaken—the existing structure/functioning of the system, leading to "lock in " (a stable configuration), or a change in orientation, or—eventually—the dissolution of the system. Hence, systems may—just as firms—be locked into a specific path of development that supports certain types of activities and constrains others. This may be seen as an advantage, as it pushes the participating firms and other actors in the system in a direction that is deemed to be beneficial. But it may also be a disadvantage, if the configuration of the system leads firms to ignore potentially fruitful avenues of exploration. The character of such processes will be affected by the extent to which the system exchanges impulses with its environment. The more open a system is for impulses from outside, the less the chance of being "locked out" from promising new paths of development that emerge outside the system. It is, therefore, important for "system managers"—such as policy makers—to keep an eye on the openness of the system, to avoid the possibility of innovation activities becoming unduly constrained by self-reinforcing path-dependency.

Another important feature of systems that has come into focus is the strong complementarities that commonly exist between the components of a system. If, in a dynamic system, one critical, complementary component is lacking, or fails to progress or develop, this may block or slow down the growth of the entire system. This is, as pointed out earlier, one of the main reasons why there is often a very considerable time lag between invention and innovation. Economic historians have commonly used concepts such as "reverse salients" and "bottlenecks" to characterize such phenomena (Hughes 1983; Rosenberg 1982). However, such constraints need not be of a purely technical character (such as, for instance, the failure to invent a decent battery, which has severely constrained the diffusion of electric cars for more than century), but may have to do with lack of proper infrastructure, finance, skills, etc. Some of the most important innovations of this century, such as electricity and automobiles (Mowery and Rosenberg 1998), were dependent on very extensive infrastructural investments (wiring and roads/distribution-systems for fuel, respectively). Moreover, to fulfil the potential of the new innovation, such investments often need to be accompanied by radical changes in the organization of production and distribution (and, more generally, attitudes: see Perez 1983, 1985; Freeman and Louçâ 2001). There are important lessons here for firms and policy makers. Firms may need to take into account the wider social and economic implications of an innovation project. The more radical an innovation is, the greater the possibility that it may require extensive infrastructural investments and/or organizational and social change to succeed. If so, the firm needs to think through the way in which it may join up with other agents of change in the private or public sector. Policy makers, for their part, need to consider what different levels of government can do to prevent "bottlenecks" to occur at the system level in areas such as skills, the research infrastructure, and the broader economic infrastructure.

1.5 How Innovation Differs

One of the striking facts about innovation is its variability over time and space. It seems, as Schumpeter (see Box 1.2) pointed out, to "cluster," not only in certain sectors but also in certain areas and time periods. Over time the centers of innovation have shifted from one sector, region, and country to another. For instance, for a long period the worldwide center of innovation was in the UK, and the productivity and income of its population increased relative to its neighboring countries, so that by the mid-nineteenth century its productivity (and income) level was 50 per cent higher than elsewhere; at about the beginning of the twentieth century the center of innovation, at least for the modern chemical and electrical technologies of the day, shifted to Germany; and now, for a long time, the worldwide center of innovation has been in the USA, which during most of the twentieth century enjoyed the highest productivity and living standards in the world. As explained by Bruland and Mowery in this volume, the rise of the US to world technological leadership was associated with the growth of new industries, based on the exploitation of economies of scale and scope (Chandler 1962, 1990) and mass production and distribution.

How is this dynamic to be explained? Schumpeter, extending an earlier line of argument dating back to Karl Marx,[17] held technological competition (competition through innovation) to be the driving force of economic development. If one firm in a given industry or sector successfully introduces an important innovation, the argument goes, it will be amply rewarded by a higher rate of profit. This functions as a signal to other firms (the imitators), which, if entry conditions allow, will "swarm" the industry or sector with the hope of sharing the benefits (with the result that the initial innovator's first mover advantages may be quickly eroded). This "swarming" of imitators implies that the growth of the sector or industry in which the innovation occurs will be quite high for a while. Sooner or later, however, the effects on growth (created by an innovation) will be depleted and growth will slow down.

To this essentially Marxian story Schumpeter added an important modification. Imitators, he argued, are much more likely to succeed in their aims if they improve on the original innovation, i.e., become innovators themselves. This is all the more natural, he continued, because one (important) innovation tends to facilitate (induce) other innovations in the same or related fields. In this way, innovation–diffusion becomes a creative process—in which one important innovation sets the stage for a whole series of subsequent innovations—and not the passive, adaptive process often assumed in much diffusion research (see Hall in this volume). The systemic interdependencies between the initial and induced innovations also imply that innovations (and growth) "tend to concentrate in certain sectors and their surroundings" or "clusters" (Schumpeter 1939: 100–1). Schumpeter, as is well known, looked at this dynamic as a possible explanatory factor behind business cycles of various lengths (Freeman and Louçâ 2001).

This simple scheme has been remarkably successful in inspiring applications in different areas. For instance, there is a large amount of research that has adapted the Marx–Schumpeter model of technological competition to the study of industrial growth, international trade, and competitiveness,[18] although sometimes, it must be said, without acknowledging the source for these ideas. An early and very influential contribution was the so-called "product-life-cycle theory" suggested by Vernon (1966), in which industrial growth following an important product innovation was seen as composed of stages, characterized by changing conditions of and location of production.[19] Basically what was assumed was that the ability to do product innovation mattered most at the early stage, in which there were many different and competing versions of the product on the market. However, with time, the product was assumed to standardize, and this was assumed to be accompanied by a greater emphasis on process innovation, scale economics, and cost-competition. It was argued that these changes in competitive conditions might initiate transfer of the technology from the innovator country (high income) to countries with large markets and/or low costs. Such transfers might also be associated with international capital flows in the form of so-called foreign direct investments (FDIs), and the theory has therefore also become known as a framework for explaining such flows (see Narula and Zanfei in this volume).

The "product-life-cycle theory," attractive as it was in its simplicity, was not always corroborated by subsequent research. While it got some of the general conjectures (borrowed from Schumpeter) right, the rigorous scheme it added, with well-defined stages, standardization, and changing competitive requirements, was shown to fit only a minority of industries (Walker 1979; Cohen 1995). Although good data are hard to come by, what emerges from empirical research is a much more complex picture,[20] with considerable differences across industrial sectors in the way this dynamic is shaped. As exemplified by the taxonomy suggested by Pavitt (see Box 1.4), exploration of such differences ("industrial dynamics") has evolved into one of the main areas of research within innovation studies (see in this volume:

Box 1.4 What is high-tech? Pavitt's taxonomy

The degree of technological sophistication, or innovativeness, of an industry or sector is something that attracts a lot of interest, and there have been several attempts to develop ways of classifying industries or sectors according to such criteria. The most widely used in common parlance is probably the distinction between "high-tech," "medium-tech," and "low-tech," although it is not always clear exactly what is meant by this. Often it is equated with high, medium, and low R&D intensity in production (or value added), either directly (in the industry itself) or including R&D embodied in machinery and other inputs. Based on this, industries such as aerospace, computers, semiconductors, telecommunications, pharmaceuticals, and instruments are commonly classified as "high-tech," while "medium-tech" typically include electrical and non-electrical machinery, transport equipment, and parts of the chemical industries. The remaining, "low-tech," low R&D category, then, comprises industries such as textiles, clothing, leather products, furniture, paper products, food, and so on (Fagerberg 1997; see Smith in this volume for an extended discussion).

However, while organized R&D activity is an important source of innovation in contemporary capitalism, it is not the only one. A focus on R&D alone might lead one to ignore or overlook innovation activities based on other sources, such as skilled personnel (engineers, for instance), learning by doing, using, interacting, and so forth. This led Pavitt (1984) to develop a taxonomy or classification scheme which took these other factors into account. Based a very extensive data-set on innovation in the UK (see Smith in this volume), he identified two ("high-tech") sectors in the economy, both serving the rest of the economy with technology, but very different in terms of how innovations were created. One, which he labeled "science-based," was characterized by a lot of organized R&D and strong links to science, while another—so-called "specialized suppliers" (of machinery, instruments, and so on)—was based on capabilities in engineering, and frequent interaction with users. He also identified a scale-intensive sector (transport equipment, for instance), also relatively innovative, but with fewer repercussions for other sectors. Finally, he found a number of industries that, although not necessarily non-innovative in every respect, received most of their technology from other sectors.

An important result of Pavitt's analysis was the finding that the factors leading to successful innovation differ greatly across industries/ sectors. This obviously called into question technology or innovation polices that only focused on one mechanism, such as, for instance, subsidies to R&D.

Ch. 14 by Malerba; Ch. 15 by VonTunzelmann and Acha; Ch. 16 by Miles). Inspired, to a large extent, by the seminal work by Nelson and Winter (see Box 1.5), research in this area has explored the manner in which industries and sectors differ in terms of their internal dynamics (or "technological regimes": see Malerba and Orsenigo 1997), focusing, in particular, on the differences across sectors in knowledge bases, actors, networks, and institutions (so called "sectoral systems": see Malerba, Ch. 14 in this volume). An important result from this research is that, since the factors that influence innovation differ across industries, policy makers have to take such differences into account when designing policies. The same policy (and policy instruments) will not work equally well everywhere.

Box 1.5 Industrial dynamics—an evolutionary interpretation

The book *An Evolutionary Theory of Economic Change* (1982) by Richard Nelson and Sidney Winter is one of the most important contributions to the study of innovation and long run economic and social change. Nelson and Winter share the Schumpeterian focus on "capitalism as an engine of change." However, building on earlier work by Herbert Simon and others (so-called "procedural" or "bounded" rationality), Nelson and Winter introduce a more elaborate theoretical perspective on how firms behave. In Nelson and Winter's models, firms' actions are guided by routines, which are reproduced through practice, as parts of the firms' "organizational memory." Routines typically differ across firms. For instance, some firms may be more inclined towards innovation, while others may prefer the less demanding (but also less rewarding) imitative route. If a routine leads to an unsatisfactory outcome, a firm may use its resources to search for a new one, which—if it satisfies the criteria set by the firm—will eventually be adopted (so-called "satisficing" behavior).

Hence, instead of following the common practice in much economic theorizing of extrapolating the characteristics of a "representative agent" to an entire population (so-called "typological thinking"), Nelson and Winter take into account the social and economic consequences of interaction within populations of heterogeneous actors (so-called "population thinking"). They also emphasize the role of chance (the stochastic element) in determining the outcome of the interaction. In the book, these outcomes are explored through simulations, which allow the authors to study the consequences of varying the value of key parameters (to reflect different assumptions on technological progress, firm behavior, etc.). They distinguish between an "innovation regime," in which the technological frontier is assumed to progress independently of firms' own activities (the "science based" regime), and another in which technological progress is more endogenous and depends on what the firms themselves do (the "cumulative" regime). They also vary the ease/difficulty of innovation and imitation.

Nelson and Winter's work has been an important source of inspiration for subsequent work on "knowledge-based firms," "technological regimes," and "industrial dynamics," and evolutionary economics more generally, to mention some important topics.

Sources: Nelson and Winter 1982; Andersen 1994; Fagerberg 2003.

1.6 INNOVATION AND ECONOMIC PERFORMANCE

The Marx–Schumpeter model was not intended as a model of industrial dynamics; its primary purpose was to explain long run economic change, what Schumpeter called "development." The core of the argument was (1) that technological competition is the major form of competition under capitalism (and firms not responding to these demands fail), and (2) that innovations, e.g. "new combinations" of existing knowledge and resources, open up possibilities for new business opportunities and future innovations, and in this way set the stage for continuing change. This perspective, while convincing, had little influence on the economics discipline at the time of its publication, perhaps because it did not lend itself easily to formal, mathematical modeling of the type that had become popular in that field. More recently, however, economists (Romer 1990), drawing on new tools for mathematical modeling of economic phenomena, have attempted to introduce some of the above ideas into formal growth models (so-called "new growth theory" or "endogenous growth theory").[21]

In developing this perspective, Schumpeter (1939) was, as noted, particularly concerned with the tendency of innovations to "cluster" in certain contexts, and the resulting structural changes in production, organization, demand, etc. Although these ideas were not well received by the economic community at the time, the big slump in economic activity worldwide during the 1970s led to renewed attention, and several contributions emerged viewing long run economic and social change from this perspective. Both Mensch (1979) and Perez (1983, 1985), to take just two examples, argued that major technological changes, such as, for instance, the ICT revolution today, or electricity a century ago, require extensive organizational and institutional change to run their course. Such change, however, is difficult because of the continuing influence of existing organizational and institutional patterns. They saw this inertia as a major growth-impeding factor in periods of rapid technological change, possibly explaining some of the variation of growth over time (e.g. booms and slumps) in capitalist economies. While the latter proposition remains controversial, the relationship between technological, organizational, and institutional change continues to be an important research issue (Freeman and Louçã 2001), with important implications both for the analysis of the diffusion of new technologies (see Hall in this volume) and the policy discourse (see Lundvall and Borras in this volume).

Although neither Marx nor Schumpeter applied their dynamic perspective to the analysis of cross-national differences in growth performance, from the early 1960s onwards several contributions emerged that explore the potential of this perspective for explaining differences in cross-country growth. In what came to be a very influential contribution, Posner (1961) explained the difference in economic growth between two countries, at different levels of economic and technological development, as resulting from two sources: innovation, which enhanced the difference, and imitation, which tended to reduce it. This set the stage for a long series of contributions, often labeled "technology gap" or "north–south" models (or approaches), focusing on explaining such differences in economic growth across countries at different levels of development (see Fagerberg 1994, 1996 for details). As for the lessons, one of the theoretical contributors in this area summed it up well when he concluded that: "Like Alice and the Red Queen, the developed region has to keep running to stay in the same place" (Krugman 1979: 262).

A weakness of much of this work was that it was based on a very stylized representation of the global distribution of innovation, in which innovation was assumed to be concentrated in the developed world, mainly in the USA. In fact, as argued by Fagerberg and Godinho in this volume, the successful catch-up in technology and income is normally not based only on imitation, but also involves innovation to a significant extent. Arguably, this is also what one should expect from the Schumpeterian perspective, in which innovation is assumed to be a pervasive phenomenon. Fagerberg (1987, 1988) identified three factors affecting differential growth rates across countries: innovation, imitation, and other efforts related to the commercial exploitation of technology. The analysis suggested that superior innovative activity was the prime factor behind the huge difference in performance between Asian and Latin American NIC countries in the 1970s and early 1980s. Fagerberg and Verspagen (2002) likewise found that the continuing rapid growth of the Asian NICs relative to other country groupings in the decade that followed was primarily caused by the rapid growth in the innovative performance of this region. Moreover, it has been shown (Fagerberg 1987; Fagerberg and Verspagen 2002) that, while imitation has become more demanding over time (and hence more difficult and/or costly to undertake), innovation has gradually become a more powerful factor in explaining differences across countries in economic growth.

1.7 What do we Know about Innovation? And what do we Need to Learn more about?

Arguably, we have a good understanding of the role played by innovation in long run economic and social change, and many of its consequences:

- The function of innovation is to introduce novelty (variety) into the economic sphere. Should the stream of novelty (innovation) dry up, the economy will settle into a "stationary state" with little or no growth (Metcalfe 1998). Hence, innovation is crucial for long-term economic growth.
- Innovation tends to cluster in certain industries/sectors, which consequently grow more rapidly, implying structural changes in production and demand and, eventually, organizational and institutional change. The capacity to undertake the latter is important for the ability to create and to benefit from innovation.
- Innovation is a powerful explanatory factor behind differences in performance between firms, regions, and countries. Firms that succeed in innovation prosper, at the expense of their less able competitors. Innovative countries and regions have higher productivity and income than the less innovative ones. Countries or regions that wish to catch up with the innovation leaders face the challenge of increasing their own innovation activity (and "absorptive capacity") towards leader levels (see Godinho and Fagerberg in this volume).

Because of these desirable consequences, policy makers and business leaders alike are concerned with ways in which to foster innovation. Nevertheless, in spite of the large amount of research in this area during the past fifty years, we know much less about why and how innovation occurs than what it leads to. Although it is by now well established that innovation is an organizational phenomenon, most theorizing about innovation has traditionally looked at it from an individualistic perspective, as exemplified by Schumpeter's "psychological" theory of entrepreneurial behavior (Fagerberg 2003). Similarly, most work on cognition and knowledge focuses on individuals, not organizations. An important exception was, of course, Nelson and Winter (1982), whose focus on "organizational memory" and its links to practice paved the way for much subsequent work in this area.[22] But our understanding of how knowledge—and innovation—operates at the organizational level remains fragmentary and further conceptual and applied research is needed.

A central finding in the innovation literature is that a firm does not innovate in isolation, but depends on extensive interaction with its environment. Various concepts have been introduced to enhance our understanding of this phenomenon, most of them including the terms "system" or (somewhat less ambitious) "network." Some of these, such as the concept of a "national system of innovation," have become popular among policy makers, who have been constrained in their ability to act by lack of a sufficiently developed framework for the design and evaluation of policy. Still, it is a long way from pointing to the systemic character of innovation processes (at different levels of analysis), to having an approach that is sufficiently developed to allow for systematic analysis and assessment of policy issues. Arguably, to be really helpful in that regard, these system approaches are in need of substantial elaboration and refinement (see the chapter by Edquist in this volume).

One obstacle to improving our understanding is that innovation has been studied by different communities of researchers with different backgrounds, and the failure of these communities to communicate more effectively with one another has impeded progress in this field. One consequence of these communication difficulties has been a certain degree of "fuzziness" with respect to basic concepts, which can only be improved by bringing these different communities together in a constructive dialogue, and the present volume should be seen as a contribution towards this aim. Different, and to some extent competing, perspectives should not always be seen as a problem: many social phenomena are too complex to be analyzed properly from a single disciplinary perspective. Arguably, innovation is a prime example of this.

Notes

1. I wish to thank my fellow editors and contributors for helpful comments and suggestions. Thanks also to Ovar Andreas Johansson for assistance in the research, Sandro Mendonça for his many creative inputs (which I unfortunately have not have been able to follow to the extent that he deserves), and Louise Earl for good advice. The responsibility for remaining errors and omissions is mine.

2. A consistent use of the terms invention and innovation might be to reserve these for the first time occurrence of the idea/concept and commercialization, respectively. In practice it may not always be so simple. For instance, people may very well conceive the same idea independently of one another. Historically, there are many examples of this; writing, for instance, was clearly invented several times (and in different cultural settings) throughout history (Diamond 1998). Arguably, this phenomenon may have been reduced in importance over time, as communication around the globe has progressed.

3. In the sociological literature on diffusion (i.e. spread of innovations), it is common to characterize any adopter of a new technology, product, or service an innovator. This then leads to a distinction between different types of innovators, depending on how quick they are in adopting the innovation, and a discussion of which factors might possibly explain such differences (Rogers 1995). While this use of the terminology may be a useful one in the chosen context, it clearly differs from the one adopted elsewhere. It might be preferable to use terms such as "imitator" or "adopter" for such cases.

4. Similarly for automobiles: while the idea of a power-driven vehicle had been around for a long time, and several early attempts to commercialize cars driven by steam, electricity, and other sources had been made, it was the incorporation of an internal combustion engine driven by low-cost, easily available petrol that made the product a real hit in the market (Mowery and Rosenberg 1998).

5. A somewhat similar distinction has been suggested by Henderson and Clark (1990). They distinguish between the components (or modules) of a product or service and the way these components are combined, e.g. the product "design" or "architecture." A change only in the former is dubbed "modular innovation," change only in the latter "architectural innovation." They argue that these two types of innovation rely on different types of knowledge (and, hence, create different challenges for the firm).

6. In fact, many economists go so far as to argue that the savings in costs, following a process innovation in a single firm or industry, by necessity will generate additional income and demand in the economy at large, which will "compensate" for any initial negative effects of a process innovation on overall employment. For a rebuttal, see Edquist 2001 and Pianta, Ch. 21 in this volume.

7. Schumpeter 1934: 66.

8. In the sociological literature on innovation, the term "reinvention" is often used to characterize improvements that occur to a product or service, while it is spreading in a population of adopters (Rogers 1995).

9. In the Community Innovation Survey (CIS) firms are asked to qualify novelty with respect to the context (new to the firm, industry or the world at large). See Smith in this volume for more information about these surveys.

10. Kim and Nelson (2000*a*) suggest the term "active imitation" for producers who, by imitating already existing products, modify and improve them.

11. For instance, in one of his last papers, he pointed out: "To let the murder out and start my final thesis, what is really required is a large collection of industrial and locational monographs all drawn up according to the same plan and giving proper attention on the one hand to the incessant historical change in production and consumption functions and on the other hand to the quality and behaviour of leading personnel" (Schumpeter 1949/1989: 328).

12. Even in cases where the project ultimately is successful in aims, entrepreneurs face the challenge of convincing the leadership of the firm to launch it commercially (which may be much more costly than developing it). This may fail if the leadership of the firm has doubts about its commercial viability. It may be very difficult for management to foresee the economic potential of a project, even if it is "technically" successful. Remember, for instance, IBM director Thomas Watson's dictum in 1948 that "there is a world market for about five computers" (Tidd et al. 1997: 60)!

13. "A unified homogenous leadership structure is effective for routine trial-and-error learning by making convergent, incremental improvements in relatively stable and unambiguous situations. However, this kind of learning is a conservative process that maintains and converges organizational routines and relationships towards the existing strategic vision ... although such learning is viewed as wisdom in stable environments, it produces inflexibility and competence traps in changing worlds" (Van de Ven et al. 1999: 117).

14. It would also imply that large countries should be expected to be more innovative than smaller ones, consistent with, for instance, the prediction of so-called "new growth" theory (Romer 1990). See Verspagen in this volume.

15. See, in particular, Ch. 10 by Granstrand (intellectual property rights), Ch. 8 by Mowery and Sampat (universities and public research infrastructure), and Ch. 9 by O'Sullivan (finance).

16. This is essentially what was suggested by Porter (1990).

17. See Fagerberg 2002, 2003 for a discussion of this "Marx–Schumpeter" model.

18. See Fagerberg (1996), Wakelin (1997), and Cantwell, Ch. 20 in this volume for overviews of some of this literature.

19. For a more recent analysis in this spirit, with a lot of empirical case-studies, see Utterback (1994).

20. Available econometric evidence suggests that innovation, measured in various ways (see Smith in this volume), matters in many industries, not only those which could be classified as being in the early stage of the product-cycle (Soete 1987; Fagerberg 1995).
21. For an overview, see Aghion and Howitt (1998). See also the discussion in Fagerberg (2002, 2003), and Ch. 18 by Verspagen in this volume.
22. For a discussion of the role of different types of knowledge in economics, including the organizational dimension, see Cowan et al. (2000) and Ancori et al. (2000).

References

AGHION, P., and HOWITT, P. (1998), *Endogenous Growth Theory*, Cambridge, Mass.: MIT Press.

ANCORI, B., BURETH, A., and COHENDET, P. (2000), "The Economics of Knowledge: The Debate about Codification and Tacit Knowledge," *Industrial Dynamics and Corporate Change* 9: 255–87.

ANDERSEN, E. S. (1994), *Evolutionary Economics, Post-Schumpeterian Contributions*, London: Pinter.

ARTHUR, W. B. (1994), *Increasing Returns and Path Dependency in the Economy*, Ann Arbor: University of Michigan Press.

BRACZYK, H. J. et al. (1998), *Regional Innovation Systems*, London: UCL Press.

BUSH, V. (1945), *Science: The Endless Frontier*. Washington: US Government Printing Office.

CARLSSON, B., and STANKIEWICZ, R. (1991), "On the Nature, Function and Composition of Technological Systems," *Journal of Evolutionary Economics* 1: 93–118.

CHANDLER, A. D. (1962), *Strategy and Structure: Chapters in the History of the American Industrial Enterprise*, Cambridge, Mass.: MIT Press.

—— (1990) *Scale and Scope: The Dynamics of Industrial Capitalism*, Cambridge, Mass.: Harvard University Press.

COHEN, W. (1995), "Empirical Studies of Innovative Activity," in P. Stoneman (ed.), *Handbook of the Economics of Innovation and Technological Change*, Oxford: Blackwell, 182–264.

*—— and LEVINTHAL, D. (1990), "Absorptive Capacity: A New Perspective on Learning and Innovation," *Administrative Science Quarterly* 35: 123–33.

COWAN, R., DAVID, P. A., and FORAY, D. (2000), "The Explicit Economics of Knowledge Codification and Tacitness," *Industrial Dynamics and Corporate Change* 9: 211–53.

DIAMOND, J. (1998), *Guns, Germs and Steel: A Short History of Everybody for the Last 13000 Years*, London: Vintage.

DOSI, G. (1988), "Sources, Procedures and Microeconomic Effects of Innovation," *Journal of Economic Literature* 26: 1120–71.

—— FREEMAN, C., NELSON, R., SILVERBERG, G., and SOETE, L. G. (eds.) (1988), *Technical Change and Economic Theory*, London: Pinter.

EDQUIST, C., HOMMEN, L., and MCKELVEY, M. (2001), *Innovation and Employment: Process versus Product Innovation*, Cheltenham: Elgar.

* Asterisked items are suggestions for further reading.

FAGERBERG, J. (1987), "A Technology Gap Approach to Why Growth Rates Differ," *Research Policy* 16: 87–99, repr. as ch. 1 in Fagerberg (2002).

—— (1988), "Why Growth Rates Differ," in Dosi et al. 1988: 432–57.

—— (1994), "Technology and International Differences in Growth Rates," *Journal of Economic Literature* 32(3): 1147–75.

—— (1995), "Is There a Large-Country Advantage in High-Tech?," NUPI Working Paper No. 526, Norwegian Institute of International Affairs, Oslo, repr. as ch. 14 in Fagerberg (2002).

—— (1996), "Technology and Competitiveness," *Oxford Review of Economic Policy* 12: 39–51, repr. as ch. 16 in Fagerberg (2002).

—— (1997), "Competitiveness, Scale and R&D," in J. Fagerberg et al., *Technology and International Trade*, Cheltenham: Edward Elgar, 38–55, repr. as ch. 15 in Fagerberg (2002).

—— (2000), "Vision and Fact: A Critical Essay on the Growth Literature," in J. Madrick (ed.), *Unconventional Wisdom: Alternative Perspectives on the New Economy*, New York: The Century Foundation, 299–320, repr. as ch. 6 in Fagerberg (2002).

*—— (2002), *Technology, Growth and Competitiveness: Selected Essays*, Cheltenham: Edward Elgar.

—— (2003), "Schumpeter and the Revival of Evolutionary Economics: An appraisal of the Literature," *Journal of Evolutionary Economics* 13: 125–59.

—— and VERSPAGEN, B. (2002), "Technology-Gaps, Innovation-Diffusion and Transformation: An Evolutionary Interpretation," *Research Policy* 31: 1291–304.

FREEMAN, C. (1987), *Technology Policy and Economic Performance: Lessons from Japan*, London: Pinter.

—— CLARK, J., and SOETE, L. G. (1982), *Unemployment and Technical Innovation: A Study of Long Waves and Economic Development*, London: Pinter.

*—— and SOETE, L. (1997), *The Economics of Industrial Innovation*, 3rd edn. London: Pinter.

—— and LOUÇÃ, F. (2001), *As Time Goes By: From the Industrial Revolutions to the Information Revolution*, Oxford: Oxford University Press.

GRANOVETTER, M. (1973), "The Strength of Weak Ties," *American Journal of Sociology* 78: 1360–80.

GRANSTRAND, O., PATEL, P., and PAVITT, K. (1997), "Multi-technology Corporations: Why They Have 'Distributed' rather than 'Distinctive Core' Competencies," *California Management Review* 39: 8–25.

HENDERSON, R. M., and CLARK, R. B. (1990). "Architectural Innovation: The Reconfiguration of Existing Product Technologies and the Failure of Established Firms," *Administrative Science Quarterly* 29: 26–42.

HOBDAY, M. (2000), "East versus Southeast Asian Innovation Systems: Comparing OEM- and TNC-led Growth in Electronics," in Kim and Nelson 2000b: 129–69.

HUGHES, T. P. (1983), *Networks of Power, Electrification in Western Society 1880–1930*, Baltimore: The Johns Hopkins University Press.

KIM, L., and NELSON, R. R. (2000a) "Introduction," in Kim and Nelson 2000b: 13–68.

—— —— (2000b), *Technology, Learning and Innovation: Experiences of Newly Industrializing Economies*, Cambridge: Cambridge University Press.

*KLINE, S. J., and ROSENBERG, N. (1986), "An Overview of Innovation," in R. Landau and N. Rosenberg (eds.), *The Positive Sum Strategy: Harnessing Technology for Economic Growth*, Washington, DC: National Academy Press, 275–304.

KRUGMAN, P. (1979), "A Model of Innovation, Technology Transfer and the World Distribution of Income," *Journal of Political Economy* 87: 253–66.

LUNDVALL, B. Å. (1988), "Innovation as an Interactive Process: From User–Producer Interaction to the National System of Innovation," in Dosi et al. 1988: 349–69.

——(ed.) (1992), *National Systems of Innovation: Towards a Theory of Innovation and Interactive Learning*, London: Pinter.

MALERBA, F., and ORSENIGO, L. (1997), "Technological Regimes and Sectoral Patterns of Innovative Activities," *Industrial and Corporate Change* 6: 83–117.

—— NELSON, R. R., ORSENIGO, L., and WINTER, S. G. (1999), "'History-friendly' Models of Industry Evolution: The Computer Industry," *Industrial Dynamics and Corporate Change* 8: 1–36.

MENSCH, G. (1979), *Stalemate in Technology*, Cambridge, Mass.: Ballinger Publishing Company.

METCALFE, J. S. (1998), *Evolutionary Economics and Creative Destruction*, London: Routledge.

*MOWERY, D., and ROSENBERG, N. (1998), *Paths of Innovation, Technological Change in 20th-Century America*, Cambridge: Cambridge University Press.

NELSON, R. R. (ed.) (1993), *National Systems of Innovation: A Comparative Study*, Oxford: Oxford University Press.

—— and WINTER, S. G. (1982), *An Evolutionary Theory of Economic Change*, Cambridge, Mass.: Harvard University Press.

*NONAKA, I., and TAKEUCHI, H. (1995), *The Knowledge Creating Company*, Oxford: Oxford University Press.

*PAVITT, K. (1984), "Patterns of Technical Change: Towards a Taxonomy and a Theory," *Research Policy* 13: 343–74.

PEREZ, C. (1983), "Structural Change and the Assimilation of New Technologies in the Economic and Social System," *Futures* 15: 357–75.

——(1985), "Micro-electronics, Long Waves and World Structural Change," *World Development* 13: 441–63.

PORTER, M. E. (1990), "The Competitive Advantage of Nations," *Harvard Business Review* 68: 73–93.

POSNER, M. V. (1961), "International Trade and Technical Change," *Oxford Economic Papers* 13: 323–41.

*ROGERS, E. (1995), *Diffusion of Innovations*, 4th edn., New York: The Free Press.

ROMER, P. M. (1990), "Endogenous Technological Change," *Journal of Political Economy* 98: S71–S102.

ROSENBERG, N. (1976), *Perspectives on Technology*, New York: Cambridge University Press.

——(1982), *Inside the Black Box: Technology and Economics*, New York: Cambridge University Press.

SCHMOOKLER, J. (1966), *Invention and Economic Growth*, Cambridge, Mass.: Harvard University Press.

SCHUMPETER, J. (1934), *The Theory of Economic Development*, Cambridge, Mass.: Harvard University Press.

——(1939), *Business Cycles: A Theoretical, Historical, and Statistical Analysis of the Capitalist Process*, 2 vols., New York: McGraw-Hill.

*——(1943), *Capitalism, Socialism and Democracy*, New York: Harper.

——(1949), "Economic Theory and Entrepreneurial History," *Change and the Entrepreneur*, 63–84, repr. in J. Schumpeter (1989), *Essays on Entrepreneurs, Innovations, Business Cycles and the Evolution of Capitalism*, ed. Richard V. Clemence, New Brunswick, NJ: Transaction Publishers, 253–61.

SCHUMPETER, R. (1954), *History of Economic Analysis*, New York: Allen & Unwin.

SHIONOYA, Y. (1997), *Schumpeter and the Idea of Social Science*, Cambridge: Cambridge University Press.

SOETE, L. (1987), "The Impact of Technological Innovation on International Trade Patterns: The Evidence Reconsidered," *Research Policy* 16: 101–30.

SWEDBERG, R. (1991), *Joseph Schumpeter: His Life and Work*, Cambridge: Polity Press.

TIDD, J., BESSANT, J., and PAVITT, K. (1997), *Managing Innovation: Integrating Technological, Market and Organizational Change*, Chichester: John Wiley & Sons.

TUSHMAN, M. L., and ANDERSON, P. (1986). "Technological Discontinuities and Organizational Environments," *Administrative Science Quarterly* 31(3): 439–65.

UTTERBACK, J. M. (1994), *Mastering the Dynamics of Innovation*, Boston: Harvard Business School Press.

VAN DE VEN, A., POLLEY, D. E., GARUD, R., and VENKATARAMAN, S. (1999), *The Innovation Journey*, New York: Oxford University Press.

VERNON, R. (1966), "International Investment and International Trade in the Product Cycle," *Quarterly Journal of Economics* 80: 190–207.

VON HIPPEL, E. (1988), *The Sources of Innovation*, New York: Oxford University Press.

WAKELIN, K. (1997), *Trade and Innovation: Theory and Evidence*, Cheltenham: Edward Elgar.

WALKER, W. B. (1979), *Industrial Innovation and International Trading Performance*, Greenwich: JAI Press.

PART I

INNOVATION IN THE MAKING

INTRODUCTION TO PART I

MOST innovations occur in firms or other types of organizations. The contributions in this section survey our current knowledge on the organizational structure and context of the process of innovation. Chapter 2, by Lazonick, provides a historical perspective on the development of innovative firms, from the small and medium-sized firms of the First Industrial Revolution through the multi-divisional diversified industrial firms of the US and Japan in the twentieth century to the current debate on the "New Economy" and network-based business models. Powell and Grodal deal more extensively with the role of networks in innovation in the subsequent chapter. Chapter 4, by Pavitt, discusses innovation processes within firms, and uses an extensive survey of the relevant literature to provide an analytical perspective on the factors affecting the performance and management of innovation within the large firm. A complementary chapter by Lam (Chapter 5) focuses on firms' experiences with organizational innovation. Finally, Chapter 6 by Smith deals with an indispensable prerequisite for the study of innovation, the measurement of innovation-related activities, particularly in firms.

CHAPTER 2

THE INNOVATIVE FIRM

WILLIAM LAZONICK

2.1 INTRODUCTION

WHAT makes a firm innovative? How have the characteristics of innovative firms changed over time? To address these questions, one requires a conceptual framework for analyzing how a firm transforms productive resources into goods and services that customers want at prices they can afford. To make this productive transformation, a firm must engage in three generic activities: strategizing, financing, and organizing. The types of strategy, finance, and organization that support the innovation process change over time and can vary markedly across industrial activities and institutional environments at any point in time. The innovative firm must, therefore, be analyzed in comparative–historical perspective. This chapter presents and illustrates a framework for analyzing the "social conditions of innovative enterprise" in the comparative–historical experiences of the advanced economies.

Section 2.2 builds upon prominent theories of the innovative firm to derive the "social conditions of innovative enterprise" framework. Section 2.3 focuses on the regional agglomerations of capabilities, now known as "Marshallian industrial districts," that, by the late nineteenth century, had enabled Britain to emerge as the world's first industrial nation. Section 2.4 provides a perspective on the emergence and growth of the US managerial corporation that propelled the US economy to international industrial leadership during the first half of the twentieth century.[1]

Over the past few decades, the greatest challenges to the US managerial corporation have come from Japan. Section 2.5 identifies the social conditions of innovative enterprise that have characterized the Japanese model, while Section 2.6 outlines the distinctive characteristics of the US New Economy firm that has gained competitive advantage in a number of critical product markets in the information and communication technology (ICT) industries. Section 2.7 draws some general conclusions from this essay's comparative–historical perspective concerning strategy, finance, and organization in the innovative firm, and the methodology for studying these phenomena.

2.2 SOCIAL CONDITIONS OF INNOVATIVE ENTERPRISE

Firms strategize when they choose the product markets in which they want to compete and the technologies with which they hope to be competitive. Firms finance when they make investments to transform technologies and access markets that can only be expected to generate revenues sometime in the future. Firms organize when they combine resources in the attempt to transform them into saleable products. To strategize, finance, and organize is not necessarily to innovate. By definition, innovation requires learning about how to transform technologies and access markets in ways that generate higher quality, lower cost products. Learning is a social activity that renders the innovation process uncertain, cumulative, and collective (O'Sullivan 2000*b*). The innovation process is uncertain because, by definition, what needs to be learned about transforming technologies and accessing markets can only become known through the process itself. By investing in learning, an innovative strategy confronts the uncertain character of the innovation process. The innovation process is cumulative when learning cannot be done all at once; what is learned today provides a foundation for what can be learned tomorrow. Investments in cumulative learning, therefore, require sustained, committed finance. The innovation process is collective when learning cannot be done alone; learning requires the collaboration of different people with different capabilities. Investments in collective learning, therefore, require the integration of the work of these people into an organization.

   What is the theory of the firm that can comprehend how strategizing, financing, and organizing can support the innovation process? Over the past century, the theoretical efforts of economists have focused mainly on the optimizing firm rather than the innovating firm. The optimizing firm takes as given technological capabilities and market prices (for inputs as well as outputs), and seeks to maximize profits on the basis of these technological and market constraints. In sharp contrast, in the attempt to generate higher quality, lower cost products than had previously been available, and thus differentiate itself from competitors in its industry, the innovating firm seeks to transform the technological and market conditions that the optimizing firm takes as "given" constraints. Hence, rather than constrained optimization, the innovating firm engages in what I call "historical transformation," a mode of resource allocation that requires a theoretical perspective on the processes of industrial and organizational change (Lazonick 2002*a*).

The distinction between the innovating and optimizing firm is implicit in the work of Alfred Marshall, whose *Principles of Economics*, published in eight editions between 1890 and 1920, placed the theory of the firm at the center of economic analysis. Although Marshall's followers used his arguments to construct the theory of the optimizing firm that remains entrenched in economics textbooks, Marshall (1961: 315) himself displayed considerable insight into the dynamics of the innovating firm, as revealed in the following passage:

An able man, assisted by some strokes of good fortune, gets a firm footing in the trade, he works hard and lives sparely, his own capital grows fast, and the credit that enables him to borrow more capital grows still faster; he collects around him subordinates of more than ordinary zeal and ability; as his business increases they rise with him, they trust him and he trusts them, each of them devotes himself with energy to just that work for which he is specially fitted, so that no high ability is wasted on easy work, and no difficult work is entrusted to unskillful hands. Corresponding to this steadily increasing economy of skill, the growth of his firm brings with it similar economies of specialized machines and plants of all kinds; every improved process is quickly adopted and made the basis of further improvements; success brings credit and credit brings success; success and credit help to retain old customers and to bring new ones; the increase of his trade gives him great advantages in buying; his goods advertise one another and thus diminish his difficulty in finding a vent for them. The increase of the scale of his business increases rapidly the advantages which he has over his competitors, and lowers the price at which he can afford to sell.

What then constrains the growth of such a firm? In *Industry and Trade*, published in 1919, Alfred Marshall acknowledged that over the previous decades the large-scale enterprise had become dominant in advanced nations such as the United States and Germany. He invoked, however, the aphorism, "shirtsleeves to shirtsleeves in three generations" (Marshall 1961: 621) to explain the limit to the growth of the firm that would prevent a small number of large firms from dominating an industry. An owner-entrepreneur of exceptional ability would found and build a successful firm. In the second generation, control would pass to descendants who could not be expected to have the capabilities or drive of the founder, and as a result the firm would grow more slowly or even stagnate. The third generation would lose touch with the innovative legacy of the first generation, and the firm would wither away in the face of new entrepreneurial competition.

Writing in the first decades of the twentieth century, Joseph Schumpeter (1934) also focused on the innovative entrepreneur who, by creating "new combinations" of productive resources, could disrupt the "circular flow of economic life as conditioned by given circumstances." In effect, Schumpeter was arguing that, through entrepreneurship, which he called the "fundamental phenomenon of economic development," innovating firms could challenge optimizing firms, and thereby drive the development of the economy. In 1911, when he first published *The Theory of Economic Development* (in German), Schumpeter, like Marshall, viewed the innovative firm as the result of the entrepreneurial work of an extraordinary individual. Over the subsequent decades, however, as Schumpeter observed the actual development of the leading economies, he came to see the large corporation as the innovating firm, engaged in what he called a process of "creative destruction"; the creation of new modes of productive transformation destroyed existing modes that had themselves been the result of innovative enterprise in the past.

In *Capitalism, Socialism, and Democracy*, first published in 1942, Schumpeter (1950: 118, 132) argued that "technological 'progress' tends, through systemization and rationalization of research and management, to become more effective and sure-footed" as it is undertaken as "the business of teams of trained specialists who turn out what is required and make it work in predictable ways." In a series of major works, Alfred Chandler (1962, 1977, 1990) documented the rise of the managerial corporation in the United States from the last decades of the nineteenth century, the evolution of its multidivisional structure from the 1920s, and the emergence of managerial enterprise in Britain and Germany. In *The Theory of the Growth of the Firm*, first published in 1959, Edith Penrose (1995) conceptualized the modern corporate enterprise as an organization that administers a collection of human and physical resources. People contribute labor services to the firm, not merely as individuals, but as members of teams who engage in learning about how to make best use of the firm's productive resources—including their own.

At any point in time, this learning endows the firm with experience that gives it productive opportunities unavailable to other firms, even in the same industry, that have not accumulated the same experience. The accumulation of innovative experience enables the firm to overcome the "managerial limit" that in the theory of the optimizing firm causes the onset of increasing costs and constrains the growth of the firm (Penrose 1995: chs. 5, 7, and 8). The innovating firm can transfer and reshape its existing productive resources to take advantage of new market opportunities. Each move into a new product market enables the firm to utilize unused productive services accumulated through the process of organizational learning. These unused productive services can provide a foundation for the growth of the firm, through both in-house complementary investments in new product development and the acquisition of other firms that have already developed complementary productive services.

From the 1980s many business school academics, working in the strategy area, cited Penrose's 1959 book as an intellectual foundation for a "resource-based" view of the firm. Resource-based theory focused on the characteristics of valuable resources that one firm possessed and that competitor firms found it difficult to imitate. Resource-based theory, however, provided no perspective on why and how some firms rather than others accumulated valuable and inimitable resources, or indeed what made these resources valuable and inimitable (see Lazonick 2002a). Independently of the resource-based perspective, however, Richard Nelson and Sidney Winter (1982) fashioned a theory of the persistence of the large industrial corporation based on organizational capabilities, characterized by tacit knowledge and embedded in organizational routines, thus adding a cumulative dimension to the theory of the firm. Drawing on a highly eclectic set of sources from a number of disciplines, Bruce Kogut and Udo Zander (1996: 502) argued that "[f]irms are organizations that represent social knowledge of coordination and learning," thus emphasizing the collective dimension in the theory of the firm.

In "Why Do Firms Differ, and How Does It Matter?" Nelson (1991: 72) argued that "it is organizational differences, especially differences in abilities to generate and gain from innovation, rather than differences in command over particular technologies, that are the source of durable, not easily imitable, differences among firms. Particular technologies are much easier to understand, and imitate, than broader firm dynamic capabilities." David Teece, Gary Pisano, and Amy Shuen (1997: 516) defined "dynamic capabilities as the firm's ability to integrate, build, and reconfigure internal and external competences to address rapidly changing environments." They also argued that the firm's strategy entails choosing among and committing to long-term paths or trajectories of competence development (Teece at al. 1997: 524). Whereas the firm's asset positions determine its competitive advantage at any point in time and its evolutionary path constrains the types of industrial activities in which a firm can be competitive, its organizational processes transform the capabilities of the firm over time.

While Teece et al. (1997: 519) stressed the importance of learning processes that are "intrinsically social and collective," their dynamic capabilities perspective lacks social content. The framework does not ask what types of people are able and willing to make the strategic investments that can result in innovation, how these strategic decision makers mobilize the necessary financial resources, and how they create incentives for those people within the firm's hierarchical and functional division of labor to cooperate in the implementation of the innovative strategy. These questions about the roles of strategizing, financing, and organizing in the innovating firm are at the center of what Mary O'Sullivan and I have called the "social conditions of innovative enterprise" perspective (Lazonick and O'Sullivan 2000; O'Sullivan 2000b; Lazonick 2002b).

This perspective asks how and under what conditions the exercise of strategic control ensures that the enterprise seeks to grow using the collective processes and along the cumulative paths that are the foundations of its distinctive competitive success. The perspective emphasizes the role of human agency in determining whether and how the enterprise accumulates innovative capability, and thus adds an explicitly social dimension to work on "dynamic capabilities." Specifically, strategic control determines how strategic decision makers choose to build on "asset positions"; financial commitment determines whether the enterprise will have the resources available to it to persist along an "evolutionary path" to the point where its accumulation of innovative capability can generate financial returns; and organizational integration determines the structure of incentives that characterize "organizational processes" that can transform individual actions and individual capabilities (including those of strategic managers) into collective learning.

Of central importance to the accumulation and transformation of capabilities in knowledge-intensive industries is the skill base in which the firm invests in pursuing its innovative strategy. Within the firm, the division of labor consists of different functional specialties and hierarchical responsibilities. At any point in time a firm's functional and hierarchical division of labor defines its skill base. In the effort to generate collective and cumulative learning, those who exercise strategic control can choose how to structure the skill base, including how employees move around and up the functional and hierarchical division of labor over the course of their careers. At the same time, however, the organization of the skill base will be constrained by both the particular learning requirements of the industrial activities in which the firm has chosen to compete and the alternative employment opportunities of the personnel for whom the firm must compete.

In cross-national comparative perspective, the skill base that enterprises employ to transform technologies and access markets can vary markedly even in the same industrial activity during the same historical era, with different innovative outcomes. Precisely because innovative enterprise depends on social conditions, the development and utilization of skill bases that occur in one institutional environment may not, at a point in time at least, be possible in another institutional environment. Moreover, even within the same industry and same nation, dynamic capabilities that yielded innovative outcomes in one historical era may become static capabilities that inhibit innovative responses in a subsequent historical era.

The innovative firm requires that those who exercise strategic control be able to recognize the competitive strengths and weaknesses of their firm's existing skill base and, hence, the changes in that skill base that will be necessary for an innovative response to competitive challenges. These strategic decision makers must also be able to mobilize committed finance to sustain investment in the skill base until it can generate higher quality, lower cost products than were previously available. As the following comparative–historical syntheses illustrate, given strategic control and financial commitment, the essence of the innovative firm is the organizational integration of a skill base that can engage in collective and cumulative learning.

2.3 THE BRITISH INDUSTRIAL DISTRICT

In last half of the nineteenth century, Britain became known as the "workshop of the world." Britain's position in the world economy owed much to its mercantile power, developed through global commerce and related wars with other leading nations over the previous centuries. Mercantilism gave British industry access to world product markets and sources of raw materials, but it was the transformation of production from the late eighteenth century that enabled Britain to emerge as the world's leading (and indeed first) industrial nation.

In the late nineteenth century, Britain's productive power resided in industrial districts that, for building machines and using them to manufacture products as varied as cloth and ships, possessed an immense accumulation of capabilities. Beyond evening courses at local "mechanics' institutes," formal vocational or professional education played no role in the development of Britain's skilled labor force. Nor did British industry make use of corporate, university, or government research labs to develop new technology. Regionally based on-the-job apprenticeship arrangements, through which craft workers passed on their skills to the next generation, constituted in effect the "national innovation system" of the world's first industrial economy.

What accounts for the importance of the craft worker for Britain's industrial leadership? While the mechanization of the factory was a central feature of the British industrial revolution—and in its time a wonder of the world—the standardization of materials and the automation of machinery that British industry achieved during its industrial revolution were, in historical retrospect, incipient. Skilled craft workers maintained critical roles in keeping imperfect machinery in motion and ensuring high levels of throughput of work-in-progress made from imperfect materials. Within the firm, experienced workers typically were responsible for training younger workers in the craft, supervising their work, and coordinating the flow of work through the production process. In some industries, the central employment relation took the form of an internal subcontract system; for example, in the cotton spinning industry, employers paid piece-rates to senior workers, known as "self-acting minders," who in turn trained, supervised, and paid time wages to junior workers known as "piecers" and "doffers." In the metalworking industries, specialized workers such as "turners" and "fitters" were generally classified as "engineers," an appellation that in the British context signified membership in the "labor aristocracy" of skilled production workers (Lazonick 1990: chs. 1–6).

The localized, on-the-job character of skill formation was the major factor underlying the growth of industrial districts that made use of particular specialized craft skills. As Alfred Marshall (1961: 271) famously put it, in the British industrial districts "mysteries of the trade become no mysteries; but are as it were in the air." In periods of strong product-market demand, the ready availability of specialized craft labor induced new specialized manufacturing firms, often founded by craft workers themselves, to set up in these districts. The growth of a district induced other firms to invest in regionally specific communication and distribution facilities for the supply of materials, the transfer of work-in-progress across vertically specialized firms, and the marketing of output.

Regional concentration encouraged vertical specialization, which in turn eased firm entry into a particular speciality, thus resulting in high levels of horizontal competition. Firms could be owned and managed by the same people; there was no need to invest in the types of managerial organization that by the late nineteenth century were becoming central to the growth of firms in the United States, Germany, and Japan. In the industrial districts, economies of scale were, as Marshall argued, external, rather than internal, to the firm.

As producers and users of machinery, craft workers constituted the prime source of innovation in a particular region. Over time they devised incremental techno-logical and organizational improvements that, through the local trade press (includ-ing workers' newspapers) as well as the movement of workers (especially trained apprentices) to new employers, diffused across firms in the district. Some specialized engineering firms distinguished themselves through in-house learning. But even the strongest of these firms—for example, the textile machinery firm of Platt Brothers based in Oldham—did no in-house R&D, and from the last half of the nineteenth century generated no significant technological innovations. Their strength resided in their employment of craft labor that could flexibly produce customized machines for many different types of users (Farnie 1990).

The importance of localized craft labor to the innovative capabilities of local firms meant that it was the industrial district, and often a particular town within a district, not the individual firm, that constituted the learning entity. At the firm level, craft workers made countless "strategic" decisions to improve products and processes. For both individual firms and the district as a whole, the fixed costs of developing this source of innovation were, in historical and comparative perspective, low. At the same time, craft-oriented employment systems encouraged a high level of utilization of the plant and equipment in place. Union bargains protected the tenure and remuneration of senior workers who, paid by the piece, were willing to work long, hard, and steady. The inducement for junior workers, typically paid time wages, was that they could eventually join the aristocracy of labor. There is evidence that, within an industrial district, those localities in which negotiated piece-rate bargains shared productivity gains between employees and employers on a stable and equitable basis saw the fastest growth in productivity and market share (Lazonick 1990: chs. 3–5; Huberman 1996).

Based on craft organization, British industrial districts were highly innovative (see also Bruland and Mowery, this volume). The fact that it was the industrial district as a whole, rather than the individual enterprise within it, that was the innovating entity gave rise to the notion that differences among firms in an industrial activity were unimportant to economic performance, and indeed that they could all be characterized by depicting a "representative firm" that optimized subject to given technological and market constraints. Within the Marshallian perspective, even innovation at the district level did not require strategic direction, since the industrial arts were "in the air." Indeed, Marshall (1919: 600–1) described the organization of the Lancashire cotton textile industry, with its high degrees of horizontal competition and vertical specialization, as "perhaps the present instance of concentrated organisation mainly automatic." Yet just as Marshall was writing these words, the cotton textile industry, which had accounted for one-quarter of British exports on the eve of World War I, entered into a long-run decline from which it never recovered, and the other major British industrial districts suffered a similar fate (Elbaum and Lazonick 1986).

From the late 1970s, however, the notion of the "Marshallian industrial district" as a driver of innovative enterprise saw an academic resurgence, based on the rapid growth during the 1960s and 1970s of many highly specialized and localized districts in what became known as "the Third Italy" (Brusco 1982; Sabel 1982; Becattini 1990). On the basis of this experience, a number of US academics, headed by Charles Sabel, Michael Piore, and Jonathan Zeitlin, posited a new model of "flexible specialization" as an alternative to mass production on the US corporate model (Piore and Sabel 1984; Sabel and Zeitlin 1985). The industrial activities of the districts of the Third Italy focused on, among other things, textiles, footwear, and light machinery, just as the British districts had done. Large numbers of vertically specialized proprietary firms in which craft labor was a prime source of competitive advantage populated each industrial activity, and many entrepreneurs had previously been craft workers.

There were, however, two important differences between the British industrial districts that Marshall had observed in the late nineteenth century and those that experienced rapid growth in the Third Italy more recently. The first difference was the extent to which in Italy collective institutions supported the innovative activities of small firms. Sebastiano Brusco (1992) has emphasized the importance of the "red" local governments in Emilia-Romagna in promoting policies to support the activities of small enterprises, and in particular in facilitating cooperatives that provided these firms with "real services" related to business administration, marketing, and training. While consumer cooperatives sprung up in the British industrial districts of the late nineteenth century, producer cooperatives were rare. The second difference, which became more evident in the 1990s, was the extent to which, in some districts and in some industries, "leading" firms could emerge, drawing on the resources of the industrial districts while, through their own internal growth, transforming the innovative capability of the districts (see, for example, Belussi 1999). In contrast, when in the first half of the twentieth century competitive challenges confronted the British industrial districts, dominant firms failed to emerge to lead a restructuring process.

2.4 THE US MANAGERIAL CORPORATION

Marshall located the limits to the growth of the firm in the problem of succeeding the original owner-entrepreneur. In *The Theory of Economic Development*, Schumpeter (1934: 156) concurred using the same aphorism as Marshall, literally clothed in different garb and specifically identified as a US phenomenon: "An American adage expresses it: three generations from overalls to overalls." Critical to this perspective were two assumptions: first, that the entrepreneur was the essence of the innovative firm, and second, that the integration of ownership and control was a necessary condition for entrepreneurship. Notwithstanding his own important study of comparative trends in industrial organization published in *Industry and Trade*, Marshall (1919) declined to recognize, as ultimately Schumpeter did, that the problem of innovative succession could be resolved by the separation of ownership and control.

Taking place during the same decades in which Marshall wrote his influential books, the separation of share ownership from strategic control was the essence of what Chandler (among others) would call "the managerial revolution" in American business. During this period Germany and Japan also experienced managerial revolutions (Chandler 1990; Chandler et al. 1997; Morikawa and Kobayashi 1986; Morikawa 1997). Many British firms, especially in the science-based chemical and electrical industries also made investments in managerial organization, but in such a constrained manner that it can hardly be said that a managerial revolution occurred in Britain during the first half of the twentieth century (Hannah 1983; Lazonick 1986; Chandler 1990; Owen 2000).

In the United States, the managerial revolution began in the 1890s in industries such as steel, oil refining, meatpacking, tobacco, agricultural equipment, telecommunications, and electric power that owner-entrepreneurs had built up over the previous decades. Wall Street (and especially the firm of J. P. Morgan) organized the merger of the leading companies, and in the process did what would later become known as "initial public offerings" (IPOs) in order to allow the owner-entrepreneurs to cash in on their ownership stakes. Many of them then retired from active management of the company. Taking their places in strategic decision-making positions were salaried managers, most of whom had themselves been recruited years or even decades earlier to help build the innovative firms that they now controlled. Hence, Marshall's "entrepreneurial" limit to the growth of the firm was overcome. By the turn of the century, the separation of ownership and control in many of the most successful industrial corporations served as a powerful inducement for bright young, and typically White, Anglo-Saxon, Protestant, men to consider careers as corporate executives (Lazonick 1986; O'Sullivan 2000*a*: ch. 3).

Also from the beginning of the twentieth century, a four-year undergraduate college degree became important for entry into managerial careers, and in 1908

Harvard University launched the first graduate school in business administration. In 1900 about 2 per cent of 18–24 year olds were enrolled in institutions of higher education; in 1930 over 7 per cent; and in 1950 over 14 per cent. By the 1920s the top managers of many large industrial corporations had college degrees. As employers of university graduates as well as beneficiaries of university research, big business took an active role in shaping the form and content of higher education to meet its needs for "knowledge assets" (Noble 1977; Lazonick 1986).

As they expanded, US industrial corporations tended to diversify into new lines of business. Capabilities developed for generating goods for one product market could be used as a basis for gaining entry to new product markets. Moreover, as companies were successful, they could use internally generated revenues to finance these new investments. Profitable US corporations generally paid ample dividends to share-holders, but they still generated enough revenues to invest for the future, including growing expenditures on R&D (Mowery and Rosenberg 1989: ch. 4).

Besides transforming technology, a critical role of the managerial organization was to gain access to product markets. Without high levels of sales, the high fixed costs of developing technology and investing in production facilities would have simply resulted in high levels of losses. The building of national transportation and communications infrastructures—themselves largely put in place by managerial enterprises—created the possibility for manufacturing enterprises to sell on mass markets. To take advantage of this opportunity, however, the industrial corporations had to make complementary investments in distribution capabilities, including sales personnel, sales offices, advertising, and in some cases even customized transportation facilities. As Chandler (1990) has shown, from the late nineteenth century, a "three-pronged" investment in production, distribution, and management was a necessary condition for the growth of the industrial enterprise.

If the social condition for the growth of the US industrial corporation was an integrated managerial organization, a distinguishing feature of the same corporation was a sharp organizational segmentation between salaried managers and what became known as "hourly" workers. This segmentation had its roots in the first half of the nineteenth century when industrial managers faced a skilled labor force that was highly mobile not only from one firm to another but also from one occupation and one locality to another. In contrast, in Britain the local pools of specialized craft labor generated by apprenticeship systems meant that employers had access to ample supplies of skilled labor, even in booms. As a result, there was much less pressure in Britain than in the United States for managers to invest in the development of skill-displacing technologies. In the United States, but not Britain, firms integrated technical specialists into their managerial organizations for precisely that purpose. Hence the emergence by the mid-nineteenth century of the distinctive "American system of manufactures" (Hounshell 1984: chs. 1–2).

The key to this system was the mass production of standardized, precision-engineered parts that could be used interchangeably in a product without the intervention of a skilled worker to make the parts fit together. As David Hounshell (1984) has shown, it took a century of investment in productive capabilities by many companies in many sectors of US industry before, during the boom of the 1920s, mass production, so defined, became a reality. The productivity of the mass-production enterprise, nevertheless, still relied upon the stable employment of "semi-skilled" production workers who tended high-throughput, and very expensive, machinery (Lazonick 1990: chs. 7–8).

During the Great Depression of the 1930s, such stable employment disappeared, leading semi-skilled workers at the major mass producers to turn to industrial unionism (Brody 1980: ch. 3). The major achievement of mass-production unionism in the United States was long-term employment security for so-called "hourly" workers, with seniority as the governing principle for internal promotion to higher pay grades and continued employment during company layoffs. In return, these unionized employees accepted unilateral managerial control over the organization of work and technological change. During the post-World War II decades, production workers enjoyed employment security and rising wages but they were not in general integrated with managerial personnel into the company's organizational learning processes.

The result was that going into the second half of the twentieth century US industrial corporations had powerful managerial organizations for developing new technology. These corporations also had devised arrangements with their unionized labor forces to ensure the high level of utilization of these technologies. In employing thousands and in some cases tens of thousands of production workers who were not integrated into the company's organizational learning processes, however, this US model of the innovative firm had a fundamental weakness that, in the 1970s and 1980s, would be exposed in international competition. The Japanese in particular would demonstrate the innovative capability that could be created by not only building highly integrated managerial organizations, as the Americans had done, but also, as a complement, developing the skills of shop-floor workers and integrating their efforts into the firm's collective learning processes.

Even the most insightful of the theories of the US managerial corporation could not, without elaboration, account for the Japanese challenge (Lazonick 2002c). Both Penrose (1995) and Chandler (1962 and 1977) focused exclusively on the managerial organization, as did the influential perspective of John Kenneth Galbraith (1967) with its notion of the "technostructure" as the essence of the modern firm. Penrose did not see that, once confronted by the Japanese challenge, the US managerial corporation would have to develop the capabilities of the shop-floor worker to make use of unused managerial resources. Chandler focused on speed or throughput as a basis for achieving economies of scale and scope, but ignored the role of the shop-floor worker in the process of transforming high fixed costs into low unit costs, and hence did not perceive an important limitation of the US managerial model (Lazonick 1990).

2.5 THE JAPANESE CHALLENGE

Within the new structure of cooperative industrial relations that emerged out the conflicts of the depression years, US industrial corporations were able to take advantage of the post-World War II boom to re-establish themselves as the world's pre-eminent producers of consumer durables such as automobiles and electrical appliances and related capital goods such as steel and machine tools. With the help of US government research support and contracts, US companies also became the leaders in the computer and semiconductor industries.

In the 1970s and 1980s, however, Japanese companies challenged the US industrial corporations in the very mass-production industries—steel, memory chips, machine tools, electrical machinery, consumer electronics, and automobiles—in which even as late as the 1960s US corporations seemed to have attained an insurmountable competitive advantage. During the 1950s and 1960s many Japanese companies had developed innovative manufacturing capabilities, often on the basis of technologies borrowed from abroad to produce mainly for the home market. As Japanese exports to the United States increased rapidly in the last half of the 1970s, many observers attributed the challenge to the lower wages and longer working hours that prevailed in Japan. By the early 1980s, however, with real wages in Japan continuing to rise, it became clear that Japanese advantage was based on superior capabilities for generating higher quality, lower cost products.

The three social institutions that, in combination, formed the foundation for Japan's remarkable success were cross-shareholding, the main bank system, and lifetime employment. Cross-shareholding provided the managers of Japanese industrial corporations with the strategic control to allocate resources to investments that could generate higher quality, lower cost products. The main bank system provided these companies with levels of financial commitment that permitted them to sustain the innovation processes until they could generate returns, first on home and then on foreign product markets. Given this financial support for strategic industries, lifetime employment enabled the companies involved to put in place a new model of hierarchical and functional integration that enabled them to mobilize broader and deeper skills bases for collective and cumulative learning (Lazonick 2001). Let us look briefly at how these institutions became embedded in the functioning of the Japanese industrial enterprise in the post-World War II decades.

In 1948 the Supreme Commander for the Allied Powers (SCAP)—the occupation authority in Japan—began the dissolution of the zaibatsu, the giant holding companies that had dominated the Japanese economy from the Meiji era of the late nineteenth century to World War II. The dissolution process not only dispossessed the families that owned the zaibatsu but also removed from office the top management layers of the zaibatsu holding companies and major affiliated firms (Morikawa

1997). Taking control of strategic decision making were "third-rank executives," primarily engineers plucked from the ranks of middle management to take leadership positions of companies whose challenge was to find non-military markets for their companies' accumulated capabilities.

With the reopening of the stock market in 1949, these young and ambitious executives feared that the new public shareholders might join forces to demand their traditional rights as owners. To defend themselves against these outside interests, the community of corporate executives engaged in the practice of cross-shareholding. Commercial banks and industrial companies took equities off the market by holding each other's shares. Though not contractual, cross-shareholding was sustained by the willingness of the entire Japanese business community to accept that one company would not sell its shareholdings of another company.[2] By 1975, according to its broadest, and most relevant, definition as stock in the hands of such stable shareholders, cross-shareholding represented 60 per cent of outstanding stocks listed on the Tokyo Stock Exchange. It peaked at 67.4 per cent in 1988, but by 2000 had declined to 57.1 per cent, mainly because the beleaguered banking sector had been forced to reduce their shareholdings.

During the "era of high-speed growth" from the early 1950s to the early 1970s, most of the financial commitment of Japanese companies came from bank loans, with the companies' debt–equity ratios often at 6 : 1 or 7 : 1. Each major industrial company had a "main bank" whose job it was to convince other banks to join it in making loans to the company and to take the lead in restructuring its client company should it fall into financial distress. Some economists (e.g. Aoki and Patrick 1994) have accorded the main banks a major role in monitoring the behavior of Japan's corporate managers. In funding the growth of Japanese companies, however, the Japanese banks were relatively passive agents of government development policy, with "overloans" being made by the Bank of Japan to its member banks for providing highly leveraged finance to growing industrial companies. Japanese banks, that is, played a critical role in providing financial commitment, but no significant role in the exercise of strategic control.

Integrated organizations of managers and workers, not financial interests, monitored the behavior of the top executives of Japanese corporations (Lazonick 1999). The main mode of achieving this organizational integration was the lifetime employment system, which extended from top executives to male (but not female) shop-floor workers. The origins of the lifetime employment system can be found in the widespread employment in industry of university graduates as salaried technical and administrative personnel during the early twentieth century (Yonekawa 1984). Some companies extended the promise of lifetime employment to shop-floor workers as well when dire economic conditions and democratization initiatives of the late 1940s had given rise to a militant labor movement. The goal of the new industrial unions was to implement "production control": the takeover of idle factories so that workers could put them into operation and earn a living (Gordon

1985). Leading companies such as Toyota, Toshiba, and Hitachi fired militant workers and created enterprise unions of white-collar (technical and administrative) and blue-collar employees. Foremen and supervisors were members of the enterprise unions, as were all university-educated personnel, for at least the first ten years of employment before they made the official transition into "management."

The most important achievement of enterprise unionism was the institutionalization of lifetime employment, a system that, while not contractually guaranteed, gave white-collar and blue-collar workers employment security, at first to the retirement age of 55, then from the 1980s to the age of 60, and currently (in transition) to the age of 65 (Sako and Sato 1997). This employment security both won the commitment of the worker to the company and gave the company the incentive to develop the productive capabilities of the worker. The system did not differ in principle from the organizational integration of technical and administrative employees that was at the heart of the US managerial revolution, except in one extremely important respect. In the United States there was a sharp segmentation between salaried managers and shop-floor workers, whereas the Japanese companies of the post-World War II decades integrated shop-floor workers into a company-wide process of organizational learning.

Through their engagement in processes of cost reduction, Japanese shop-floor workers were continuously involved in a more general process of improvement of products and processes that, by the 1970s, enabled Japanese companies to emerge as world leaders in factory automation (Jaikumar 1989). By the early 1990s the stock of robots in Japanese factories was over seven times that of the United States. Also of great importance was the ability of Japanese manufacturers to eliminate waste in production; by the late 1970s, for example, Japan's competitive advantage in television sets was not in labor costs or even scale economies but in a savings of materials costs (Owen 2000: 278; Fagerberg and Godinho in this volume). This productive transformation became particularly important in international competition in the 1980s as Japanese wages approached the levels of those in North America and Western Europe and, especially from 1985, as the value of the Japanese yen dramatically strengthened. During the 1980s and 1990s, influenced by not only Japan's export performance but also the impact of Japanese direct investment in North America and Western Europe, many Western companies sought, with varying degrees of success, to implement Japanese high-quality, low-cost mass-production methods.

During the 1980s most Western analyses of the sources of Japanese competitive advantage focused on the hierarchical integration of the shop-floor worker into the organizational learning process. By the early 1990s, however, as Japanese companies captured higher value-added segments of the products markets in which they competed, the emphasis shifted to the role of "cross-functional management," "company-wide quality control," or "concurrent engineering" in generating not only lower cost but also higher quality products within highly accelerated product development cycles. Much of the discussion of functional integration focused on its role in "new product development" in international comparative perspective, with, as Clark and Fujimoto (1991) showed for the automobile industry, the US managerial corporation performing quite poorly.

Given that the innovative power of the US industrial corporation resided in its integrated managerial organization, why should it have suffered from functional segmentation in competition with the Japanese? One reason was that, given the hierarchical segmentation of shop-floor activities from organizational learning processes in US companies, US engineers were not forced to communicate across their disciplines to solve "real-world" manufacturing problems. Another had to do with the increasing interfirm mobility of US engineers from the 1960s—mobility that, as we shall see, was related to the rise of the "New Economy" high-tech firm. The prospects for interfirm mobility gave scientists and engineers an interest in developing their reputations among their peers within their particular area of specialization, even if it detracted from integrating their specialist knowledge across functional areas within the particular firm for which they were working. By contrast, in the Japanese firm both the hierarchical integration of managers and workers and low levels of interfirm mobility of engineering personnel fostered functional integration.

The evolution of the semiconductor industry provides a vivid example of the competitive power, but also the limits, of Japanese organizational integration. From the late 1970s the Japanese mounted a formidable competitive challenge to US producers in dynamic random access memory (DRAM) chips, forcing most US companies, including Intel, to withdraw from the market after 1985. Already a powerhouse in semiconductors before the Japanese challenge, Intel reemerged even stronger in the 1990s as the leader in microprocessors, a product in which it was the pioneer in the early 1970s and for which during the 1980s it secured the franchise for the IBM PC and the subsequent IBM clones (Burgelman 1994).

Organizational integration was critical to the Japanese challenge in DRAMs. As Daniel Okimoto and Yoshio Nishi (1994) have shown, the most critical interactions in product and process development in Japanese semiconductor companies were between personnel in divisional R&D labs and factory engineering labs, with engineering capability being concentrated in the factory labs. They argue that in Japan "hands-on manufacturing experience . . . is almost a requirement for upward career and post-career mobility [whereas] [i]n the United States, by contrast, manufacturing engineers carry the stigma of being second-class citizens" (Okimoto and Nishi 1994: 195).

Value added in microprocessors is in the design that determines the use of the product, an activity for which US skill bases in semiconductors were more suited. Value added in memory chips is in process engineering that reduces defects and increases chip yields, an activity for which Japanese skill bases in semiconductors were more suited. By the 1980s Japanese companies such as Fujitsu, Hitachi, and

NEC were able to achieve yields in the production of DRAMs that were 40 per cent higher than the best US companies.

In the 1990s the Japanese economy as a whole has stagnated, to the point where many Western observers now blame its unique institutional framework, still largely intact, for its lack of innovation. Yet, in industries such as electronics and automobiles, Japanese companies such as Sony and Toyota, among many others, remain leading innovators in those types of products in which, as during the previous decades, their integrated skill bases gave them international competitive advantage. The main microeconomic problems in the Japanese economy are to be found in the financial system and, relatedly, institutions for creation of new innovative firms.

During the boom of the 1980s the leading Japanese manufacturing companies were able to reduce their reliance on bank debt, just as the banks were awash with cash to lend. The banks then channeled funds into speculative investments in land and stocks, thus fuelling the "bubble economy" of the late 1980s. When the bubble burst in 1990, the banks were saddled with mountains of bad debt. Although most of this bad debt has now been written off, the banks remain in fragile condition because most of their loans are being made to smaller companies that do not have anything close to the growth potential that was realized by many Japanese companies in the previous eras of high-speed growth and export expansion (Lazonick 1999). "Growth potential," however, is not exogenous to the "social conditions of innovative enterprise," as illustrated by the emergence of more powerful modes of strategy, finance and organization in the rise of the "New Economy" model of the innovative firm in the United States.

2.6 THE NEW ECONOMY MODEL

During the 1970s and 1980s while Japanese enterprises were challenging established US managerial corporations in many industries in which they had been dominant, there was a resurgence of the US information and communications technology (ICT) industries, providing the foundation for what by the last half of the 1990s became known as the "New Economy." Historically, underlying the emergence of the New Economy were massive post-World War II investments by the US government, in collaboration with research universities and industrial corporations, in developing computer and communications technologies.

By the end of the 1950s, this combined business–government investment effort had resulted in not only the first generation of computers, with IBM as the leading firm, but also the capability of imbedding integrated electronic circuits on a silicon chip, with Fairchild Semiconductor and Texas Instruments in the forefront of creating the technology that would become the standard of the semiconductor industry. Through the early 1960s the US government provided virtually all of the demand for semiconductors. From the second half of the 1960s, however, a growing array of commercial opportunities for electronic chips induced the creation of semiconductor startups. A new breed of venture capitalist, many with prior managerial or technical experience in the semiconductor industry, backed so many semiconductor startups clustered in the region around Stanford University that by the early 1970s the district was dubbed "Silicon Valley." Innovation in semiconductors, and especially the development of the microprocessor—in effect a computer on a chip—created the basis for the emergence of the microcomputer industry from the late 1970s, which in turn resulted in the enormous growth of an installed base of powerful "hosts" in homes and offices that made possible the Internet revolution of the 1990s.

As AnnaLee Saxenian (1994) has shown, intense, and often informal, learning networks that transcended the boundaries of firms contributed to the success of Silicon Valley. Like the Marshallian industrial districts of a century earlier, there is no doubt that, in Silicon Valley, "the mysteries of the trade . . . were in the air." But in its strategy, finance, and organization, the New Economy business model that emerged in Silicon Valley differed significantly from the Marshallian industrial district. Of particular importance was the extent to which in Silicon Valley organizational learning occurred within the firm, enabling some particularly innovative firms that grew to employ tens of thousands of employees to drive the development of the region. In its early stages this organizational learning tended to be backed by venture capital, a mode of finance that through its success in Silicon Valley from the 1960s evolved into an industry in its own right. Also of great importance in supporting the development of technology and the education of personnel available to firms in this high-tech industrial district were state funding and universities, institutions that for a century had been central to the US managerial model.

The founders of new ICT firms were typically engineers who had gained specialized experience in existing ICT firms, although in some cases they were university faculty members intent on commercializing their academic knowledge. While some of these entrepreneurs came from existing Old Economy companies, where it was often difficult for their new ideas to get internal backing, New Economy companies themselves became increasingly important as a source of new entrepreneurs who left their current employers to start a new firm (Gompers et al. 2003). Typically the founding entrepreneurs of a New Economy startup sought committed finance from venture capitalists with whom they shared not only ownership of the company but also strategic control. Besides sitting on the board of directors of the new company, the venture capitalists would generally recruit professional managers, who would be given company stock along with stock options, to lead the transformation of the firm from a new venture to a going concern. This stock-based compensation gave these managers a powerful financial incentive to develop the innovative capabilities of the company to the point where it could do an IPO or private sale to an established company. But, both before and after making this transition, their tenure with, and value to, the company depended on their managerial capabilities, not their fractional ownership stakes.

Key to making this transition from new venture to going concern was the organizational integration of an expanding body of technical and managerial "talent." Stock options became an important mode of compensation, usually as a partial substitute for cash salaries, for attracting these highly mobile people to the startup and retaining their services. The underlying stock would become valuable if and when they took the form of publicly traded shares. Shortening the expected period between the launch of a company and its IPO was the practice of most venture-backed high-tech startups of going public on the NASDAQ exchange (founded in 1971), with its much less stringent listing requirements than the Old Economy New York Stock Exchange. If and when the firm did an IPO or was acquired by another publicly listed company, the venture capitalists could sell their shareholdings on the stock market, thereby exiting from their investments in the firm, while entrepreneurs could also transform some or all of their ownership stakes into cash. With the company's stock being publicly traded, employees who exercised their stock options could easily turn their shares into cash.

During the 1980s and 1990s the liberal use of stock as a compensation currency, not only for top executives as had been the case in Old Economy companies since the 1950s, but also for a broad base of non-executive personnel became a distinctive feature of New Economy firms. For example, Cisco Systems, which grew from about 200 employees at the time of its IPO in 1990 to 38,000 employees in 2001, awarded stock options to all of its employees, so that by 2001 stock options outstanding accounted for over 14 per cent of the company's total stock outstanding. Since Cisco did hardly any of its own manufacturing—another distinctive characteristic of many New Economy "systems integrators"—the people in the skill base to whom these options were awarded were almost all highly educated employees who were potentially highly mobile on the labor market.

Besides using their own stock as a compensation currency, during the 1990s some New Economy companies grew large by using their stock, instead of cash, to acquire other, smaller and typically younger, New Economy firms in order to gain access to new technologies and markets. Cisco mastered this growth-through-acquisition strategy; from 1993 through 2002 Cisco made seventy-eight acquisitions (forty-one of which were during 1999–2000, the peak years of the New Economy boom), with stock providing the currency for over 98 per cent of the total value of these acquisitions.

At the same time Cisco conserved cash by paying no dividends, a mode of financial commitment that also distinguished New Economy from Old Economy companies. As a result, Cisco's astonishing growth in the 1990s occurred without the company taking on any long-term debt. Nevertheless, with the bursting of the New Economy bubble from mid-2000, Cisco spent billions of dollars repurchasing its own stock to support its sagging stock price (Carpenter et al. 2003). Even during the boom, when stock prices were rising, the extent to which New Economy companies issued stock to make acquisitions and compensate employees meant that some of them spent billions of dollars on stock repurchases; during 1997–2000, for example, Intel's stock repurchases totalled $18.8 billion and Microsoft's $13.4 billion. By way of comparison, over these years Intel's total expenditures on R&D were $14.2 billion, while Microsoft's were $11.2 billion.

As in the cases of Intel, Microsoft, and Cisco, by the end of the twentieth century a number of New Economy companies had grown to be formidable growing concerns (Lazonick 2004). In 2002 the top 500 US-based companies by sales included twenty ICT firms founded no earlier than 1965 that had been neither spun-off from nor merged with an Old Economy firm. These twenty companies had revenues ranging from $35.4 billion for Dell Computer to $3.0 billion for Computer Associates International, with an average of $10.4 billion. Their headcounts ranged from 78,700 for Intel to 8,100 for Qualcomm, with an average of 30,084, up from an average for the same twenty companies of 6,347 in 1993. Nine of these twenty companies (and seven of the top ten) were based in Silicon Valley, another two in Southern California, and the other nine in eight states around the country. Compaq Computer, the forty-sixth largest US company in 2001 with $33.6 billion in sales and 70,950 employees, would have been high up on this list in 2002 had it not been acquired by Hewlett-Packard.

Many of these large New Economy companies have become important contributors to the patenting activity of US-based corporations. Samuel Kortum and Josh Lerner (2000) have shown that in the first half of the 1980s a sharp decline in patenting by US corporations was counterbalanced by a massive increase in early-stage venture-capital disbursements. But from the last half of the 1980s patenting picked up again, in part because it became important to the competitive strategy of high-growth New Economy firms. In 2001 Intel was eighteenth in the number of US patents issued to all companies, and seventh among US-based companies. Ahead of Intel were not only Old Economy companies such as IBM, Lucent Technologies, General Electric, and Hewlett-Packard but also two much smaller, but still sizeable, New Economy semiconductor companies, Micron Technology, founded in 1978 in Idaho, in fourth place, and Advanced Micro Devices (AMD), founded in Silicon Valley in 1969, in fourteenth place. In 2002 AMD was the 535th largest US company by sales and had 12,146 employees, while Micron was 554th and employed 18,700.

Innovative New Economy companies have tended to grow large by upgrading and expanding their product offerings within their main lines of business, and thus far at least have not engaged in the indiscriminate diversification into unrelated technologies and markets that characterized, and ultimately undermined the performance of, many leading Old Economy companies in the 1960s and 1970s. At the same time, New Economy companies have become less vertically integrated than Old Economy companies because equipment manufacturers such as Cisco, Dell, and Sun Micro-systems have focused their investment strategies on activities that require organiza-tional learning in their core competencies, while outsourcing activities that, as is the case with semiconductor fabrication, are too expensive and complex to be done in-house, or, alternatively, as is the case with printed circuit board assembly, have become routine. Some of the largest ICT companies in the United States are upstream electronics components suppliers, most of which are New Economy firms. Among the top 1000 US companies by sales in 2002 were eleven semicon-ductor companies, with a total employment of 212,354, ranging from Intel with its 78,700 employees to Nvidia (a specialist producer of graphics processors founded in 1993) with 1,513 employees. The world's five largest contract manufacturers— Flextronics, Solectron, Sanmina-SCI, Celestica, and Jabil Circuit—to whom equip-ment manufacturers outsource the mass production of printed circuit boards and other components, employed a total of 260,000–270,000 people at the begin-ning of 2003.

The severe downturn in the ICT industries in 2001 and 2002 raised questions about the sustainability of the New Economy model. A major weakness of the New Economy model lay in the huge personal gains, often amounting to tens of millions and even hundreds of millions of dollars, that top executives could reap from stock-based rewards in a volatile stock market. When stock prices were rising, executives had strong personal incentives to allocate resources (or give the appearance of doing so) in ways that encouraged the speculative market. Many of these allocative deci-sions undermined the innovative capabilities of the firms over which these execu-tives exercised control (Carpenter et al. 2003). When stock prices began falling, the same executives had strong personal incentives to cash in quickly by selling stock, so that they made immense fortunes (in most instances without breaking the law) even as their companies lost money and, in many cases, struggled to survive (Gimein et al. 2002).

A major problem for some of these companies was the way in which the use of stock as a combination and compensation currency in the New Economy boom affected the role of the stock market as a source of cash (O'Sullivan 2003). Seventy years earlier, in the stock market boom of the late 1920s, US corporations had sold stock at speculative prices to pay down debt or bolster their treasuries, thus making them less financially vulnerable when the boom turned to bust. In the boom of the late 1990s corporations did not take advantage of the speculative market by selling stock; if anything, these companies purchased stock to support their already inflated stock prices. While employees, and particularly high-level executives, benefited from these stock price increases, their companies were weakened financially, as became painfully evident for many ICT companies from mid-2000 when the stock market turned down.

2.7 Understanding the Innovative Firm: Implications for Theory

This chapter has illustrated that the social characteristics of the innovative firm have varied markedly over time and across institutional environments. To study the innovative firm in abstraction from the particular social conditions that enable it to generate higher quality, lower cost products is to forgo an understanding of why it became innovative in the first place and how its innovative capabilities may be rendered obsolete. A comparative–historical analysis enables us to learn from the past and provide working hypotheses for ongoing research.

First, the comparative–historical experience of the innovative firm suggests that, contrary to a common belief that has persisted since the time of Marshall, the form of firm ownership is not the critical issue for understanding the type of strategic control that supports innovative enterprise. Critical are the abilities and incentives of those managers who exercise strategic control. Whether they are majority owners of the firm, state employees, or employees of publicly listed companies, we need to know where and how these strategic managers gained the experience to allocate resources to the innovation process, and the conditions under which their personal rewards depend on the firm's innovative success.

Second, the most fundamental, if by no means the only, source of financial commitment for the innovative firm is to be found in those funds that are generated by the firm itself. When bank finance is used to leverage financial commitment, it requires close relations between financial institutions and innovative firms, as for example in the Japanese model. In certain times and places, the stock market can provide some well-positioned firms with financial commitment. But as a financial institution, the fundamental role of the stock market is to provide liquidity, not commitment. It enables owner-entrepreneurs and venture capitalists to cash out of their investments, and it enables households to diversify their savings portfolios so that they can (hopefully) tap into the yields of the stock market without having to devote time and effort to understanding the innovative capabilities of the companies that have listed their securities on it.

Third, while strategic control and financial commitment are essential to the innovative firm, it is organizational integration that determines the innovative capability that the firm actually possesses. The types of organizational integration that result in innovation vary across industries and institutional environments as well as over time. The hierarchical and functional divisions of labor that, when integrated into learning organizations, generated innovation in the past cannot necessarily be expected to do so in the future when faced with changes in technology, markets, and competition—changes which to some extent successful innovation in itself brings about.

When a society's previously innovative firms no longer generate innovative outcomes, there will be pressures on these firms to reallocate resources from investments in existing skill bases to investments in new types of organizations. Such organizational restructuring does not always occur smoothly or successfully, as the historical experiences of British industrial districts, US managerial corporations, and Japanese enterprise groups have shown. Precisely because the innovative firm is a social organization, the reallocation of its resources is a social process in which different groups of people can have very different interests. An understanding of the changing organization of the innovative firm is important for understanding not only how a society innovates but also how a society copes with processes of social disruption in which the gains of some may be the losses of others.

In a theory of innovative enterprise, strategy, finance, and organization are interlinked as a dynamic process with learning as an outcome. To fully comprehend the innovative firm, there is a need to understand the actual learning processes: the relation between tacit knowledge and codified knowledge, between individual capabilities and collective capabilities, and between what is learned at a point in time and how that learning cumulates over time. The prevailing social conditions of innovative enterprise provide the context for those learning processes, shaping the types of learning that are attempted, the extent to which these processes are sustained, and the ways in which people interact both cognitively and behaviorally. The influence of the social context is manifested by the functional and hierarchical integration of skill bases that, as this essay has illustrated, can vary dramatically across industries and institutional environments as well as over time.

A theory of innovative enterprise must be based on an understanding of comparative–historical experience that is broad and deep enough to evoke confidence that the assumptions and relations forming the substance of the theory capture the essence of the reality to which the theory purports to be relevant. The development of relevant theory requires an iterative approach in which theoretical postulates are derived from the study of the historical record and the resultant theory is used to analyze history as an ongoing and unfolding process (Lazonick 2002a). The intellectual challenge is to integrate theory and history.

As Edith Penrose (1989: 11) perceptively put it in an article written late in her career:

"Theory" is, by definition, a simplification of "reality" but simplification is necessary in order to comprehend it at all, to make sense of "history". If each event, each institution, each fact, were really unique in all aspects, how could we understand, or claim to understand, anything at all about the past, or indeed the present for that matter? If, on the other hand, there are common characteristics, and if such characteristics are significant in the determination of the course of events, then it is necessary to analyse both the characteristics and their significance and "theoretically" to isolate them for that purpose.

If we need theory to make sense of history, so we also need history to make sense of theory. As Penrose concluded: "universal truths without reference to time and space are unlikely to characterise economic affairs."

NOTES

1. For a comparison of managerial corporations of European nations such as Britain, France, Germany, and Italy with the US model, see Lazonick 2003.
2. When in financial distress, a company might raise cash by selling some of its cross-shareholdings to other companies at the going market price but with an understanding that the shares would be repurchased, also at the going market price, if and when its financial condition improved.

REFERENCES

AOKI, M., and PATRICK, H. (eds.) (1994), *The Japanese Main Bank System: Its Relevance for Developing and Transforming Economies*, Oxford: Oxford University Press.

BECATTINI, G. (1990), "The Marshallian Industrial District as a Socio-Economic Notion," in F. Pyke, G. Becattini, and W. Sengenberger (eds.), *Industrial Districts and Inter-Firm Cooperation in Italy*, International Institute for Labour Studies, 37–51.

BELUSSI, F. (1999), "Path-Dependency versus Industrial Dynamics: An Analysis of Two Heterogeneous Districts," *Human Systems Management* 18: 161–74.

BRODY, D. (1980), *Workers in Industrial America: Essays on the Twentieth Century Struggle*, Oxford: Oxford University Press.

BRUSCO, S. (1982), "The Emilian Model: Productive Decentralisation and Social Integration," *Cambridge Journal of Economics* 6: 167–84.

—— (1992), "Small Firms and the Provision of Real Services," in F. Pyke and W. Sengenberger (eds.), *Industrial Districts and Local Economic Regeneration*, International Institute for Labour Studies, 177–96.

BURGELMAN, R. (1994), "Fading Memories: A Process Theory of Strategic Business Exit in Dynamic Environments," *Administrative Science Quarterly* 39(1): 24–56.

*CARPENTER, M., LAZONICK, W., and O'SULLIVAN, M. (2003), "The Stock Market and Innovative Capability in the New Economy: The Optical Networking Industry," *Industrial and Corporate Change* 12(5): 963–1034.

CHANDLER, A. (1962), *Strategy and Structure: Chapters in the History of the American Industrial Enterprise*, Cambridge, Mass.: MIT Press.

*—— (1977), *The Visible Hand: The Managerial Revolution in American Business*, Cambridge, Mass.: Harvard University Press.

—— (1990), *Scale and Scope: The Dynamics of Industrial Capitalism*, Cambridge, Mass.: Harvard University Press.

*—— AMATORI, F., and HIKINO, T. (eds.) (1997), *Big Business and the Wealth of Nations*, Cambridge: Cambridge University Press.

CLARK, K., and FUJIMOTO, T. (1991), *Product Development Performance: Strategy, Organization and Management in the World Auto Industries*, Cambridge, Mass.: Harvard Business School Press.

* Asterisked items are suggestions for further reading.

*ELBAUM, B., and LAZONICK, W. (eds.) (1986), *The Decline of the British Economy*, Oxford: Oxford University Press.

FARNIE, D. (1990), "The Textile Machinery-Making Industry and the World Market, 1870–1960," *Business History* 32(4): 150–70.

GALBRAITH, J. K. (1967), *The New Industrial State*, Boston: Houghton Mifflin.

GIMEIN, M., DASH, E., MUNOZ, L., and SUNG, J. (2002), "You Bought. They Sold." *Fortune* 146(4): 64–72.

GOMPERS, P., LERNER, J., and SCHARFSTEIN, D. (2003), "Entrepreneurial Spawning: Public Corporations and the Genesis of New Ventures, 1986–1999," NBER Working Paper 9816, July.

GORDON, A. (1985), *The Evolution of Labor Relations in Japan: Heavy Industry, 1853–1955*, Cambridge, Mass.: Harvard University Press.

*HANNAH, L. (1983), *The Rise of the Corporate Economy: The British Experience*, 2nd edn., London: Methuen.

HOUNSHELL, D. (1984), *From the American System to Mass Production, 1800–1932*, Baltimore: Johns Hopkins University Press.

HUBERMAN, M. (1996), *Escape from the Market: Negotiating Work in Lancashire*, Cambridge: Cambridge University Press.

JAIKUMAR, R. (1989), "Japanese Flexible Manufacturing Systems: Impact on the United States," *Japan and the World Economy* 1(2): 113–43.

KOGUT, B., and ZANDER, U. (1996), "What Firms Do? Coordination, Identity, and Learning," *Organization Science* 7: 502–18.

KORTUM, S., and LERNER, J. (2000), "Assessing the Contribution of Venture Capital to Innovation," *Rand Journal of Economics* 31(4): 674–92.

LAZONICK, W. (1986), "Strategy, Structure, and Management Development in the United States and Britain," in K. Kobayashi and H. Morikawa (eds.), *Development of Managerial Enterprise*, Tokyo: University of Tokyo Press, 101–46.

—— (1990), *Competitive Advantage on the Shop Floor*, Cambridge, Mass.: Harvard University Press.

—— (1999), "The Japanese Economy and Corporate Reform: What Path to Sustainable Prosperity?," *Industrial and Corporate Change* 8(4): 607–33.

—— (2001), "Organizational Learning and International Competition: The Skill-Base Hypothesis," in W. Lazonick and M. O'Sullivan (eds.), *Corporate Governance and Sustainable Prosperity*, Basingstoke: Palgrave, 37–77.

—— (2002a), "Innovative Enterprise and Historical Transformation," *Enterprise & Society* 3(1): 35–54.

—— (2002b), "Innovative Enterprise, The Theory of," in M. Warner, (ed.), *International Encyclopedia of Business and Management*, 2nd edn., Stamford, Conn.: Thomson Learning, 3055–76.

—— (2002c), "The US Industrial Corporation and The Theory of the Growth of the Firm," in Christos Pitelis (ed.), *The Growth of the Firm: The Legacy of Edith Penrose*, Oxford: Oxford University Press, 249–77.

—— (2003), "The Social Foundations of Innovative Enterprise," Franco Momigliano Lecture, Instituto per la Cultura e la Storia d'Impresa, Terni, Italy, July 10.

—— (2004), "Corporate Restructuring," in S. Ackroyd, R. Batt, P. Thompson, and P. Tolbert (eds.), *The Oxford Handbook of Work and Organization*, Oxford: Oxford University Press (forthcoming).

LAZONICK, W., and O'SULLIVAN, M. (2000), "Perspectives on Corporate Governance, Innovation, and Economic Performance," Report prepared for the project on Corporate Governance, Innovation, and Economic Performance under the Targeted Socio-Economic Research Programme of the European Commission, June (www.insead.edu/cgep).

LESLIE, S., and KARGON, R. (1996), "Selling Silicon Valley: Frederick Terman's Model for Regional Advantage," *Business History Review*, 70(4): 435–72.

MARSHALL, A. (1919), *Industry and Trade*, London: Macmillan.

—— (1961), *Principles of Economics*, 9th (variorum) edn., London: Macmillan.

MORIKAWA, H. (1997), "Japan: Increasing Organizational Capabilities of Large Industrial Enterprises, 1880s–1980s," in A. Chandler, F. Amatori, and T. Hikino (eds.), *Big Business and the Wealth of Nations*, Cambridge: Cambridge University Press: 307–35.

—— and KOBAYASHI, K. (eds.) (1986), *Development of Managerial Enterprise*, Tokyo: University of Tokyo Press.

*MOWERY, D., and ROSENBERG, N. (1989), *Technology and the Pursuit of Economic Growth*, Cambridge: Cambridge University Press.

NELSON, R. (1991), "Why Do Firms Differ, and How Does It Matter?," *Strategic Management Journal* 12, Special Issue: 61–74.

—— and WINTER, S. (1982), *An Evolutionary Theory of Economic Change*, Cambridge, Mass.: Harvard University Press.

NOBLE, D. (1977), *America by Design: Science, Technology, and the Rise of Corporate Capitalism*, New York: Knopf.

OKIMOTO, D., and NISHI, Y. (1994), "R&D Organization in Japanese and American Semiconductor Firms," in M. Aoki and R. Dore (eds.), *The Japanese Firm: The Sources of Competitive Strength*, Oxford: Oxford University Press: 178–208.

*O'SULLIVAN, M. (2000*a*), *Contests for Corporate Control: Corporate Governance and Economic Performance in the United States and Germany*, Oxford: Oxford University Press.

—— (2000*b*), "The Innovative Enterprise and Corporate Governance," *Cambridge Journal of Economics* 24(4): 393–416.

—— (2003), "The Stock Market as a Source of Cash in the US Corporation," INSEAD working paper.

*OWEN, G. (2000), *From Empire to Europe: The Decline and Revival of British Industry since the Second World War*, New York: HarperCollins.

PENROSE, E. (1989), "History, the Social Sciences and Economic 'Theory,' with Special Reference to Multinational Enterprise," in A. Teichova, M. Lévy-Leboyer, and H. Nussbaum (eds.), *Historical Studies in International Corporate Business*, Cambridge: Cambridge University Press, 7–13.

*—— (1995), *The Theory of the Growth of the Firm*, 3rd edn., Oxford: Oxford University Press [first published 1959].

PIORE, M., and SABEL, C. (1984), *The Second Industrial Divide: Possibilities for Prosperity*, Basic Books.

SABEL, C. (1982), *Work and Politics*, Cambridge: Cambridge University Press.

—— and ZEITLIN, J. (1985), "Historical Alternatives to Mass Production: Politics, Markets and Technology in Nineteenth-Century Industrialization," *Past and Present* 108: 133–76.

*SAKO, M., and SATO, H. (eds.) (1997), *Japanese Labour and Management in Transition: Diversity, Flexibility, and Participation*, London: Routledge.

SAXENIAN, A. (1994), *Regional Advantage: Culture and Competition in Silicon Valley and Route 128*. Cambridge, Mass.: Harvard University Press.

SCHUMPETER, J. (1934), *The Theory of Economic Development*, Cambridge, Mass.: Harvard University Press.

—— (1950), *Capitalism, Socialism, and Democracy*, 3rd edn., New York: Harper.

TEECE, D., PISANO, G., and SHUEN, A. (1997), "Dynamic Capabilities and Strategic Management," *Strategic Management Journal* 18(7): 509–33.

YONEKAWA, S. (1984), "University Graduates in Japanese Enterprises before the Second World War," *Business History* 26(3): 193–218.

CHAPTER 3

NETWORKS OF INNOVATORS

WALTER W. POWELL

STINE GRODAL

3.1 INTRODUCTION

In February of 2001, two rival consortia published rough draft (roughly 90 per cent complete) sequences of the human genome in *Nature* and *Science*. The "public" Human Genome Project consisted of five key institutions and eleven collaborators,[1] supported by the US National Institutes of Health, Department of Energy, and the Wellcome Trust in the United Kingdom. The rival "private" consortia, led by the biotech firm Celera, included both commercial firms and academic researchers from the University of California, Penn State, Case Western, Johns Hopkins, Cal Tech, Yale, Rockefeller, as well as scientists in Spain, Israel, and Australia. These projects have been acclaimed for their remarkable scientific achievement; they were also the product of considerable organizational innovation. In contrast to the Manhattan Project or Project Apollo, both of which were hierarchically organized, national projects, the Human Genome Project (HGP) and the Celera team were pluralist, multiorganizational, multinational confederations. These two groups were intensely rivalrous, but collaborated intensively within their own groups (Lambright 2002). HGP involved management by two government agencies and a private British

We are grateful to David Mowery and Jan Fagerberg for their careful readings of earlier drafts.

foundation that coordinated activities in government labs, universities, and nonprofit institutes in the US and England. As the lead firm, Celera's organization was more focused, but its research team included scientists and state-of-the-art equipment at private firms, public and private universities, and nonprofit institutes in four countries.

Both projects were organized as large-scale networks, and their rivalry spurred each side to engage in a high-stakes learning race. While the cost, scale, and distributed nature of these projects may have been unusual, the form of organization—collaboration across multiple organizational boundaries and institutional forms—is no longer rare. Indeed, many analysts have noted that the model of networks of innovators has become commonplace over the past two decades (Powell 1990; Rosenbloom and Spencer 1996; Roberts and Liu 2001; Chesbrough 2003).

Collaboration among ostensible rivals was once regarded as a provisional or transitional step taken to enter new markets, spread risks, or to share early stage R&D costs (Mowery 1988). Such forays were often followed by mergers as the transitory activities became incorporated inside the boundaries of the firm. Recent studies suggest, however, that various forms of interorganizational partnerships are now core components of corporate strategy. Even where these linkages endure for relatively lengthy periods of time, they do not entail vertical integration (Gomes-Casseres 1996; Hagedoorn 1996; Noteboom 1999; Ahuja 2000a). Contemporary studies of industrial performance are replete with reports of a significant upsurge in various types of interorganizational collaboration. While these collaborations can take a number of forms (including research consortia, joint ventures, strategic alliances, and subcontracting) and span a wide range of key functions, a National Research Council analysis of trends in industrial research and development (R&D) suggests that the innovation process has undergone the most significant transformation over the past decade (Merrill and Cooper 1999). In a survey of the period 1960–98, Hagedoorn (2002) finds a sharp growth in R&D collaborations, beginning in the late 1970s and continuing through the mid-1990s.

A National Research Council assessment of eleven US-based industries, purposefully diverse in character and technology but all resurgent in the 1990s, observes in every sector an increased reliance on external sources of R&D, notably universities, consortia and government labs, and greater collaboration with domestic and foreign competitors, as well as customers in the development of new products and processes (Mowery 1999: 7). Other surveys also point to the enhanced centrality of interorganizational collaboration, especially in R&D. For example, National Science Foundation data show a marked increase in the number of international alliances between US and Western European countries between 1980 and 1994; but by the mid-1990s, the formation rates for intranational alliances linking US firms with their domestic competitors outpace international linkages (National Science Board 1998). The former collaborations were motivated largely by concerns with market access, while the latter focus more on the development of new technologies.

Similarly, there is now ample research illustrating the growing links between US firms and universities (Powell and Owen-Smith 1998), and greater involvement by firms and government labs in research joint ventures (Link 1996, 1999). In the realm of science, Hicks and Katz (1996) find that research papers are much more likely to be co-authored and involve authors with multiple institutional affiliations that span universities, government, and industry. Distributed networks of practice are the organizing basis for many technical communities, suggesting both that sources of knowledge are now more widely dispersed and that governance mechanisms are emerging to orchestrate distributed knowledge. The open source software movement is but one highly visible example of this trend (O'Mahony 2002; Weber 2003), which illustrates how advances in information technology have greatly facilitated virtual networks. In short, as Mowery (1999: 9) observes, "the diversity of institutional actors and relationships in the industrial innovation process has increased considerably." Complex networks of firms, universities, and government labs are critical features of many industries, especially so in fields with rapid technological progress, such as computers, semiconductors, pharmaceuticals, and biotechnology.

Our goal in this chapter is to assess the state of scholarly research on the role of networks in the innovation process. We begin with a review of the factors that have triggered the increased salience of networks. We discuss different types of networks, distinguishing between networks that are based more on contractual or market considerations, and those that are based on less formal, and more primordial relationships, such as common membership in a technological community or a regional economy. We then turn to a discussion of the analytical leverage provided by the tools of network analysis. This stream of research, which spans sociology, social psychology, organizational behavior, and business strategy, highlights key distinctions between highly clustered, dense networks, steeped in overlapping ties and high in trust, and weak-tie networks, that provide access to novel, non-redundant information. We next review a number of empirical studies of the contribution of networks to the innovative output of firms. We take up the issue of knowledge transfer, examining how the codification of knowledge can shape what is transmitted through networks. We briefly discuss the governance of networks, and then conclude with an assessment of what types of organizations and settings derive the greatest impact on innovation from participation in networks.

Research on the relationship between networks and innovation is a relatively recent area of inquiry. While there is a good deal of work underway, direct analyses measuring the impact of interfirm networks on performance are limited. Much of the extant research focuses on the effects of networks on patenting, access to information, and the generation of novel ideas. Moreover, the studies often examine high-tech industries, where investment in R&D is pronounced. Attention to the consequences of network ties for the financial performance of firms is relatively rare.

3.2 WHY HAVE NETWORKS GROWN IN IMPORTANCE?

The advantages of a heterogeneous group of contacts are well established in both social theory and network analysis. A strong tradition of theory and research, running from Simmel (1954) to Merton (1957) to Granovetter (1973) to Burt (1992), makes abundantly clear that there are informational, status, and resource advantages to having broad and diverse social circles. Below we review an array of recent empirical studies that demonstrate how interorganizational relationships lead to various benefits with respect to information diffusion, resource sharing, access to specialized assets, and interorganizational learning. In science and technology-based fields, the advantages that accrue from diverse sources of information and resources are considerable. Not surprisingly, then, as the commercialization of knowledge has assumed greater importance in economic growth, collaboration across organizational boundaries has become more commonplace. Interorganizational networks are a means by which organizations can pool or exchange resources, and jointly develop new ideas and skills. In fields where scientific or technological progress is developing rapidly, and the sources of knowledge are widely distributed, no single firm has all the necessary skills to stay on top of all areas of progress and bring significant innovations to market (Powell and Brantley 1992; Powell, Koput, and Smith-Doerr 1996; Hagedoorn and Duysters 2002). In such settings, networks can become the locus of innovation, as the creation of knowledge is crucial to improving competitive position.

Collaborative networks have long been central to the production process in craft-based industries (Eccles 1981), in industrial districts (Brusco 1982; Piore and Sabel 1984), and in fields such as aerospace where assembly depended upon key inputs from diverse participants. The growth of knowledge-intensive industries has heightened the importance of networks in R&D as well as product development and distribution. A persistent finding from a diverse set of empirical studies is that internal R&D intensity and technological sophistication are positively correlated with both the number and intensity of strategic alliances (Freeman 1991; Hagedoorn 1995).

For organizations in rapidly developing fields, heterogeneity in the portfolio of collaborators allows firms to learn from a wide stock of knowledge. Organizations with broader networks are exposed to more experiences, different competencies, and added opportunities (Beckman and Haunschild 2002). Such access creates an environment in which "creative abrasion," the synthesis that is developed from multiple points of view, is more likely to occur. In this view, "innovation occurs at the boundaries between mind sets, not within the provincial territory of one knowledge and skill base" (Leonard-Barton 1995: 62). By having access to a more varied set of activities, experiences, and collaborators, companies broaden the resource and knowledge base that they can draw on. By developing more multiplex ties with individual partners, either through pursuing multiple collaborations or expanding an existing R&D partnership into downstream development, companies increase the points of contact between them. When relationships are deepened, greater commitment and more thorough knowledge sharing ensue. Organizations with multiple and/or multifaceted ties to others are likely to have developed better protocols for the exchange of information and the resolution of disputes (Powell 1998). Parties that develop a broader bandwidth for communication are, in turn, more capable of transferring complex knowledge. In science-driven fields such as biotechnology, organizations that develop ties to different kinds of organizations and carry out multiple types of activities with these organizations are central players in industry networks (Powell et al. 2004). These centrally positioned organizations are both capable of pulling promising new entrants into the network and collaborating with a wide assortment of incumbents. Moreover, research shows that in biotechnology, organizations lacking such connections fail to keep pace and fall by the wayside (Powell et al. 2005).

3.3 VARIETIES OF NETWORKS

The literature on networks emphasizes that they are most pronounced in the domain between the flexibility and autonomy of markets and the force and control of organizational authority (Powell 1990). Networks thus combine some of the incentive structures of markets with the monitoring capabilities and administrative oversight associated with hierarchies (Mowery, Oxley, and Silverman 1996). For our purposes, we include networks based on formal contractual relations, such as subcontracting relationships, strategic alliances or participation in an industry-wide research consortium, and informal ties, based on common membership in a professional or trade association, or even a looser affiliation with a technological community.

One can differentiate networks with respect to their duration and stability, as well as whether they are forged to accomplish a specific task or evolve out of pre-existing bonds of association. Networks vary from short-term projects to long-term relationships, and the different temporal dimensions have important implications for governance. Some networks are hierarchical, monitored by a central authority; while others are more heterarchical, with distributed authority and strong self-organizing features. Grabher and Powell (2004) focus on temporal stability and forms of governance to differentiate four key types: informal networks (based on shared experience); project networks (short-term combinations to accomplish specific tasks); regional networks (where spatial propinquity helps sustain a common community); and business networks (purposive, strategic alliances between two parties). These types do not represent essentialist categories; rather they may overlap and interweave with one another. Consider these forms as useful coordinates to locate networks with respect to different combinatory elements.

Several key concepts provide potent analytical tools that apply across different types of networks and permit assessment of their effects. First, consider the differences between strong and weak ties (Granovetter 1973). In interpersonal terms, a strong tie is a person with whom you interact on a regular basis, while a weak tie is an acquaintance, or a friend of a friend. Strong ties are important for social support, but much of the novel information that a person receives comes from weak ties. Strong ties are based on common interests, consequently most information that is passed reinforces existing views. Weak ties introduce novelty in the form of different ideas or tastes, and by introducing new information they are, for example, invaluable in job searches and other circumstances where a small amount of new information is highly useful. Weak ties have a longer reach, but a much narrower bandwidth than strong ties. The latter are more cohesive, and often prove to be more effective at the exchange of complex information. Figure 3.1 illustrates the difference between strong and weak ties.

Much of the research on interfirm networks extrapolates from interpersonal relations.[2] In general, this is a plausible analytical move; however, it elides the question of whether relationships at the firm level are dependent on ongoing interpersonal ties, and whether the business relationship would be harmed or severed if the key participants were to depart. The extent to which interorganizational ties are contingent upon relations among individuals is a key question for scholarly research, as well as a critical challenge for business strategy (Gulati 1995; Powell 1998).

A second notable contrast is the distinction between networks as bridges and as structural holes (Burt 1992). Bridges are points of connection between parties that

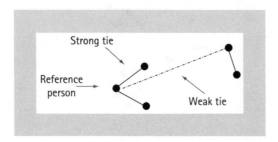

Fig. 3.1 Strong and weak ties

lack ties, such as when A knows B, and C knows B but not A. B is the bridge between A and C, thus the gateway to a linkage between A and C. Granovetter (1973) argued that bridges are the links that make weak ties possible. Burt (1992) deepened the argument by moving from the who question (i.e., which position in a network is best situated) to the question of how certain structural arrangements generate benefits and opportunities. He coined the term "structural holes" as a potential connection between clusters of units that are not connected. The possibility of making such a connection provides leverage, or opportunities for arbitrage. Those positioned to take advantage of structural holes can broker gaps in the social structure. See Figure 3.2 for illustration.

There is debate as to whether strong or weak ties, or bridges or structural holes, offer greater opportunities for innovation (Ahuja 2000*a*; Ruef 2002). Clearly, variation in network structures is associated with different content in relationships. Strong ties between two parties may restrict information gathering in terms of the breadth of search, but the information that is exchanged is "thick," or detailed and rich. Weak ties are thinner and less durable, but provide better access to non-redundant information. There is also disagreement as to whether networks can be designed or "pruned" to produce "optimal" shapes, without triggering repercussions. Whether location in a network is highly malleable or not, position in a network both empowers and constrains opportunities.

A third point of contrast is between networks formed intentionally across a market interface to accomplish a task and emergent networks that grow out of ongoing relationships. The former may be considered an instrumental or strategic relation, while the latter stems from more primordial relations, such as common ethnicity, friendship, or location. These different starting points matter, but in the fluid world of networks, the point of origin does not fix the evolution of a relationship. Consider two cases. There is a global trend toward vertical disaggregation in manufacturing, as firms are relying on suppliers for design and component inputs

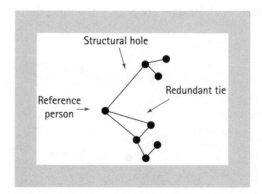

Fig. 3.2 Structural holes and redundant ties

in a variety of industries (see Womack, Jones, and Roos 1990, for autos; McKendrick, Doner, and Haggard 2000, for disk drives). Often these outsourcing decisions are driven by the need to reduce costs, save time, and enhance flexibility, while the large firm concentrates on those activities in which it has some form of competitive advantage. But as many analysts have noted, there is no natural stopping point in this relation (Sabel 1994; Helper, McDuffie, and Sabel 2000). The subcontractor can end up involved in design issues, doing critical R&D, or become central to efforts to improve quality. What began as a choice to outsource can, in some circumstances, become either a deep, mutually dependent collaboration or a highly ambiguous and opportunistic partnership. Helper et al. (2000) and Dyer and Nobeoka (2000) illustrate the marked trend for automobile subcontracting to evolve into interdependent, bilateral relationships.

Or consider the contemporary life sciences, where many R&D partnerships emerge out of ongoing intellectual relationships—co-authorships, mentor–mentee relationships, and common training (Murray 2002). These informal personal relationships may, however, come to involve significant intellectual property in the form of patents, and thus become highly formalized contractual agreements between organizations. We offer these examples as illustrations that networks forged out of strategic purposes can take on strong relational elements, while more personal ties can become contractual and highly specified. While it is possible to assign networks to either a transcational (i.e., based on a consideration of business opportunities without regard to prior social relations) or relational (i.e., embedded in ongoing social relationships) category, it is inappropriate to assume that relations remain fixed. As networks evolve, there is considerable give and take. Intense competition can render calculative strategic alliances more embedded, while the prospect of great financial reward can turn a "handshake" relationship between individuals into a formal legal linkage between firms.

Figure 3.3 provides a typology of different forms of networks, with the horizontal axis representing degree of purposiveness, ranging from informal to contractual. The vertical axis represents the extent of embeddedness, varying from open, episodic, or fluid to recurrent, dense connections among a fairly closed group (Granovetter, 1985). We illustrate each of the four cells with examples of types of innovation networks. In the lower left cell we place informal networks, such as a scientific invisible college, that emerge out of shared experience or common interest. Although these relations tend to be temporary and short-lived, the gray arrows are intended to show that these informal linkages can evolve into formal business alliances or more enduring primordial relations, where participation is more constant and less fluid. The primordial network in the upper left cell is characterized by a common social identity, continuous participation, and close ties. All these features are often found in professional networks, craft-based occupations, ethnic communities, and industrial districts. The upper right cell is typified by involvement in a common project. Membership in such a network is typically restricted and often

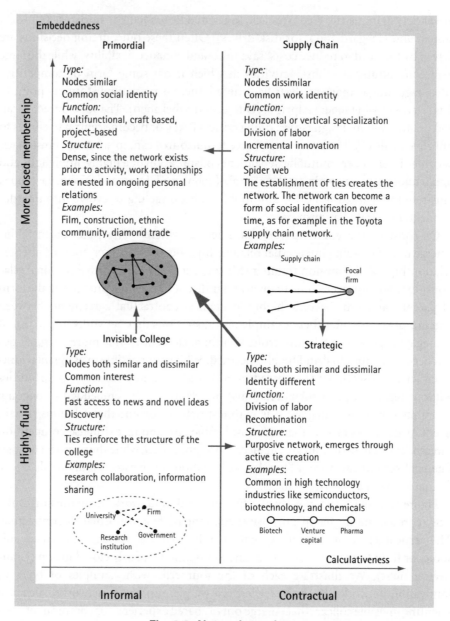

Fig. 3.3 Network typology

governed by a "lead" firm. A supply-chain network or a large construction project are apt examples. The gray arrows are meant to illustrate that supply-chain relations can evolve into either occupational communities and industrial districts or into formal business partnerships. The most purposive and instrumental type of network is the strategic alliance, represented by the lower right cell. We turn now to a discussion of the relationship between these forms and the innovation process.

3.4 EMPIRICAL STUDIES OF THE ROLE OF NETWORKS IN INNOVATION

3.4.1 Formal Ties

Most empirical studies of the relationship between networks and innovation focus on formal ties established among organizations. This stream of research documents a strong positive relationship between alliance formation and innovation, across such diverse industries as chemicals (Ahuja 2000a), biotechnology (Powell et al. 1996, 1999; Walker, Kogut, and Shan 1997; Baum, Calabrese, and Silverman 2000), telecommunications (Godoe 2000), and semiconductors (Stuart 1998, 2000). The diversity of the research contexts suggests the effects of network structure may be generalizable. Nevertheless, most research has focused on high technology industries, and uses patents as a proxy for innovation. More direct measures of innovative outputs are needed. Some of the important themes that emerge from this research highlight specific tie characteristics, technological uncertainty, and network evolution. In addition, researchers have emphasized the increased benefits in the form of resources and knowledge that alliances provide to entrepreneurial firms.

Tie characteristics. One line of research focuses on how different types of ties influence the benefits derived from alliances. Vinding (2002) identified 548 Danish manufacturing firms that developed one or more new products over a two-year period. In interviews with a subset of the companies, he finds that the impact of a collaboration on innovation is significantly related to both the type of partner and the pattern of previous collaborative relationships. The importance of prior interaction with partners points to the significance of relationship building, and how elements such as trust and cognitive understandings require time to develop. Domestic partners were found to have a greater positive impact on innovative performance than foreign partners, possibly due to the higher costs, both psychological and financial, associated with more distant collaborations. Vinding's research emphasizes benefits derived from strong local ties. Similarly, in a ten-year (1980–90) case study of the R&D portfolio of a Norwegian telecommunications organization, Godoe (2000) reports comparable results with respect to strong ties. His analysis suggests that radical innovations were more likely to emerge from intimate and prolonged interaction. But in the Norwegian case, the affiliations were not local, but instead based on membership in international telecommunications associations.

Powell et al. (1999) emphasize that experience with collaboration and centrality in the network derived from a diverse set of ties are important determinants of innovation among biotechnology firms over the period 1988–99. Their analyses suggest that centrality and experience resulted in more patenting. The most consequential connections in terms of patenting were R&D partnerships. The diversity of network ties also had a positive influence on rates of patenting. Powell et al. (1999) found that while network experience had a positive influence on patenting, the rate of increase diminished with additional experience, suggesting possible declining returns to network connectivity. The question of whether there are limits to connectivity needs to be investigated further.[3] These results suggested a "cycles of learning" process in which R&D collaborations generate attention that attracts other partners, who collaborate in developing novel ideas. This enhanced diversity of affiliations increases a firm's experience at managing collaborations and transferring knowledge, and increases their centrality in the industry network. Greater centrality is associated with a higher rate of patenting, and both centrality and higher volumes of patenting trigger subsequent R&D partnerships, restarting the cycle for centrally placed firms.

Most research has looked at the presence or absence of a formal collaboration. Ahuja (2000a), however, developed a more nuanced analysis including both direct and indirect ties, and the level of indirectness. Drawing data from a sample of 97 firms in the global chemicals industry, he used the number of patents as a measure of innovative output, while collaboration was measured through formal ties. More distant connections through affiliates of partners were coded as weak or indirect ties. The results show that both direct and indirect ties have a positive influence on innovation, though the impact of indirect ties is smaller than the impact of direct ties. The number of direct ties also negatively moderates the impact of indirect ties. In contrast to Burt's (1992) arguments about the arbitrage opportunities available through non-redundant contacts, Ahuja shows that a network with many structural holes can reduce innovative output, as measured by rates of patenting. A key advantage of close-knit networks may be due to their superior ability to transfer tacit knowledge (Van Wijk, Van den Bosch, and Volberda 2003). In an analysis of the exchange of information across project teams in a large multinational computer company, Hansen (1999) also illustrates that complex knowledge is transferred most easily through tightly knit networks.

Entrepreneurial firms. One active area of research concerns the effects of networks on survival chances of newly founded firms. Larson's (1992) ethnographic study of how a startup firm grew and prospered by drawing on external resources and support for key business functions illustrates how relationships are forged and sustained as startup firms grow. While not explicitly looking at innovative output, Larson added insight into the signal importance of networks in obtaining resources necessary to fuel a startup firm's success. Shan, Walker, and Kogut (1994) examined whether biotechnology startup firms' cooperative relationships with other firms had a positive effect on patenting. Their results offer support for the argument that collaborative relationships increased innovation, because formal cooperative relationships explained innovative output, while innovative output did not account for the pattern of alliances. The salience of alliances for young and small firms is further emphasized in Stuart's (2000) study of innovation in the semiconductor industry. His dataset includes 150 firms, followed by the consultancy firm Dataquest over the period 1985–91. Drawing on sales figures, patterns of strategic alliances, and patenting activity, Stuart shows that firms possessing technologically sophisticated alliance partners patented at a substantially greater rate than those that lacked such ties. Firms establishing strategic alliances with large partners also grew at a higher rate than firms without access to such partners. The returns to networks with regard to patenting were greatest for both young and small firms.

Baum, Calabrese, and Silverman (2000) pursue a similar question, asking how the composition of a startup firm's alliance portfolio affects its performance. Using data on 142 biotechnology firms founded in Canada between 1991 and 1996, they find a positive effect of alliance formation on startup innovation. Network efficiency, defined as the diversity of information and capabilities per alliance, showed a large positive effect on the number of biotech patents. Alliances with direct competitors had a negative effect on innovation, however. These results were moderated when the rival biotechnology firm had a larger share of the relevant market or if the rival biotechnology firm was highly innovative. Of the various performance measures used, the number of patents and the volume of R&D expenditures were most significantly influenced by rates of alliance formation.

Network dynamics. Drawing on Cohen and Levinthal's (1990) ideas about absorptive capacity, Powell et al. (1996) argued that firms utilize external collaborations to stay abreast in rapidly developing technological fields. But organizations cannot be passive recipients of new knowledge. "What can be learned is crucially affected by what is already known" (Powell et al. 1996: 120). To understand the "news" generated externally, organizations have to make "news" internally. In this fashion, the rate of acquisition of skills and resources from the outside is closely linked to the generation of expertise internally. In their work on the global biotechnology industry, they find that firms that develop experience at managing collaborative R&D relationships garner faster access to centrally positioned organizations. As experience at collaborating grows, firms widen the network of organizations with whom they partner. As a firm's experience with collaboration and its diversity of partners increase, the more central and visible the firm becomes in the industry. This centrality leads, in turn, to growth in the size of the firm, and to the ability to coordinate more alliances, creating a feedback cycle. This cycle of learning has been shown to be associated with positive financial performance (Powell et al. 1999), and a greater ability to collaborate with diverse kinds of organizations, which permits firms to retain a leadership position in the industry (Powell et al. 2005).

The general picture that emerges from research in organizational sociology and business strategy is one in which networks and innovation constitute a virtuous cycle. External linkages facilitate innovation, and at the same time innovative outputs attract further collaborative ties. Both factors stimulate organizational growth, and appear to enhance further innovation. Ahuja (2000b) and Stuart (2000), for example, demonstrate that firms with many prior patents are more likely to form alliances than firms lacking patents, suggesting a recursive process of innovation and growth in which collaborative ties play a central role. Further attention needs to be given, however, to such issues as the effects of the duration of linkages, experience with collaboration, and the consequences of broken ties on rates of innovation.

Technological uncertainty. Two additional aspects of the innovation process involve the relationship between strategy and alliance formation and the level of technological uncertainty in the field. Eisenhardt and Schoonhoven (1996) studied the population of semiconductor firms launched in the US between 1978 and 1985, and found that the more risk-taking a company's strategy, the more alliances a company formed. One explanation is that as firms gain credibility for developing pioneering technologies, access to financial and other resources for developing innovative technology is secured through alliances. An alternative view is that alliances are necessary to share the attendant risks in high-velocity environments. Sarkar, Echambadi, and Harrison's (2001) analysis of managers in a range of high-tech industries revealed that an active strategy of alliance formation enhanced performance, as measured by market share, sales growth, market development, and product innovation. They also report that managers who perceived the environment as more uncertain were more likely to pursue alliances. In addition, smaller firms derived more value from network linkages than larger firms, presumably because smaller companies viewed the technological landscape as more uncertain.

Rosenkopf and Tushman (1998) examined the role of technical communities in the flight simulation industry, where cooperative technical organizations play a critical role in developing standards and advancing the state of the art. In a study covering the years 1958–92, they found long periods marked by incremental change punctuated by shorter eras of ferment. They show that the rate of founding of technical networks increases during periods of discontinuity, and stabilizes into core "cliques" when ferment declines and a dominant design emerges. Subsequent technological change disrupts dominant cliques, and triggers the formation of new networks, restarting the cycle. Thus, both technological strategy and industry evolution are linked to patterns of network formation, with external networks assuming greater importance during periods of technical discontinuity and for firms with more risk-taking strategies. The importance of industry technical committees in standards setting has also been emphasized in the computer industry (Farrell and Salomer 1988) and videocassette recorders (Cusumano et al. 1992).

The overall conclusion of this group of studies is that networks provide access to more diverse sources of information and capabilities than are available to firms lacking such ties, and, in turn, these linkages increase the level of innovation inside firms. Younger and smaller firms may benefit more from collaborative relationships than do larger firms. Most notably, firms with a central location within networks generate more innovative output. Both direct and indirect ties provide a positive contribution to innovation, but the effect of indirect ties is moderated by the prevalence of direct ties. The evidence for the benefits of structural holes is not uniform; where structural holes might be beneficial is in the search for new information, but the knowledge transfer process appears to be facilitated by closer-knit networks. From the view of the dynamics of collaboration, successful external relations appear to beget more ties, which fuel firm growth and innovation. Clearly, there are limits to this cycle, but research has not addressed this question in depth thus far.

The majority of the studies reviewed in this section have been carried out using patents as the dependent variable and formal relationships as the independent variable. Patents provide a measure of novelty that is externally validated through the patent examination process, hence they are a useful indicator of knowledge creation (Griliches 1990). But patents have some limitations. Some kinds of innovations are not patented, and there is variation in the extent of patenting across industries. (See Chapter 14 by Malerba on inter-industry variation in innovation processes.) On the other hand, the focus of many of these studies—semiconductors, chemicals, biotechnology—is in fields where patenting is commonplace, and competitors in these sectors are active patentors. The attention to these high-technology industries raises questions, however, as to the generalizability of the results to other less knowledge-intensive industries.

One study that speaks to differences across industries is Rowley, Behrens, and Krackhardt's (2000) analysis of strong and weak ties in the steel and semiconductor industries. This study made a notable effort to distinguish between strong affiliations, where alliances entailed significant resource commitments and regular interactions, and more "arm's-length" transactions, where there was a rapid exchange, and the relationship was characterized by less frequency and depth. For example, equity alliances, joint ventures, and R&D partnerships were categorized as strong ties, while licensing, patent agreements, and marketing relations were classified as weak ties. Recognizing that weak ties serve as bridges to novel information, while strong ties are useful for both social control and the exchange of tacit knowledge, they find divergent results. In the steel industry strong ties are positively associated with performance; while in semiconductors weak ties are more efficacious. They suggest these findings reflect the importance of search and product innovation in semiconductors, and a focus on improvements in the production process for steel.

Much of the research on buyer–supplier relations and subcontracting has focused on more traditional industries, such as automobiles or textiles. To be sure, these industries make considerable use of technological advances, but they are less science-driven. As a consequence, the sources of relevant knowledge are not as widely dispersed. Strong ties thus tend to predominate over weak ties. But the content of those ties can evolve, changing from contractual to relational. Consider subcontracting relations, particularly one of the more notable examples—the Toyota auto production network. Researchers stress how the density of overlapping ties that connect this chain of production facilitates knowledge sharing, mutual learning, and fast responsiveness (Dyer 1996; Dyer and Nobeoka 2000). But the trust and reciprocity that characterize this dense network are the outcome of a long developmental process. In the 1950s and early 1960s, when Japanese firms competed on the basis of lowest cost, relationships with subcontractors were hierarchical and asymmetric. As firms increasingly competed on the basis of quality and innovation, however, complex multitiered supply relationships underwent significant change. These relationships can remain hierarchical in two key respects: the larger lead firm often has a significant financial stake in a supplier or affiliate, and it initiates the production process. But the asymmetry has been sharply reduced. Suppliers, in an effort to remain competitive, make significant investments in new equipment, constantly upgrade workers' skills, and take on more critical aspects of the assembly process (Helper et al. 2000). In turn, the larger firms offer long-term contracts, share employees and provide technical assistance, and make financial investments to fund equipment upgrades.

Too explicit a focus on formal, contractual linkages, however, neglects the myriad informal ties that connect organizations. All kinds of informal interactions take place between organizations, including participation in ad hoc industry committees, or executive education programs, conferences, trade association activities, and the like. Personnel mobility and common educational backgrounds may also foster informal linkages across firms. Such informal connections may be the basis on which more formal, contractual alliances are forged. Indeed, the success of formal affiliations may hinge on the strength of informal ties. Thus we turn to a discussion of noncontractual relations.

3.4.2 Informal Ties

Informal patterns of affiliation have long been a central topic in sociology and anthropology, where studies of friendship networks, advice and referral networks, and communities are common. There is also a well-established strand of research in organization theory that points out how informal relations within organizations are often not closely aligned with formal authority (Dalton 1959; Blau 1963). A small line of work focuses on the impact of informal networks in large, multinational companies (Ghoshal and Bartlett 1990; Hansen 1999). Relatively few studies, however, link informal ties to the innovation process, and there is scant research on informal interorganizational relations.

Scholars have often argued that the sharing of complex information is enhanced by embedded ties, which suggests that informal ties have the potential to make a significant contribution to innovation. There is a strong sense among researchers that informal relations undergird formal ties. Powell et al. (1996) argue that, in the life sciences, "beneath most formal ties lie a sea of informal ties." Nevertheless, many organizations are largely unaware of the extent to which formal activities are buttressed by informal connections (Cross, Borgatti, and Parker 2002).

One of the key studies of informal networks among firms was Von Hippel's (1987) work on the sharing of proprietary information among US steel mini-mill produc- ers. Based on interviews with plant managers and other engineers with direct knowledge of manufacturing processes, he found that the trading of proprietary knowledge with both cooperating and rival firms was commonplace. He was initially surprised that proprietary knowledge was so "leaky," but he came to recognize that information exchange was highly reciprocal and conditioned on expectations that requests for help would be met. Much of the information that was shared focused on production problems, matters of pollution control and safety, and issues dealing with industry-wide concerns. But when relationships among engineers in rival firms were particularly close, more proprietary information was exchanged. Von Hippel also found that engineers had strong norms of membership in a professional community that cut across firms, and that information trading was a means to secure reputation and status in that community. He provides numerous examples of how the sharing of complex information by engineers contributed to the productiv- ity of mini-mills.

The cluster of individuals that share a similar set of skills and expertise has been dubbed a "community of practice" (Wenger 1998), or a "network of practice" (Brown and Duguid 2001). Similar in some respects to a technical community, or a sophisticated hobby club, these loose groups are engaged in related work practices, though they do not necessarily work together. Such fluid groups are important to the circulation of ideas. Saxenian (1994) observed ample sharing of proprietary know- ledge among engineers in Silicon Valley, many of whom have as strong a commit- ment to their peers within the same occupational group as to their companies. Saxenian argues that informal knowledge sharing, widely institutionalized as a professional practice in Silicon Valley, is one of the crucial factors contributing to its fertile innovative climate. Cohen and Fields (1999) stress that professional ties in Silicon Valley are forged in complex collaborations between entrepreneurs, scien- tists, firms, and associations, focused on the pursuit of innovation and its commer- cialization. This collaborative process generates and refines the intangible raw material of technical change—ideas.

Kreiner and Schultz (1993) analyze the importance of informal ties through in- depth interviews with university researchers and industry research directors in the Danish biotech field. They stress that successful collaborative R&D alliances within the Danish biotech industry are often based on informal ties. A barter economy, where materials, laboratory tests, chemicals, etc. are exchanged, was pervasive in this sector. They show that norms of sharing information on the frontier of research aid in the formation of more formal networks. As in the mini-mills, information exchange is not under managerial control, even though such reciprocal flows can be channeled by managerial actions.

Many studies of informal relationships stress the significance of trust. Tsai and Ghoshal (1998) studied the association between intrafirm networks and innovation in fifteen business units of a multinational electronics company. They found, not surprisingly, that social ties led to a higher degree of trustworthiness among business units. Trust increased resource-exchange and combination between the business units, which contributed to product innovation. The importance of trust also looms large in Uzzi's (1997) analysis of the difference between "arm's-length" ties ("a deal in which costs are everything") and embedded ties ("you become friends with these people—business friends. You trust them and their work. They're part of the family"). Uzzi conducted interviews and ethnographic observations at twenty-three women's better-dress firms in the New York City apparel industry. His study is notable not only for the quality of his data, but also for his attention to the performance consequences of different kinds of exchange relations. Uzzi found that organizational performance increases with the use of embedded ties to network partners, as these ties were superior at conveying complex, context-dependent knowledge. He argued, however, that a balance between a firm's embedded ties and a firm's arm's-length ties needed to be struck, because a network structure comprising only arm's-length ties or embedded ties decreased organizational performance.

The significance of a balanced network structure, mixing formal and informal affiliations, is also emphasized in Ruef's (2002) analysis of entrepreneurship. He found that individuals positioned in heterogeneous networks, comprising both strong and weak ties, are more likely to be regarded as innovative by peers, in comparison to entrepreneurs in more homogeneous networks. Rosenkopf has found a similar interweaving of formal and informal relations in her research on industry-wide expert communities in the areas of flight simulation and mobile phones. Rosenkopf, Metiu, and George (2001) analyze joint participation by cellular service firms in technical committees, finding that such membership facilitates subsequent formal interfirm alliance formation. The effect of participation in technical committees decreases when firms have already established prior alliances, suggesting that the effect of informal ties is more catalytic when firms do not already have established alliance partners.

3.4.3 Multi-Party Relationships

Most studies of networks and innovation have examined either dyadic ties, or a focal firm in the context of an overall network. Rosenkopf and Tushman's (1998) work on expert communities emphasized the importance of studying multi-party relationships that connect technical professionals across organizations. Akera (2001) uses archival data to map the importance of the IBM user group Share in the early days of computing, offering a portrait of a large multi-party collaboration. Share was formed soon after IBM's release of the first mainframe computer in 1953, to enable IBM's costumers to swap programs, and collaborate on programming so that duplicative effort was avoided. Some of the main innovations that came out of Share provided the basis for both systems programming and operating systems, which today form the backbone of modern computer use. Share conveyed information about hardware changes that substantially improved the design of IBM computers and peripherals. One of the main contributions of the Share network was the creation of technical standards, but a second-order effect was also the diffusion of knowledge across companies and between users.

A modern analogue to Share is the Linux community, founded as a group of users trying to develop an alternative to the operating system supplied by Microsoft. The network of Linux programmers has proven effective in developing software in a highly distributed fashion. In the beginning, most programmers had never met each other and only knew each other virtually—by the username they used when coding. The Linux community has a very modest organizational structure, relying on a combination of interpersonal networks and an individual's reputation as a skilled programmer to serve as the admission ticket to the network (O'Mahony 2002; Weber 2003). One difference between the Share network and the Linux network is that corporate interests drove the Share network, while Linux has been primarily driven by the end users. Nevertheless, in both cases, the decomposability of programming tasks is an important factor in facilitating distributed networks.

Another large network that has been widely studied is the network of scientists, often termed invisible colleges (Crane 1972). An invisible college is an informal network of researchers who form around a common problem or paradigm. By studying invisible colleges, Crane (1972) hoped to understand how knowledge grows and how the structure of scientific communities affects the expansion of knowledge. There are now numerous studies of scientific networks, mapping the structure of co-authorship and citations (Newman 2003 provides a good overview), though few attend explicitly to the issue of innovation. David (2001), however, develops a formal model to show that the liberal sharing of knowledge within the scientific community is a major driver of scholarly innovation. One of the historic characteristics of scientific communities is that information and research results have been distributed openly among members of the relevant community. The shift toward increasing research commercialization by universities has led some scholars to question whether the innovative benefits of invisible colleges will persist, or if commercial interests will block informal knowledge sharing among scientists (Powell and Owen-Smith 1998; Owen-Smith 2003). Chapter 8 by Mowery and Sampat on university–industry interfaces offers a more detailed discussion of the role of universities in the innovation process.

Another line of research that has attended to scientific and technological networks, dubbed Actor–Network Theory (ANT), examines how particular definitions or configurations of science and technology triumph over alternative conceptions. Actor–Network Theory is unique in its treatment of artifacts and technologies, as well as people and organizations, as members of a network (Callon 1998; Latour 1987). The primary contribution of ANT to the relationship between networks and innovation is to show that not only can networks facilitate innovation, but they also constrain it by determining the kind of innovations produced, their subsequent interpretation, and their final use (Callon 2002).

A related line of work that looks at networks as systems of activity is the markets-as-networks approach, developed by Scandinavian marketing researchers (Håkansson and Snehota 1995). This approach examines the multiple relationships among organizations, and shows how these different aspects of interorganizational relationships transform and evolve over time. For example, supplier networks may change frequently, with different elements of production being either outsourced or insourced (Waluszewski 1995). The resources that are exchanged among the partners in a production network are constantly changing. What determines whether an entity is a resource depends on the situation, and its use in combination with other resources. Resources are, thus, always polyvalent in both use and value. The participants in a production network, both individually and collectively, develop bonds characterized by trust and commitment. These bonds also have an organizing effect on networks, as they shape the identities of actors, and account for different levels of commitment among participants. This rich vein of qualitative research has not explicitly focused on innovation, however.

The various studies of multi-party networks tend to emphasize the processual aspects of collaboration. This attention to content is welcome, but it sometimes comes at the expense of measuring the output of relationships, particularly how the sharing and processing of information by members of a network can determine the generation of novelty. A fuller understanding of the innovation process needs to examine the topic of information sharing, a subject to which we now turn.

3.5 KNOWLEDGE TRANSFER

The role of knowledge transfer is clearly central to the innovation process. Research has highlighted two different aspects of the knowledge-transfer process, each of which influences innovation, albeit in different respects. One explanation for the exchange of information through networks emphasizes the importance of complementary assets in the division of innovative labor (Mowery, Oxley, and Silverman 1996). If firm A is good at producing a specific component and firm B is capable of using that component to produce an engine, they collaborate in a joint production in which their capabilities reinforce one another. In biotechnology, for example, small firms with close ties to university scientists may excel at drug development, but lack the skills and resources to manage or fund costly clinical trials. By working closely with a research hospital and an established firm that has a limited pipeline of new medicines, the parties collaborate in a division of labor that is mutually rewarding, and can result in the participants learning from one another, and accomplishing tasks they could not do individually.

A second form of knowledge sharing occurs when existing information within a network is recombined in novel ways. Indeed, novelty is often the unanticipated result of reconfiguring existing knowledge, problems, and solutions (Nelson and Winter 1982; Fleming and Sorenson 2001). As a consequence of such collisions or transpositions, firms can generate something they were unable to create on their own. Both forms of knowledge transfer depend on some manner of successful exchange of ideas, however.

An oft-used distinction is drawn between tacit and explicit knowledge (Cowan, David, and Foray 2000). Interest in tacit knowledge stems from Polanyi's (1956) argument that we frequently know a good deal more than we can express verbally. Explicit knowledge is highly codified, as in blueprints, recipes, manuals, or in the form of training. Tacit knowledge lacks such extensive codification (Nonaka and Takeuchi 1995). Valuable, productive knowledge often demands considerable effort to acquire, and such knowledge is frequently altered in the process of acquisition and application. Perhaps the most vivid example is the continuing effort of US automakers to acquire, understand, and implement the Japanese system of lean production (Womack, Jones, and Roos 1990; Dyer and Nobeoka 2000). Knowledge of complex production technologies is rarely obtained in a fully digestible form; understanding inevitably entails learning by using. The distinction between codified and tacit is key because the latter demands considerably more trial-and-error learning to apply the new knowledge in a different setting.

Many studies point to the relatively easy transferability of explicit knowledge in contrast to tacit knowledge. Simonin (1999) shows that knowledge transfer within alliances is negatively affected by both the nature of knowledge and differences in organizational culture. He observes important differences in knowledge exchange between long and short-lived alliances. Older alliances develop a common language and shared mental models between partners, suggesting a learning curve within alliances where the negative effects of lack of experience and knowledge complexity subside as the alliance matures. Thus, as an alliance ages and participants develop relationship-specific understanding, there is the opportunity to convey more subtle forms of information more effectively. Complex tacit knowledge can become more explicit as partners develop a wider bandwidth of communication.

If knowledge tacitness is a limiting factor in the transfer of knowledge, then the cost of transferring knowledge is proportional to the type of knowledge transferred. Easily transferred knowledge is widely dispersed at a low cost (Boisot 1998), thus the likelihood that explicit knowledge contains novel elements that would lead to innovation is lower. On the other hand, when knowledge is very "sticky" (von Hippel 1998) and contains a large tacit component, the degree of difficulty and the costs of transfer are high. Consequently, the expected gains realized from this information are uncertain, as the cost of obtaining information may exceed its value. This suggests that when knowledge involves a moderate level of complexity, the benefits derived from transfer may be greatest. Figure 3.4 suggests a hypothetical inverted U-shaped relationship between innovation and codification. Here we assume that there is variability in the cost of information transfer, and that the greatest value may be derived when novel ideas are transmitted without too much difficulty.

Szulanski's (1996) analysis of the transfer of internal benchmarking efforts in eight companies suggests key dimensions along which knowledge transfer can be

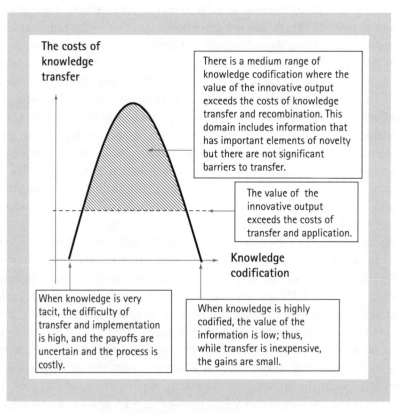

The costs of knowledge transfer

There is a medium range of knowledge codification where the value of the innovative output exceeds the costs of knowledge transfer and recombination. This domain includes information that has important elements of novelty but there are not significant barriers to transfer.

The value of the innovative output exceeds the costs of transfer and application.

Knowledge codification

When knowledge is very tacit, the difficulty of transfer and implementation is high, and the payoffs are uncertain and the process is costly.

When knowledge is highly codified, the value of the information is low; thus, while transfer is inexpensive, the gains are small.

Fig. 3.4 Knowledge codification and innovation

distinguished. He demonstrates that relationships between sender and receiver are important, in that both parties need mutual awareness of state-of-the-art practices. Obviously, communication is critical to information exchange. But even when relationships function well, some knowledge is causally ambiguous, or sticky, and thus not easily transferred. Moreover, information exchange is hindered when the parties have differential levels of absorptive capacity, that is, the ability to recognize the value of new information, assimilate it, and apply it to commercial ends (Cohen and Levinthal 1990). This capacity is essential to innovative capability. For example, internal success with R&D and R&D expenditures positively affect a firm's ability to exploit the opportunities presented in external relations (Cohen and Levinthal 1990; Powell et al. 1996).

The productive transfer of knowledge is also essential when two or more organizations are able to combine their different capabilities, and create a product or service that they would not be able to construct on their own. A good illustration is the Italian motorcycle industry where the locus of innovation is broadly dispersed, and through cooperation, the participants benefit from specialization and variety generation (Lipparini, Lorenzoni, and Zollo 2001). Because all the participants provide valuable inputs, there is a high commitment to knowledge generation. The lead firms develop relational capacities aimed at pooling the skills of specialized participants, helping the overall flow of information and resources in the network. A parallel analysis of the Italian packaging machinery industry stresses the creation of a supplier network in which specialized roles are highly complementary (Lorenzoni and Lipparini 1999). Over time, managers of the core companies developed a specialized supplier network and each participant focused on a narrow, but highly competitive set of core competencies. This network structure enabled every step of the supply chain to specialize in improvements of their specific component, thereby increasing the responsiveness of the participants to market conditions.

3.6 GOVERNANCE AND INCENTIVE ISSUES

Many studies of interfirm networks draw data from a single point in time and thus do not examine how collaborations unfold over time. Even studies that do look at dynamics tend to do so from the perspective of a dyad. Initially, the choice by a young firm of whom to partner with is often driven by resource needs. As both firms and the network mature, various dyadic choices increasingly reflect structural properties of the network. Thus, the existing network structure shapes search behavior. Consequently, networks both enable information to become knowledge and determine the nature of knowledge (Kogut 2000). But we do not, as yet, have a parallel understanding of the management and governance of networks to accompany analyses of structure and topology. Concerns with how the parties in a relationship adapt to changing circumstances, or attend to the incentives to adjust the relationship to make improvements remain largely unexamined.

Not surprisingly, then, many studies assume that as long as a relationship persists, the participants are achieving their goals, and view termination as a sign of failure. Such a view misses the point that the relationship may have completed its goal or outlived its usefulness. Inkpen and Ross (2001) find evidence that the termination of alliances might not signify failure, but simply be a sign of the conclusion of collaborative activity (for example, a new product is launched). Nevertheless, some collaborations seem to persist even though they have stagnated or outlived their usefulness. Inkpen and Ross suggest several reasons why interfirm alliances may persist beyond their optimal duration.[4] Alliances can be difficult and expensive to form; thus, once established, there may be reluctance to disband them. Moreover, as more firms pursue alliances, a bandwagon effect is created and many firms jump on it out of fear of being left without a partner. There are also challenges and costs associated with managing a partnership. If the relationship is poorly coordinated, the costs can outweigh the benefits. Alliances can also become synonymous with a firm's values, making them difficult to discontinue. Finally, there may be costs associated with closing an alliance. All of these factors may contribute to alliances existing beyond the period when they create value for a firm.

Several studies have pointed to problems of stagnation that can occur in some long-term associations. Although this work does not deal directly with rates of innovation, focusing instead on viability and survival, the general point is apt. When the participants in a network become too tightly knit and information circulates only among a small group, networks can become restrictive and ossified. Information that cycles back and forth only among the same participants can lead to lock-in or sclerosis. When networks turn inward and become restricted in terms of access to new members, the possibility of "group think" increases. Grabher's (1993) study of steelmaking in the Ruhr illustrates how a highly cohesive, homogeneous region became so overembedded that no producers opted for alternative strategies. This cognitive lock-in eventually led to the decline of steelmaking in Germany. Powell (1985) shows how the failure of editors to renew and update their networks leads to a decline in the quality of a book publisher's list and reputation. Portes and Sensenbrenner (1993) illustrate how ethnic community networks can become restrictive and subject successful entrepreneurs to ostracism when they deviate too much from community standards. In sum, the ties that bind economic actors together can become the ties that blind, thwarting recognition of preferable alternatives.

3.7 Summary

Interorganizational networks have grown considerably in importance over recent decades. Networks contribute significantly to the innovative capabilities of firms by exposing them to novel sources of ideas, enabling fast access to resources, and enhancing the transfer of knowledge. Formal collaborations may also allow a division of innovative labor that makes it possible for firms to accomplish goals they could not pursue alone. Research on alliances documents that investments in mutual learning and a portfolio of diverse collaborations are associated with increased patenting. While patenting is only an input to the innovation process, the strength of these empirical results highlights the importance of access to heterogeneous contacts.

We have argued that the nature of knowledge, conceptualized in terms of tacitness or explicitness, is an important factor in determining whether members of a network can effectively share information and skills. Networks that are rooted in a division of innovative labor logic may find it easier to transfer tacit knowledge in the form of finished inputs, while networks involved in the co-creation of novel ideas may succeed or fail on the basis of their ability to convey and transfer ideas that are not easily codified.

Another central challenge to networks of innovators is developing the capacity to simultaneously enhance the flow of information among current participants and be open to new entrants. The twin tests of increasing cohesion within the network and recognizing promising sources of new ideas are difficult to surmount. Some research suggests that a mixture of strong and weak ties affords the proper blend of reliability and novelty.

Much more research is needed, however, to ascertain how mixtures of thick, reliable affiliations can be combined with novel linkages to newcomers. Moreover, since some affiliations are essentially person to person ties, a greater understanding is needed of how relations among individuals are aggregated to productive relations between corporate actors. Or, in cases where this issue is neglected, as in studies of contractual relations between firms where the informal relational underpinnings are not analyzed, the opportunity to examine the intertwining of the careers of individuals and the strategies of firms is missed. Equally important, the network literature has not focused explicitly on different measures of innovative output, whether it is new products or services, new modes of organizing production, or more rapid response to competitive demands. The standard measures are based on either patents, an input to the innovation process, or problem solving, without sufficient attention to either the timeliness or the optimality of the solution. The network literature is still relatively young, however, so it may be premature to expect such sophisticated answers. We look forward to future research that offers a more compelling analysis of the specific ways in which networks shape innovative outputs.

Notes

1. The five that formed the core of the human sequencing component were the Whitehead Institute, Washington University, Baylor College of Medicine, the Joint Genome Institute (a cluster of three national laboratories at the Department of Energy), and the Sanger Institute in England.
2. Notable exceptions include Ahuja 2000*a*, Rowley et al. 2000, Powell et al. 2005.
3. Owen-Smith and Powell (2003) found that US universities with strong ties to a limited set of commercial partners had "fertile" patent portfolios, with fertility measured by the impact of patent citations. Those universities with few ties also had les fertile patents and patented much less. The optimal strategy for research universities, with respect to patent volume and impact, appears to be one of diverse ties to a wide array of industrial partners. Diversity mitigates possible capture from too close relations with commercial firms.
4. Inkpen and Ross (2001) do not specify how to measure the appropriate duration of relationships. Instead, they assume that some relationships become stale and dutiful over time, and no longer generate benefits that outweigh the costs of sustaining them.

References

*Ahuja, G. (2000*a*), "Collaboration Networks, Structural Holes, and Innovation: A Longitudinal Study," *Administrative Science Quarterly* 45: 425–55.

——(2000*b*), "The Duality of Collaboration: Inducements and Opportunities in the Formation of Interfirm Linkages," *Strategic Management Journal* 21: 317–43.

Akera, A. (2001), "The IBM User Group, Share," *Technology and Culture* 42: 710–36.

Baum, J. A. C., Calabrese, T., and Silverman, B. S. (2000), "Don't Go It Alone: Alliance Network Composition and Startups' Performance in Canadian Biotechnology," *Strategic Management Journal* 21: 267–94.

Beckman, C., and Haunschild, P. (2002), "Network Learning: The Effects of Partner's Heterogeneity of Experience on Corporate Acquisitions," *Administrative Science Quarterly* 47: 92–1724.

Blau, P. M. (1963), "Consultation Among Colleagues," in *Dynamics of Bureaucracy*, Chicago: University of Chicago Press, 157–69.

Boisot, M. H. (1998), *Knowledge Assets: Securing Competitive Advantage in the Information Economy*, Oxford: Oxford University Press.

Brown, J. S., and Duguid, P. (2001), "Knowledge and Organization: A Social Practice Perspective," *Organization Science* 12(2): 198–213.

Brusco, S. (1982), "The Emilian Model: Productive Decentralization and Social Integration," *Cambridge Journal of Economics* 6: 167–84.

*Burt, R. S. (1992), *Structural Holes*, Cambridge, Mass.: Harvard University Press.

Callon, M. (1998), "Actor-Network Theory—The Market Test," in J. Law and J. Hassard (eds.), *Actor Network Theory and After*, Cambridge, Mass.: Blackwell, 186–96.

* Asterisked items are suggestions for further reading.

—— (2002), "From Science as an Economic Activity to Socioeconomics of Scientific Research," in P. Mirowski and E. M. Sent (eds.), *Science Bought and Sold*, Chicago: University of Chicago Press.

CHESBROUGH, H. (2003), *The Era of Open Innovation*, Boston: Harvard Business School Press.

COHEN, S., and FIELDS, G. (1999), "Social Capital and Capital Gains in Silicon Valley," *California Management Review* 41(2): 108–30.

COHEN, W., and LEVINTHAL, D. (1990), "Absorptive Capacity: A New Perspective on Learning and Innovation," *Administrative Science Quarterly* 35: 128–52.

COWAN, R., DAVID, P. A., and FORAY, D. (2000), "The Explicit Economics of Knowledge Codification and Tacitness," *Industrial and Corporate Change* 9: 211–53.

CRANE, D. (1972), *Invisible Colleges: Diffusion of Knowledge and Scientific Communities*, Chicago: University of Chicago Press.

CROSS, R., BORGATTI, S., and PARKER, A. (2002), "Making Invisible Work Visible," *California Management Review* 44(2): 25–46.

CUSUMANO, M., MYLONADIS, Y., and ROSENBLOOM, R. (1992), "Strategic Maneuvering and Mass-Market Dynamics: The Triumph of VHS over Beta," *Business History Review* 66: 51–94.

DALTON, M. (1959), "Power Struggles in the Line," in *Men Who Manage*, New York: Wiley, 71–109.

DAVID, P. A. (2001), "Cooperation, Creativity, and Closure in Scientific Research Networks: Modeling the Simpler Dynamics of Epistemic Communities," Conference at Centre Saint-Gobain, Paris.

*DYER, J. H. (1996), "Specialized Supplier Networks as a Source of Competitive Advantage: Evidence from the Auto Industry," *Strategic Management Journal* 17(4): 271–91.

—— and NOBEOKA, K. (2000), "Creating and Managing a High-Performance Knowledge-Sharing Network: The Toyota Case," *Strategic Management Journal* 21: 345–67.

ECCLES, R. (1981), "The Quasifirm in the Construction Industry," *Journal of Economic Behavior and Organization* 2: 335–57.

EISENHARDT, K. M., and SCHOONHOVEN, C. B. (1996), "Resource-based View of Strategic Alliance Formation: Strategic and Social Effects in Entrepreneurial Firms," *Organization Science* 7: 136–50.

FARRELL, J., and SALOMER, G. (1988), "Coordination through Committees and Markets," *RAND Journal of Economics* 19(2): 235–52.

FLEMING, L., and SORENSON, O. (2001), "Technology as a Complex Adaptive System: Evidence from Patent Data," *Research Policy* 30: 1019–39.

*FREEMAN, C. (1991), "Networks of Innovators: A Synthesis of Research Issues," *Research Policy* 20: 499–514.

GALUNIC, D. C., and RODAN, S. (1998), "Resource Recombinations in the Firm: Knowledge Structures and the Potential for Schumpeterian Innovation," *Strategic Management Journal* 19: 1193–201.

GHOSHAL, S., and BARTLETT, C. (1990), "The Multinational Corporation as an Interorganizational Network," *Academy of Management Review* 15: 561–85.

GODOE, H. (2000), "Innovation Regimes, R&D and Radical Innovations in Telecommunications," *Research Policy* 29: 1033–46.

GOMES-CASSERES, B. (1996), *The Alliance Revolution: The New Shape of Business Rivalry*, Cambridge, Mass.: Harvard University Press.

GRABHER, G. (1993), "The Weakness of Strong Ties: The Lock-in of Regional Development in the Ruhr Area," in G. Grabher (ed.), *The Embedded Firm*, London: Routledge.

—— and POWELL, W. W. (2004), "Introduction," in *Critical Studies in Economic Institutions: Networks*, London: Edward Elgar.

*GRANOVETTER, M. (1973), "The Strength of Weak Ties," *American Journal of Sociology* 78: 1360–80.

—— (1985), "Economic Action and Social Structure: The Problem of Embeddedness," *American Journal of Sociology* 91: 481–510.

GRILICHES, Z. (1990), "Patent Statistics as Economic Indicators: A Survey," *Journal of Economic Literature* 28: 1661–707.

GULATI, R. (1995), "Does Familiarity Breed Trust? The Implications of Repeated Ties for Contractual Choice in Alliances," *Academy of Management Journal* 38: 85–112.

HAGEDOORN, J. (1995), "Strategic Technology Partnering During the 1980s: Trends, Networks, and Corporate Patterns in Non-Core Technologies," *Research Policy* 24: 207–31.

—— (1996), "Trends and Patterns in Strategic Technology Partnering since the Early Seventies," *Review of Industrial Organization* 11: 601–16.

—— (2002), "Inter-Firm R&D Partnerships: An Overview of Major Trends and Patterns since 1960," *Research Policy* 31: 477–92.

——, and DUYSTERS, G. (2002), "External Sources of Innovative Capabilities: The Preference for Strategic Alliances or Mergers and Acquisitions," *Journal of Management Studies* 39(2): 167–88.

HÅKANSSON, H., and SNEHOTA, I. (1995), *Developing Relationships in Business Networks*, London: Routledge.

HANSEN, M. T. (1999), "The Search–Transfer Problem: The Role of Weak Ties in Sharing Knowledge across Organization Subunits," *Administrative Science Quarterly* 44: 82–111.

HELPER, S., MacDUFFIE, J. P., and SABEL, C. (2000), "Pragmatic Collaborations: Advancing Knowledge While Controlling Opportunism," *Industrial and Corporate Change* 9(3): 443–89.

HICKS, D. M., and KATZ, J. S. (1996), "Where is Science Going?" *Science, Technology and Human Values* 21(4): 379–406.

INKPEN, A. C., and ROSS, J. (2001), "Why do Some Strategic Alliances Persist beyond their Useful Life?" *California Management Review* 44: 132–48.

KOGUT, B. (2000), "The Network as Knowledge: Generative Rules and the Emergence of Structure," *Strategic Management Journal* 21: 405–25.

KREINER, K., and SCHULTZ, M. (1993), "Informal Collaboration in R&D. The Formation of Networks across Organizations," *Organization Studies* 14: 189–209.

LAMBRIGHT, W. H. (2002), "Managing 'Big Science': A Case Study of the Human Genome Project," Washington, DC: Pricewaterhouse Coopers Endowment for the Business of Government.

LARSON, A. (1992), "Network Dyads in Entrepreneurial Settings: A Study of the Governance of Exchange Processes," *Administrative Science Quarterly* 37: 76–104.

LATOUR, B. (1987), *Science in Action*, Cambridge, Mass.: Harvard University Press.

LEONARD-BARTON, D. (1995), *Wellsprings of Knowledge: Building and Sustaining the Sources of Innovation*, Boston: HBS Press.

LINK, A. (1996), "Research Joint Ventures: Patterns from Federal Register Filings," *Review of Industrial Organization* II (October): 617–28.

—— (1999), "Public/Private Partnerships in the United States," *Industry and Innovation* 6(2): 191–217.

LIPPARINI, A., LORENZONI, G., and ZOLLO, M. (2001), "Dual Network Strategies: Managing Knowledge-based and Efficiency-based Networks in the Italian Motorcycle Industry," Working paper.

LORENZONI, G., and LIPPARINI, A. (1999), "The Leveraging of Interfirm Relationships as a Distinctive Organizational Capability," *Strategic Management Journal* 20: 317–38.

McKENDRICK, D. G., DONNER, R. F., and HAGGARD, S. (2000), *From Silicon Valley to Singapore: Competitive Advantage in the Hard Disk Drive Industry*, Stanford, Calif.: Stanford University Press.

MERRILL, S. A., and COOPER, R. S. (1999), "Trends in Industrial Research and Development: Evidence from National Data Sources," in *Securing America's Industrial Strength*, Washington, DC: National Academy Press, 99–116.

MERTON, R. K. (1957), "The Role-Set: Problems in Sociological Theory," *British Journal of Sociology* 8: 110–20.

MOWERY, D. C. (ed.) (1988), *International Collaborative Ventures in US Manufacturing*, Cambridge, Mass.: Ballinger.

—— (1999), "America's Industrial Resurgence? An Overview," in *U.S. Industry in 2000: Studies in Competitive Performance*. Washington, DC: National Academy Press, 1–16.

—— OXLEY, J. E., and SILVERMAN, B. S. (1996), "Strategic Alliances and Interfirm Knowledge Transfer," *Strategic Management Journal* 17: 77–91.

MURRAY, F. (2002), "Innovation as Co-evolution of Scientific and Technological Networks: Exploring Tissue Engineering," *Research Policy* 31: 1389–403.

National Science Board (1998), *Science and Engineering Indicators—1998*. Arlington, Va.: National Science Foundation.

NELSON, R., and WINTER, S. (1982), *An Evolutionary Theory of Economic Change*, Cambridge, Mass.: Harvard University Press.

NEWMAN, M. (2003), "The Structure and Function of Complex Networks," *SIAM Review* 45: 167–256.

NONAKA, I., and TAKEUCHI, N. (1995), *The Knowledge-Creating Company*, New York: Oxford University Press.

NOTEBOOM, B. (1999), *Inter-Firm Alliances: Analysis and Design*, London: Routledge.

O'MAHONY, S. (2002), *The Emergence of a New Commercial Actor: Community Managed Software Projects*, Ph.D. dissertation, Department of Management Science and Engineering, Stanford University.

OWEN-SMITH, J. (2003), "From Separate Systems to a Hybrid Order: Accumulative Advantage Across Public and Private Science at Research One Universities," *Research Policy* 32(6): 1081–104.

—— and POWELL, W. W. (2003), "The Expanding Role of University Patenting in the Life Sciences," *Research Policy* 32(9): 1695–711.

PIORE, M., and SABEL, C. (1984), *The Second Industrial Divide*, New York: Basic Books.

POLANYI, M. (1956), *Personal Knowledge—Towards a Post-Critical Philosophy*, Chicago: University of Chicago Press.

PORTES, A., and SENSENBRENNER, J. (1993), "Embeddedness and Immigration: Notes on the Social Determinants of Economic Action," *American Journal of Sociology* 98: 1320–50.

POWELL, W. W. (1985), *Getting Into Print: The Decision-Making Process in Scholarly Publishing*, Chicago: University of Chicago Press.

POWELL, W. W. (1990), "Neither Market Nor Hierarchy: Network Forms of Organization," in B. M. Straw and L. L. Cummings (eds.), *Research in Organizational Behavior*, Greenwich, Conn.: JAI Press, 12: 295–336.

—— (1996), "Inter-Organizational Collaboration in the Biotechnology Industry," *Journal of Institutional and Theoretical Economics* 120(1): 197–215.

—— (1998), "Learning from Collaboration: Knowledge and Networks in the Biotechnology and Pharmaceutical Industries," *California Management Review* 40(3): 228–40.

—— and BRANTLEY, P. (1992), "Competitive Cooperation in Biotechnology: Learning Through Networks?" in R. Eccles and N. Nohria (eds.), *Networks and Organizations*, Boston: Harvard University Press.

*—— KOPUT, K. W., and SMITH-DOERR, L. (1996), " Interorganizational Collaboration and the Locus of Innovation in Biotechnology," *Administrative Science Quarterly* 41(1): 116–45.

—— —— and OWEN-SMITH, J. (1999), "Network Position and Firm Performance," in S. Andrews and D. Knoke (eds.), *Research in the Sociology of Organizations* Greenwich, Conn.: JAI Press, 16: 129–59.

—— and OWEN-SMITH, J. (1998), "Commercialism in Universities: Life Sciences Research and Its Linkage with Industry," *Journal of Policy Analysis and Management* 17 (2): 253–77.

——, WHITE, D. R., KOPUT, K., and OWEN-SMITH, J. (2005), "Network Dynamics and Field Evolution: The Growth of Interorganizational Collaboration in the Life Sciences," *American Journal of Sociology*.

ROBERTS, E. B., and LIU, W. K. (2001), "Ally or Acquire? How Technology Leaders Decide," *Sloan Management Review* 43: 26–34.

ROSENBLOOM, R. S., and SPENCER, W. J. (1996), "The Transformation of Industrial Research," *Issues in Science and Technology* 12(3): 68–74.

*ROSENKOPF, L., and TUSHMAN, M. (1998), "The Coevolution of Community Networks and Technology: Lessons from the Flight Simulation Industry," *Industrial and Corporate Change* 7: 311–46.

—— METIU, A., and GEORGE, V. P. (2001), "From the Bottom Up? Technical Committee Activity and Alliance Formation," *Administrative Science Quarterly* 46: 748–72.

ROWLEY, T., BEHRENS, D., and KRACKHARDT, D. (2000), "Redundant Governance Structures: An Analysis of Structural and Relational Embeddedness in the Steel and Semiconductor Industries," *Strategic Management Journal* 21: 369–86.

RUEF, M. (2002), "Strong Ties, Weak Ties and Islands: Structural and Cultural Predictors of Organizational Innovation," *Industrial and Corporate Change* 11: 427–49.

SABEL, C. F. (1994), "Learning by Monitoring: The Institutions of Economic Development," in N. Smelser and R. Swedberg (eds.), *Handbook of Economic Sociology*, Princeton: Princeton University Press, 137–65.

SARKAR, K. B., ECHAMBADI, R., and HARRISON, J. S. (2001), "Alliance Entrepreneurship and Firm Market Performance," *Strategic Management Journal* 21: 369–86.

*SAXENIAN, A. (1994), *Regional Advantage: Culture and Competition in Silicon Valley and Route 128*. Cambridge, Mass.: Harvard University Press.

SHAN, W., WALKER, G., and KOGUT, B. (1994), "Interfirm Cooperation and Startup Innovation in the Biotechnology Industry," *Strategic Management Journal* 15: 387–94.

SIMMEL, G. (1954), *Conflict and the Web of Group Affiliations*, Glencoe, Ill.: The Free Press.

SIMONIN, B. L. (1999), "Ambiguity and the Process of Knowledge Transfer in Strategic Alliances," *Strategic Management Journal* 20: 595–623.

STUART, T. E. (1998), "Network Positions and Propensities to Collaborate," *Administrative Science Quarterly* 43: 668–98.

—— (2000), "Interorganizational Alliances and the Performance of Firms: A Study of Growth and Innovation Rates in a High-technology Industry," *Strategic Management Journal* 21: 791–811.

SZULANSKI, G. (1996), "Exploring Internal Stickiness: Impediments to the Transfer of Best Practice with the Firm," *Strategic Management Journal* 17: 27–43.

TSAI, W., and GHOSHAL, S. (1998), "Social Capital and Value Creation: The Role of Intrafirm Networks," *Academy of Management Journal* 41: 464–76.

*UZZI, B. (1997), "Social Structure and Competition in Interfirm Networks: The Paradox of Embeddedness," *Administrative Science Quarterly* 42: 464–76.

VAN WIJK, R., VAN DEN BOSCH, F. A. J., and VOLBERDA, H. W. (2003), "Knowledge and Networks," in M. Easterby-Smith and M. A. Lyles (eds.), *Handbook of Organizational Learning and Knowledge Management*, Oxford: Blackwell.

VINDING, A. L. (2002), "Interorganizational Diffusion and Transformation of Knowledge in the Process of Product Innovation," Ph.D. Thesis, Aalborg University.

VON HIPPEL, E. (1987), "Cooperation between Rivals: Informal Know-how Trading," *Research Policy* 16: 291–302.

—— (1998), "Economics of Product Development by Users: The Impact of 'Sticky' Local Information," *Management Science* 44: 629–44.

*WALKER, G. B., KOGUT, B., and SHAN, W. (1997), "Social Capital, Structural Holes and the Formation of an Industry Network," *Organization Science* 8: 109–25.

WALUSZEWSKI, A. (1995), "Glulam," in H. Håkansson and I. Snethota (eds.), *Developing Relationships in Business Networks*, London: Routledge.

WEBER, S. (2003), *The Success of Open Source*, Cambridge, Mass.: Harvard University Press.

WENGER, E. (1998), *Communities of Practice*, New York: Cambridge University Press.

WOMACK, J. P., JONES, D. T., and ROOS, D. (1990), *The Machine That Changed the World: The Story of Lean Production*, New York: Harper & Row.

CHAPTER 4

INNOVATION PROCESSES

KEITH PAVITT

4.1 INTRODUCTION[1]

THIS chapter concerns innovation processes within firms, focusing mainly on innovation within large corporations in advanced countries.[2] What do we know about the historical evolution of these innovation processes and about the key challenges facing "innovation managers" within modern industrial corporations?[3] The chapter draws on empirical studies of innovation processes, bearing in mind the difficulties for generalization posed by the highly contingent nature of innovation. Section 4.2 presents a short introduction to the many theories and empirical studies of innovation[4] and suggests a simple framework for disaggregating the many innovation activities which take place at the firm level. Three broad, overlapping subprocesses (not stages) of innovation are identified: the production of knowledge; the transformation of knowledge into artifacts—by which we mean products, systems, processes, and services; and the continuous matching of the latter to market needs and demands.[5] Sections 4.3 through 4.5 examine key aspects of each of these three subprocesses, showing how each has evolved historically and why they pose such difficult problems for innovation-related managers, entrepreneurs, researchers, and workers. The chapter identifies these management difficulties and points to some of the strategies firms have deployed to meet these challenges.

4.2 CORPORATE INNOVATION PROCESSES

Since there is more than one process of innovation, there is no easy way to organize this chapter. Innovation processes differ in many respects according to the economic sector, field of knowledge, type of innovation, historical period and country concerned. They also vary with the size of the firm, its corporate strategy or strategies, and its prior experience with innovation. In other words, innovation processes are "contingent."[6] These difficulties are compounded by the fact that there is no widely accepted theory of firm-level processes of innovation that satisfactorily integrates the cognitive, organizational, and economic dimensions of innovation processes in firms. Economists tend to concentrate on the economic incentives for, and the effects of, innovation (largely ignoring what happens in between). Organizational specialists focus on the structural and procedural correlates of innovative activities and processes, sociologists on their social determinants and consequences. Managerial specialists address the practices most likely to lead to competitive success, psychologists may examine the phenomenon of creativity or the ways in which people's vision is restricted to one or other set of opportunities. Rich accounts and data resources have come from these and other lines of work over the past decades. Empirical evidence and theoretical understanding have been amassed by historians, from survey researchers and those concerned with bibliometrics, patenting, and other quantifiable dimensions of innovation. A growing number of "innovation studies" shows little allegiance to any particular discipline, and widely disparate theories and methods coexist in relevant journals and handbooks.

Joseph Schumpeter is considered a pioneer in the economic analysis of innovation, having concentrated more effort on this topic than any other economist in the first half of the twentieth century. His insights have guided the subsequent development of the field, and helped to explicate the vital role of innovation in growth and competitiveness. But Schumpeter's early work in particular (see Fagerberg 2003; Fagerberg, Ch. 1 in this volume), stressed the role of individuals, rather than organizations, in the innovation process, highlighting the character and determination of outstanding individuals, and defining innovations as "Acts of Will" rather than "Acts of Intellect." His interpretation reflected the nature of the evidence available to him at the time. As the authors of the SAPPHO study (Rothwell et al. 1974) found, the vast majority of earlier studies of successful and unsuccessful innovation were personal memoirs and anecdotes of the exploits of individual scientists, inventors or managers and contained little or no systematic comparison or analysis. Project SAPPHO and many similar, subsequent studies were in part attempts to overcome the "individualist" bias of personal memoirs. They set out to examine a wider set of organizational factors, as well as the skills and experience of a wider range of individuals participating in each innovation.

In order to understand this rich but potentially confusing mosaic of knowledge about innovation processes, I suggest the following general framework:

(1) Innovation processes involve the exploration and exploitation of opportunities for new or improved products, processes or services, based either on an advance in technical practice ("know-how"), or a change in market demand, or a combination of the two. Innovation is therefore essentially a matching process. The classic paper on this subject is Mowery and Rosenberg (1979).

(2) Innovation is inherently uncertain, given the impossibility of predicting accurately the cost and performance of a new artifact, and the reaction of users to it. It therefore inevitably involves processes of learning through either experimentation (trial and error) or improved understanding (theory). Some (but not all) of this learning is firm-specific. The processes of competition in capitalist markets thus involve purposive experimentation through competition among alternative products, systems, processes, and services and the technical and organizational processes that deliver them.

In organizing the evidence, it is useful to divide innovation into three, partially overlapping processes, consistent with the two features described above.[7] Each process is closely associated with contributions from particular academic disciplines. Each process also has undergone major historical transformations as the process of innovation has evolved.

- *The production of scientific and technological knowledge*: a major trend, since the industrial revolution, has been for the production of scientific and technological knowledge to have become increasingly specialized, by discipline, by function and by institution. History and social studies of science, technology and business have contributed significantly to our understanding of this transformation.

- *The translation of knowledge into working artifacts*: in spite of the explosive growth in scientific knowledge in recent years, theory remains an insufficient guide to technological practice. This reflects an underlying trend for growing complexity of technological artifacts,[8] and in the knowledge bases underpinning them. Technological and business history have made major contributions here as have the cognitive sciences more recently.

- *Responding to and influencing market demand*: this involves a continual process of matching working artifacts with users' requirements. The nature and extent of the opportunities to transform technological knowledge into useful artifacts vary amongst fields and over time, and determine in part the nature of products, users and methods of production. In the competitive capitalist system, corporate technological and organizational practices co-evolve with markets. Social change and innovations in marketing and market research have contributed to complex problems and equally complex solutions to the challenge of matching technological opportunities with market needs and organizational practices. These processes are central concerns of scholars in management, economics and marketing studies.

I now discuss the implications of change over time in the structure and nature of each of these features of innovation processes. Such change presents considerable challenges to the modern innovation manager and to the corporation as a whole.

4.3 The Production of Scientific and Technological Knowledge

All the improvements in machinery, however, have by no means been the inventions of those who had occasion to use the machines. Many... have been made by the makers of the machines, when to make them became the business of a peculiar trade: and some by... those who are called philosophers, or men of speculation, whose trade is not to do anything but to observe everything: and who, upon that account are often capable of combining together the powers of the most distant and dissimilar objects.... *Like every other employment... it is subdivided into a number of different branches, each of which affords occupation to a peculiar tribe or class of philosophers; and this subdivision of employment in philosophy, as well as in every other business, improves dexterity and saves time.* (Smith 1937 (orig. 1776): 8, my italics)

Adam Smith's identification of the benefits of specialization in the production of knowledge has been amply confirmed by experience. Professional education, the establishment of laboratories, and improvements in techniques of measurement and experimentation have increased the efficiency of discovery, invention, and innovation. Increasingly difficult problems can be tackled and solved.[9] New and useful fields of knowledge have been developed, punctuated by the periodic emergence of fields with rapid rates of technological advance and offering rich opportunities for commercial exploitation. For example, progress in such activities as metal cutting and forming, or the use of new power sources, has been informed by and has drawn upon physics, chemistry, and biology, as well as a variety of related engineering disciplines. Today, the coordination of increasing specialization remains a fundamental task of the large corporation.

Three forms of corporate specialization have developed in parallel. First is the development in large manufacturing firms of R&D laboratories specialized in the production of knowledge for commercial exploitation. Second is the development of a myriad of small firms providing continuous improvements in specialized producers' goods. A third trend of specialization is the changing "division of labor" between private knowledge developed and applied in business firms, and public knowledge developed and disseminated by universities and similar institutions. Taken together, all these forms of specialization have combined to make a heterogeneous and path-dependent pattern of technical change that places great demands on corporations for coordination of processes within their boundaries and between these organizations and others external to the firm. These processes of change, examined below in more detail, have intensified and broadened the challenges facing managers, "intrapreneurs," and entire corporations.[10]

4.3.1 Functional Specialization and Integration: Industrial R&D Laboratories

A major source of innovation in the twentieth century was the industrial R&D laboratory, which remains important in the twenty-first century. It emerged first in Germany in the chemical industry and in the USA in the electrical industry, for two reasons. It was part of the more general process of functional specialization of the large manufacturing firm (Mowery 1995; see also the chapter by Bruland and Mowery in this volume), which itself emerged from the exploitation of economies of scale and speed made possible by radical innovations in materials processing and forming, and in power sources (Chandler 1977). But the industrial research laboratory also provided these firms with a means for exploiting the rich veins of useful knowledge emerging from fundamental advances in chemistry and physics. In addition, these new in-house laboratories served as a "monitoring post" for established firms seeking to acquire new technologies from other firms.

Mowery (1995) has shown for the USA that a growing proportion of industrial R&D in the twentieth century was integrated within large manufacturing firms, rather than in independent companies. Until about 10 years ago, business-funded R&D in all OECD countries was almost exclusively performed within innovating firms (not only manufacturing firms, since telecommunications and some other services have long undertaken R&D). Mowery explained this lack of vertical disintegration by the difficulties of writing contracts for an activity whose output is uncertain and idiosyncratic, and pointed out that integration of R&D within the firm reflected important operating advantages as well. Thus competitive advantage can be gained by the effective combination of specialized and often tacit knowledge across functional boundaries, within the individual firm. Accumulated firm-specific experience is very important.[11]

A robust conclusion emerging from research on innovation processes is that one of the most important factors differentiating successful from unsuccessful innovation has been the degree of collaboration and feedback between product design and other corporate functions, especially manufacturing and marketing within the firm (Rothwell 1992). Many product designs turn out to be technically difficult (even impossible) to manufacture, and/or fail to take into account often elementary user requirements (Forrest 1991).

From a corporate strategy perspective, the importance of such intrafirm collaboration has increased the importance of cross-functional integration spanning departmental boundaries. Japanese automobile firms pioneered the use of "heavyweight" project managers, empowered to control resources across the firm, reporting directly to the senior management team at the same level as the departmental manager (Clark and Fujimoto 1992). These project managers were in turn linked to innovation processes within customers and key suppliers, enabling fast, project-based innovation. These "heavyweight" project managers occasionally clash with functional bosses, some of whom are unwilling to "give up" control over their resources and object to project-led management. Many large firms now provide formal training in project management to their professional project manager, covering such issues as management of fast-moving project teams responsible for integrating research outputs, conceptual and detailed design, and various engineering functions, while at the same time responding to changing or emerging customer requirements during the production process.

Many writers stress the importance of personal contacts and exchanges across functions within the firm to deal with tacit elements of both product design and its successful transfer to manufacture and market. There is no perfect or foolproof process for ensuring effective coordination. Indeed, so called "best practice" can be positively harmful when its application is taken too far. Excessive use of "heavyweight" project managers can lead to the loss of such benefits from integration as economies of scale, and cost reductions from the use of common components (e.g. in automobile development; see Leonard-Barton 1995). Firms can find it hard to decide what to do with a heavyweight product manager (and the associated staff) when a product development is failing. Failure to grasp this difficult nettle can lead to the problem of "escalation" or an inability to terminate a failing project.[12] Managing the trade-offs between project and functional management, and overcoming the inherent difficulties in project-based management, present major difficulties to senior technology managers.

4.3.2 Technological Convergence and Vertical Disintegration in Production Techniques

Even in industries with heavy investments in product innovation, however, some "vertical disintegration" (outsourcing of specific activities to supplier firms) in manufacturing process innovation has occurred since the nineteenth century, often stimulated by technological advances. Rosenberg (1976) highlighted the emergence of specialized machine tool firms in the nineteenth-century US economy as a result of advances in metal cutting and metal-forming techniques that produced technological "convergence" in operations that were common to a number of manufacturing processes. For example, boring accurate circular holes in metal was an operation common to the making both of small arms and of sewing machines. Although the skills associated with such machining operations were often craft-based and tacit, their outputs could be codified and standardized. The demand for such common operations grew sufficiently that markets for machines to perform them became large enough to sustain the growth of specialized firms who designed and made such machines. Large manufacturing customers could therefore buy machines that incorporated the latest improvements (drawing on the feedback from many users) and were far superior to what they could do by themselves.

Similar processes of technological convergence and vertical disintegration have been common features of twentieth-century capitalist development (see Table 4.1). New opportunities have emerged from breakthroughs that have created applications spanning multiple product groups. Examples include: materials shaping and forming, properties of materials, common stages of continuous processes, storage and manipulation of information for controlling various business functions such as manufacturing operations and design.

Lundvall (1988) and other writers show that the links between the (often small) firms providing these specialized production inputs and their (mainly large) customers are often "relational" rather than arm's-length, and include considerable exchange of information and personnel related to the development, operation and improvement of the specialized inputs. Managing the outsourcing of these critical inputs has become a major challenge to managers of large firms (Quinn 2000). For

Table 4.1 Examples of technological convergence and vertical disintegration

Underlying technological advance	Technological convergence	Vertical disintegration
Metal cutting & forming	Production operations	Machine Tool Makers
Chemistry & metallurgy	Materials analysis & testing	Contract Research
Chemical engineering	Process control	Instrument Makers *Plant Contractors*
Computing	Design Repeat operations	CAD Producers Robots Makers
New Materials	Building Prototypes	Rapid Prototyping Firms
ICT	Application software	Knowledge-Intensive
	Production systems	Business Services Contract manufacture

example, logistics and ICT systems often differ between component suppliers and integrator companies, creating serious (albeit often technically simple) problems for communication and transactions among these firms. More fundamentally, the choice of which activities to outsource and which to retain in-house defines the "core competence" of the modern corporation, molding the boundaries of the firm (Hamel and Prahalad 1994; Davies et al. 2001).

4.3.3 Industrial Linkages with Universities

As innovative activities in business firms have become more professionalized, and university research more specialized, universities now play an important role in providing the trained researchers for firms in some sectors to perform their innovative activities. At the same time, firms have found it important to have effective processes in order to benefit from progress in those longer-term research programs in universities in fields that have possible impacts on their current and future activities.

The range of interactions between firms and universities is considerable. At one extreme, there is something close to the so-called (but relatively rare) "linear model." Here, fundamental research by a university scientist leads to a discovery, its practical importance is recognized by a business firm, which may collaborate with the university scientist in order to exploit it. This happens most often in science-based industries including the chemical, biotechnology, and pharmaceutical sectors, where the focus is on the discovery of interesting and useful synthetic molecules.

At the other extreme, the provision of trained researchers, familiar with the latest research techniques and integrated in international research networks, is important to firms. It is ranked by many industrialists as the greatest benefit provided by universities (Martin and Salter 1996). Thus, even if university research in mechanical engineering has fewer direct applications than research in chemistry, it still provides mechanical engineers trained (for example) in those simulation and modelling techniques that are increasingly important in the design and development of automobiles and aero-engines.

In between are a variety of other, often complementary, processes which have to be managed in order to link university research with industrial innovation, including direct industrial funding of university research, university-based consultants, and exchanges of research personnel. Three common features of university–firm links emerge from the literature.

(1) The importance of personal and often informal contacts. Informal relationships give practitioners entry points into the academic world, people who they can ask about where the important developments lie and who the relevant people are. Such relationships give researchers insight into the

problems that are confronting industry, how the leading edge of corporate practice is developing, and so on. The informal relationships can result in formal outputs that can in turn trigger more informal contacts. For instance, industrial publications in the scientific literature can be seen as signals to the wider academic community of fields and problems of industrial interest that would benefit from more intense personal exchanges (Hicks 1995).

(2) Much university research that is useful to industrialists also is valued by academics. Some academics erect a demarcation between industrially rele- vant applied work and more fundamental research, and such a demarcation may apply to the work of some university groups playing a role in technology transfer to local businesses. But it does not apply to a good deal of more advanced research activities. US-based studies by Mansfield (1995) and Narin et al. (1997) suggest that a high proportion of industrially significant research is publicly funded, performed in the academically prestigious research uni- versities, and published in high quality academic journals.

(3) The practical benefits of most university research emerge from processes that are roundabout and indirect. Probably the most frequent contribution is the provision of graduates trained by leading researchers, and often conversant with emergent research methods and approaches. Such individuals can be "carriers" of new theoretical insights, new techniques and observations, and new skills, all of which industrial firms find difficult to provide themselves. Over time, they will turn these capabilities to solving the problems that their employers face—or provide support by means of forming spin-offs or joining consultancies that provide innovation-supporting services to industrial clients.

Over the past twenty years, governments have begun to expect greater direct useful- ness from university research. Often this has been supported by empirically ques- tionable assumptions and theories,[13] or an incomplete understanding of the indirect benefits actually valued by industrialists. For entirely different reasons, certain fields of university research—many fields of biotechnology and some of software and related activities—now provide an increasing stream of inventions with potential industrial application. These are reflected in increases in university licensing activity, in university-founded spin-off firms, and increases in private funding of university research. In Chapter 8, Mowery and Sampat discuss the nature and implications of these recent developments.

These university–industry relationships can be extremely difficult for firms to manage.[14] Managers often complain that universities operate on extended "time lines" with little regard for the urgent deadlines of business. Therefore, they argue, universities should not be placed on the critical path of any important projects. Universities, in turn, sometimes find themselves in the invidious position of being viewed as a low-cost performer of industrial projects, often with the encouragement of government and research council programs of "technology transfer." At worst, these programs may focus on the short-term needs of industry at the expense of the long term quality of universities' basic research, graduate training, and support for experimentation and critical questioning (Salter et al. 2000). Such technology transfer programs often are based, implicitly or explicitly, on a version of the linear model of innovation in which universities (and other public sector organizations) perform the science/basic research, which generates innovations "for" industry to take up in its engineering, manufacturing, and marketing.

4.3.4 Heterogeneity in Innovation Processes

These characteristics of industrial innovation contribute to inter-industry heterogeneity in the structure and management of the processes of innovation, as well as change over time in the identities of the major actors in these processes. The dominant sources of technical change in the twentieth century were—and still are—large manufacturing firms exploiting different fields of specialized knowledge, with in-house R&D laboratories, together with a myriad of small firms providing specialized producer goods. Over time, the mix of firms and technologies has changed, reflecting the appearance of innovative opportunities generated by the different rates of growth of knowledge in various specialized fields. These characteristics and heterogeneity have major implications for the study of innovation processes.

- Given the increasingly specialized and professionalized nature of the knowledge on which they are based, manufacturing firms are path-dependent. Where they search for the future is heavily conditioned by what they have learned to do in the past.[15] As we shall see below, path-dependency reflects the conservatism of professional and functional groups. It also results from cognitive limits in individuals' knowledge about technologies, markets, and the opportunities presented by developments in each. Thus, for example, it is difficult if not impossible to convert a traditional textile firm into one making and selling semiconductors. The recent experience of Vivendi suggests it may also be difficult to transform a national water company into an international ICT and media giant, however enthusiastic some managers and financiers were about this.
- Firms that specialize in different products and related technological fields are likely to stress different features of innovation processes, reflecting the nature of the fields of knowledge on which they depend. Thus, for contemporary automobile firms, effective feedback between product design and manufacture is more important than feedback between product design and university research. For a pharmaceutical firm, the reverse is likely to be the case. These patterns reflect the direct usefulness of university research in the field, and other factors, such as the complexity of manufacturing processes. Users are likely to be an important source of innovations for small innovating firms providing producer goods; but users'

role in innovation is likely to be low where large firms are selling to a mass market of users who lack strong technological capabilities (e.g. many consumer products).

- Innovation processes will differ greatly between large and small firms in other respects. Innovations in large firms involve a larger number of people in specialized functions, with shifting responsibilities over time. Innovation processes in large firms are also more likely to involve recognizable procedures, whether formal or informal. In small firms, there are fewer resources to apply to such issues. Decisions related to the recognition of opportunities, the allocation of resources, and the coordination of functional activities, are more likely to reflect the competencies and behavior of senior managers.[16]

The heterogeneity and contingent nature of innovation means that there can be no simple "best practice" innovation model for firms or managers to follow. Each firm proceeds on the basis of its prior experience and the technological trajectories evident in the specific industry or product group. But the lack of global "best practices" should not be taken to mean that innovation strategy does not matter, nor that good management cannot make a difference to firms' productivity, market share, or profitability.

4.4 TRANSFORMATION OF KNOWLEDGE INTO WORKING ARTIFACTS

Scientific advances enable the creation of artifacts of increasing complexity, embodying an increasing number of subsystems and components, and drawing on a broadening range of fields of specialized knowledge. This increasing system complexity is one consequence of the growing specialization in knowledge production. It has resulted in both better understanding of cause–effect relationships, and better and cheaper methods of experimentation. These advances have reduced the costs of technological search, enabling greater complexity in terms of the number of components, parts or molecules that can be successfully embodied in a new product or service. Developments within ICT are accelerating this trend: digitalization opens options for more complex systems, and simulation techniques reduce the costs of experimentation (Pavitt and Steinmueller 2001).

Managers involved in transforming S&T knowledge into products, systems, and services need to be aware of four sets of specific trends in their industries. These are: (*a*) technology trajectories and scientific theories; (*b*) where relevant, government-funded R&D programs; (*c*) systems integration; and (*d*) techniques and approaches to managing uncertainty. I deal with each of these issues below.

4.4.1 Keeping Technological Practice (not too far) ahead of Scientific Theory

In spite of the spectacular increases in scientific knowledge over the past 200 years, theory remains an insufficient guide to technological practice. This is partly because of the increasing complexity of physical artifacts and the knowledge bases that underpin them. The importance of practice is reflected in the continuing dominance in industrial R&D laboratories of development activities—the design, building and testing of specific artifacts—compared to research in fields on which they are based. Constant (2000) describes this as technology advancing through the recursive practices of scientists and engineers, involving "alternate phases of selection and of corroboration by use.... The result is strongly corroborated foundational knowledge: knowledge that is implicated in an immense number and variety of designs embodied in an even larger population of devices, artifacts, and practices, that is used recursively to produce new knowledge" (p. 221).

Continuous innovation requires constant improvement in methods of technological search, but technical complexity cannot run too far ahead of scientific understanding.[17] The feedbacks in both directions—between improvements in scientific understanding and improvements in technical performance—have been well documented by historians and others in areas such as aerodynamics and thermodynamics.[18] Mahdi (2002) has recently developed a taxonomy of technological search that depends on three factors: (1) the degree to which technological problems can be decomposed into simpler sub-tasks; (2) the level of understanding of cause–effect relations; and (3) the costs of experimentation with possible solutions.

Advances in the technologies of measurement and manipulation of the increasingly small, are a major source of improvements in technological search. This has been the case in the past few decades in molecular biology and materials, both of which have opened major new opportunities for technical change.[19] ICT can also reduce the costs of search and selection. Major advances in large-scale computing and simulation technology have reduced the costs of exploring alternative technical configurations, and have created opportunities for increasingly complex systems made possible through the digitalization of data of all sorts (Pavitt and Steinmueller 2001). Innovation managers and engineers involved in transforming knowledge into working artifacts today need to be aware of specialized ICT trends in their own industries, as well as new measurement and manipulation techniques elsewhere that themselves frequently involve the application of advanced ICTs.

Nightingale (2000) has shown that these mechanisms have radically changed experimental techniques in the pharmaceutical industry during the past ten years. There has been a shift towards more fundamental science, for example, linking biochemical mechanisms to the expression of genes; and there has been a much greater use of simulations (involving models, extensive data banks, etc.) to conduct virtual experiments complementary to real ones. A third related development has been the use of high throughput screening techniques.[20]

4.4.2 Government-Funded Programs

Technological activities directly financed by governments have sometimes been of major importance in opening and exploiting innovative opportunities. Successes include ICT in the USA, where military-related programs played a major role in the early development of computers, semiconductors, software, and the Internet. Military programs have also had important technological spin-offs into civil aviation in the USA, while governments in Japan and France have successfully supported the development of high-speed trains. But there have also been many disappointments. Policies to support the development of civilian nuclear power have on the whole not been successful, nor have those for the support of high-volume, low-cost residential construction (See Eads and Nelson 1971, as well as the case studies in Nelson 1982). More recently, policies to encourage the development of renewable energy technologies have met with mixed success. And controversy surrounds the achievements of a whole series of EU programs.

Since the 1980s government programs for "pre-competitive" collaborative R&D in Europe (e.g. ESPRIT and Eureka), the USA (e.g. Sematech) and Japan (e.g. the 5G ICOT Program) have proliferated. Thus, most major firms are presented with opportunities for participation in such programs. Firms require methods for evaluating their potential contribution to corporate goals, the financial and organizational costs of participating, the risks involved in not participating, and the ways in which government programs can complement or fit into the overall corporate strategy (Floyd 1997).

It is difficult to generalize from the recent history of government support for industrial innovation. All such programs involve technical lobbies successfully putting pressure on governments for financial support, often in fields related closely to military applications, or (often large-scale) infrastructure, such as transport, energy, housing, and communications. This process can lead to neglect of commercial constraints and to premature commitments to particular designs. Economists highlight the opportunity costs of these programs; but government support can also speed up critical technological learning at a time when purely private markets are not ready to take the risks. The early development of ICT in the USA suggests the importance of diversity and experimentation in government support for technological progress. But would this have worked for the development of high-speed trains, where the costs of experiments are much higher and technical change is more incremental? And, as we shall see in the next section, everyone makes mistaken assumptions about future developments in a complex and fast-changing world.

4.4.3 Multi-Technology Firms, Modularization and Systems Integration

In addition to increasingly complex artifacts, specialization in knowledge production has increased the range of fields of knowledge that contribute to the design of each product. Compare the original mechanical loom with the many fields of specialized knowledge—electrical, aerodynamic, software, materials—that are now embodied in the contemporary design. Or observe the way in which modern automobiles increasingly integrate plastic and other new materials, as well as electronic and software control systems.

Firms designing these increasingly complex products find it difficult to master advances in all the fields that they embody. Hence the growing importance of modular product architectures, where component interfaces are standardized, and interdependencies amongst components are decoupled. This enables the outsourcing of design and production of components and subsystems, within the constraints of overall product (or system) architecture.

Technological convergence has also provided opportunities for further vertical disintegration between product design and manufacture. Sturgeon (1999), for example, describes recent growth in contract manufacturing in electronics, in which specialized firms take over product design from other firms, and assume responsibility for detailed engineering and manufacture. The technological convergence here is based on increasing automation of routine operations (e.g. component insertion), and on growing use of standard software tools.

Contract manufacturing is growing in other industries as well,[21] giving rise to "modular production networks." These modules are defined by distinct breaks in the value chain at points where information regarding product specifications can be highly formalized. This occurs within functionally specialized value chain nodes, where activities tend to be highly integrated and based on tacit linkages. Between these nodes linkages depend on the transfer of codified information.

At first sight, these recent changes might appear to point to a neatly specialized system for the production of innovations, with product and systems designers, their subcontractors for components and subsystems, and their manufacturers, working together through arm's-length market relations—a trend foreseen by Sturgeon (1999). But this neglects the consequences of important distinctions that need to be made between the properties of artifacts, the knowledge on which they are based, and the degree to which such knowledge can be transformed into codified information (Granstrand et al. 1997).

Briefly stated, the development and production of increasingly complex artifacts are, as we have seen, based on the integration of an increasing number of fields of specialized knowledge. These fields advance at different speeds and their progress cannot be tracked solely by monitoring codified information. The division of labour between companies in production thus cannot be mirrored by an equivalent division of labour in knowledge (Brusoni et al. 2001). Some overlap between companies in knowledge competencies is necessary to deal with the transfer of tacit knowledge, to manage unforeseen consequences of systemic complexity, and to resolve imbalances between components that are liable to result from their uneven rates of technical change. Similarly arm's-length relations between firms will not be as effective as forms of "loose coupling" with periodic bouts of integration, when systems architectures and the tasks of component suppliers are redefined by firms specializing increasingly in systems design and integration.

4.4.4 "Managing" Uncertainty?

Specialized R&D and related activities in business firms have become institutionalized and predictable sources of discoveries, inventions, innovations, and improvements. However, the process of innovation is complex, involving many variables whose properties and interactions (and economic usefulness) are understood imperfectly. As a consequence, firms are not able to explain fully and predict accurately either the technical performance of major innovations, or their acceptability to potential users (or in some cases even who the potential users are). They cannot accurately predict the technical and commercial outcomes of their own innovative activities, nor those of other firms.[22] On average, research scientists and engineers tend to be over-optimistic about the costs, benefits, and time periods of their proposed projects, and about market demand for the products resulting from them. But there is typically great variation in the ratio of *ex post* outcomes to *ex ante* estimates of investment or profit within any corporate portfolio of projects (Freeman 1982; Mansfield 1995). Indeed, commercially unsuccessful projects often account for a disproportionate share of corporate R&D spending (Griliches 1990).

Business firms (and others) rarely are capable of defining the full array of possible uses that may emerge for their innovations, especially radical ones. Examples of inaccurate predictions about what turned out later to be spectacularly successful technologies and innovations are legion. Early twentieth-century pioneers of radio communication conceived it as a system of point-to-point communications, particularly between naval vessels—it was only much later that the much larger market for mass radio communications was recognized. After World War II, the founder of IBM foresaw a world market for computers in single figures. For the more recent period between the 1960s and the 1980s, Schnaars and Berenson (1986) concluded that only about half the major new product families announced in the USA turned out to be commercially successful. The recent experience of inaccurate forecasts of the potential markets for various generations of the mobile phone, and various functions associated with these (the unexpected success of text messaging, for instance) is equally instructive.

Corporate managers therefore face severe difficulties in deciding how to deal with innovative activities, which have some of the elements of conventional investment activities, but for which severe uncertainty means that continuous feedback from the market, past experience and experimentation are essential. In practice, top-down corporate visions can be a poor guide to innovation strategies. Ericsson's success in opening up mobile telephony began with initiatives from middle-level technical management, rather than from the top. In the academic and business literatures, the failures of top-down visions are easily forgotten and the successes oversimplified. For example, Prahalad and Hamel (1990) tell the story of Canon's successful diversification from optics and precision mechanics into electronics technology, and from cameras into photocopying and computer peripheral products; but they ignore the firm's failed diversification into recording products and electronic calculators (Sandoz 1997).

The broad differences between search and selection activities have been recognized for a long time—in practice, with the distinction between corporate and divisional R&D activities; and in theory with the distinction between "knowledge building" and "strategic positioning" on the one hand, and "business investment" on the other (Mitchell and Hamilton, 1988). However, as the recent history of corporate R&D shows, maintaining balance and linkages between the two is not an easy task. Briefly stated, there is no one best way of evaluating the costs and benefits of corporate R&D expenditures *ex ante*. Rule-based systems fail, because they inevitably simplify, and may therefore neglect what turn out to be important factors in a complex system. Judgement-based systems fail, because of the impossibility of quickly distinguishing good judgement from good luck. One consequence is periodic swings in fashions and management practices. These often reflect struggles for influence between financially trained managers, who tend to prefer rule-based systems, and those who are technically trained and prefer to rely on technical judgements.

4.5 MATCHING OF ARTIFACTS, ORGANIZATIONAL PRACTICES AND MARKET DEMAND

The matching of products, processes, systems, and services (and organizational practices) with actual and potential market demand is a major responsibility for managers overseeing innovation in the successful corporation. The corporation builds on its accumulated knowledge of product and process technologies, of organizational practices, and of users' needs to carry out this function. Responding to (and creating) market needs and demands, as well as matching organizational practices with technological opportunities, involves dealing with disruptive change. This disruptive change interacts with one of the negative consequences of specialization: namely, the potential for tribal warfare over the old and the new between specialized functions and disciplines within the firm.

4.5.1 Matching Technology and Organizational Practices with Market Needs

Chandler (1977) has shown that the rise in the USA at the end of the nineteenth century of the large, multi-unit firm, and of the coordinating function of professional middle managers, depended critically on the development of the railroads, coal, the telegraph, and continuous flow production. Similarly, the later development of the multidivisional firm reflected in part the major opportunities for product diversification in the chemical industry opened up by breakthroughs in synthetic organic chemistry. Although it is commonplace today to argue that technologies and organizational practices do, or should, co-evolve with market demands, there is some risk of "technological determinism" in Chandler's argument that this process largely involves the adaptation of corporate organizational practices to emerging market needs and technological opportunities.

Technical advances often precede organizational and market advances, among other reasons because of the firmer knowledge base and lower costs of experimentation associated with technical, as opposed to market or organizational, innovation. This does not mean that technology imposes one organizational "best way" or even a clear strategy towards the marketplace. Variety in the characteristics of technologies, their continuous change and uncertain applications also produces variety and experimentation in organizational and marketing practices. But this variety and change does not mean that "anything goes" in either organizational or marketing terms. It may be practically impossible for a firm that wishes to remain competitive to resist making use of new technologies and knowledge in its own future product and process development—unless it wishes to become a niche producer like the manufacturers of clockwork watches and analogue record players. But moving into new technologies without appropriate changes in terms of skills and training, divisions of labour and interrelations between parts of the organization, can be even more costly. Or consider the firm's investment decisions. A firm applying conventional cost–benefit analysis and strict cost controls to all of these decisions will not prosper in the long term in a competitive market governed by the exploitation of a rich, varied and rapidly advancing body of technological knowledge.[23]

The empirical literature summarized in the first two columns in Table 4.2 highlights the organizational and marketing practices that must be made consistent with key features of technologies:

- External linkages with potential customers, and with the important sources of knowledge and skills.
- Internal linkages in the key functional interfaces for experimentation and learning.
- The centralization of resource allocation and monitoring activities needs to be consonant with the costs of technological and market experimentation.

Table 4.2 Matching corporate technology and organizational practices with market needs and demands

Corporate technology →	Matching organizational and marketing practices →	Dangers in radical technological change
Inherent characteristics		
1. Richness of opportunities	1a. Allocating resources for exploring options 1b. Matching technologies with product markets	1a. Greater opportunities not matched by resources for exploring options 1b. Matching opportunities missed in the marketplace
2. Costs of specific experiments	2. Degree of centralization in decision making	2. Reduced cost of experiments not matched by decentralization or market testing
Supporting skills and networks		
1. Specific sources of external knowledge	1. Participation in specific professional knowledge networks	1. Difficulties in recognizing & joining new knowledge networks
2. Accumulated knowledge of specific customers' demands, distribution channels, production methods, supply chains.	2. Learning and improving in key functions and across key functional interfaces	2a. Difficulties in recognizing & responding to new customers' demands, distribution channels, production methods, supply chains 2b. Difficulties in recognizing new key functional interfaces
3. Skill and Expertise Base	3. Anticipating and organizing for the necessary communication and gatekeeper skills	3. Scepticism and resistance from established or potentially obsolescent professional and functional groups

- Criteria for resource allocation need to be consonant with levels of technological and market opportunity.
- Alignment of professional groups, who possess power and control, with fields of future opportunity.

The richness of the technological and market opportunities and the scale of technical experiments should determine the appropriate share of resources allocated to technological search, as well as the degree of centralization and fluidity in organization structures. Supporting skills and networks will define the specific competencies to be accumulated, professional networks to be joined and key functions and functional interfaces within and across which learning must take place within the firm.

The particular circumstances of the individual firm and project will obviously define the basic skills required for commercial innovation. But the discussion in this chapter leads to a clear-cut conclusion: in addition to specialist skills, "gatekeeper" skills and general communication skills now are more important almost everywhere. People who are capable of communicating across organizational barriers, disciplinary barriers, and professional barriers can be invaluable. In very small firms it may be satisfactory for one or two key individuals to possess the unique combination of required skills. In larger firms, the specific requirements may be hard to anticipate. There is no single managerial planning prescription.

Differences amongst technologies are reflected in differences in organizational and marketing practices. For example, both pharmaceutical and consumer electronics firms see rich technological and market opportunities, and thus devote substantial resources to technological search. The much higher costs of experimentation in pharmaceuticals mean that drug firms tend to have centralized and formal procedures for launching new products, whilst in consumer electronics the situation is more likely to be decentralized and informal. Similarly, as already mentioned, both pharmaceutical and automobile companies have centralized decision structures, but the former stress interfaces between corporate R&D and public research in biomedical fields, whilst the latter concentrate on links between R&D and production.

4.5.2 Coping with Radical Change

The past 200 years have seen periodic step-jumps in technological understanding and performance in specific fields. During the past several decades, these discontinuous advances have been more often than not based on major scientific breakthroughs. "Radical" innovations have reduced considerably the costs of key economic inputs, and have therefore been widely adopted and become the catalysts for major structural changes in the economy. They include steam power, electricity, motorization, synthetic materials, and radio communications (Freeman and Louçã 2001). The most celebrated contemporary example, of course, is the massive and continuing reductions in the costs of storing, manipulating, and transmitting information brought about by improvements in ICT.

Each wave of radically new technologies has been associated with the growth of firms that have mastered the new technologies and have pioneered in the development and commercialization of related products, processes, and services. In the current jargon of corporate strategy, these firms have developed core competencies in the new technologies, which have become a distinctive and sustainable competitive advantage.

Ever since Schumpeter associated the advent of revolutionary technologies with "waves of creative destruction," there has been debate about the relative role of incumbent large firms and new entrants in exploiting them. Over the past twenty years, most of the analytical writing has been stacked against incumbents, although recent empirical studies point to evidence in favour of both (Methe et al. 1996). Over time, the weight of the argument has shifted somewhat away from emphasis on the difficulties facing incumbents in mastering new fields of technological knowledge (Cooper and Schendel 1976; Tushman and Anderson 1986; Utterback 1994).

More recent work has emphasized the difficulties faced by incumbent firms that must adapt established organizational practices to seize the opportunities opened by revolutionary technological changes. Examples include the organizational consequences of changes in product architectures (Henderson and Clark 1990), resistance from groups with established competencies (Leonard-Barton 1995; Tripsas and Gavetti 2000), and the unexpected emergence of new markets (Christensen 1997; Levinthal 1998).

Contrary to a widespread assumption, the nature and directions of radical new technological opportunities are easily recognized by the technically qualified: for example, miniaturization, compression, and digitalization are key trajectories in ICT. The result in this case is that a growing number of large firms, in a growing number of industries, are now technically active in ICT (Granstrand et al. 1997; Mendonçã 2000). However, the difficult, costly and uncertain task is that of combining radically new technical competencies with existing technical competencies and organizational practices, many of which may be threatened or must be changed in order to exploit potential market opportunities. Experimentation and diversity therefore are necessary for exploration of the directions and implications of radical technological changes, but also for assessing the implications of these changes for products, markets and organizational practices.

The third column of Table 4.2 identifies some reasons why such experiments may fail in incumbent firms. Some are a consequence of the need to modify competencies or organizational practices; and some of the inevitable uncertainties in the early stages of radically new technologies. The likelihood that established firms will fail increases with the number of practices and competencies that need to be changed.

Here a comparison between the conclusions of two recent industry studies is instructive. Klepper and Simons (2000) have shown that firms already established in making radios were subsequently the most successful in the newly developing colour TV market. On the other hand, Holbrook and his colleagues (2000) have shown that none of the firms established in designing and making thermionic valves was subsequently successful in semiconductors.

With the benefit of hindsight, we can see that success in semiconductors required more changes in technological competencies, organizational practices, and market experimentation amongst incumbents, than did success in colour TV. The valve firms required new competencies and networks in quantum physics, a much stronger interface between product design and very demanding manufacturing technology, and the ability to deal with new sorts of customers (computer makers and the military, in addition to consumer electronics firms). For the radio firms, the shift to color TV required basically the same technological competencies, augmented by well-known screen technologies. Otherwise, the customers and distribution channels remained unchanged, as did the key networks and linkages both inside and outside the firm.

According to Chandler's (1997: 76) so-called "continuity" thesis, the population of incumbent large firms has remained stable in recent times, because of their accumulated skills and resources in adopting new technologies and adapting to them. This thesis has been challenged by Louçã and Mendonça (1999) and by Freeman and Louçã (2001: 340–55), who argue that, on the contrary, a cohort of new large firms continues to join the population of incumbent large firms with each new wave of technical change. Only a minority of the largest firms was able to remain at the top through several waves. This suggests that the micro-level evidence considered in this Section, especially the factors listed in the third (right-hand) column of Table 4.2, have had significant consequences for the evolution of industry structure.[24] Radical technological change, and the success or failure of incumbent firms in adapting to it, thus may have important consequences for structural change in the economy as a whole.

Firms in the vanguard of developing and exploiting radically new technologies must be distinguished from the more numerous firms who adopt and integrate the new technologies with their current activities. For these firms, in-house competencies in the new technologies are background: in other words, necessary for the effective adoption of advances made outside the firm. Paradoxically, the very fact that radically new technologies allow step-jump reductions in the costs of a key input simultaneously makes their adoption a competitive imperative and an unlikely source for sustained competitive advantage among adopting firms. For example, in the past many factories had no choice but to adopt coal and steam—and later electricity—as a source of power, given their cost and other advantages. The same is true today for many ICT-based management practices. But in neither case were these revolutionary advances by themselves a source of sustainable competitive advantage for the adopting firms. Much of the emphasis by writers on corporate strategy—like Barney (1991) and Porter (1996)—on the importance of establishing a distinctive and sustainable advantage accordingly does not, and cannot, apply to the major transformations now inevitably happening in many companies through the adoption of ICT. Their framework can help understand CISCO (a major US supplier of equipment for the Internet), but does not help much with TESCO (a major UK supermarket chain, increasingly using the Internet).

4.5.3 Tribal Warfare

In his enumeration of the potential advantages of increasing specialization in knowledge production, Adam Smith describes the various scientific disciplines as "tribes." Certainly an important element in contemporary processes of innovation is "tribal" conflict between different professional groups with specialized knowledge. Some such conflicts have been touched upon above: financial versus technological competencies in the evaluation of R&D programs; technical versus marketing competencies in product development. But perhaps the most important is the potential resistance of a company's current top managers and technical staff (rooted in the successes of the past), to the introduction of new specialized competencies and methods, reflecting potential opportunities for tomorrow.

The difficulties in introducing the new, in the face of the tried and tested old, were spelled out long ago:

It must be considered that there is nothing more difficult to carry out, nor more doubtful of success, nor more dangerous to handle, than to initiate a new order of things. For the reformer has enemies in all those who profit from the old order of things, and only lukewarm defenders in all those who would profit by the new order, this lukewarmness arising partly from fear of their adversaries . . . and partly from the incredulity of mankind, who do not truly believe in anything new until they have had actual experience of it. Thus it arises that on every opportunity for attacking the reformer, his opponents do so with the zeal of partisans, the others only defend him half-heartedly.   (Machiavelli 1950: 21–2)

Well-documented contemporary examples of this process include IBM's early reluctance to enter the personal computer market and Polaroid's early commitment but subsequent failure to develop a business based on digital imaging (Tripsas and Gavetti 2000). In both cases, a company with the technical resources to develop the new technology, failed to do so, due to resistance and scepticism from the established power structures. In these cases, it can plausibly be argued that yesterday's "core competencies," became today's "core rigidities" (Leonard-Barton 1995).

But the new does not always turn out to be better than the old. Conservative resistances in oil companies to investments in nuclear power in the 1970s turn out to have been largely justified. So, more recently, was scepticism about the dot.com boom. And in the light of IBM's subsequent success in systems integration and software, resistance to heavy commitment to the PC could yet be seen as beneficial in the longer run. This is what makes decision making about radical innovations so difficult. The political battle for influence often involves one-sided and distorted analyses, reflecting the interests of specific disciplines and functions. Crucial factors and key uncertainties may be ignored, consciously or otherwise. The successes and failures only become clear well after the smoke of battle has cleared.

For today's corporate manager there can be no simple tools or model to neutralize the uneasy, politicized task of dealing with radical innovations. Good judgement, experience, trial and error learning remain the only feasible "toolkit" available to today's innovative corporations.

4.6 CONCLUSIONS

Despite spectacular improvements in the scientific knowledge base, and slower but steady improvements in organizational know-how, innovation processes are neither tidy, nor easy to delineate or manage. Increasing specialization in the production of artifacts and in knowledge has also increased levels of complexity—in artifacts themselves, in the knowledge on which they are based, and in the organizational forms and practices for their development and commercial exploitation. As a consequence—and contrary to some of the predictions of Schumpeter (1962) and Penrose (1959):

- innovations—especially radical innovations—remain unpredictable in their technical and commercial outcomes;
- technical entrepreneurship is not a general-purpose management skill; at a time of radical breakthroughs and new opportunities it is in large part specific to a particular technological field and often to a particular place;
- major innovation decisions are a largely political process, often involving professional groups advocating self-interested outcomes under conditions of uncertainty (i.e. ignorance), rather than balanced and careful estimates of costs, benefits and measurable risk.

As a consequence, established large firms have sometimes found it difficult to deal with the radically new. In the future, they will confront new challenges. Increasing complexities in products, systems and the underlying knowledge base are leading firms to experiment with modular product architectures and greater use of ICTs and the outsourcing of component design and production. Large innovating firms are therefore likely to become less self-sufficient in their processes, not more so.

Finally, increasing specialization in the production of artifacts, and their underlying knowledge bases, has made innovative processes increasingly path-dependent. As a consequence, several aspects of innovation processes are contingent on sector, firm, and technology field. These include: the knowledge base underlying innovative opportunities; the links between scientific theory and technological practice; possibilities for knowledge-based diversification; methods of research budget allocation; degree of centralization; and the critical skills, interfaces and networks that need to be developed. Only two innovation processes remain generic: coordinating and integrating specialized knowledge, and learning under conditions of uncertainty.

Notes

1. Keith Pavitt had prepared several drafts of this chapter, but passed away before completing it. Amendments to the original have been made by Christopher Freeman, Mike Hobday, Ian Miles, and the editors.
2. This chapter focuses mainly on large firms within the USA, Europe, and Japan. It does not consider issues arising in connection with entrepreneurial attitudes and motivations, such as their willingness to take risks (Drucker (1985) examines some of the links between entrepreneurship and innovation, and Roberts (1991) and Oakey (1995) examine the process of innovation within SMEs). For studies of innovation processes within firms based in the industrializing countries (or "latecomer firms") see Hobday (1995), Kim (1997), and Fagerberg and Godinho in this volume.
3. Arguably, "innovation managers" (in practice, if not formally identified by job title) can be found at all levels of the firm (for an overview of innovation management in firms see Tidd et al. 2001). Hamel (2000) examines "strategic innovation" (or innovation in strategy approach) while Schonberger (1982) and Robinson (1991) deal with *kaizen* (or continuous improvements) to current vintages of capital equipment and organization.
4. For critical assessments of firm-level models of innovation, ranging from the linear model to chain-link models and more recent interactive/contingent models, see Rothwell (1992), Forrest (1991) and Mahdi (2002).
5. The term "stages" is avoided here, as it implies linearity. Research has consistently shown that the process of innovation within the firm is anything but linear (Kline and Rosenberg 1986; Tidd et al. 2001; Van der Ven et al. 1999). The three sub-processes of innovation, although distinctive, overlap considerably and often occur concurrently.
6. For a review of models of innovation and the importance of contingency, see Mahdi (2002: ch. 2).
7. As noted earlier, most innovation processes are overlapping and intertwined, terms like "stages" or "phases" impose an unrealistic linearity on the various innovation processes.
8. Recall that by artifacts we mean services and systems as well as more tangible material artifacts such as items of equipment.
9. The classic texts on this are Rosenberg (1974), Price (1984), and Mowery and Rosenberg (1989).

10. See Pinchot (1997) for a discussion of the intrapreneur, that is, an innovator operating within a large corporation.

11. See Georghiou et al. (1986).

12. For a discussion of escalation using examples from the UK stock exchange and other major ICT project failures, see Flowers (1996). In the public sector, or state-dependent industries, the problem can be acute. The experience of many countries that established agencies to develop nuclear power, for instance, suggests that these are often extremely difficult to restrain.

13. Three very different lines of argument have been mobilized over recent years. First is the view that the output of university research is a free good available to everybody—as assumed by orthodox economics. In contrast, a second view argues that publicly funded university research is a form of conspicuous intellectual consumption reflecting techno-logical and economic achievements, but not contributing to them (as suggested by Kealey 1996). A more nuanced argument is the "mode 2" claim that the locus of useful scientific discovery is moving from universities to "contexts of application" (as persuasively argued, if somewhat overstated, by Gibbons et al. 1994).

14. Universities also face management challenges here, of course—not least in reconciling the established structures of academic recognition with the contexts of industrially relevant research.

15. Georghiou et al. (1986).

16. Community Innovation Survey data (see the chapter by Smith in this volume for further discussion) are now beginning to provide us with systematic data on many aspects of the innovative effort and sources of innovative ideas for firms of different types and in different sectors.

17. For example, in the absence of theory and/or cheap methods for constructing and testing prototypes, the costs of search and selection become prohibitively high—see Martin (2000) on why Japanese swords did not improve over a period of more than 500 years.

18. See e.g. Rosenberg and Nelson (1994) on the origins of the engineering disciplines in US universities.

19. A similar conclusion has been reached by Becker and Murphy (1992). They argue that the degree of specialization in tasks is limited not by the extent of the market, as in Smith's famous formulation, but by the costs of coordinating specialized activities. These coordination costs are reduced by increases in general knowledge.

20. In addition, some prestigious academic institutions, such as Stanford University in the USA, and the Ecole des Mines in Paris, are developing research programs in "bio-inform-atics," growing out of the challenges to the complex results of the Human Genome project.

21. Sturgeon lists apparel and footwear, toys, data processing, offshore oil drilling, home furnishings and lighting, semiconductor fabrication, food processing, automotive parts, brewing, enterprise networking, and pharmaceuticals. In addition, Prencipe (1997) demonstrates that outsourcing has increased in the production of aircraft engine com-ponents.

22. Rosenberg (1994) has pointed out that in the nineteenth century, the Western Union turned down an opportunity to purchase Bell's patent for the telephone, which it regarded as an inferior product to the telegraph.

23. See e.g. the history of the UK General Electric Company under Arnold Weinstock (Aris 1998).

24. Note also, the first point in this column (1a). This relates also to the attitude and behavior of the financial system. Some recent work on financial capital has returned to the original Schumpeterian emphasis on "credit creation" for the finance of innovation at various stages of the successive technological revolutions. These factors which affect the growth, composition and fluctuation of demand and hence the influence of demand upon innovation at the firm level are further considered in the chapters by Verspagen and O'Sullivan in this volume.

References

ARIS, S. (1998), *Arnold Weinstock and the Making of GEC,* London: Aurum Press.

BARNEY, J. (1991), "Firm Resources and Sustained Competitive Advantage," *Journal of Management* 17: 99–120.

BECKER, G., and MURPHY, K. (1992), "The Division of Labour, Co-ordination Costs, and Knowledge," *The Quarterly Journal of Economics* 107: 1137–60.

*BRUSONI, S., PRENCIPE, A., and PAVITT, K. (2001), "Knowledge Specialization, Organizational Coupling and the Boundaries of the Firm: Why Firms Know More Than They Make" *Administrative Science Quarterly* 46 (4): 597–621.

CHANDLER, A. (1977), *The Visible Hand: The Managerial Revolution in American Business,* Cambridge, Mass.: Belknap.

—— (1997), "The United States: Engines of Economic Growth in the Capital Intensive and Knowledge Intensive Industries," in A. Chandler, F. Amatori, and T. Hikino (eds.), *Big Business and the Wealth of Nations,* Cambridge: Cambridge University Press.

CHRISTENSEN, C. (1997), *The Innovator's Dilemma,* Boston: Harvard University Press.

CLARK, K., and FUJIMOTO, T. (1992), *Product Development Performance,* Boston: Harvard Business School Press.

CONSTANT, E. (2000), "Recursive Practice and the Evolution of Technological Knowledge," in J. Ziman (ed.), *Technological Innovation as an Evolutionary Process,* Cambridge: Cambridge University Press.

COOPER, A., and SCHENDEL, D. (1976), "Strategic Responses to Technological Threats," *Business Horizons* (Feb.), 61–9.

DAVIES, A., TANG, P., BRADY, T., HOBDAY, M., RUSH, H., and GANN, D. (2001), *Integrated Solutions: The New Economy between Manufacturing and Services,* Brighton: SPRU.

DRUCKER, P. F. (1985), *Innovation and Entrepreneurship,* New York: Harper & Row.

EADS, G., and NELSON, R. R. (1971), "Government Support of Advanced Civilian Technologies: Power Reactors and the Supersonic Transport," *Public Policy* (Summer): 405–28.

FAGERBERG, J. (2003), "Schumpeter and the Revival of Evolutionary Economics: an Appraisal of the Literature," *Journal of Evolutionary Economics* 13: 125–59.

FLOYD, C. (1997), *Managing Technology for Corporate Success,* Hampshire: Gower.

FLOWERS, S. (1996), *Software Failure: Management Failure: Amazing Stories and Cautionary Tales,* Chichester: Wiley.

* Asterisked items are suggestions for further reading.

FORREST, J. E. (1991), "Models of the Process of Technological Innovation," *Technology Analysis and Strategic Management* 3(4): 439–52.

*FREEMAN, C. (1982), *The Economics of Industrial Innovation*, London: Frances Pinter.

——and LOUÇÃ, F. (2001), *As Time Goes By: From the Industrial Revolutions to the Information Revolution*, Oxford: Oxford University Press.

GEORGHIOU, L., METCALFE J. S., GIBBONS M., RAY, T., and EVANS, J. (1986), *Post-Innovation Performance*, Basingstoke and New York: MacMillan.

GIBBONS, M., LIMOGES, C., NOWOTNY, H., SCHWARTZMAN, S., SCOTT, P., and TROW, M. (1994), *The New Production of Knowledge, the Dynamics of Science and Research in Contemporary Societies*, London: Sage.

*GRANSTRAND, O., PATEL, P., and PAVITT, K. (1997), "Multi-technology Corporations: Why They Have 'Distributed' rather than 'Distinctive Core' Competencies," *California Management Review* 39: 8–25.

GRILICHES, Z. (1990), "Patent Statistics as Economic Indicators," *Journal of Economic Literature* 28: 1661–707.

HAMEL, G. (2000), *Leading the Revolution*, Boston: Harvard Business School Press.

——and PRAHALAD, C. K. (1994), *Competing for the Future*, Boston: Harvard Business School Press.

*HENDERSON, R., and CLARK, K. (1990), "Architectural Innovation: the Reconfiguration of Existing Product Technologies and the Failure of Established Firms," *Administrative Sciences Quarterly*, 35: 9–30.

*HICKS, D. (1995), "Published Papers, Tacit Competencies and Corporate Management of the Public/private Character of Knowledge," *Industrial and Corporate Change* 4: 401–24.

HOBDAY, M. (1995), *Innovation in East Asia: the Challenge to Japan*, Aldershot: Edward Elgar.

*HOLBROOK, D., COHEN, W., HOUNSHELL, D., and KLEPPER, S. (2000), "The Nature, Sources, and Consequences of Firm Differences in the Early History of the Semiconductor Industry," *Strategic Management Journal* 21: 1017–41.

KEALEY, T. (1996), *The Economic Laws of Scientific Research*, London: MacMillan.

KIM, L. (1997), *Imitation to Innovation: the Dynamics of Korea's Technological Learning*, Boston: Harvard Business School Press.

KLEPPER, S., and SIMONS, K. (2000), "Dominance by Birthright: Entry of Prior Radio Producers and Competitive Ramifications in the U.S. Television Receiver Industry," *Strategic Management Journal* 21: 997–1016.

*KLINE, S. J., and ROSENBERG, N. (1986), "An Overview of Innovation," in R. Landau and N. Rosenberg (eds.), *The Positive Sum Strategy: Harnessing Technology for Economic Growth*, Washington, DC: National Academy Press, 275–304.

LEONARD-BARTON, D. (1995), *Wellsprings of Knowledge*, Boston: Harvard Business School Press.

LEVINTHAL, D. (1998), "The Slow Pace of Rapid Technological Change: Gradualism and Punctuation in Technological Change," *Industrial and Corporation Change* 7: 217–47.

LOUÇÃ, F., and MENDONÇÃ, S. (1999), "Steady Change: The 200 Largest US Manufacturing Firms in the Twentieth Century," Working Paper No. 14/99, CISEP-ISEG, UTL, Lisbon.

LUNDVALL, B. (1988), "Innovation as an Interactive Process: from User–Producer Interaction to the National System of Innovation," in G. Dosi, C. Freeman, R. Nelson, G. Silverberg, and L. Soete (eds.), *Technical Change and Economic Theory*, London: Frances Pinter.

MACHIAVELLI, N. (1950 edn.), *The Prince*, New York: Modern Library College Editions.

MAHDI, S. (2002), "Search Strategy in Product Innovation Process: Theory and Evidence from the Evolution of Agrochemical Lead Discovery Process," D.Phil. Thesis, Unpublished, SPRU, University of Sussex.

MANSFIELD, E. (1995), *Innovation, Technology and the Economy*, Aldershot: Edward Elgar.

MARTIN, G. (2000), "Stasis in Complex Artefacts," in J. Ziman (ed.), *Technological Innovation as an Evolutionary Process*, Cambridge: Cambridge University Press.

MARTIN, B., and SALTER, A. (1996), "The Relationship between Publicly Funded Basic Research and Economic Performance: a SPRU Review," *Report for HM Treasury*, SPRU, University of Sussex.

MENDONÇÃ, S. (2000), *The ICT Component of Technological Diversification*, M.Sc. dissertation, SPRU, University of Sussex.

METHE, D., SWAMINATHAN, A., and MITCHELL, W. (1996), "The Underemphasized Role of Established Firms as Sources of Major Innovations," *Industrial and Corporate Change* 5: 1181–203.

*MITCHELL, G., and HAMILTON, W. (1988), "Managing R&D as a Strategic Option," *Research Technology Management* 31: 15–24.

MOWERY, D. (1995), "The Boundaries of the U.S. Firm in R&D," in N. R. Lamoreaux and D. M. G. Raff (eds.), *Coordination and Information: Historical Perspectives on the Organization of Enterprise*, Chicago: University of Chicago Press of NBER.

—— and ROSENBERG, N. (1979), "The Influence of Market Demand upon Innovation: a Critical Review of Some Recent Empirical Studies," *Research Policy* 8: 102–53.

—— —— (1989), *Technology and the Pursuit of Economic Growth*, Cambridge: Cambridge University Press.

NARIN, F., HAMILTON, K., and OLIVASTRO, D. (1997), "The Increasing Linkage Between U.S. Technology and Public Science," *Research Policy* 26: 317–30.

NELSON, R. R. (ed.) (1982), *Government and Technical Progress: A Cross-Industry Comparison*, New York: Pergamon.

NIGHTINGALE, P. (2000), "Economies of Scale in Experimentation: Knowledge and Technology in Pharmaceutical R&D," *Industrial and Corporate Change* 9: 315–59.

OAKEY, R. (1995), *High-Technology New Firms*, London: Paul Chapman.

PAVITT, K., and STEINMUELLER, W. (2001), "Technology in Corporate Strategy: Change, Continuity and the Information Revolution," in A. Pettigrew, H. Thomas and R. Whittington (eds.), *Handbook of Strategy and Management*, London: Sage.

PENROSE, E. T. (1959), *The Theory of the Growth of the Firm*, Oxford: Basil Blackwell.

PINCHOT, G. (1997), "Innovation Through Intrapreneuring," ch. 26 in R. Katz (ed.), *The Human Side of Managing Technological Innovation*, New York: Oxford University Press.

PORTER, M. (1996), "What is Strategy?" *Harvard Business Review*, Nov./Dec.: 61–78.

PRAHALAD, C. K., and HAMEL, G. (1990), "The Core Competence of the Corporation," *Harvard Business Review*, 90(3): 79–91.

PRENCIPE, A. (1997), "Technological Competencies and Product's Evolutionary Dynamics: a Case Study from the Aero-engine Industry," *Research Policy* 25: 1261–76.

PRICE, D. DE SOLLA (1984), "The Science/Technology Relationship, the Craft of Experimental Science, and Policy for the Improvement of High Technology Innovation," *Research Policy* 13(1), 3–20.

QUINN, J. B. (2000), "Outsourcing Innovation: The New Engine of Growth," *Sloan Management Review* (Summer), 13–28.

ROBERTS, E. (1991), *Entrepreneurs in High Technology: Lessons from MIT and Beyond*, Oxford: Oxford University Press.

ROBINSON, A. (1991), *Continuous Improvement in Operations*, Cambridge, Mass.: Productivity Press.

ROSENBERG, N. (1974), "Science, Invention, and Economic Growth," *Economic Journal* 84(333), 90–108.

—— (1976), "Technological Change in the Machine Tool Industry, 1840–1910," in *Perspectives on Technology*, Cambridge: Cambridge University Press.

—— (1994), *Exploring the Black Box: Technology, Economics and History*, Cambridge: Cambridge University Press.

—— and NELSON, R. (1994), "American Universities and Technical Advance in Industry," *Research Policy* 23: 323–48.

ROTHWELL, R. (1992), "Successful Industrial Innovation: Critical Success Factors for the 1990s," *Research Policy* 22(3), 221–39.

*—— FREEMAN, C., HORSLEY, A., JERVIS, V., ROBERTSON, A., and TOWNSEND, J. (1974), "SAPPHO Updated—Project SAPPHO Phase II," *Research Policy* 3(3), 258–91.

SANDOZ, P. (1997), *Canon*, London: Penguin.

SALTER, A., D'ESTE, P., MARTIN, B., GEUNA, A., SCOTT, A., PAVITT, K., PATEL, P., and NIGHTINGALE, P. (2000), *Talent Not Technology: Publicly Funded Research and Innovation in the UK*, CVCP, SPRU, and HEFCE.

SCHNAARS, S., and BERENSON, C. (1986), "Growth Market Forecasting Revisited: A Look Back at a Look Forward," *California Management Review* 28: 71–88.

SCHONBERGER, R. (1982), *Japanese Manufacturing Technique: Nine Hidden Lessons in Simplicity*, New York: Free Press.

SCHUMPETER, J. (1962), *Capitalism Socialism and Democracy*, 3rd edn., New York: Harper Torchbooks Edition (originally published 1942 by Harper and Brothers).

SMITH, A. (1937), *An Inquiry into the Nature and Causes of the Wealth of Nations*, New York: Modern Library Edition.

STURGEON, T. J. (1999), "Turn-key Production Networks: Industry Organization, Economic Development, and the Globalization of Electronics Contract Manufacturing," Unpublished Ph.D. dissertation, University of California at Berkeley.

*TIDD, J., BESSANT, J., and PAVITT, K. (2001), *Managing Innovation: Integrating Technological, Market and Organizational Change*, 2nd edn., Chichester: Wiley.

TRIPSAS, M., and GAVETTI, G. (2000), "Capabilities, Cognition and Inertia: Evidence from Digital Imaging," *Strategic Management Journal* 21: 1147–61.

TUSHMAN, M. and ANDERSON, P. (1986), "Technological Discontinuities and Organizational Environments," *Administrative Science Quarterly*, 31: 439–65.

UTTERBACK, J. (1994), *Mastering the Dynamics of Innovation*, Boston, Mass.: Harvard Business School Press.

VAN DE VEN, A., POLLEY, D. E., GARUD, R., and VENTKATARAMAN, S. (1999), *The Innovation Journey*, New York: Oxford University Press.

CHAPTER 5

ORGANIZATIONAL INNOVATION

ALICE LAM

5.1 INTRODUCTION

ORGANIZATIONAL creation is fundamental to the process of innovation (Van de Ven et al. 1999). Innovation constitutes part of the system that produces it. The system is itself "organization" or "organizing," to put it in Weick's (1979) term. The ability of an organization to innovate is a precondition for the successful utilization of inventive resources and new technologies. Conversely, the introduction of new technology often presents complex opportunities and challenges for organizations, leading to changes in managerial practices and the emergence of new organizational forms. Organizational and technological innovations are intertwined. Schumpeter (1950) saw organizational changes, alongside new products and processes, as well as new markets as factors of "creative destruction."

In a general sense, the term "organizational innovation" refers to the creation or adoption of an idea or behavior new to the organization (Daft 1978; Damanpour and Evan 1984; Damanpour 1996). The existing literature on organizational innovation is indeed very diverse and not well integrated into a coherent theoretical framework. The phenomenon of "organizational innovation" is subject to different interpretations within the different strands of literature. The literature can be broadly classified into three different streams, each with a different focus and a set of different questions which it addresses. Organizational design theories focus predominantly on the link between structural forms and the propensity of an organization to innovate (Burns and Stalker 1961; Lawrence and Lorsch 1967; Mintzberg 1979). The unit of analysis is the organization and the main research aim is to identify the structural characteristics of an innovative organization, or to determine the effects of organizational structural variables on product and process innovation. This strand of literature has been most influential and well integrated into the literature on technological innovation (e.g. Teece 1998). Theories of organizational cognition and learning, by contrast, tend to focus on the micro-level process of how organizations develop new ideas for problem solving. They emphasize the cognitive foundations of organizational innovation which is seen to relate to the learning and organizational knowledge creation process (Agyris and Schon 1978; Nonaka 1994; Nonaka and Takeuchi 1995). This camp of research provides a micro-lens for understanding the capacity of organizations to create and exploit new knowledge necessary for innovative activities. A third strand of research concerns organizational change and adaptation, and the processes underlying the creation of new organizational forms. Its main focus is to understand whether organizations can overcome inertia and adapt in the face of radical environmental shifts and technological changes, and whether organizational change occurs principally at the population level through selection (e.g. Hannan and Freeman 1977; 1984; Romanelli and Tushman 1994). In this context, innovation is considered as a capacity to respond to changes in the external environment, and to influence and shape it (Burgleman 1991; Child 1997).

While there are important empirical overlaps between these three different strands of research, they remain theoretically distinct. The separation of these research streams has prevented us from developing a clear view of "organizational innovation," and of how its different dimensions are interrelated.[1] This chapter seeks to understand the interaction between organization and innovation from the three different but interdependent perspectives. Section 5.2 examines the relationship between organizational structures and innovation, drawing on the various strands of work in organizational design theories. Section 5.3 looks at organizational innovation from the micro-level perspective of learning and organizational knowledge creation. It argues that organizations with different structural forms vary in their patterns of learning and knowledge creation, engendering different types of innovative capabilities. Section 5.4 discusses organizational adaptation and change, focusing on whether and how organizations can overcome inertia in the face of discontinuous technological changes and radical shifts in environmental conditions. The chapter concludes by discussing the limitations and gaps in the existing literatures, and the areas for future research.

5.2 Organizational Structure and Innovation

Conventional research on organizational innovativeness has explored the determinants of an organization's propensity to innovate. Although researchers have analysed the influence of individual, organizational, and environmental variables (Kimberley and Evanisko 1981; Baldridge and Burnham 1975), most of the research has focused on organizational structure (Wolfe 1994). Within the field of organizational design theories, there has been a long tradition of investigating the links between environment, structures, and organizational performance. Several studies have shown how certain organizational structures facilitate the creation of new products and processes, especially in relation to fast changing environments. The work of micro-economists in the field of strategy also emphasizes the superiority of certain organizational forms within particular types of business strategies and product markets (Teece 1998). More recently, there has been a significant shift in the focus of theoretical enquiry away from purely formal structures towards a greater interest in organizational processes, relationships, and boundaries (Pettigrew and Fenton 2000). The growing influence of economic sociology and the introduction of "network" concepts into the organizational design field denotes such a shift. The relationship between network structure and innovation is dealt with by Powell and Grodal (Ch. 3 in this volume).

5.2.1 Contingency Theory: Context, Structure, and Organizational Innovativeness

The classical theory of organizational design was marked by a preoccupation with universal forms and the idea of "one best way to organize." The work of Weber (1947) on the bureaucracy and of Chandler (1962) on the multidivisional form, was most influential. The assumption of "one best way" was, however, challenged by research carried out during the 1960s and 1970s under the rubric of contingency theory which explains the diversity of organizational forms and their variations with reference to the demands of context. Contingency theory argues that the most "appropriate structure" for an organization is the one that best fits a given operating contingency, such as scale of operation (Pugh et al. 1969; Blau 1970), technology (Woodward 1965; Perrow 1970) or environment (Burns and Stalker 1961; Lawrence and Lorsch 1967). This strand of research and theory underpins our understanding of the relationships between the nature of the task and technological environments, structure and performance. However, only some of the studies deal specifically with the question of how structure is related to innovation.

Burns and Stalker's (1961) polar typologies of "mechanistic" and "organic" organizations (see Box 5.1) demonstrate how the differences in technological and market environment, in terms of their rate of change and complexity, affect organ-

---

Box 5.1 Burns and Stalker: mechanistic and organic structures

Burns and Stalker (1961) set out to explore whether differences in the technological and market environments affect the structure and management processes in firms. They investigated twenty manufacturing firms in depth, and classiffied environments into "stable and predictable" and "unstable and unpredictable." They found that firms could be grouped into one of the two main types, mechanistic and organic forms, with management practices and structures that Burns and Stalker considered to be logical responses to environmental conditions.

The *Mechanistic Organization* has a more rigid structure and is typically found where the environment is stable and predictable. Its characteristics are:

(*a*) tasks required by the organization are broken down into specialized, functionally differentiated duties and individual tasks are pursued in an abstract way, that is more or less distinct from the organization as a whole;

(*b*) the precise definition of rights, obligations, and technical methods is attached to roles, and these are translated into the responsibilities of a functional position, and there is a hierarchical structure of control, authority, and communication;

(*c*) knowledge of the whole organization is located exclusively at the top of the hierarchy, with greater importance and prestige being attached to internal and local knowledge, experience, and skill rather than that which is general to the whole organization;

(*d*) a tendency for interactions between members of the organization to be vertical, i.e. between superior and subordinate.

The *Organic Organization* has a much more fluid set of arrangements and is an appropriate form to changing environmental conditions which require emergent and innovative responses. Its characteristics are:

(*a*) individuals contribute to the common task of the organization and there is continual adjustment and redefinition of individual tasks through interaction with others;

(*b*) the spread of commitment to the organization beyond any technical definition, a network structure of control authority and communication, and the direction of communication is lateral rather than vertical;

(*c*) knowledge may be located anywhere in the network, with this ad hoc location becoming the center of authority and communication;

(*d*) importance and prestige attach to affiliations and expertise valid in industrial and technical and commercial milieu external to the firm.

Mechanistic and organic forms are polar types at the opposite ends of a continuum and, in some organizations, a mixture of both types could be found.

Source: Burns and Stalker (1961).

izational structures and innovation management. Their study found that firms could be grouped into one of the two main types: the former more rigid and hierarchical, suited to stable conditions; and the latter, a more fluid set of arrangements, adapting to conditions of rapid change and innovation. Neither type is inherently right or wrong, but the firm's environment is the contingency that prompts a structural response. Related is the work of Lawrence and Lorsch (1967) on principles of organizational differentiation and integration and how they adapt to different environmental conditions, including the market, technical-economic and the scientific sub-environments, of different industries. Whereas Burns and Stalker treat an organization as an undifferentiated whole that is either mechanistic or organic, Lawrence and Lorsch recognize that mechanistic and organic structures can coexist in different parts of the same organization owing to the different demands of the functional sub-environments. The work of these earlier authors had a profound impact on organizational theory and provided useful design guidelines for innovation management. Burns and Stalker's model remains highly relevant for our understanding of the contemporary challenges facing many organizations in their attempts to move away from the mechanistic towards the organic form of organizing, as innovation becomes more important and the pace of environmental change accelerates. Lawrence and Lorsch's suggestion that mechanistic and organic structures can coexist is reflected in the contemporary debate about the importance of developing hybrid modes of organizations—"ambidextrous organizations"—that are capable of coping with both evolutionary and revolutionary technological changes (Tushman and O'Reilly 1996; see also Sections 5.4.2 and 5.4.3).

Another important early contribution is the work of Mintzberg (1979) who synthesized much of the work on organizational structure and proposed a series of archetypes that provide the basic structural configurations of firms operating in different environments. In line with contingency theory, he argues that the successful organization designs its structure to match its situation. Moreover, it develops a logical configuration of the design parameters. In other words, effective structuring requires consistency of design parameters and contingency factors. The "configurational hypothesis" suggests that firms are likely to be dominated by one of the five pure archetypes identified by Mintzberg, each with different innovative potential: simple structure, machine bureaucracy, professional bureaucracy, divisionalized form, and adhocracy. The characteristic features of the archetypes and their innovative implications are shown in Table 5.1. The main thrust of the argument is that bureaucratic structures work well in stable environments but they are not innovative and cannot cope with novelty or change. Adhocracies, by contrast, are highly organic and flexible forms of organization capable of radical innovation in a volatile environment (see also Section 5.3.3).

Contingency theories account for the diversity of organizational forms in different technological and task environments. They assume that as technology and

Table 5.1 Mintzberg's structural archetypes and their innovative potentials

Organization archetype	Key features	Innovative potential
Simple structure	An organic type centrally controlled by one person but can respond quickly to changes in the environment, e.g. small start-ups in high-technology.	Entrepreneurial and often highly innovative, continually searching for high-risk environments. Weaknesses are the vulnerability to individual misjudgement and resource limits on growth.
Machine bureaucracy	A mechanistic organization characterized by high level of specialization, standardization, and centralized control. A continuous effort to routinize tasks through formalization of worker skills and experiences, e.g. mass production firms.	Designed for efficiency and stability. Good at dealing with routine problems, but highly rigid and unable to cope with novelty and change.
Professional bureaucracy	A decentralized mechanistic form which accords a high degree of autonomy to individual professionals. Characterized by individual and functional specialization, with a concentration of power and status in the "authorized experts." Universities, hospitals, law, and accounting firms are typical examples.	The individual experts may be highly innovative within a specialist domain, but the difficulties of coordination across functions and disciplines impose severe limits on the innovative capability of the organization as a whole.
Divisionalized form	A decentralized organic form in which quasi-autonomous entities are loosely coupled together by a central administrative structure. Typically associated with larger organizations designed to meet local environmental challenges.	An ability to concentrate on developing competency in specific niches. Weaknesses include the "centrifugal pull" away from central R&D towards local efforts, and competition between divisions which inhibit knowledge sharing.
Adhocracy	A highly flexible project-based organization designed to deal with instability and complexity. Problem-solving teams can be rapidly reconfigured in response to external changes and market demands. Typical examples are professional partnerships and software engineering firms.	Capable of fast learning and unlearning; highly adaptive and innovative. However, the unstable structure is prone to short life, and may be driven over time toward the bureaucracy (see also Section 5.3.3).

Sources: Mintzberg (1979); Tidd et al. (1997: 313–14); Lam (2000).

product markets become more complex and uncertain, and task activities more heterogeneous and unpredictable, organizations will adopt more adaptive and flexible structures, and they will do so by moving away from bureaucratic to organic forms of organizing. The underlying difficulties in achieving the "match," however, are not addressed in this strand of research. Contingency theories neglect the possibility that the factors identified as most important in this theory are susceptible to different interpretations by organizational actors (Daft and Weick 1984), and ignores the influence of other factors such as managerial choice (Child 1972; 1997) or institutional pressures (Powell and DiMaggio 1991). These aspects will be discussed in Sections 5.3 and 5.4.

5.2.2 Industrial Economics: Strategy, Structure, and the Innovative Firm

The work of micro-economists in the field of strategy considers organizational structure as both cause and effect of managerial strategic choice in response to market opportunities. Organizational forms are constructed from the two variables of "strategy" and "structure." The central argument is that certain organizational types or attributes are more likely to yield superior innovative performance in a given environment because they are more suited to reduce transaction costs and cope with alleged capital market failures. The multidivisional, or M-form, for example, has emerged in response to increasing scale and complexity of enterprises and is associated with a strategy of diversification into related product and technological areas (Chandler 1962). It can be an efficient innovator within certain specific product markets, but may be limited in its ability to develop new competencies.

The theory of "the innovative enterprise" developed by Lazonick and West (1998) is rooted in the Chandlerian framework, inasmuch as it focuses on how strategy and structure determine the competitive advantage of the business enterprise. It also builds on Lawrence and Lorsch's (1967) conceptualization of organizational design problems as differentiation and integration. This theory postulates that, over time, business enterprises in the advanced economies have to achieve a higher degree of "organizational integration" in order to sustain competitive advantage. Japanese firms are said to have gained a competitive advantage in industries such as electronics and automobiles over the USA because of their superior organizational capacity for integrating shop-floor workers and enterprise networks, enabling them to plan and coordinate specialized divisions of labor and innovative investment strategies. Lazonick and West also argue that those US firms (e.g. Motorola and IBM) that have been able to sustain their competitive advantage also benefit from a high degree of organizational integration. The "organizational integration" hypothesis directs our attention to the social structure of the enterprise and its internal cohesiveness as a critical determinant of corporate strategy and innovative performance. But this interpretation itself is insufficiently attentive to the contingency viewpoint—the Japanese model of organizational integration works well in established technological fields in which incremental innovation is important, but not necessarily in rapidly developing new fields where radical innovation is vital for competitiveness.

Teece (1998) explains the links between firm strategy, structure, and the nature of innovation by specifying the underlying properties of technological innovation and then proposing a related set of organizational requirements of the innovation process. His framework suggests that both the formal (governance modes) and informal (cultures and values) structures, as well as firms' external networks, powerfully influence the rate and direction of their innovative activities. Based on four classes of variables including firm boundaries, internal formal structure, internal informal structure (culture), and external linkages, the author identifies four archetypal corporate governance modes: multiproduct integrated hierarchy, highflex silicon valley type, virtual corporation, and conglomerate. He argues that different organizational arrangements are suited to different types of competitive environments and differing types of innovation. Teece (1998: 156–7) illustrates the argument by distinguishing between two main types of innovation, namely "autonomous" and "systemic" innovation, and matching them with different organizational structures. An autonomous innovation is one that can be introduced to the market without massive modification of related products and processes. An example is the introduction of power steering which did not initially require any significant alternatives to the design of cars or engines. This can often be advanced rapidly by smaller autonomous structures, such as "virtual" firms, accomplishing necessary coordination through arm's-length arrangements in the open market. By contrast, the move to front-wheel drive required the complete redesign of many automobiles in the 1980s. This type of change is systemic innovation which favors integrated enterprises because it requires complex coordination amongst various subsystems, and hence is usually accomplished under one "roof." These propositions, however, have yet to be empirically verified (Teece 1998: 146–7).

The work by micro-economists highlights the interaction between market and organizational factors in shaping innovative performance, although it devotes little attention to the internal dynamics and social processes within organizations. Many of the empirical predictions within this literature on the relationship between firm strategy and structure, and innovative performance have yet to be verified, and pose intriguing opportunities for future research.

5.3 Organizational Cognition, Learning and Innovation

5.3.1 The Cognitive Foundations of Organizational Innovation

The "structural perspectives" discussed above treat innovation as an output of certain structural features; but some organizational researchers regard innovation as "a process of bringing new, problem-solving, ideas into use" (Amabile 1988; Kanter 1983). Mezias and Glynn (1993: 78) define innovation as "nonroutine, significant, and discontinuous organizational change that embodies a new idea that is not consistent with the current concept of the organization's business." This approach defines an innovative organization as one that is intelligent and creative (Glynn 1996; Woodman et al. 1993), capable of learning effectively (Senge 1990; Agyris and Schon 1978), and creating new knowledge (Nonaka 1994; Nonaka and Takeuchi 1995). Cohen and Levinthal (1990) argue that innovative outputs depend on the prior accumulation of knowledge that enables innovators to assimilate and exploit new knowledge. From this perspective, understanding the role of cognition and organizational learning in fostering or inhibiting innovation becomes crucially important.

The cognitively oriented literature in organization and management research is rooted in cognitive psychology and analyzes the various intervening mental processes that mediate responses to the environment (see Hodgkinson 2003). The terms "cognition" or "cognitive" refer to the idea that individuals develop mental models, belief systems, and knowledge structures that they use to perceive, construct, and make sense of their worlds and to make decisions about what actions to take (Weick 1979, 1995; Walsh 1995). Individuals are limited in their ability to process the complex variety of stimuli contained in the external environment (Simon's "bounded rationality" problem), and hence they develop "mental representations" to filter, interpret, and reconstruct incoming information which, under certain circumstances may form the basis of creative ideas and new insights, but may also lead to biases and inertia. The psychological literature has focused predominantly on the information processing consequences of mental models. Organization and management researchers have extended the analysis to the group and organizational levels. Their analysis suggests that organizations develop collective mental models and interpretative schemes which affect managerial decision making and organizational action. Organizational cognition differs from individual cognition because it encompasses a social dimension. Thus much of the research has focused on the socio-cognitive connectedness, and seeks to account for the social processes in the formation of collective cognition and knowledge structures.

The idea that organizations can think and act collectively, and serve as a repository of organized knowledge has stimulated much research on organizational learning and knowledge creation. This work has sought to understand how social inter- action and group dynamics within organizations shape collective intelligence, learning, and knowledge generation, and yields important insights into the micro- dynamics underpinning the innovative capability of organizations. It has also examined how shared mental models or interpretive schemes affect organizational adaptiveness. On the positive side, some argue that shared interpretative schemes facilitate an organization's capacity to process and interpret information in a purposeful manner, promote organizational learning and collective problem solv- ing, and thus enhance its adaptive potential (Fiol 1993; Brown and Duguid 1991). Other studies suggest that organizational interpretative schemes can create "blind spots" in organizational decision making and block organizational change (Shri- vasta and Schneider 1984; Shrivastava et al. 1987). The paradox seems to be that organizational cognition can be at once enabling and crippling, like two sides of the same coin.

Viewing organizational innovation from the cognitive perspective shifts our analysis from organizational structures and systems to the processes of organiza- tional learning and knowledge creation. The analysis below suggests that organiza- tions with different structural forms vary in their patterns of learning and knowledge creation, giving rise to different types of innovative capabilities. Organizational boundaries and the social context of learning influence an organization's cognitive vision and its capacity for radical change and innovation.

5.3.2 Organizational Learning and Knowledge Creation: Shared Context and Collective Learning

Innovation can be understood as a process of learning and knowledge creation through which new problems are defined and new knowledge is developed to solve them. Central to theories of organizational learning and knowledge creation is the question of how organizations translate individual insights and knowledge into collective knowledge and organizational capability. While some researchers argue that learning is essentially an individual activity (Simon 1991; Grant 1996), most theories of organizational learning stress the importance of collective knowledge as a source of organizational capability. Collective knowledge is the accumulated know- ledge of the organization stored in its rules, procedures, routines and shared norms which guide the problem-solving activities and patterns of interaction among its members. Collective knowledge resembles the "memory" or "collective mind" of the organization (Walsh and Ungson 1991). It can either be a "stock" of knowledge stored as hard data or represent knowledge in a state of "flow" emerging from interaction. Collective knowledge exists between rather than within individuals. It can be more or less than the sum of the individuals' knowledge, depending on the mechanisms that translate individual into collective knowledge (Glynn 1996). Both individuals and organizations are learning entities. All learning activities, however, take place in a social context, and it is the nature and boundaries of the context that make a difference to learning outcomes.

Much of the literature on organizational learning points to the importance of social interaction, context, and shared cognitive schemes for learning and knowledge creation (Nonaka 1994; Agyris and Schon 1978; Lave and Wenger 1991; Brown and Duguid 1991, 1998). This builds on Polanyi's (1966) idea that a large part of human knowledge is subjective and tacit, and cannot be easily codified and transmitted independent of the knowing subject. Hence its transfer requires social interaction and the development of shared understanding and common interpretative schemes.

Nonaka's theory of organizational knowledge creation is rooted in the idea that shared cognition and collective learning constitute the foundation of organizational knowledge creation (Nonaka 1994; Nonaka and Takeuchi 1995). At the heart of the theory is the idea that tacit knowing constitutes the origin of all human knowledge, and organizational knowledge creation is a process of mobilizing individual tacit knowledge and fostering its interaction with the explicit knowledge base of the firm. Nonaka argues that knowledge needs a context to be created. He uses the Japanese word "ba," which literally means "place," to describe such a context. "Ba" provides a shared social and mental space for the interpretation of information, interaction, and emerging relationships that serves as a foundation for knowledge creation. Participating in a "ba" means transcending one's limited cognitive perspective or social boundary to engage in a dynamic process of knowledge sharing and creation. In a similar vein, the notion of "community of practice" developed in the work of Lave and Wenger (1991), Wenger (1998), and Brown and Duguid (1991; 1998), suggests that organizational members construct their shared identities and perspectives through "practice," that is shared work experiences. Practice provides a social activity in which shared perspectives and cognitive repertoires develop to facilitate knowledge sharing and transfer. Hence, the work group provides an important site where intense learning and knowledge creation may develop. The group, placed at the intersection of horizontal and vertical flows of knowledge within the organization, serves as a bridge between the individual and organization in the knowledge creation process. Nonaka's theory stresses the critical role of the semi-autonomous project teams in knowledge creation. Much of the recent literature on new and innovative forms of organization also focuses on the use of decentralized, group-based structure as a key organizing principle.

Many organizational and management researchers regard the firm as a critical social context where collective learning and knowledge creation takes place. Nonaka and Takeuchi (1995) talk about the "knowledge creating company." Argyris and

Schon (1978) suggest that an organization is, at its root, a cognitive enterprise that learns and develops knowledge. "Organizational knowledge" essentially refers to the shared cognitive schemes and distributed common understanding within the firm that facilitate knowledge sharing and transfer. It is similar to Nelson and Winter's (1982) concept of "organizational routines": a kind of collective knowledge rooted in shared norms and beliefs that aids joint problem solving and capable of supporting complex patterns of action in the absence of written rules. The notion of "core competence" (Prahalad and Hamel 1990) implies that the learning and knowledge creation activities of firms tend to be cumulative and path-dependent. Firms tend to persist in what they do because learning and knowledge are embedded in social relationships, shared cognition, and existing ways of doing things (Kogut and Zander 1992). Several authors have analyzed how collective learning in technology depends on firms' cumulative competences and evolves along specific trajectories (Dosi 1988; Pavitt 1991). Thus, the shared context and social identity associated with strong group-level learning and knowledge accumulation processes may constrain the evolution of collective knowledge. Firms may find it difficult to unlearn past practices and explore alternative ways of doing things. Levinthal and March (1993) argue that organizations often suffer from "learning myopia," and have a tendency to sustain their current focus and accentuate their distinctive competence, what they call falling into a "competency trap." The empirical research by Leonardo-Barton (1992) illustrates how firms' "core capabilities" can turn into "core rigidities" in new product development.

An inherent difficulty in organizational learning is the need to maintain an external boundary and identity while at the same time keeping the boundary sufficiently open to allow the flow of new knowledge and ideas from outside. March (1991) points out that a fundamental tension in organizational learning is balancing the competing goals of "the exploitation of old certainties" and the "exploration of new possibilities." Whereas knowledge creation is often a product of an organization's capability to recombine existing knowledge and generate new applications from its existing knowledge base, radically new learning tends to arise from contacts with those outside the organization who are in a better position to challenge existing perspectives and paradigms. Empirical research has suggested that sources of innovation often lie outside an organization (von Hippel 1988; Lundvall 1992). External business alliances and network relationships, as well as using new personnel to graft new knowledge onto the existing learning systems, are important mechanisms for organizational learning and knowledge renewal in an environment characterized by rapid technological development and disruptive changes. The "dynamic capability" perspective argues that the long-term competitive perform-ance of the firm lies in its ability to build and develop firm-specific capability and, simultaneously, to renew and reconfigure its competences in response to an environ-ment marked by "creative destruction" (Teece and Pisano 1994). Thus, a fundamen-tal organizational challenge in innovation is not simply to maintain a static balance between exploitation and exploration, or stability and change, but a continuous need to balance and coordinate the two dynamically throughout the organization.

5.3.3 Two Alternative Models of Learning and Innovative Organizations: "J-form" vs. "Adhocracy"

While all organizations can learn and create knowledge, their learning patterns and innovative capabilities vary. During the past decade, a large literature has discussed new organizational models and concepts designed to support organizational learning and innovation. These models include "high performance work systems" or "lean production" (Womack et al. 1990), pioneered by Japanese firms in the automobile industry; and the "N-form corporation" (Hedlund 1994) and "hypertext organization" (Nonaka and Takeuchi 1995). More recent concepts such as "cellular forms" (Miles et al. 1997), "modular forms" (Galunic and Eisenhardt 2001) and "project-based networks" (DeFillippi 2002) reflect the growth of flexible and adaptive forms of organization with a strategic focus on entrepreneurship and radical innovation in knowledge-intensive sectors of the economy. These studies highlight the different ways in which firms seek to create learning organizations capable of continuous problem solving and innovation. Very few studies explain the nature of the learning processes underpinning these structural forms, the types of innovative competences generated and the wider institutional context within which this organizational learning is embedded.

A closer examination of the literature on new forms suggests that the various models of innovative organizations can be broadly classified into two polar ideal types, namely, the "J-form" and "adhocracy". The former refers to an organization which is good at cumulative learning and derives its innovative capabilities from the development of organization-specific collective competences and problem-solving routines. The term J-form is used because its archetypal features are best illustrated by the "Japanese type" of organizations, such as Aoki's (1988) model of the "J-firm", and Nonaka and Takeuchi's (1995) "knowledge-creating companies." Adhocracy (Mintzberg, 1979), by contrast, tends to rely more upon individual specialist expertise organized in flexible market-based project teams capable of speedy responses to changes in knowledge and skills, and integrating new kinds of expertise to generate radical new products and processes. Mintzberg's term is used here to capture the dynamic, entrepreneurial, and adaptive character of the kind of organization typified by Silicon Valley type companies (Bahrami and Evans 2000). Both the "J-form" and "adhocracy" are learning organizations with strong innovative capabilities, but they differ markedly in their structural forms, patterns of learning, and the type of innovative competences generated.

The J-form organization relies on knowledge that is embedded in its operating routines, team relationships, and shared culture. Learning and knowledge creation within the J-form takes place within an "organizational community" that incorporates shop-floor skills in problem solving, and intensive interaction and knowledge sharing across different functional units. The existence of stable organizational careers rooted in an internal labor market provide an incentive for organizational members to commit to organizational goals and to develop firm-specific problem-solving knowledge for continuous product and process improvement. New knowledge is generated through the fusion, synthesis, and combination of the existing knowledge base. The J-form tends to develop a strong orientation towards pursuing an incremental innovation strategy and to do well in relatively mature technological fields characterized by rich possibilities of combinations and incremental improvements of existing components and products (e.g. machine-based industries, electronics components, and automobiles). But the J-form's focus on nurturing organizationally embedded, tacit knowledge and its emphasis on continuous improvement in such knowledge can inhibit learning radically new knowledge from external sources. The disappointing performance of Japanese firms in such fields as software and biotechnology during the 1990s may constitute evidence of the difficulties faced by "J-form firms" in entering and innovating in rapidly developing new technological fields (Lam 2002; Whitley 2003; see also Box 5.2).

The adhocracy is an organic and adaptive form of organization that is able to fuse professional experts with varied skills and knowledge into adhoc project teams for solving complex and often highly uncertain problems. Learning and knowledge creation in an adhocracy occurs within professional teams that often are composed of employees from different organizatons. Careers are usually structured around a series of discrete projects rather than advancing within an intra-firm hierarchy. The resulting project-based career system is rooted in a relatively fluid occupational labor market which permits the rapid reconfiguration of human resources to align with shifting market requirements and technological changes. The adhocracy has a much more permeable organizational boundary that allows the insertion of new ideas and knowledge from outside. This occurs through the recruitment of new staff, and the open professional networks of the organizational members that span organizational boundaries. The adhocracy derives its competitive strength from its ability to reconfigure the knowledge base rapidly to deal with high levels of technical uncertainty, and to create new knowledge to produce novel innovations in emerging new industries. It is a very adaptive form of organization capable of dynamic learning and radical innovation. However, the fluid structure and speed of change may create problems in knowledge accumulation, since the organization's competence is embodied in its members' professional expertise and market-based know-how which are potentially transferable. The adhocracy is subject to knowledge loss when individuals leave the organization. Starbuck (1992: 725), for instance, talks

Box 5.2 Japan: an example of organizational community model of learning

The Japanese economy is characterized by a high level of cooperation and organizational integration. This occurs through extensive long-term collaboration between firms in business groups and networks. Additionally, integration within large firms is particularly strong. Japanese social institutions and employment practices foster the close involvement of shop-floor workers in the development of organizational capability. The successful state education system and large, company-driven networks equip the majority of workers with a high level of skills that employers respect and so can rely on them to contribute usefully to innovation activities. The internal labor market system is characterized not only by long-term attachment but also by well-organized training and job rotation schemes. These practices promote continuous skills formation through learning-by-doing and systematic career progression (Lam 1996; 1997). Hence, a strong organizational capacity to accumulate knowledge and learn incrementally. Over the past three decades, Japanese firms have gained international competitive advantage in those industries such as transport equipment, office machines, consumer electronics, electronic components for computing equipment, and telecommunication hardware. The strength of Japan in these sectors stems from the capability of firms to develop highly flexible production systems through the close integration of shop-floor skills and experience, the tight linkages between R&D, production, and marketing, and a unique innovation strategy based on continual modification and upgrading of existing components and products (Womack et al. 1991). Conversely, organization-specific and path-dependent learning have constrained Japan's success in a number of leading-edge technological fields. Japan finds it harder to excel in sectors which do not exclusively rely on incremental upgrading of system components (e.g. aerospace; supercomputers) and those in which fast-paced radical innovations are crucial for success (e.g. pharmaceuticals and biotechnology). The human-network-based interaction and internal tacit knowledge transfer appear to be less effective in coordinating systems involving complex interactions among components. The organizational community model of learning limits the development of highly specialized scientific expertise, and makes it difficult to adopt radically new skills and knowledge needed for radical learning in emerging new technological fields.

about the "porous boundaries" of this type of organizations and points out that they often find it hard to keep unique expertise exclusive.

The long-term survival of this loose, permeable organizational form requires the support of a stable social infrastructure rooted in a wider occupational community or localized firm networks. The example of high-technology firms in Silicon Valley highlights the importance for the "adhocracy" of supportive local labor markets and other external institutions typically included in analyses of national, sectoral and regional innovation systems (Saxenian 1996; Bahrami and Evans 2000; Angels 2000; see also Ch. 7 in this volume by Edquist, Ch. 11 by Asheim and Gertler, and Ch. 14 by Malerba, as well as Box 5.3).

Box 5.3 Silicon Valley: an example of professional team model of learning

Silicon Valley has been an enormously successful and dynamic region characterized by rapid innovation and commercalization in the fast growing technological fields. The core industries of the region include microelectronics, semiconductors, computer networking, both hardware and software, and more recently biotechnology. Firms operating in these industries undergo frequent reconfiguration and realignment in order to survive in a constantly changing environment marked by incessant innovation. The availability of a large pool of professional experts with known reputations in particular fields enables firms to quickly reconstitute their knowledge and skill base in the course of their innovative endeavours. The rapid creation of new start-up firms focusing on novel innovative projects, and the ease with which project-based firms are able to assemble and reassemble their teams of highly skilled scientists and engineers to engage in new innovative activities are central to the technological and organizational dynamism of the region. The high rate of labor mobility and extensive hiring and firing creates a permissive environment for entrepreneurial start-ups and flexible reconfiguration of project teams and knowledge sources. Labor mobility within the context of a region plays a critical role in the generation of professional networks and facilitates the rapid transmission of evolving new knowledge, a large part of which may be tacit. Such a regionally based occupational labor market provides a stable social context and shared industrial culture needed to ensure the efficient transfer of tacit knowledge in an interfirm career framework. The shared context and industry-specific values within the regional community ensure that tacit knowledge will not be wasted when one changes employers, and this gives the individual a positive incentive to engage in tacit "know-how" learning (Deffillipi and Arthur 1996). A regionally based labor market and networks of firms create a stable social structure to sustain collective learning and knowledge creation within and across firm boundaries. The creation of a wider social learning system amplifies the learning and innovative capability of the individual firms locating within the system. It provides an anchor of stability for fostering and sustaining the innovative capability of the adhocracy.

Although firms in the high-technology sectors are under intense pressure to learn faster and organize more flexibly, evidence thus far suggests that complete adhocracies remain rare. Adhocracies are usually confined to organizational subunits engaged in creative work (e.g. "skunk work" adhocracies) (Quinn 1992: 294–5), or knowledge-intensive professional service fields (e.g. law, management consultancies, software engineering design) where the size of the firm is generally relatively small, enabling the whole organization to function as an interdependent network of project teams (DeFillippi 2002). Attempts by large corporations to adopt the adhocracy mode have proved to be difficult to sustain in the long run. An illustration is the case of Oticon, the Danish manufacturer of hearing aids, which adopted a radical form of project-based organization (described as the "spaghetti organization") to stimulate entrepreneurship and innovation but only to find itself giving way to a more traditional matrix organization a decade later (Foss 2003; see Box 5.4). Elsewhere, the most successful examples of adhocracies are found in regionally based industrial communities, as in the case of Silicon Valley, and other high-technology clusters. There, the agglomeration of firms creates a stable social context and shared

---

Box 5.4 Oticon: the rise and decline of the "spaghetti organization"

Oticon, a Danish electronics producer, is one of the world market leaders in hearing aids. The company became world famous for radical organizational transformation in the early 1990s, and has been treated as an outstanding example of the innovative benefits that a radical project-based organization may generate (Verona and Ravasi 2003). The "spaghetti organization," as it has come to be known, refers to a flat, loosely coupled, project-based organization characterized by ambiguous job boundaries and extensive delegation of task and project responsibilities to autonomous teams. The adoption of the radical structure in 1990 represented a dramatic break from the traditional hierarchical, functional-based organization that the company had relied upon in the past.

The background to the implementation of the spaghetti organization was the loss of competitive advantage that Oticon increasingly realized during the 1980s. Although for decades the company had played a leading role in the hearing aids industry, at the end of the 1980s its products largely depended on a mature and declining technology. The advent of digital technology had gradually led to a shift in the technological paradigm during the 1980s and Oticon was losing ground to its major competitors. In 1990, the company underwent extensive restructuring in response to the crisis. The spaghetti organization was introduced, aiming at developing a more creative and entrepreneurial organization. The radical reorganization had immediate and strong performance effects, resulting in a series of remarkable innovations during the 1990s. Despite this success, the spaghetti organization was partially abandoned from about 1996 and was gradually superseded by a more stable, traditional matrix organization.

The study by Foss (2003) suggests that the Oticon spaghetti organization had encountered severe problems of coordination and knowledge sharing between projects because of the fluid and adhocratic nature of project assignments, and difficulties in ensuring employee commitments to projects. More notably, Foss argues that the spaghetti organization, as an "internal hybrid" (i.e. the infusion of elements of market autonomy and flexibility into a hierarchy), was inherently unstable partly because of the motivational problems caused by "selective intervention." Attempts by management selectively to intervene in project selection and coordination became increasingly at odds with the official rhetoric that stresses self-organization. The mounting frustration among employees eventually led to the retreat from the radical spaghetti organization.

Although the Oticon experiment is widely considered as a success story of organizational innovation, the partial retreat from the spaghetti organization illustrates the inherent difficulties of sustaining a complete adhocracy.

Sources: Foss (2003); Verona and Ravasi (2003).

cognitive framework to sustain collective learning and reduce uncertainty associated with swift formation of project teams and organizational changes. An important item for future research is a clear identification of the population of "adhocracies" in different industries and regions of the global "knowledge-based economy." Current work on this organizational type consists largely of case studies and anecdotes.

5.3.4 The Social Embeddedness of Organizations and their Innovative Capabilities

Although competitive pressures are felt by nearly all organizations in the advanced economies, the emergence and structure of new organizational forms are affected by their particular institutional contexts. A large literature contrasts the patterns of innovation and technological change in different countries and attributes these differences to national institutional frameworks and the ways in which they shape organizational forms and innovative competences (Whitley 2000, 2003; Hollingsworth 2000). The "varieties of capitalism" framework, for example, makes a stylized contrast between coordinated (CME) and liberal market economies (LME). It highlights how differences in labor market organization, training systems, and societal norms and values governing business and economic relationships encourage firms to organize and coordinate their skills and knowledge resources differently to pursue distinctive innovation strategies (Soskice 1999; Hall and Soskice 2001).

Much of the work adopting the "varieties of capitalism" perspective argues that "coordinated market economies" such as Japan and Germany have developed institutions that encourage long-term employment and business relationships, facilitating the development of distinctive organizational competences conducive to continuous but incremental innovation. The J-form organization is facilitated by this type of institutional context. Conversely, "liberal market economies" like the US and UK are better able to foster adhocracies in rapidly emerging new industries through radical innovation. The more permissive institutional environment associated with the US and UK facilitates high labor mobility between firms, and reconfiguration of new knowledge and skills within flexible forms of organization to support risky entrepreneurial activities. In addition to labor markets, other institutional features such as education systems and financial markets also shape the development of skills and innovative competences of firms (Lam 2000; Casper 2000; see also O'Sullivan, this volume). The linkages among institutions, organizations, and innovation are more complex than the simplified stylized contrast between J-form and adhocracy suggests. What the polar-type contrast suggests is that the ability of firms to develop different patterns of learning and innovative competences is contingent upon the wider social context, and that institutional frameworks affect how firms develop and organize their innovative activities in different societies. Societal institutions create constraints on and possibilities for firms to develop different types of organizations and innovative competences, giving rise to distinctive national innovative trajectories.

5.4 ORGANIZATIONAL CHANGE AND INNOVATION

Organizational theories have long considered the ways in which organizations evolve and adapt to their environments, including the influence of technological change on the evolution of organizations (see Tushman and Nelson 1990). A core debate concerns whether organizations can change and adapt to major discontinuous technological change and environmental shifts, or whether radical change in organizational forms occurs principally at the population level through the process of selection (Lewin and Volberda 1999). This literature includes at least three broad views on the nature of organizational adaptation and change. Organizational ecology and institutional theories, as well as evolutionary theories of the firm, emphasize the powerful forces of organizational inertia and argue that organizations respond only slowly and incrementally to environmental changes. This strand of work focuses on the way environments select organizations, and how this selection process creates change in organizational forms. A second view, the punctuated equilibrium model, proposes that oganizations evolve through long periods of incremental and evolutionary change punctuated by discontinuous or revolutionary change. It sees organizational evolution as closely linked to the cyclical pattern of technological change. The punctuated model regards organizational transformation as a discontinuous event occurring over a short period of time. The third perspective, which might be described as strategic adaptation, argues that organizations are not always passive recipients of environmental forces but also have the power to influence and shape the environment. The strategic adaptation perspective stresses the role of managerial action and organizational learning, and the importance of continuous change and adaptation in coping with environmental turbulence and uncertainty.

The following sections examine their main arguments and relevance to our understanding of the relationships between organizational change and innovation.

5.4.1 Incremental/Evolutionary View
of Organizational Change

Organization population ecologists (e.g. Hannan and Freeman 1977; 1984) argue that individual organizations seldom succeed in making radical changes in strategy and structure in the face of environmental turbulence because they are subject to strong inertial forces. Such forces are inherent in the established structures of the organization which represent relatively fixed repertoires of highly reproducible routines. While giving organizations reliability and stability, these routines also make them resistant to change. As a result, organizations respond relatively slowly to threats and opportunities in the environment. Organizational ecology theories posit that adaptation of organizational structures within an industry occurs principally at the population level, with new organizations replacing the old ones that fail to adapt.

The institutional perspective on organizations also emphasizes the stability and persistence of organizational forms in a given population or field of organizations (DiMaggio and Powell 1983; Zucker 1987). A major source of resistance to change arises from the normative embeddedness of an organization within its institutional context. Organizations are socially defined and operate within a web of values, norms, rules, and beliefs and taken-for-granted assumptions that they represent values, interests, and cognitive schemas of organizational and institutional actors which are hard to change (Hinings et al. 1996). In this view, organizational change consists largely of constant reproduction and reinforcing of existing modes of thought and organization (Greenwood and Hinings 1996). In other words, organizational change is usually convergent change that occurs within the parameter of an existing archetype, rather than revolutionary change which involves moving from one archetype to another.[3]

Evolutionary theories of the firm (Nelson and Winter 1982) also argue that organizations are subject to inertial forces. Organizations accumulate know-how and tacit knowledge in the course of their development, and the resulting organizational routines and skills become core competences and are difficult to change. Evolutionary theories regard organizational change as a product of the search for new practices in the neighborhood of an organization's existing practices, that is, "local search," and thus organizational routines and skills change only slowly and incrementally.

In the face of environmental change, new entrants within the industry may displace the established organizations that cannot adapt fast enough; new organizational forms thus tend to evolve and develop from the entrepreneurial activities of new firms. This viewpoint is consistent with the widespread argument in the literature on technological innovation that it is usually new firms which pioneer novel forms of organization to take full advantage of radical changes in technology (Schumpeter 1950; Aldrich and Mueller 1982). However, the relative importance of new entrants versus established organizations in developing new forms of organizing is partly shaped by the scale and pace of environmental change. Some evidence suggests that the effects of technological change on organizational evolution depend on whether the new technology destroys or enhances the competences of existing organizations (Tushman and Anderson 1986; Henderson and Clark 1990). The general observation is that new entrants play a much more significant role in organizational evolution in the face of "competence-destroying" technological innovations; while established organizations are in a better position to initiate changes to adapt to "competence-enhancing" technological changes.

The ability of an organization to adapt to technological change is thus influenced by the speed at which new competences and skills can be developed to match the demands of the new technologies. This is another reason to expect the institutional context to play an important role in shaping the dynamics of organizational change, for reasons noted above. New firms have played a much more prominent role in capitalizing on the new opportunities opened by radical technological changes in the United States than in other industrial economies because of the flexibility of professional labor markets and venture capital markets. In coordinated market economies such as Japan or Germany, new firms are not created as quickly because of the inflexibility of the labor market and relative absence of venture capital. As a result, established organizations may have more time to create new organizational structures and competences to adapt to technological changes. The relative importance of selection versus adaptation as a mechanism underlying the creation of new organizational forms thus may vary between different contexts. Ecology and evolutionary theories of organizational change have tended not to take these contextual factors into account.

5.4.2 Punctuated Equilibrium and Discontinuous Organizational Transformation

In contrast, the punctuated equilibrium model proposes that organizations are capable of initiating revolutionary structural change during periods of environmental turbulence. It depicts organizations as evolving through relatively long periods of stability (equilibrium periods) in their basic patterns of activity that are punctuated by relatively short bursts of fundamental change (revolutionary periods) (Gersick 1991; Romanelli and Tushman 1994). It argues that organizations will typically accomplish fundamental transformations in short, discontinuous bursts of change involving most or all key domains of organizational activity. These include changes in strategy, structure, power distribution, and control systems. Punctuated equilibrium theorists argue that the common state of organizations is one of stability and inertia, and as a result, these "revolutionary periods" provide rare opportunities for organizations to break the grip of structural and cultural inertia. In this view, organizations are most likely to introduce radical changes in times of performance crisis or when they are confronted with disruptive environmental conditions such as radical competence destroying new technologies (Anderson and Tushman 1990). A number of empirical studies based on company histories (e.g. Tushman, Newman, and Romanelli 1986; Romanelli and Tushman 1994) show that in many organizations fundamental organizational transformations occur according to the patterns predicted by the punctuated model. Other studies (e.g. Miller and Friesen 1982; Virany, Tushman, and Romanelli 1992) show that organizations that were able to drastically transform themselves perform better than those that changed incrementally. However, most of the empirical evidence supporting the radical transformative mode of organizational change was based on retrospective archival studies of surviving companies. This approach does not permit analysis of the dynamics of the change process, and fails to account for unsuccessful transitions.

The punctuated model also suggests that the underlying dynamics of technological change influence patterns of organizational evolution. This argument builds on the technology cycle model developed by Anderson and Tushman (1990) which proposes that technological progress is characterized by relatively long periods of incremental, competence-enhancing innovation devoted to elaboration and improvement in dominant design. These periods of increasing consolidation and organizational alignment are punctuated by radical, competence-destroying technological discontinuities which pose fundamental challenges and strategic opportunities for organizations. The implication of the technology cycle concept is that the competitive environment repeatedly changes over time, and successful organizations accordingly have to initiate periodic discontinuous or revolutionary change to accommodate changing environmental conditions. A fundamental challenge facing organizations is to develop diverse competences and capabilities to shape and deal with the technology cycle. Tushman and O'Reilly (1996; 1999) argue that firms operating in the turbulent technological environment need to become "ambidextrous", that is, capable of simultaneously pursuing both incremental and discontinuous technological changes.[4]

The punctuated model provides important insights into patterns of organizational evolution and their relationship to the underlying dynamics of technological change, but it is largely descriptive. This model assumes that new organizational forms would emerge during periods of radical, discontinuous change; but fails to address the crucial question of how organizational actors create new forms during the revolutionary period. The model also does not address the long-term prospects for survival of the new organizational forms that emerge during the revolutionary period.

5.4.3 Strategic Adaptation and Continuous Change

Theories of strategic organizational adaptation and change focus on the role of managerial action and strategic choice in shaping organizational change (Child 1972; 1997; Burgleman 1991). They view the evolution of organizations as a product of actors' decisions and learning, rather than the outcome of a passive environmental selection process. Organizational agents are seen as enjoying a kind of "bounded autonomy." According to Child (1997: 60), organizational action is bounded by the cognitive, material, and relational structures internal and external to the organization, but at the same time it impacts upon those structures. Organizational actors, through their actions and "enactment" (Weick 1979), are capable of redefining and modifying structures in ways that will open up new possibilities for future action. As such, the strategic choice perspective projects the possibility of creativity and innovative change within the organization.

Many strategic adaptation theorists view organizational change as a continuous process encompassing the paradoxical forces of continuity and change, rather than an abrupt, discontinuous, episodic event described by the punctuated equilibrium model. Continuity maintains a sense of identity for organizational learning (Weick 1996; Kodama 2003), and provides political legitimacy and increase the acceptability of change among those who have to live with it (Child and Smith 1987). Burgleman's (1983; 1991) study of the Intel Corporation illustrates how the company successfully evolved from a memory to a microprocessor company by combining the twin elements of continuity and change for strategic renewal. Burgleman argues that consistently successful organizations use a combination of "induced" and "autonomous" processes in strategy making to bring about organizational renewal. According to the author, the induced process develops initiatives that are within the scope of the organizations' current strategy and build on existing organizational learning (i.e. continuity). In contrast, the autonomous process concerns initiatives that emerge outside of the organization and provide the opportunities for new organizational learning (i.e. change). These twin processes are considered vital for successful organizational transformation. In a similar vein, Brown and Eisenhardt (1997) note that continuous organizational change for rapid product innovation is becoming a crucial capability for firms operating in high-velocity industries with short product cycles. Based on detailed case studies of multi-product innovations in six firms in the computer industry, the authors conclude that continuous change and product innovations are supported by organizational structures that can be described as "semi-structures," a combination of "mechanistic" and "organic" features, that balance order and chaos. More notably, the authors identify "links in time" that force simultaneous attention and linkages among past, present and future projects as essential to change processes. The key argument is that links in time create the direction, continuity, and tempo of change to support fast pace adaptation in an uncertain and volatile environment.

Most strategic adaptation theories assume that organizational adaptation can occur through incremental and frequent shifts, and that new organizational forms and discontinuous transformation can be brought about by such processes. This strand of research highlights the importance of firm-level adaptation and internal organizational processes in the creation of new organizational forms. Once again, however, most studies of strategic adaptation present retrospective studies of successful organizational adaptation. They tend to focus on organizational restructuring and transformation within prevailing organizational forms and are not specifically concerned with the creation of new organizational forms (Lewin and Volberda 1999). We remain in need of a theory to account for how and under what conditions managerial action and organizational learning is connected to the emergence of new organizational forms.

5.5 Conclusion

The relationship between organization and innovation is complex, dynamic, and multilevel. The existing literature is voluminous and diverse. This chapter has sought to understand the nature of the relationship from three different but interdependent perspectives: (a) the relationship between organizational structural forms and innovativeness; (b) innovation as a process of organizational learning and knowledge creation; and (c) organizational capacity for change and adaptation. Although there are potentially important overlaps and interconnections between these different aspects of the relationships, the different strands of research have remained separate and there is no single coherent conceptual framework for understanding the phenomenon of "organizational innovation." This is partly due to the great conceptual ambiguity and confusion surrounding the term "organizational innovation." Our review of the existing literature reveals no consensus definition of the term "organizational innovation." Different researchers have used the term to describe different aspects of the relationships between organization and innovation. Indeed, the concept has been used in a rather loose and slippery manner in many writings and some authors are coy about stating definitions. Perhaps this conceptual indeterminacy reflects the fact that "organizational innovation" embraces a very wide range of phenomena. Much work remains to be done if we are to understand how the different dimensions fit together.

This large literature has advanced our understanding of the effects of organizational structure on the ability of organizations to learn, create knowledge, and generate technological innovation. We know relatively less, however, about how internal organizational dynamics and actor learning interact with technological and environmental forces to shape organizational evolution. It remains unclear how and under what conditions organizations shift from one structural archetype to another, and the role of technological innovation in driving the process of organizational change is also obscure. Progress in these areas will require greater efforts to bridge the different levels of analysis and multidisciplinary research to add insight and depth beyond one narrow perspective.

At present, research on organizational change and adaptation is fragmented: the different levels of analysis are disconnected and often rooted in different theoretical paradigms that use different research methods. Thus, ecology and evolutionary theorists have sought to understand the dynamic relationship between innovation and organizational evolution at the population or industry levels using retrospective historical data, while organizational and management researchers tend to examine the process of adaptation at the level of individual organizations, mostly based on cross-sectional case studies. The former is rooted in a structuralist deterministic paradigm whereas the latter takes into account actor choice and intentionality. The disconnection between these two different levels of analysis has meant that we continue to treat selection and adaptation as two separate processes in organizational evolution, whereas in reality new forms of organization emerge from the dynamic interaction between the two processes (Lewin and Volberda 1999). The biggest challenge for researchers is to bridge the wide gulf between ecology/evolutionary theories (dealing with organizational evolution and external forces of change) and strategic choice and learning theories (focusing on actor choice, interpretation, and group dynamics within organizations). A useful avenue for future research would consider how organizational choice and evolutionary processes interact to facilitate organizational change and innovation. This will require longitudinal research on organizational adaptation in "real time," as distinct from retrospective historical case studies (Lewin et al. 1999).

Another factor that inhibits major theoretical progress in the field is the failure of researchers in the fields of innovation and organizational studies to work more closely together. Although innovation scholars have long recognized the importance of the organizational dimension of innovation, many innovation studies continue to be dominated by an economic approach that allows little room for the analysis of creative change and innovation within the organization itself. By contrast, researchers in the field of organizational studies who have developed a rich literature on organizational cognition, learning, and creativity rarely relate their work explicitly to innovation. As a result, this stream of work which offers great potential for understanding the micro-dynamics of organizational change and innovation remains outside the main arena of innovation studies. The bulk of the existing research on the relationship between organization and innovation continues to focus on how technology and market forces shape organizational outcomes and treat organizations primarily as a vehicle or facilitator of innovation, rather than as innovation itself. For example, we tend to assume that technological innovation triggers organizational change because it shifts the competitive environment and forces organizations to adapt to the new set of demands. This deterministic view neglects the possibility that differences in organizational interpretations of, and responses to, external stimuli can affect the outcomes of organizational change. The literature in organizational cognition argues that the environment is equivocal and changes in the environment creates ambiguity and uncertainty which prompts the organization to embark on a cycle of environmental scanning, interpretation, and learning (Daft and Weick 1984; Greve and Taylor 2000). The scanning and search process may lead to new interpretative schemata and organizational action which could be an important source of innovative organizational change. Treating the organization as an interpretation and learning system directs our attention to the important role of internal organizational dynamics, actor cognition, and behavior in shaping the external environment and outcomes of organizational change.

Another promising direction for future research recognizes that organizational innovation may be a necessary precondition for technological innovation, rather than treating this process uniformly as a response to external forces, and focuses on the processes of internal organizational reform and transformation that are necessary to create such preconditions. This requires that scholars take greater account of the role of endogenous organizational forces such as capacity for learning, values, interests, and power in shaping organizational evolution and technological change. This is an area where organization and management researchers could make a significant contribution by placing a greater emphasis on rigorous empirical research and theory building.

Notes

1. The term "organizational innovation" is ambiguous. Some authors use it to refer to the broad meaning of "innovation or innovative behaviour in organizations" (Slappendel 1996; Sorensen and Stuart 2000), or "organizational adoption of innovations" (Kimberley and Evanisko 1981; Damanpour and Evan 1984; Damanpour 1996). Within these broad meanings, the dependent variable "innovation" is defined to encompass a range of types, including new products or process technologies, new organizational arrangements or administrative systems. The main aim of these studies has been to identify a range of individual, organizational, and environmental variables that affect an organization's propensity to adopt an innovation. Others (e.g. Pettigrew and Fenton 2000) use the term in a more restrictive way simply to refer to innovation in organizational arrangements. Here the dependent variable is new organizational practices or organizational forms. Innovation may refer to the widespread adoption by organizational population of an organizational innovation, or merely some novel combination of organizational processes or structures not previously associated. There is a tendency for authors in this camp to equate organizational innovation to organizational change or development,

assuming that change in itself is necessarily innovative, without making an explicit link between organizational change and technological innovation.

2. For a detailed analysis of the interaction between institutions and organizations in innovation systems, see Edquist and Johnson (1992) and Hollingsworth (2000).

3. Institutional theorists accept that radical, innovative change would be possible in newly emerging sectors (e.g. biotechology) where the organizational fields are "illformed" and there is no stipulated template for organizing (Greenwood and Hinings 1996).

4. According to Tushman and O'Reilly (1996; 1999), ambidextrous organizations are ones that can sustain their competitive advantage by operating in multiple modes simultaneously—managing for short-term efficiency by emphasizing stability and control, and for long-term innovation by taking risks. Organizations that operate in this way develop multiple, internally inconsistent architectures, competences, and cultures, with built-in capabilities for efficiency, consistency, and reliability on the one hand, and experimentation and improvisation on the other. During periods of incremental change, organizations require units with relatively formalized roles, responsibilities, functional structures, and efficiency-oriented cultures that emphasize teamwork and continuous improvement. By contrast, during periods of ferment—times that can generate architectural and discontinuous innovation—organizations require entrepreneurial "skunkworks" types of units. These units are relatively small, have loose decentralized product structures, experimental cultures, loose work processes, strong entrepreneurial and technical competences. Examples of companies that have successfully developed ambidextrous organizations include Hewlett-Packard, Johnson and Johnson, and ABB (Asea Brown Boveri), as well as such large Japanese companies as Canon and Honda.

References

ALDRICH, H. E., and MUELLER, S. (1982), "The Evolution of Organizational Forms: Technology, Coordination and Control," in B. M. Staw and L. L. Cummings (eds.), *Research in Organizational Behaviour*, Greenwich, Conn.: JAI Press, 4: 33–87.

AMABILE, T. M. (1988), "A Model of Creativity and Innovation in Organizations," in N. M. Staw and L. L. Cummings (eds.). *Research in Organizational Behaviour*, Greenwich, Conn.: JAI Press, 10: 123–67.

ANDERSON, P., and TUSHMAN, M. L. (1990), "Technological Discontinuities and Dominant Designs: a Cyclical Model of Technological Change," *Administrative Science Quarterly* 35(4): 604–33.

ANGELS, D. P. (2000), "High-Technology Agglomeration and the Labour Market: The Case of Silicon Valley," in K. Martin (ed.), *Understanding Silicon Valley: The Anatomy of an Entrepreneurial Region*, Stanford: Stanford University Press, 125–89.

AOKI, M. (1988), *Information, Incentives and Bargaining in the Japanese Economy*, Cambridge: Cambridge University Press.

*ARGYRIS, C., and SCHON, D. (1978), *Organizational Learning: A Theory of Action Perspective*, Reading, Mass.: Addison-Wesley.

* Asterisked items are suggestions for further reading.

BAHRAMI, H., and EVANS, S. (2000), "Flexible Recycling and High-Technology Entrepreneurship," in K. Martin (ed.), *Understanding Silicon Valley: The Anatomy of an Entrepreneurial Region*, Stanford: Stanford University Press, 166–89.

BALDRIDGE, J. V., and BURNHAM, R. A. (1975), "Organizational Innovation: Individual, Organizational, and Environmental Impacts," *Administrative Science Quarterly* 20(2): 165–76.

BLAU, P. M. (1970), "A Formal Theory of Differentiation in Organizations," *American Sociological Review* 35(2): 201–18.

*BROWN, J. S., and DUGUID, P. (1991), "Organizational Learning and Communities of Practice: Towards a Unified View of Working, Learning and Innovation," *Organization Science* 2(1): 40–57.

—— —— (1998), "Organizing Knowledge," *California Management Review* 40(3): 90–111.

BROWN, S. L., and EISENHARDT, K. M. (1997), "The Art of Continuous Change: Complexity Theory and Time-Paced Evolution in Relentlessly Shifting Organizations," *Administrative Science Quarterly* 42(1): 1–34.

BURGLEMAN, R. A. (1983), "A Model of the Interaction of Strategic Behaviour, Corporate Context, and the Concept of Strategy," *Academy of Management Review* 8(1): 61–70.

—— (1991), "Intraorganizational Ecology of Strategy Making and Organizational Adaptation: Theory and Research," *Organization Science*, 2(3): 239–62.

*BURNS, T., and STALKER, G. M. (1961), *The Management of Innovation*, London: Tavistock.

CASPER, S. (2000), "Institutional Adaptiveness, Technology Policy and the Diffusion of New Business Models: The Case of German Biotechnology," *Organization Studies* 21: 887–914.

CHANDLER, A. D. (1962), *Strategy and Structure: Chapters in the History of the American Industrial Enterprise*, Cambridge, Mass.: MIT Press.

CHILD, J. (1972), "Organizational Structure, Environment and Performance—the Role of Strategic Choice," *Sociology* 6(1): 1–22.

—— (1997), "Strategic Choice in the Analysis of Action, Structure, Organizations and Environment: Retrospect and Prospect," *Organization Studies* 18(1): 43–76.

—— and SMITH, C. (1987), "The Context and Process of Organizational Transformation—Cadbury Limited in its Sector," *Journal of Management Studies* 24: 565–93.

COHEN, W. M., and LEVINTHAL, D. A. (1990), "Absorptive Capacity: A New Perspective on Learning and Innovation," *Administrative Science Quarterly* 35: 123–38.

DAFT, R. L. (1978), "A Dual-Core Model of Organizational Innovation," *Academy of Management Review* 21: 193–210.

—— and LEWIN, A. (1993), "Where Are the Theories for New Organizational Forms? An Editorial Essay," *Organization Science* 4(4): i–vi.

*—— and WEICK, K. E. (1984), "Toward a Model of Organizations as Interpretation Systems," *The Academy of Management Review* 9(2): 284–95.

DAMANPOUR, F. (1996), "Organizational Complexity and Innovation: Developing and Testing Multiple Contingency Models," *Management Science* 42(5): 693–716.

—— and EVAN, W. M. (1984), "Organizational Innovation and Performance: The Problem of Organizational Lag," *Administrative Science Quarterly* 29: 392–402.

DEFILLIPI, R. (2002), "Organization Models for Collaboration in the New Economy," *Human Resource Planning* 25(4): 7–19.

—— and ARTHUR, M. B. (1996), "Boundarlyess Contexts and Careers: A Competency-Based Perspective," in M. B. Arthur and D. M. Rousseau (eds.). *The Boundaryless Career: A New*

Employment Principle for a New Organizational Era, New York: Oxford University Press, 116–31.

DiMaggio, P. J., and Powell, W. W. (1983), "The Iron Cage Revisited: Institutional Isomorphism and Collective Rationality in Organizational Fields," *American Sociological Review* 48: 147–60.

Dosi, G. (1988), "Sources, Procedures, and Microeconomic Effects of Innovation," *Journal of Economic Literature* 26: 1120–71.

Edquist, C., and Johnson, B. (1997), "Institutions and Organizations in Systems of Innovation," in C. Edquist (ed.), *Systems of Innovation: Technologies, Institutions and Organizations*, London: Pinter, 41–63.

Fiol, C. M. (1993), "Consensus, Diversity, and Learning in Organizations," *Organization Science* 5: 403–20.

Foss, N. J. (2003), "Selective Intervention and Internal Hybrids: Interpreting and Learning from the Rise and Decline of the Oticon Spaghetti Organization," *Organization Science* 14(3): 331–49.

Galunic, D. C., and Eisenhardt, K. M. (2001), "Architectural Innovation and Modular Corporate Forms," *Academy of Management Journal* 44(6): 1229–49.

Gersick, C. J. G. (1991), "Revolutionary Change Theories: A Multilevel Exploration of the Punctuated Paradigm," *Academy of Management Review* 16(1): 10–36.

Glynn, M. A. (1996), "Innovative Genius: A Framework for Relating Individual and Organizational Intelligence to Innovation," *Academy of Management Review* 21(4): 1081–111.

Grant, R. M. (1996), "Toward a Knowledge-Based Theory of the Firm," *Strategic Management Journal* 17: 109–22.

*Greenwood, R., and Hinings, C. R. (1996), "Understanding Radical Organizational Change: Bringing Together the Old and New Institutionalism," *Academy of Management Review* 21(4): 1022–54.

Greve, H. R., and Taylor, A. (2000), "Innovations as Catalysts for Organizational Change: Shifts in Organizational Cognition and Change," *Administrative Science Quarterly* 45: 54–80.

Hall, P., and Soskice, D. (eds.) (2001), *Varieties of Capitalism: The Institutional Foundations of Comparative Advantage*, Oxford: Oxford University Press.

Hannan, M. T., and Freeman, J. H. (1977), "The Population Ecology of Organizations," *American Journal of Sociology* 82(5): 929–63.

*———— (1984), "Structural Inertia and Organizational Change," *American Sociological Review* 49(2): 149–64.

Hedlund, G. (1994), "A Model of Knowledge Management and The N-Form Corporation," *Strategic Management Journal* 15: 73–90.

Henderson, R. M., and Clark, R. B. (1990), "Architectural Innovation: The Reconfiguration of Existing Product Technologies and the Failure of Established Firms," *Administrative Science Quarterly* 29: 26–42.

Hinings, C. R., Thibault, L., Slack, T., and Kikulis, L. M. (1996), "Values and Organizational Structure," *Human Relations* 49(7): 885–916.

Hodgkinson, G. P. (2003), "The Interface of Cognitive and Industrial, Work and Organizational Psychology," *Journal of Occupational and Organizational Psychology* 76(1): 1–24.

Hollingsworth, J. R. (2000), "Doing Institutional Analysis: Implications for the Study of Innovations," *Review of International Political Economy* 7(4): 595–644.

KANTER, R. M. (1983), *The Change Masters*, New York: Simon & Schuster.

KIMBERLY, J. R., and EVANISKO, M. J. (1981), "Organizational Innovation: The Influence of Individual, Organizational, and Contextual Factors on Hospital Adoption of Technological and Administrative Innovations," *The Academy of Management Journal* 24(4): 689–713.

KODAMA, M. (2003), "Strategic Innovation in Traditional Big Business: Case Studies of Two Japanese Companies," *Organization Studies* 24(2): 235–68.

KOGUT, B., and ZANDER, U. (1992), "Knowledge of the Firm, Combinative Capabilities, and the Replication of Technology," *Organization Science* 3(3): 383–97.

LAM, A. (1996), "Engineers, Management and Work Organization: a Comparative Analysis of Engineers' Work Roles in British and Japanese Electronics Firms," *Journal of Management Studies* 33(2): 183–212.

——(1997), "Embedded Firms, Embedded Knowledge: Problems of Collaboration and Knowledge Transfer in Global Cooperative Ventures," *Organization Studies* 18(6): 973–96.

*——(2000), "Tacit Knowledge, Organizational Learning, Societal Institutions: an Integrated Framework," *Organization Studies* 21(3): 487–513.

——(2002), "Alternative Societal Models of Learning and Innovation in the Knowledge Economy," *International Social Science Journal* 17(1): 67–82.

LAVE, J., and WENGER, E. (1991), *Situated Learning: Legitimate Peripheral Participation*. New York: Cambridge University Press.

LAWRENCE, P. R., and LORSCH, J. W. (1967), "Differentiation and Integration in Complex Organizations," *Administrative Science Quarterly* 12: 1–47.

LAZONICK, W., and WEST, J. (1998), "Organizational Integration and Competitive Advantage," in G. Dosi et al. (eds.). *Technology, Organization, and Competitiveness*, Oxford: Oxford University Press.

LEONARD-BARTON, D. (1992), "Core Capabilities and Core Rigidities: A Paradox in Managing New Product Development," *Strategic Management Journal* 13: 111–25.

LEVINTHAL, D. A., and MARCH, J. G. (1993), "The Myopia of Learning," *Strategic Management Journal* 14: 95–112.

*LEWIN, A. Y., and VOLBERDA, H. W. (1999), "Prolegomena on Coevolution: a Framework for Research on Strategy and New Organizational Forms," *Organization Science* 10(5): 519–34.

LEWIN, A. Y., LONG, C. P., and CARROLL, T. N. (1999), "The Co-evolution of New Organizational Forms," *Organization Science* 10: 535–50.

LUNDVALL, B.-A. (ed.) (1992), *National Systems of Innovation: Towards a Theory of Innovation and Interactive Learning*, London: Pinter.

MARCH, J. G. (1991), "Exploration and Exploitation in Organizational Learning," *Organization Science* 2: 71–87.

MEZIAS, S. J., and GLYNN, M. A. (1993), "The Three Faces of Corporate Renewal: Institution, Revolution, and Evolution," *Strategic Management Journal* 14: 77–101.

MILES, R. E., SNOW, C. C., MATHEWS, J. A., MILES, G., and COLEMAN, H. J. Jr. (1997), "Organizing in the Knowledge Age: Anticipating the Cellular Form," *Academy of Management Executive* 11(4): 7–20.

MILLER, D., and FRIESEN, P. H. (1982), "Structural Change and Performance: Quantum versus Piecemeal-Incremental Approaches," *Academy of Management Journal* 25(4): 867–92.

MINTZBERG, H. (1979), *The Structuring of Organization*, Englewood Cliffs, NJ: Prentice Hall.

NELSON, R. R., and WINTER, S. G. (1982), *An Evolutionary Theory of Economic Change*, Cambridge, Mass.: Belknap Press.

NONAKA, I. (1994), "A Dynamic Theory of Organizational Knowledge Creation," *Organization Science* 5: 14–37.

*—— and TAKEUCHI, H. (1995), *The Knowledge Creating Company*, New York: Oxford University Press.

PAVITT, K. (1991), "Key Characteristics of the Large Innovating Firm," *British Journal of Management* 2: 41–50.

PERROW, C. (1970), *Organizational Analysis*, London: Tavistock.

PETTIGREW, A. M., and FENTON, E. M. (eds.) (2000), *The Innovating Organization*, London: Sage Publications.

POLANYI, M. (1966), *The Tacit Dimension*, New York: Anchor Day Books.

POWELL, W. W., and DiMaggio, P. J. (eds.) (1991), *The New Institutionalism in Organizational Analysis*, Chicago: University of Chicago Press.

PRAHALAD, C. K., and HAMEL, G. (1990), "The Core Competence of the Corporation," *Harvard Business Review* (May/June): 79–91.

PUGH, D. S., HICKSON, D. J., and HININGS, C. R. (1969), "The Context of Organization Structures," *Administrative Science Quarterly* 14: 47–61.

QUINN, J. B. (1992), *Intelligent Enterprise: A Knowledge and Service Based Paradigm for Industry*, New York: The Free Press.

ROMANELLI, E., and TUSHMAN, M. L. (1994), "Organizational Transformation as Punctuated Equilibrium: An Empirical Test," *Academy of Management Journal* 37(5): 1141–66.

SAXENIAN, A. (1996), "Beyond Boundaries: Open Labour Markets and Learning in the Silicon Valley," in M. B. Arthur and D. M. Rousseau (eds.), *The Boundaryless Career: A New Employment Principle for a New Organizational Era*, New York: Oxford University Press, 23–39.

SCHUMPETER, J. (1950), "The Process of Creative Destruction," in J. Schumpeter (ed.). *Capitalism, Socialism and Democracy*, 3rd edn., London: Allen and Unwin.

SENGE, P. (1990), *The Fifth Discipline: the Art and Practice of the Learning Organization*, New York: Doubleday.

SHRIVASTAVA, P., and SCHNEIDER, S. (1984), "Organizational Frame of Reference," *Human Relations* 37(10): 795–809.

—— MITROFF, I., and ALVESSON, M. (1987), "Nonrationality in Organizational Actions," *International Studies of Management and Organization* 17: 90–109.

SIMON, H. A. (1991), "Bounded Rationality and Organizational Learning," *Organization Science* 2: 125–34.

SLAPPENDEL, C. (1996), "Perspective on Innovation in Organizations," *Organization Studies* 17(1): 107–29.

SORENSEN, J. B., and STUART, T. E. (2000), "Age, Obsolescence, and Organizational Innovation," *Administrative Science Quarterly* 45(1): 81–112.

SOSKICE, D. (1999), "Divergent Production Regimes: Coordinated and Uncoordinated Market Economies in the 1980s and 1990s," in H. Kitschelt, P. Lange, G. Marks, and J. Stephens (eds.). *Continuity and Change in Contemporary Capitalism*, Cambridge: Cambridge University Press, 101–34.

STARBUCK, W. H. (1992), "Learning by Knowledge-Intensive Firms," *Journal of Management Studies* 29(6): 713–40.

TEECE, D. J. (1998), "Design Issues for Innovative Firms: Bureaucracy, Incentives and Industrial Structure," in A. D. Chandler Jr., P. Hagstrom, and O. Solvell (eds.). *The Dynamic Firm*, Oxford: Oxford University Press, 134–65.

—— and PISANO, G. (1994), "The Dynamic Capabilities of Firms: an Introduction," *Industrial and Corporate Change* 3(3): 537–56.

TIDD, J., BESSANT, J., and PAVITT, K. (1997), *Managing Innovation*, Chichester: John Wiley & Sons.

*TUSHMAN, M. L., and ANDERSON, P. (1986), "Technological Discontinuities and Organizational Environments," *Administrative Science Quarterly* 31(3): 439–65.

—— and NELSON, R. R. (1990), "Introduction: Technology, Organizations and Innovation," *Administrative Science Quarterly* 35(1): 1–8.

—— and O'REILLY, C. A. III (1996), "Ambidextrous Organizations: Managing Evolutionary and Revolutionary Change," *California Management Review* 38(4): 8–30.

—— —— (1999), "Building Ambidextrous Organizations: Forming Your Own 'Skunk Works'," *Health Forum Journal* 42(2): 20–3.

—— NEWMAN, W. H., and ROMANELLI, E. (1986), "Convergence and Upheaval: Managing the Unsteady Pace of Organizational Evolution," *California Management Review* 29(1): 29–44.

VAN DE VEN, A., POLLEY, D., GARUD, S., and VENKATARAMAN, S. (1999), *The Innovation Journey*, New York: Oxford University Press.

VERONA, G., and RAVASI, D. (2003), "Unbundling Dynamic Capabilities: an Exploratory Study of Continuous Product Innovation," *Industrial and Corporate Change* 12(3): 577–606.

VIRANY, B., TUSHMAN, M. L., and ROMANELLI, E. (1992), "Executive Succession and Organizational Outcomes in Turbulent Environments: An Organization Learning Approach," *Organization Science* 3: 72–91.

VON HIPPEL, E. (1988), *The Sources of Innovation*, New York: Oxford University Press.

WALSH, J. P. (1995), "Managerial and Organizational Cognition: Notes From a Trip Down Memory Lane," *Organization Science* 6(3): 280–321.

—— and UNGSON, G. R. (1991), "Organizational Memory," *Academy of Management Review* 16: 57–91.

WEBER, M. (1947), *The Theory of Social and Economic Organization*, Glencoe, Ill.: The Free Press.

WEICK, K. E. (1979), *The Social Psychology of Organizing*, 2nd edn., Reading, Mass.: Addison-Wesley.

—— (1995), *Sensemaking in Organizations*, Thousand Oaks, Calif.: Sage.

—— (1996), "The Role of Renewal in Organizational Learning," *International Journal of Technology Management* 11(7–8): 738–46.

WENGER, E. (1998), *Communities of Practice: Learning, Meaning, and Identity*, New York: Cambridge University Press.

WHITLEY, R. (2000), "The Institutional Structuring of Innovation Strategies: Business Systems, Firm Types and Patterns of Technical Change in Different Market Economies," *Organization Studies* 21(5): 855–86.

—— (2003), "The Institutional Structuring of Organizational Capabilities: the Role of Authority Sharing and Organizational Careers," *Organization Studies* 24(5): 667–95.

WOLFE, B. (1994), "Organizational Innovation: Review, Critique and Suggested Research Directions," *Journal of Management Studies* 31: 405–31.

WOMACK, J. P., JONES, D. T., and ROOS, D. (1990), *The Machine that Changed the World*, New York: Rawson Associates.

WOODMAN, R. W., SAWYER, J. E., and GRIFFIN, R. W. (1993), "Toward a Theory of Organizational Creativity," *Academy of Management Review* 18(2): 293–321.

WOODWARD, J. (1965), *Industrial Organization, Theory and Practice*, London: Oxford University Press.

ZUCKER, L. G. (1987), "Institutional Theories of Organizations," *Annual Review of Sociology* 13: 443–64.

CHAPTER 6

MEASURING INNOVATION

KEITH SMITH

6.1 INTRODUCTION[1]

It is sometimes suggested that innovation is inherently impossible to quantify and to measure. This chapter argues that while this is true for some aspects of innovation, its overall characteristics do not preclude measurement of key dimensions of processes and outputs. An important development has been the emergence of new indicators of innovation inputs and outputs, including economy-wide measures that have some degree of international comparability. Following sections discuss first some broad issues in the construction and use of science, technology, and innovation (STI) indicators, then turn (briefly) to the strengths and weaknesses of current indicators, particularly R&D and patents. Final sections cover recent initiatives focusing on the conceptualization, collection, and analysis of direct measures of innovation.

New rather than "traditional" indicators are emphasized here because, as Kenneth Arrow remarked many years ago, "too much energy has gone into squeezing the last bit of juice out of old data collected for different purposes relative to the design of new types of data," a point echoed by Zvi Griliches: "far too little fresh economics data is collected" (Arrow 1984: 51; Griliches 1987: 824). Innovation data producers have responded to this kind of challenge. The most important development has been new survey-based indicators, especially the *Community Innovation Survey* (CIS), which has been carried out three times in all EU Member States. The basic format of CIS has diffused to many other countries (including Canada, Australia, Hungary, Brazil, Argentina, and China). Has this effort been justified? In answering this much depends on the quality of analysis these surveys make possible, so the final section discusses the rapidly growing research and publication efforts deriving from CIS.

6.2 THE CONCEPTUAL BACKGROUND: MEASUREMENT ISSUES

Measurement implies commensurability: that there is at least some level on which entities are qualitatively similar, so that comparisons can be made in quantitative terms.

An immediate problem is that innovation is, by definition, novelty. It is the creation of something qualitatively new, via processes of learning and knowledge building. It involves changing competences and capabilities, and producing qualitatively new performance outcomes. This may lead to new product characteristics that are intrinsically measurable in some way—new lift/drag aspects of an aircraft wing, for example, or improved fuel efficiency of an engine. However, such technical measurement comparisons are only rarely meaningful across products. More generally, innovation involves multidimensional novelty in aspects of learning or knowledge organization that are difficult to measure or intrinsically non-measurable. Key problems in innovation indicators therefore concern the underlying conceptualization of the object being measured, the meaning of the measurement concept, and the general feasibility of different types of measurement. Problems of commensurability are not necessarily insoluble, but a main point arising from recent work is the need for care in distinguishing between what can and what cannot be measured in innovation.

Quite apart from the problem of whether novelty can be measured, a fundamental definitional issue is what we actually mean by "new" (see Ch. 1 by Fagerberg in this volume). Does an innovation have to contain a basic new principle that has never been used in the world before, or does it only need to be new to a firm? Does an innovation have to incorporate a radically novel idea, or only an incremental change? In general, what kinds of novelty count as an innovation? These issues of commensurability and novelty are basic problems for all S&T indicators—R&D in particular—but have been most explicitly addressed in the development of direct innovation indicators.

6.3 THEORIES OF INNOVATION AND THEIR USE IN INDICATOR DEVELOPMENT

Although statistics are often treated as though their meanings are transparent, they always rest on some kind of (usually implicit) conceptual foundations. The system of national accounts, for example, derives from Keynesian macroeconomic concepts that seek to identify components of aggregate demand. R&D data has a complex background in the scientification of innovation—the notion that acts of research and discovery underpin innovation (Laestadius 2003). These conceptual foundations are rarely considered when indicators are used. Such issues are complicated by the fact that some key S&T indicators are by-products of other processes—legal procedures (as with patents), or academic institutions (as with bibliometrics, which rest on publishing conventions).

What kinds of ideas have formed the conceptual foundations of innovation indicators? An important figure here has been Nathan Rosenberg, whose work quite explicitly affected the OECD's *Innovation Manual* (OECD 1992, 1997). (This manual is usually called the *Oslo Manual* because much of the drafting and expert meetings on it occurred there.) First, Rosenberg challenged the notion of research-based discovery as a preliminary phase of innovation. Second, he challenged the idea of separability between innovation and diffusion processes, pointing out that most diffusion processes involve long and cumulative programs of post-commercialization improvements (see Rosenberg 1976 and 1982). Perhaps his best-known contribution, with Steven Kline, has been the so-called chain-link model of innovation, which stresses three basic aspects of innovation (Kline and Rosenberg 1986):

- innovation is not a sequential (linear) process but one involving many interactions and feedbacks in knowledge creation
- innovation is a learning process involving multiple inputs
- innovation does not depend on invention processes (in the sense of discovery of new principles), and such processes (involving formal R&D) tend to be undertaken as problem-solving within an ongoing innovation process rather than an initiating factor

The work of Rosenberg alone, and of Rosenberg and Kline, has at least two important implications for indicator development. The first is that novelty implies not just the creation of completely new products or processes, but relatively small-scale changes in product performance which may—over a long period—have major technological and economic implications. A meaningful innovation indicator should therefore be able to pick up such change. The second is the importance of non-R&D inputs to innovation—design activities, engineering developments and experimentation, training, exploration of markets for new products, etc. So there is a need for input indicators that reflect this input variety and its diverse distributions across activities.

The CIS effort has in general been informed by ideas from recent innovation research. One in particular should be mentioned, especially because it has had a strong impact on research using the new data. This is the idea that innovation relies on collaboration and interactive learning, involving other enterprises, organizations, and the science and technology infrastructure. Data gatherers have been concerned to explore the networking dimension of innovation, and this has been an important conceptual issue in survey design (see Howells 2000, for an overview of research on this topic).

6.3.1 Existing and New Indicators: What Can Be Measured, and What are the Limitations?

What does it mean to measure qualitatively diverse phenomena? Clearly this is a serious problem for R&D data. Research is a knowledge-creating process for which both activities and outcomes are radically incommensurable—there is no meaningful way to assess the dissimilar actions and events that feed into research, let alone to compare the increments to knowledge that follow from research. This problem cannot be overcome—it can only be circumvented by carefully specifying aspects of the research process that are in some serious sense measurable. The solution adopted by the framers of the *Frascati Manual* (the OECD's operating statistical manual for R&D data collection) has been to write definitions of research-comprising activities, and then seek data on either expenditure or personnel resources devoted to such activities. The measurement concept for R&D is therefore economic in character, and the datasets that result are collections of economic indicators compatible with industrial datasets, and indeed with the national accounts.[2]

This approach to measurement has also been taken with innovation surveys. The problem is that innovation is usually conceptualized in terms of ideas, learning, and the creation of knowledge (moreover knowledge creation of a far wider character than research), or in terms of competences and capabilities. As with "research," innovation is a multidimensional process, with nothing clearly measurable about many aspects of the underlying process. Most modern innovation theory rests on some kind of "resource-based" theory of the firm, in which firms create physical and intangible assets that underpin capabilities (see Lazonick in this volume). Innovative learning can be seen as change in the knowledge bases on which capabilities rest. Neither learning, nor the capabilities which result, seem to be measurable in any direct way. However, just as "research" can be captured via expenditures on certain activities, or by the use of time by certain research personnel, so learning processes can to some extent be captured by activities such as design, training, market research, tooling up, etc. Expenditure on such activities can in principle be measured (of course the practice may be difficult, since some of these innovation-related activities are not straightforwardly reflected in the accounting procedures of firms). On the output side, the question is whether capability outcomes can be measured by some tangible change in physical or economic magnitudes. Once again there are also potential measurement areas—experience (with pilot or experimental surveys in the 1980s) showed that firms can identify changes in their product mixes, and can estimate sales from new or changed products (Smith 1992). So it is possible to define product change, in terms of construction, use of materials, technical attributes, or performance characteristics, and then to look at the place of (differently) changed products in the sales of the firm. These considerations lead to expenditure measures of inputs to innovation, and sales measures of outputs of innovation. These economic measures of innovation are clearly analogous to the measurement of research. This similarity in approach incidentally suggests that it makes no sense to use R&D data while rejecting the use of more direct innovation data.

6.4 Current Major Indicators

This section outlines the major established indicators that have been used for innovation analysis, and provides a brief guide to further analysis of them. There are three broad areas of indicator use in STI analysis: first, R&D data; second, data on patent applications, grants and citations; and third, bibliometric data (that is data on scientific publication and citation).

In addition to this there are three other important classes of indicators:

- technometric indicators, which explore the technical performance characteristics of products (see e.g. Saviotti 1996 and 2001 for a theoretical view of this, and Grupp 1994 and 1998 for analysis and empirical specifications);
- synthetic indicators developed for scoreboard purposes mainly by consultants (see World Economic Forum 2003);
- databases on specific topics developed as research tools by individuals or groups (such as the large firm database used by Pavitt and Patel, or the MERIT-CATI database on technological collaboration developed by John Hagedoorn, or the DISKO surveys on technological collaboration emanating from the University of Ålborg (see Patel and Pavitt 1997 and 1999, Hagedoorn and Schakenraad 1990, and—for extensive reporting on the use of collaboration data—OECD 2001).

Box 6.1 Bibliometric data

Bibliometric analysis, meaning the analysis of the composition and dynamics of scientific publication and citation, revolves around the Science Citation Index and the Institute for Scientific Information database. The Institute for Scientific Information (ISI) was founded in 1958, and acquired by Thomson Business Information—a subsidiary of the Thomson Corporation—in 1992. The ISI National Science Indicators database currently contains publication and citation statistics from more than 170 countries, and 105 subfields in the sciences, social sciences, and arts and humanities, representing approximately 5,500 journals in the sciences, 1,800 in the social sciences, and 1,200 in the arts and humanities.

The following discussion concentrates on R&D and patents, since bibliometric analysis relates primarily to the dynamics of science rather than innovation (see Moed et al. 1995, and Kaloudis 1997 for reviews of the state of the art).

6.4.1 Research and Development (R&D) Statistics and Indicators

By far the longest-standing area of data collection is R&D.

The key OECD document for the collection of R&D statistics is the *Standard Practice for Surveys of Research and Experimental Development*, better known as the *Frascati Manual*. The first edition was the result of an OECD meeting of national experts on R&D statistics in Frascati, Italy, in 1963. The manual has been continuously monitored and modified through the years: the current version of the manual, the *Frascati Manual 2002*, is the seventh edition (OECD 2002). The Manual defines R&D as comprising both the production of new knowledge and new practical applications of knowledge: R&D is conceived as covering three different kinds of activities: basic research, applied research, and experimental development—these categories are distinguished in terms of their distance from application.

It is often difficult to draw the dividing line between what should be counted as R&D and what should be excluded: "The basic criterion for distinguishing R&D from related activities is the presence in R&D of an appreciable element of novelty and the resolution of scientific and/or technological uncertainty, i.e. when the solution to a problem is not readily apparent to someone familiar with the basic stock of commonly used knowledge and techniques in the area concerned" (OECD 2002: 33). Education and training in general is not counted as R&D. Market research is excluded. There are also many other activities with a scientific and technological base that are kept distinct from R&D. These include such industrial activities related to innovation as acquisition of products and licenses, product design, trial production, training and tooling up, unless they are a component of research, as well as the acquisition of equipment and machinery related to product or process innovations.

R&D is often classified according to multiple criteria, and data is collected in highly detailed forms. Beyond the distinction between basic research, applied research and development the data is classified into sector of performance: business enterprise, government, higher education, and private non-profit. It also distinguishes between sources of finance, both domestic and international. Then there is classification by socio-economic objectives, and a further classification by fields of research. These detailed classifications are usually ignored both by policy analysts and researchers, who tend to focus on gross expenditure only (at industry or country level), thereby missing most of the really interesting detail in the data. For example, a major issue is that, when looking at R&D by fields of research, ICT (information and communications technologies) turns out to be the largest single category in all countries that classify R&D data in this way. However most of the ICT research is actually performed outside the ICT sector, in the form of systems and software development by users.[3] On the one hand, this raises interesting questions about the cross-industry significance of the ICT sector; but there are also questions about the extent to which such activity should be classified as R&D at all. Concerns have also been expressed about whether the R&D definitions are comprehensible to firms (especially SMEs), and whether or not there is systematic undercounting of small-firm R&D (Kleinknecht, Montfort, and Brouwer 2002).

R&D data is always constrained as an innovation indicator by the fact that it measures an input only (Kleinknecht et al. 2002). However, R&D also has fundamental advantages. These include the long period over which it has been collected, the detailed subclassifications that are available in many countries, and the relatively good harmonization across countries. Unfortunately a great deal of the literature consists essentially of an attempt to match aggregate R&D measures across time and across sectors or countries to some measure of productivity (see Griffith, Redding, and Van Reenen (2000) for a very thorough recent example; Dowrick (2003) is a recent survey of this very large literature). However this research effort is limited in two senses—on the one hand it tends to imply (along with the new growth theory, incidentally) that R&D is the primary source of productivity growth, and on the other it fails to exploit the basic complexity of the data that is actually available. The disaggregation processes that are possible with R&D data continue to offer rich and unexploited opportunities for researchers.

6.4.2 Policy Pitfalls: The Use and Misuse of R&D Indicators

It is worth saying something about the pitfalls of R&D as a policy indicator, especially via the most widely-used indicator, that of "R&D Intensity." This is the ratio of R&D expenditure to some measure of output. For a firm, it is usually the R&D/Sales ratio. For an industry or a country it is the ratio of business expenditure on R&D (often known as BERD) to total production or value added. For a country it is usually gross expenditure on R&D (GERD) to GDP.

The R&D/GDP ratio is used in two primary ways. First, it is used to characterize industries—high BERD/GDP ratios for an industry are held to identify high-technology activities. Second, a high GERD/GDP ratio for a country is often believed to indicate technological progressiveness and commitment to knowledge creation (see Godin 2004 for an account of the historical background to these notions).

For countries, there is a distribution of GERD/GDP intensities, as Table 6.1 indicates. Both analysts and policy makers often treat a particular place in the ranking, or the OECD average, or some particular GERD/GDP ratio as desirable in itself. So Canada, for example, has the objective of raising its ranking to fifth in the OECD table; Norway has the target of reaching the OECD average for GERD/GDP; and the EU as a whole has a target of reaching a GERD/GDP ratio of 3 per cent (it could be argued that this target dominates EU technology policy making at the present time). But what is the indicator really telling us?

A basic problem is that R&D intensity depends on the industrial mix. Currently the OECD uses a four-tier model to classify industries, in which the basic criterion is the BERD/Production ratio:

high-tech industries		> 5%	R&D/Production
medium high-tech industries	> 5%	> 3%	R&D/Production
low-tech industries	> 3%	> 1%	R&D/ Production
low-tech industries	> 1%	> 0%	R&D/ Production

Since industries vary considerably in their BERD/GDP ratios, the aggregate BERD/GDP ratio may simply be an effect of that fact that industrial structures are different across countries. A country or region with large high-R&D industries will naturally have a higher aggregate BERD/GDP ratio than one with most of its activities in low-R&D industries. These structural issues largely explain the differences in R&D intensities across large and smaller economies (Sandven and Smith 1997). The question then is, does a specific industrial structure really matter? This is question for debate, which cannot be addressed let alone settled here (it is interestingly explored in Pol et al. 2002); however the desirability of specific industrial structures is the real issue underlying use of this aggregate indicator, though it is rarely explicitly discussed. It is worth noting also that within an industry there tends to be a wide distribution of R&D intensities among firms, so it is common to find

Table 6.1 GERD/GDP ratios across countries

Country	GERD/GDP 2000	Percentage point deviation from OECD mean
Sweden	3.65 (1999)	1.40
Finland	3.40	1.15
Japan	2.98	0.73
United States	2.72	0.47
Korea	2.65	0.40
Germany	2.49	0.29
France	2.18	−0.07
Netherlands	1.94	−0.26
Canada	1.87	−0.33
United Kingdom	1.85	−0.35
Austria	1.84	−0.36
Norway	1.65	−0.6
Australia	1.53	−0.72
Ireland	1.15	−1.1
Italy	1.07	−1.18
New Zealand	1.03	−1.22
Spain	0.94	−1.31
Greece	0.67	−1.58
Total OECD	2.25	

Source: OECD, Main Science and Technology Indicators Database, accessed August 2003.

high-R&D firms in low-R&D industries and vice versa (Hughes 1988 discusses the intra-industry distributions using UK data).

An important recent modification of this indicator has been the addition of "acquired technology," calculated as the R&D embodied in capital and intermediate goods used by an industry, and computed via the most recent input–output table. The method for calculating acquired R&D is to assume that the R&D embodied in a capital good is equal to the capital good's value multiplied by the R&D intensity of the supplying industry. The most recent year for which relevant input–output data is generally available is 1990. The overall structure of the classification as currently used can be seen in Table 6.2, which shows direct R&D intensities for the main industrial groups for 1997, plus the proportion of acquired to direct R&D for 1990, the last year for which it was calculated.

Table 6.2 Classification of industries based on R&D intensity

	ISIC Rev 3	Direct R&D Intensity 1997	Acquired R&D intensity as % of direct R&D intensity, 1990
High technology Industries			
Aircraft and spacecraft	353	12.7	15
Pharmaceuticals	2423	11.3	8
Office, accounting and computing machinery	30	10.5	25
Radio, television and communications equipment	32	8.2	17
Medical, precision and optical instruments	33	7.9	29
Medium–high-technology industries			
Electrical machinery and apparatus	31	3.8	42
Motor vehicles and trailers	34	3.5	29
Chemicals	24 exc 2423	2.6	18
Railroad and transport eqpt. n.e.c.	352+359	2.8	88
Machinery and eqpt n.e.c.	29	1.9	104
Medium–low-technology industries			
Coke, refined petroleum products and nuclear fuel	23	0.8	30
Rubber and plastic products	25	0.9	127
Other non-metallic mineral products	26	0.9	285
Building and repairing of ships and boats	351	0.7	200
Basic metals	27	0.7	289
Fabricated metals products	28	0.6	133
Low-technology industries			
Manufacturing n.e.c. and recycling	36–37	0.4	n.a.
Wood, pulp, paper, paper products, printing and publishing	20–22	0.3	167
Food products, beverages and tobacco	15–16	0.4	267
Textiles, textile products, leather and footwear	17–19	0.3	250

Sources: OECD, *Science, Technology and Industry Scoreboard 1999: Benchmarking Knowledge-Based Economies* (Paris:OECD 1999), Annex 1, p. 106; OECD, *Science, Technology and Industry Scoreboard 2001: Towards a Knowledge-Based Economy*, Annex 1.1, pp. 13–139.

Note: The ISIC classification was revised in 1996, though changes were relatively minor. 1990 data has been reassigned to the most relevant Rev 3 category.

Table 6.2 shows that "acquired technology" as a proportion of direct R&D rises dramatically as we move from high- to low-technology industries. This suggests that technology intensity is likely to be very sensitive to how the measurement of acquired technology is carried out. For example: suppose we assume that when a firm buys a machine it acquires not a proportion of the R&D that went into the machine (corresponding to the R&D/output ratio) but all of it? In other words, purchasing a computer gives the customer access to all of the R&D that was used to produce it—this assumption seems to be compatible with the knowledge externality ideas of the new growth theory (for an overview see Verspagen 1992, see also Verspagen in this volume). Making this assumption would significantly alter the rankings of technology intensity in Table 6.2 by improving the position of industries with substantial use of R&D embodied in capital goods. Another point to make here is that so-called low-technology industries do not create or access knowledge via direct R&D, and the classification is in effect biased against all industries that employ non-R&D methods of knowledge creation (Hirsch-Kreinson et al. 2003). So the indicator has drawbacks at the levels of countries, industries and firms; there are therefore pitfalls in the uncritical use of this apparently simple indicator.

6.4.3 Patent Data

A patent is a public contract between an inventor and a government that grants time-limited monopoly rights to the applicant for the use of a technical invention (see Iversen 1998 for a good review). The patentee must first demonstrate a non-obvious advance in the state of the art after which the inventor enters into a binding relationship with the state: in general, the inventor contracts to reveal detailed information about the invention in return for limited protection against others using that invention for the time and geographical area for which the contract is in force. In terms of the concessions made by the parties, there is a trade-off between the disclosure of detailed information by the inventor against the permission of limited monopoly by the state. In this sense, the patent-system is designed as an incentive-mechanism for the creation of new economically valuable knowledge and as a knowledge-dissemination mechanism to spread this information. There has been a prolonged debate about whether the patent system would be worth creating if we did not have it (the usual answer is no), and whether—since we do have it—it should be abolished (again the usual answer is no), or whether a reward system would be superior (again, no).[4]

In general the patent system gathers detailed information about new technologies into a protracted public record of inventive activity, which is more or less continuous. This gives it striking advantages as an innovation-indicator. These include:

- Patents are granted for inventive technologies with commercial promise (i.e. innovation).
- The patent system systematically records important information about these inventions.
- The patent system collates these technologies according to a detailed and slow-to-change classification system.
- The patent system systematically relates the invention to relevant technologies, and also provides links (via citations) to relevant technical and scientific literature.
- The patent system is an old institution, providing a long history (see Granstrand in this volume)—it is the only innovation indicator extending back over centuries, and this means that it is possible to use patents to explore quantitative issues over very long periods (see Bruland and Mowery in this volume).
- The data is freely available.

The major sources of patent data are the records of the US Patent Office and the European Patent Office. Recent years have seen major increases in patenting activity, as Figure 6.1 shows. The causes of this rise are an important issue: there does seem to be growth of patenting extending back at least fifteen years, possibly signifying acceleration of innovation efforts, or changes in strategic behavior by firms; however, the rise may also be shaped by significant reductions in patent costs. (An analysis of the issues here can be found in Hall and Ziedonis 2001; see also Kortum and Lerner 1999.)

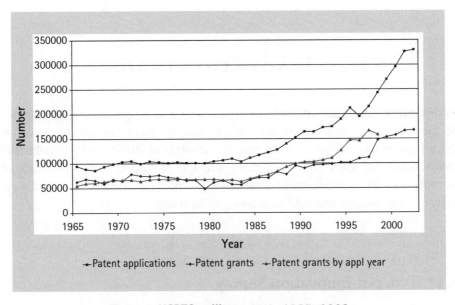

Fig. 6.1 USPTO utility patents 1965–2002

Source: Hall 2003.

Patents also of course have weaknesses, the most notable of which is that they are an indicator of invention rather than innovation: they mark the emergence of a new technical principle, not a commercial innovation. Many patents refer to inventions that are intrinsically of little technological or economic significance. More generally, Kleinknecht et al. have argued that

It is obvious that the patent indicator misses many non-patented inventions and innovations. Some types of technology are not patentable, and, in some cases, it is still being debated whether certain items (e.g. new business formulae on the internet) can be patented. On the other hand, what is the share of patents that is never translated into commercially viable products and processes? And can this share be assumed to be constant across branches and firm size classes? Moreover in some cases patent figures can be obscured by strategic behavior: a firm will not commercialize the patent but use it to prevent a competitor patenting and using it.   (Kleinknecht et al. 2002: 112)

But taking such qualifications into account, the analysis of patent data has proven very fruitful. Important achievements include the mapping of inventive activity over long time periods (Macleod 1988; Sullivan 1990); assessing the impacts of economic factors on the rate of invention (Schmookler 1971); the elucidation of the complexity of technological knowledge bases in large firms (Patel and Pavitt 1999); the use and roles of science in industrial patenting (Narin and Noma 1985; Meyer 2000); the mapping of inter-industry technology flows (Scherer 1982); the analysis of spillovers of knowledge using patent citations (Jaffe, Henderson, and Trajtenberg 1993) and the analysis of patent values (Hall, Jaffe, and Trajtenberg 2001).

6.5 NEW INNOVATION INDICATORS

Recent years have seen attempts to create new and better-designed indicators focused directly on innovation: for example, the European Commission has supported large-scale efforts to overcome the absence of direct data on industrial innovation—and there have been other attempts to improve our knowledge of outputs, sources, instruments and methods of innovation (recent discussions are Hansen 2001; Guellec and Pattinson 2001; Smith 2002).

6.5.1 Types of Innovation Survey

Innovation surveys divide into two basic types: those that focus on firm-level innovation activity, asking about general innovation inputs (both R&D and non-

R&D) and outputs (usually of product innovations), and those that focus on significant technological innovations (usually identified through expert appraisal, or through new product announcements in trade journals or other literature). Sometimes the first of these approaches is called a "subject" approach, since it focuses on the innovating agent; the latter is referred to as the "object" approach, since it focuses on the objective output of the innovation process, on the technology itself (Archibugi and Pianta 1996). Both approaches can and do incorporate attempts to explore aspects of the innovation process itself: sources of innovative ideas, external inputs, users of innovation, and so on. Both approaches define an innovation in the Schumpeterian sense, as the commercialization of a new product or process. However the object approach tends to focus on significantly new products, while the subject approach includes small-scale, incremental change.

6.5.2 The "Object" Approach to Innovation Indicators

Perhaps the most important example of the "object" approach is the SPRU database, developed by the Science Policy Research Unit at the University of Sussex, which collected information on major technical innovations in British industry, covering sources and types of innovation, industry innovation patterns, cross-industry linkages, regional aspects, and so on.[5] The SPRU approach used a panel of about 400 technical experts, drawn from a range of institutions, to identify major innovations across all sectors of the economy, from 1945 through to 1983. The database covered a total of about 4,300 innovations. An important related database is the US Small Business Administration database, covering innovations introduced to the market by small firms in the US in one year, 1982. This was constructed through an examination of about one hundred trade, engineering, and technology journals— a major study by Acs and Audretsch (1990) has been based upon it. In addition there is a range of smaller literature-based surveys—based on searches of trade literature—that have been undertaken in recent years: the Netherlands, Austria, Ireland, and the UK for example—Kleinknecht and Bain (1993) and Kleinknecht (1996) report the results from this work.

This type of approach has a number of strong advantages. Technology-oriented approaches have the merit of focusing on the technology itself, and allow a form of external assessment of the importance of an innovation—the fact that an innovation is recognized by an expert or a trade journal makes the counting of an innovation somewhat independent of personal judgements about what is or is not an innovation. Both expert-based and literature-based approaches can be backward looking, thus giving a historical perspective on technological development.

But the approach also has weaknesses. The very fact that innovations must pass a test of significance—that is, must be sufficiently innovative to be publicized in trade journals or the general press—also imparts a sample selection bias to the exercise. In effect what these surveys cover is an important subset of the population of innovations: those that are new to an industry. What gets lost is the population of innovation outputs which are "routine," incremental, part of the normal competitive activity of firms, yet not strikingly new enough to be reported.

6.5.3 Results from "Object" Studies

One of the most important results of work using the SPRU database was to show the existence of quite different types of innovative activity across different types of industry. In a pioneering study, Pavitt (1984) distinguished between four basic firm types, which he called "science based," "scale intensive," "specialized suppliers," and "supplier dominated." He showed that these categories of firms were characterized by differences in sources of technology, types of users, means of appropriation, and typical firm size. This work was among the first to really demonstrate empirically the importance of technological diversity within the economy, with important implications for the design of R&D policy in circumstances where firms have very different technology creation patterns. Other work with the SPRU database has emphasized the inter-sectoral flow of innovations (using the important data on first users of innovations within the dataset), and gave an early empirical insight into the complexity of what is now called the system of innovation (Pavitt, 1983; Robson et al. 1988). Geroski (1994: 19) has summarized these intersectoral flows as shown in Figure 6.2 where the key result is the importance of the three major engineering sectors (mechanical engineering, instruments and electronic engineering) in terms of the flow of innovations into other sectors. But it is important to note also the importance of flows within this broad engineering complex.

6.5.4 The "Subject" Approach and the Community Innovation Survey

In the early 1990s, the OECD attempted to synthesize the results of earlier trial innovation surveys, and to develop a manual that might form the basis of a common practice in this field. A group of experts was convened, and over a period of approximately fifteen months developed a consensus on an innovation manual which became known as the *Oslo Manual* (OECD 1992).

The European Commission, in a joint action between Eurostat and DG-Enterprise, followed up the OECD initiative in 1992–3, implementing the *Community Innovation Survey*. CIS was an innovative action in a number of respects. First, it

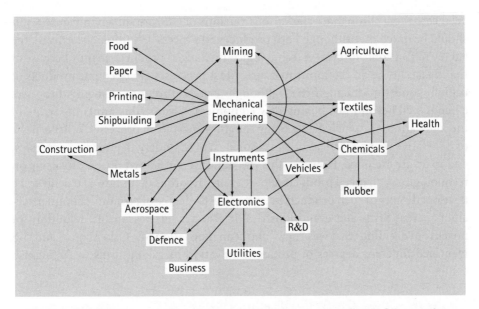

Fig. 6.2 The SPRU innovation database: The intersectoral flow of innovations

Source: Geroski (1994).

was a large-scale attempt to collect internationally comparable direct measures of innovation outputs. Second, it collected data at a highly disaggregated level and made this data available in disaggregated form to analysts. The survey has now been carried out three times, most recently in 2002; in that year the survey covered approximately 140,000 European firms.

CIS, in its various versions, developed and incorporated data on the following topics:

- expenditure on activities related to the innovation of new products (R&D, training, design, market exploration, equipment acquisition and tooling-up etc). There is therefore a unique focus on non-R&D inputs to the innovation process;
- outputs of incrementally and radically changed products, and sales flowing from these products;
- sources of information relevant to innovation;
- technological collaboration;
- perceptions of obstacles to innovation, and factors promoting innovation.

In terms of definitions, the CIS followed the *Oslo Manual* in a number of crucial respects. Firstly, it focused on technological innovation, particularly in products. But it then defined different categories of change, asking firms to assign the product range of the firm to these different categories. The CIS also asked firms to estimate the proportions of sales which were coming from: new or radically changed products, from products which had been changed in minor ways, or from unchanged products. The definitions of technological innovation used in CIS-2, which have been consistent throughout the various versions of CIS, are shown in Figure 6.3. It should be noted that although both product and process definitions are offered, the survey in fact concentrates on technologically changed products, mainly because of the availability of an economic measure. Most processes are of course products of capital goods-producing firms, although expenditure on changing processes extends well beyond just buying new equipment. Clearly, this limits the scope of the innovations on which data is being sought—apart from processes, other aspects of innovative change, such as organizational change, underlying learning processes, and so on are excluded. However, this was done for considered reasons: focusing on technologically changed products allows a fairly rigorous definition of change to be developed. Sales of such products permit at least a degree of economic commensurability across firms and even industries. It also permits reasonable definitions of novelty: in deciding what was "new" about an innovation, the Oslo Manual and CIS identified different degrees of product innovation by asking firms to distinguish

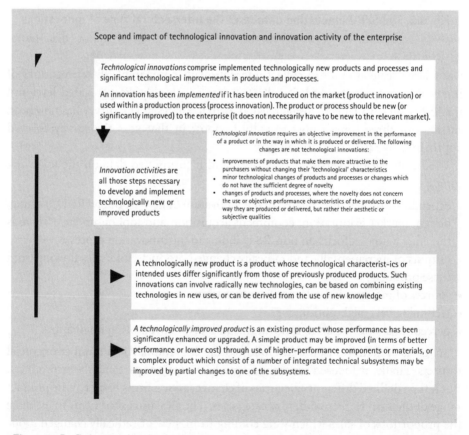

Fig. 6.3 Defining technological innovation—Community Innovation Survey (CIS)

Source: CIS-2 Questionnaire.

between sales of products new to the firm only, products new to the industry, or products that were wholly new. So although the *Oslo Manual*/CIS approach constrains innovation to the field of the technological, it does so in a way that allows a consistency between the concepts of change, novelty, and commensurability. Without such consistency, survey methods are not appropriate.

6.5.5 Innovation Activities and their Measurement

A second feature of the *Oslo Manual* and of CIS was the attempt to estimate expenditures on categories of innovation activity other than R&D. Six main categories of innovation activities were identified, and the basic structure of the questions and definitions was as shown in Figure 6.4. The basic idea here was that firms invest in a wide range of non-R&D activities, resulting in both tangible and intangible assets, and that these are likely to vary across firms and industries. The categories here are drawn closely from Kline and Rosenberg (1986), which provides the general conceptual foundation. But it can easily be seen that there are likely to be problems: these are complex categories, in an area where firms do not necessarily keep separate or detailed records. In practice, in the first round of the CIS, there were many firms who did not respond to the questions which were asked on this topic, and many who were clearly able to answer only in terms of broad estimates. But there are strong interfirm variations—some firms operate project management systems that permit accurate answers in this area, and the data quality seems to have improved over time.

 One of the important results to have emerged from this part of CIS is that capital expenditure related to innovation is the largest single component of innovation expenditure across all sectors (Evangelista et al. 1998). This emphasizes the importance of the embodied R&D in capital and intermediate goods, discussed above.

6.5.6 CIS: Some Main Results

What have we learned so far from attempts to measure and map innovation? In this section we look at some of the results that have emerged from a range of studies using CIS. The literature using innovation survey data is growing rapidly at the present time, and it falls into three broad categories.

Descriptive overviews of data results at national level. These studies are usually written for policy makers, and typically consist of tables and charts, accompanied by commentary, showing results such as the distribution of innovation expenditures and their differences across industries, proportions of firms introducing product or

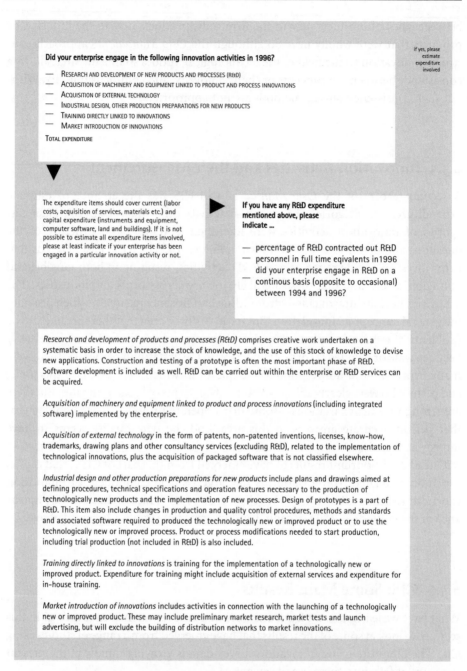

Did your enterprise engage in the following innovation activities in 1996?

if yes, please estimate expenditure involved

— RESEARCH AND DEVELOPMENT OF NEW PRODUCTS AND PROCESSES (R&D)
— ACQUISITION OF MACHINERY AND EQUIPMENT LINKED TO PRODUCT AND PROCESS INNOVATIONS
— ACQUISITION OF EXTERNAL TECHNOLOGY
— INDUSTRIAL DESIGN, OTHER PRODUCTION PREPARATIONS FOR NEW PRODUCTS
— TRAINING DIRECTLY LINKED TO INNOVATIONS
— MARKET INTRODUCTION OF INNOVATIONS

TOTAL EXPENDITURE

The expenditure items should cover current (labor costs, acquisition of services, materials etc.) and capital expenditure (instruments and equipment, computer software, land and buildings). If it is not possible to estimate all expenditure items involved, please at least indicate if your enterprise has been engaged in a particular innovation activity or not.

If you have any R&D expenditure mentioned above, please indicate ...

— percentage of R&D contracted out R&D
— personnel in full time eqivalents in1996
— did your enterprise engage in R&D on a continous basis (opposite to occasional) between 1994 and 1996?

Research and development of products and processes (R&D) comprises creative work undertaken on a systematic basis in order to increase the stock of knowledge, and the use of this stock of knowledge to devise new applications. Construction and testing of a prototype is often the most important phase of R&D. Software development is included as well. R&D can be carried out within the enterprise or R&D services can be acquired.

Acquisition of machinery and equipment linked to product and process innovations (including integrated software) implemented by the enterprise.

Acquisition of external technology in the form of patents, non-patented inventions, licenses, know-how, trademarks, drawing plans and other consultancy services (excluding R&D), related to the implementation of technological innovations, plus the acquisition of packaged software that is not classified elsewhere.

Industrial design and other production preparations for new products include plans and drawings aimed at defining procedures, technical specifications and operation features necessary to the production of technologically new products and the implementation of new processes. Design of prototypes is a part of R&D. This item also include changes in production and quality control procedures, methods and standards and associated software required to produced the technologically new or improved product or to use the technologically new or improved process. Product or process modifications needed to start production, including trial production (not included in R&D) is also included.

Training directly linked to innovations is training for the implementation of a technologically new or improved product. Expenditure for training might include acquisition of external services and expenditure for in-house training.

Market introduction of innovations includes activities in connection with the launching of a technologically new or improved product. These may include preliminary market research, market tests and launch advertising, but will exclude the building of distribution networks to market innovations.

Fig. 6.4 Resources devoted to innovation activities in 1996

process innovations, the distribution of different types of new product sales across industries, major patterns of technological collaboration, perceptions of obstacles to innovation, and data on objectives of innovation. These studies tend to be important not just in reaching policy makers, but in emphasizing some robust results which emerge from this data—in particular the conclusion that innovation is pervasively distributed across modern economies, and that non-R&D inputs to innovation are particularly important in non-high-tech sectors. In some cases these reports are sophisticated productions—the German reports, for example, rest on a substantial panel dataset, and the Canadian analytical effort (similar to but not identical with CIS) is very wide ranging indeed (Janz et al. 2002, and Statistics Canada: www.statcan.ca).[6] Most EU countries produce these reports and Eurostat in addition produces a Europe-wide overview (Eurostat 2004).

Analytical studies sponsored by the European Commission. The European Innovation Monitoring System (within DG-Enterprise) has sponsored twenty-five specific studies addressing a wide range of questions arising from the innovation data. These cover, for example, Europe-wide surveys of innovation expenditure patterns, innovation outputs across Europe, studies of links between innovation and employment patterns, and sectoral studies (pharmaceuticals, telecoms, pulp and paper, machinery, machine tools, service sector innovation, spin-offs, and regional impacts). Most of these studies are substantial pieces of work, often book length. An overview of the full range of material is provided in Appendix 6.2 to this chapter (reports are accessible via the European Innovation Monitoring System on the EU's CORDIS website: www.cordis.lu)

Econometric or statistical studies of innovation. The innovation survey data has a more or less unique feature, which is that it is available in a highly disaggregated form (as so-called "micro-aggregated" data). This makes possible a wide range of micro-level studies of innovation processes and their effects, and the research opportunities this provides are being exploited rather vigorously at the present time. Publication in this field has been building rapidly, in the form of books (e.g. Thuriaux, Arnold, and Couchot 2001; Kleinknecht and Mohnen 2002, and Gault 2004), articles, journal special issues (such as *STI Review* 27 (2001), and a forthcoming special issue of *Economics of Innovation and New Technology*), and so on. The book edited by Thurieaux et al. collects no less than thirty-one chapters on various empirical aspects of innovation using primarily CIS data. These covered methodological issues, the extension of the CIS approach to services, micro analysis of innovation and firm performance, innovation and employment, innovation in traditional industries, regional innovation, and the use of indicators in policy decision making.

By far the most rapidly growing area of publication is in scholarly journals. A non-exhaustive review of journals in 2002–4 reveals eighteen CIS-based publications. These articles are briefly summarized in Appendix 6.1 to this chapter. Studies focus on such topics as determinants of innovation, innovation and firm performance analysis, diversity (both in innovation patterns and firm performance outcomes), the role of science in innovation, sectoral performance (such as employment impacts), inter-firm collaboration and innovation performance, as well as regional and country studies, and methodological issues. There is every sign that this pace of publication will continue in years ahead. This is a rapid growth in publication, and it is worth noting that it is occurring not only in the front-line journals of innovation studies, but also in the heart of the economic mainstream (notably Mairesse and Mohnen in *American Economic Review*).

Space limitations prevent a detailed overview of the results from the work described above, but some robust conclusions that seem to have emerged from the literature as a whole are as follows:

- Innovation is prevalent across all sectors of the economy—it is not confined to high-tech activities, and so-called low-tech activities contain high proportions of innovating firms, and often generate high levels of sales from new and changed products (SPRU, 1996; European Commission 2001).[7]
- R&D is by no means the most important innovation input. In all sectors, across all countries, investment in capital equipment related to new product introduction is the major component of innovation expenditure, suggesting the need to focus on the knowledge elements embodied in such items (STEP, 1997; Evangelista et al. 1998; Evangelista 1999).
- Across all sectors and countries innovation inputs and outputs are distributed highly asymmetrically—small proportions of firms account for large proportions of innovation outputs as measured by the CIS.
- Collaboration is widespread among innovating firms, to such an extent that it appears almost a *sine qua non* for innovation activity. This result from CIS has led to a range of specific subsidiary surveys, which have generated deeper detail and have confirmed the importance of collaboration suggested by the CIS surveys (see OECD 2001 for papers on this).
- Extension of the CIS format to service sector activities is illustrative but problematical, and deserves more attention (Djellal and Gallouj 2001; Tether and Miles 2001; Ch. 16 by Miles in this volume).
- There continue to be significant differences in collection methodologies and response rates across countries, implying that the data appears to be much better suited to within-sector micro studies than to cross-country macro comparisons.

6.6 CONCLUSION

While the CIS is clearly a step forward in terms of the type and volume of innovation data that is available, it is of course open to criticism. Most criticisms focus on the definitional restrictions in CIS with respect to innovation inputs and outputs, and on whether an approach that was originally adopted for manufacturing is extendable to services. On the output side, the decisions made concerning the technological definitions of change obviously limit the forms of innovation that can be studied: it seems to be the case that CIS works well for manufactures, but not for the extremely heterogeneous services sector and its often intangible outputs. The analyses of Djellal and Gallouj (2001) and Tether and Miles (2001) suggest the need for quite different approaches to data gathering on services. In defence of the CIS approach it can be argued that it is, and was intended to be, manufacturing-specific and that extension to services would always be problematic. Similar problems arise with other non-technological aspects of innovation, such as organizational change (see Lam, this volume, for an overview of organizational innovation). It is very unclear whether CIS, or indeed any other survey-based method, can grasp the dimensions of this. The challenge for those who would go beyond this is whether they can generate definitional concepts, survey instruments, and collection methodologies that make sense for other sectors or other aspects of innovation.

On the side of R&D and non-R&D innovation inputs, it is generally unclear just how much of a firm's creative activity is captured by the types of innovation outputs that CIS measures. Arundel has pointed out that "When we talk about a firm expending a great deal of effort on innovation, we are not only speaking of financial investments, but of the use of human capital to think, learn and solve complex problems and to produce qualitatively different types of innovations" (Arundel 1997: 6). This point cannot be argued with, but again the question arises as to what can be done with survey questionnaires and what cannot. If we want to explore complex problem solving, for example, then it is doubtful whether a survey instrument is the right research tool at all. Perhaps an underlying issue here is the long-standing tension between statistical methods, with their advantages of generality but lack of depth, versus case study methods, which offer richness at the expense of generalizability.

Nevertheless it is reasonable to conclude that this data source is proving itself with researchers. Both formal evaluations of CIS as well as data tests by researchers have been broadly positive to the quality of the data flowing from the survey (Aalborg University 1995). One of the positive features of CIS is that survey definition and construction, collection methodologies, and general workability have been subjected to a degree of evaluation, critique, and debate that goes far beyond anything that has been carried out with other indicators (see Arundel et al. 1997, for one contribution to the critical development of CIS). This process is continuing, with both positive and negative potential outcomes. On the positive side, the data source may continue to be improved; on the negative, too much may be asked of this approach. But the real achievement is that CIS has produced results that have not been possible with other data sources, and there is no doubt more to come as researchers master the intricacies of the data. In fact empirical studies using CIS data may well be the most rapidly growing sub-field of publication within innovation studies at the present time. An interesting feature of the publications using CIS is the breadth of work being done—the data is being used for public presentations, for policy analyses, and for a wide range of scholarly research. It was argued above that researchers have yet to make full use of the richness of R&D data, and this applies even more to the existing survey-based innovation data. This source will continue to offer considerable scope to researchers in years ahead: issues such as innovation and firm performance, the use of science by innovating firms, the roles of non-R&D inputs, and the employment impacts of innovation are among likely areas of development.

This chapter has concentrated on the Community Innovation Survey, but future developments are unlikely to rely on this source alone. One possible trend is for greater integration of existing data sources, and this can already be seen in multi-indicator approaches to such issues as national competitiveness. Another likely trend is for the continued development of new survey instruments aligned to specific needs, along the lines of the DISKO surveys on interfirm collaboration (OECD, 2001). Such developments are much to be welcomed as Innovation Studies seeks to generalize its propositions beyond the limits of the case study method.

APPENDIX 6.1

Recent (2002 onwards) journal publications using CIS data

Author(s)	Data source	Topic
Cox, Frenz, and Prevezer (2002)	CIS-2	Distinguishing high- and low-tech industries
Evangelista and Savona (2002)	CIS 1 and 2	Employment impacts of innovation in service sector
Hesselman (2002)	CIS-2	Methodological issues and response patterns
Hinloopen (2003)	CIS-1 and CIS-2	Determinants of innovation performance at firm level across Europe
Inzelt (2002)	CIS-2	Service sector innovation in Hungary
Kleinknecht et al (2002)	CIS-2	Indicator choice and biases
Lööf and Heshmati (2004)	CIS-2	Innovation and firm performance
Lööf and Heshmati (2002)	CIS-2	Performance diversity and innovation

Mairesse and Mohnen	CIS-1	Determinants of innovation at firm level
Mohnen and Horeau (2003)	CIS-2	University–industry collaboration
Mohnen, Mairesse, and Dagenais (2003)	CIS-1	Expected vs. actual innovation output levels
Nascia and Perani (2002)	CIS-1	Diversity of innovation patterns in Europe
Quadros et al. (2001)	Brazilian innovation survey	Innovation in San Paulo region
Sellenthin and Hommen (2002)	CIS-2	Innovation patterns in Swedish industry
Tether (2002)	CIS-2	Innovation and inter-firm collaboration
Tether and Swann (2003)	CIS-3	Role of science in innovation
Van Leeuwen and Klomp (2004)	CIS-2	Innovation and multi-factor productivity

APPENDIX 6.2

Publications using CIS data sponsored by the European Commission

Publications are listed in chronological order, by topic and instituion.

Evaluation of the Community Innovation Survey (CIS)—Phase 1,
Aalborg University (Denmark), 1995

Europe's Pharmaceutical Industry: An Innovation Profile (CIS),
SPRU (UK), 1996

Innovation Outputs in European Industry (CIS),
SPRU (UK), 1996

Innovation in the European Food Products and Beverages Industry (CIS),
IKE (Denmark) and SPRU (UK), 1996

Technology Transfer, Information Flows and Collaboration (CIS),
Manchester School of Management & University of Warwick (UK), 1996

The Impact of Innovation on Employment: Alternative interpretations and results of the Italian CIS,
University of Rome "La Sapienza" (Italy), 1996

Innovation in the European Chemical Industry (CIS),
WZB (Germany) 1996

Publications are listed in deronological order, by topic and instituion. (cont.)

Innovation in the European Telecom Equipment Industry (CIS),
MERIT (Netherlands), 1996

Innovation Activities in Pulp, Paper and Paper Products in Europe (CIS),
STEP Group (Norway), 1996

The Impact of Innovation in Employment in Europe—An Analysis Using CIS Data,
Centre for European Economic Research/ZEW (Germany), 1996

Computer and Office Machinery—Firms' external growth & technological diversification:
*   analysis during CIS,*
CESPRI (Italy) 1997

Innovation Expenditures in European Industry: analysis from CIS,
STEP Group (Norway), 1997

Manufacture of Machinery and Electrical Machinery (CIS),
Centre for European Economic Research/ZEW (Germany), 1997

Innovation Measurements and Policies: Proceedings of International Conference,
20–21 May 1996, Luxembourg

Analysis of CIS 2 Data on the Impact of Innovation on the Pharmaceuticals and
*   Biotechnology Sector,*
SOFRES (Belgium), 2001

Analysis of CIS 2 Data on the Impact of Innovation on Growth in the Sector of Office
*   Machinery and Computer Manufacturing,*
SOCINTEC (Spain), 2001

Analysis of CIS 2 Data on the Impact of Innovation on Growth in Manufacturing of Machinery
*   and Equipment and of Electrical Equipment,*
STEP Group (Norway), 2001

Analysis of CIS Data on the Role of NTBFs, Spin-offs and Innovative Fast Growing SMEs in the
*   Innovation Process,*
Institute for Advanced Studies and Johanneum Research (Austria), 2001

Innovation and the Acquisition and Protection of Competencies,
MERIT (Netherlands), 2001

Analysis of Empirical Surveys on Organisational Innovation and Lessons for Future Community
*   Innovation Surveys,*
Fraunhofer Institute (Germany), 2000

Regional Patterns of Innovation: the Analysis of CIS 2 Results and Lessons from other
*   Innovation Surveys,*
STEP S.A.S (Italy), 2000

Use of Multivariate Techniques to Investigate the Multidimensional Aspects of Innovation,
University of Newcastle Upon Tyne (ISRU) (UK), 2000

Statistics on Innovation in Europe,
European Commission, 2001

Analysis of CIS 2 Data on Innovation in the Service Sector,
Manchester University (UK), 2000

"Innovation and enterprise creation: Statistics and indicators," Proceedings of the International
    Conference, 23

Notes

1. I would like to thank Ian Miles, Bart Verspagen, and Richard Nelson for comments on an earlier draft, and in particular Bronwyn Hall for comments and advice. None are implicated in the outcome, of course.
2. The question of what can be measured is an issue with all economic statistics. For example, the national accounts do not cover all economic activity (in the sense of all human activity contributing to production or material welfare). They incorporate only activity that leads to a measurable market outcome or financial recompense. This tends to leave out economic activity such as domestic work, mutual aid, child rearing, and the informal economy in general. Those services that are measured not by the value of output but by the compensation of inputs also provide problems for measurement of output and productivity.
3. In both Australia and Norway, each of which collects data by field of research for all industrial sectors, roughly 25 per cent of all R&D is in ICT.
4. An excellent overview of the literature on these and other patent issues can be found on the website of Bronwyn Hall: http://emlab.berkely.edu/users/bhhall See also Granstrand in this volume.
5. For analyses using the SPRU database, see e.g. Pavitt 1983, 1984; Robson et al. 1988; the most recent sustained analytical work using the SPRU database is Geroski 1994.
6. Canada is a leading site of policy-related indicator work at the present time—see e.g. the outstanding work of the Canadian Science and Innovation Indicators Consortium which can be found at the website given above.
7. On innovation in low-tech industries, see Ch. 15 by von Tunzelmann and Acha in this volume.

References

Aalborg University (1995), *Evaluation of the Community Innovation Survey (CIS)—Phase 1*, 2 vols., Report to the European Innovation Monitoring System.

Acs, Z., and Audretsch, D. (1990), *Innovation and Small Firms*, Cambridge, Mass: MIT Press.

*Archibugi and Pianta (1996), "Innovation surveys and patents as technology indicators: the state of the art," in OECD, *Innovation, Patents and Technological Strategies*, Paris: OECD.

Arrow, K. J. (1984), "Statistical Requirements for Greek Economic Planning," *Collected Papers of Kenneth J. Arrow*, vol. 4: *The Economics of Information*, Oxford: Blackwell.

Arundel, A. (1997), "Why Innovation Measurement Matters," in A. Arundel and R. Garrelfs (eds.), *Innnovation Measurement and Policies*, EU Luxembourg: EIMS 94/197.

Arundel, A., Patel, P., Sirilli, G., and Smith, K. (1997), *The Future of Innovation Measurement in Europe: Concepts, Problems and Practical Directions*, IDEA Paper No 3, STEP Group Oslo.

* Asterisked items are suggestions for further reading.

BIFFL, G., and KNELL, M. (2001), "Innovation and Employment in Europe in the 1990s," *WIFO Working Paper* 169/2001.

BROUWER, E., and KLEINKNECHT, A. (1997), "Measuring the Unmeasurable: A Country's Expenditure on Product and Service Innovation," *Research Policy* 25: 1235–42.

—— —— (1999), "Innovative Output and a Firm's Propensity to Patent: An Exploration of CIS Micro-Data," *Research Policy* 28: 615–24.

COX, H., FRENZ, M., and PREVEZER, M. (2002), "Patterns of Innovation in UK Industry: Exploring the CIS Data to Contrast High and Low Technology Industries," *Journal of Interdiciplinary Economics* 13(1–3): 267–304.

*DJELLAL, F., and GALLOUJ, F. (2001), "Innovation surveys for Service Industries: A Review," in Thurieaux, Arnold, and Couchot 2001: 70–6.

DOWRICK, S. (2003), "A Review of the Evidence on Science, R&D and Productivity," Working Paper, Department of Economics, Australian National University.

EUROPEAN COMMISSION (1994), *The European Report on Science and Technology Indicators 1994* (EUR 15897), Luxembourg.

—— (1997), *Second European Report on Science and Technology Indicators 1997* (EUR 17639), Luxembourg.

EUROPEAN COMMISSION (2001), Statistics on Innovation in Europe (EUR KS-32-00-895-EN-1), Luxembourg.

—— (2003), *Third European Report on Science and Technology Indicators*, Luxembourg.

EUROSRAT (2004), *Innovation in Europe. Results for the EU, Iceland and Norway: Data 1998–2001*, (Luxembourg: Office for one official publications of the European Communities)

—— SANDVEN, T., SIRILLI, G., and SMITH, K. (1998), "Measuring Innovation in European Industry," *International Journal of the Economics of Business* 5(3): 311–33.

—— and SAVONA, M. (2002), "The Impact of Innovation on Employment in Services: Evidence from Italy," *International Review of Applied Economics* 16(3): 309–18.

EVANGELISTA, R. (1999), *Knowledge and Investment*, Cheltenham: Edward Elgar.

GAULT, F. (ed.) (2004), *Understanding Innovation in Canadian Industry*, Montreal and Kingston: McGill-Queen's University Press.

GEROSKI, P. (1994), *Market Structure, Corporate Performance and Innovative Activity*, Oxford: Clarendon Press.

GODIN, B. (2004), "The Obsession with Competitiveness and its Impact on Statistics: The Construction of High Technology Indicators," *Research Policy* (forthcoming).

GRIFFITH, R., REDDING, S., and VAN REENEN, J. (2000), "Mapping the Two Faces of R&D: Productivity Growth in a Panel of OECD Industries," *LSE Centre for Economic Performance Discussion Paper*, 2457: 1–74.

GRILICHES, Z. (1987), "Comment," Washington, DC: Brookings Papers on Economic Activity, 3.

GRUPP, H. (1994), "The Measurement of Technical Performance of Innovations by Technometrics and its Impact on Established Technology Indicators," *Research Policy* 23: 175–93.

*—— (1998), *Foundations of the Economics of Innovation, Theory, Measurement and Practice*, Cheltenham: Elgar.

GUELLEC, D., and PATTINSON, B. (2001), "Innovation Surveys: Lessons from OECD Countries' Experience," *STI Review* 27: 77–102.

HAGEDOORN, K., and SCHAKENRAAD, J. (1990), "Inter-firm Partnerships and Co-operative Strategies in Core Technologies," in C. Freeman and L. Soete (eds.), *New Explorations in the Economics of Technological Change*, London: Pinter.

HALL, B. H. and Ziedonis, R. H. (2002), "The determinants of patenting in the U.S. semiconductor industry, 1980–1994", *Rand Journal of Economics*, 32: 101–28.

——, Jaffe, A., and Trajtenberg, M., 2001, "The NBER Patent Citations Daa file: Lessons, Insights and Methodological Tools", NBER Working Paper No. 8498.

*HANSEN, J. A. (2001), "Technological Innovation Indicators: A Survey of Historical Development and Current Practice," in M. P. Feldmann and A. Link (eds.), *Innovation Policy in the Knowledge-Based Economy*, Dordrecht: Kluwer, 73–103.

HESSLEMAN, L. (2002), "A Description of Responses to the UK Community Innovation Survey 2," *The Journal of Interdiciplinary Economics* 13(1–3): 243–66.

HINLOOPEN, J. (2003), "Innovation Performance across Europe," *Economics of Innovation and New Technology* 12(2): 145–61.

HIRSCH-KREINSEN, H., JACOBSEN, D., LAESTADIUS, S., and SMITH, K. (2003), "Low Tech Industries and the Knowledge Economy: State of the Art and Research Challenges," *Working Paper 2003:10*, Dept of Industrial Economics and Management, Royal Institute of Technology (KTH), Stockholm.

HOLBROOK, J. (1991), "The Influence of Scale Effects on International Comparisons of R&D Expenditure," *Science and Public Policy* 18(4): 259–62.

HOWELLS, J. (2000), "Innovation Collaboration and Networking: A European Perspective," in Science Policy Support Group, *European Research, Technology and Development. Issues for a Competitive Future*, London.

HUGHES, K. (1988), "The Interpretation and Measurement of R&D Intensity: A Note," *Research Policy* 17: 301–7.

INZELT, A. (2002), "Attempts to Survey Innovation in the Hungarian Service Sector," *Science and Public Policy* 29(5): 367–83.

IVERSEN, E. (1998), "Patents," in K. Smith (ed.), *Science, Technology and Innovation Indicators—a Guide for Policymakers*, IDEA Report 5, STEP Group Oslo.

JAFFE, A., and Henderson, R., and Trajtenberg, M., 1997, "University Versus Corporate Patents: A Window on the Basicness of Invention", *Economics of Innovation and New Technology*, 19–50.

JANZ, N., et al. (2002), *Innovation Activities in the German Economy. Report on Indicators from the Innovation Survey 2000*, Mannheim: ZEW.

KALOUDIS, A. (1998), "Bibliometrics," in K. Smith (ed.), *Science, Technology and Innovation Indicators—a Guide for Policymakers*, IDEA Report 5, STEP Group Oslo.

*KLEINKNECHT, A. (ed.) (1996), *Determinants of Innovation: The Message From New Indicators*, London: Macmillan.

—— and BAIN, D. (eds.) (1993), *New Concepts in Innovation Output Measurement*, London: Macmillan.

—— and MOHNEN, P. (eds.) (2002), *Innovation and Firm Performance. Econometric Explorations of Survey Data*, Hampshire and New York: Palgrave.

—— VAN MONTFORT, K., and BROUWER, E. (2002), "The Non-Trivial Choice Between Innovation Indicators," *Economics of Innovation and New Technology* 11(2): 109–21.

KLINE, S., and ROSENBERG, N. (1986), "An Overview of Innovation," in R. Landau (ed.), *The Positive Sum Strategy: Harnessing Technology for Economic Growth*, Washington: National Academy Press, 275–306.

KORTUM, S., and LERNER, J. (1999), "What is Behind the Recent Surge in Patenting?" *Research Policy* 28: 1–22.

LAESTADIUS, S. (2003), "Measuring Innovation in the Knowledge Economy," Paper presented to Pavitt Conference on Innovation, SPRU, Sussex, 13 Nov 2003.

LÖÖF, H., and HESHMATI, A. (2002), "Knowledge Capital and Performance Heterogeneity: A Firm-Level Innovation Study," *International Journal of Production Economics* 76(1): 61–85.

——— —— (2004), "On the Relationship between Innovation and Performance: A Sensitivity Analysis," *Economics of Innovation and New Technology* 13(1–2): forthcoming.

MCLEOD, C. (1988), *Inventing the Industrial Revolution: The English Patent System, 1660–1800*, Cambridge: Cambridge University Press.

MAIRESSE, J., and MOHNEN, P. (2001), "To be or not to be Innovative: An Exercise in Measurement," *STI Review* 27: 103–28.

*—— —— (2002), "Accounting for Innovation and Measuring Innovativeness: An Illustrative Framework and an Application," *American Economic Review* 92(2): 226–30.

*MEYER, M. (2000), "Does Science Push Technology? Patents Citing Scientific Literature," *Research Policy* 29: 409–34.

MOED, H. F., DE BRUIN, R. E., and VAN LEEUWEN, Th. N. (1995), "New Bibliometric Tools for the Assessment of National Research Performance: Database Description, Overview of Indicators and First Application," *Scientometrics* 33: 381–422.

MOHNEN, P., and HOAREAU, C. (2003), "What Types of Enterprise Forge Close Links with Universities and Government Labs? Evidence from CIS-2," *Managerial and Decision Economics* 24: 133–45.

—— MAIRESSE, J., and DAGENAIS, M. (2004), "Innovativeness: A Comparison across Seven European Countries," *Economics of Innovation and New Technology* 13(1–2): forthcoming.

NARIN, F., and NOMA, E. (1985), "Is Technology Becoming Science?" *Scientometrics* 3–6: 369–81.

NASCIA, L., and PERANI, G. (2002), "Diversity of Innovation in Europe," *International Review of Applied Economics* 16(3): 277–94.

OECD (1992, rev. 1997), *Innovation Manual: Proposed Guidelines for Collecting and Interpreting Innovation Data (Oslo Manual)*, Paris: OECD, Directorate for Science, Technology and Industry.

—— (1996), *Technology, Productivity and Job Creation*, Paris: OECD.

—— (2001), *Innovating Networks: Collaboration in National Innovation Systems*, Paris: OECD.

—— (2002), *The Measuerment of Scientific and Technological Actvities. Proposed Standard Practice for Surveys on Research and Experimetnal Development: Frascati Manual*, Paris: OECD.

PATEL, P., and PAVITT, K. (1997), "The Technological Competencies of the World's Largest Firms, Complex and Path-Dependent, but not Much Variety," *Research Policy* 26: 141–56.

*—— —— (1999), "The Wide (and Increasing) Spread of Technological Competencies in the World's Largest Firms: a Cahllenge to Conventional Wisdom," in A. Chandler et al. (eds.), *The Dynamic Firm: The Role of Technology, Strategy, Organization and Regions*, Oxford: Oxford University Press.

PAVITT, K. (1983), "Some Characteristics of Innovation Activities in British Industry," *Omega* 11.

*—— (1984), "Sectoral Patterns of Technological Change: Towards a Taxonomy and a Theory," *Research Policy* 13: 343–73.

POL, E., CARROLL, P., and ROBERTSON, P. (2002), "A New Typology for Economic Sectors with a View to Policy Implications," *Economics of Innovation and New Technology* 11(1): 61–76.

QUADROS, R., FURTADO, A., BERNADES, R., and ELIANE, F. (2001), "Technological Innovation in Brazilian Industry: An Assessment Based on the San Paulo Innovation Survey," *Technological Forecasting and Social Change* 67: 203–19.

ROBSON, M., TOWNSEND, J., and PAVITT, K. (1988), "Sectoral Patterns of Production and Use of Innovations in the UK: 1945–1983," *Research Policy* 17(1): 1–15.

ROSENBERG, N. (1976), *Perspectives on Technology*, Cambridge: Cambridge University Press.

—— (1982), *Inside the Black Box: Technology and Economics*, Cambridge: Cambridge University Press.

SANDVEN, T., and SMITH, K (1997), "Understanding R&D Indicators: Effects of Differences in Industrial Structure and Country Size," IDEA Paper 14, STEP Group Oslo.

SAVIOTTI, P. P. (1996), *Technological Evolution, Variety and the Economy*, Cheltenham: Elgar.

—— (2001), "Considerations about a Production System with Qualitative Change," in J. Foster and J. Stanley Metcalfe (eds.), *Frontiers of Evolutionary Economics. Competition, Self-Organization and Innovation Policy*, Aldershot: Edward Elgar, 197–227.

SCHERER, F. (1982), "Inter-industry Technology Flows in the United States," *Research Policy* 11(4): 227–45.

SCHMOOKLER, J. (1971), "Economic Sources of Inventive Activity," in N. Rosenberg (ed.), *The Economics of Technological Change*, London: Pelican, 117–36.

SELLENTHIN, M., and HOMMEN, L. (2002), "How Innovative is Swedish Industry? A Factor and Cluster Analysis of CIS II," *International Review of Applied Economics* 16(3): 319–32.

SMITH K. (1992), "Technological Innovation Indicators: Experience and Prospects," *Science and Public Policy* 19(6): 24–34.

—— (2001), "Innovation Indicators and the Knowledge Economy: Concepts, Results and Challenges," in Thurieaux, Arnold, and Couchot (eds), Luxembourg: European Commission, 14–25.

SPRU (1996), *Innovation Outputs in European Industry (CIS)*, Report to the European Innovation Monitoring System.

STEP GROUP (1997), *Innovation Expenditures in European Industry: Analysis from CIS*, Report to the European Innovation Monitoring System.

SULLIVAN, R. (1990), "The Revolution of Ideas: Widespread Patenting and Invention During the Industrial Revolution", *Journal of Economic History* 50(2): 340–62.

TETHER, B. (2002), "Who Cooperates for Innovation and Why: An Empirical Analysis," *Research Policy* 31(6): 947–67.

—— and MILES, I. (2001), "Surveying Innovation in Services—Measurement and Policy Interpretation Issues," in Thurieaux, Arnold, and Couchot (2001).

—— and SWANN, G. M. P. (2003), "Services, Innovation and the Science Base: An Investigation Into the UK's 'System, of Innovation' Using Evidence from the Third Community Innovation Survey," paper to CNR/Univerity of Urbino Workshop on Innovation In Europe.

*THURIEAUX, B., ARNOLD, E., and COUCHOT, C. (eds.) (2001), *Innovation and Enterprise Creation: Statistics and Indicators*, Luxembourg: European Commission (EUR 17038).

VAN LEEUWEN, G., and KLOMP, L. (2004), "On the Contribution of Innovation to Multifactor Productivity," *Economics of Innovation and New Technology* 13(1–2): forthcoming.

VERSPAGEN, B. (1992), "Endogenous Innovation in Neo-classical Growth Models: A Survey," *Journal of Macroeconomics* 631–62.

WORLD ECONOMIC FORUM (2003), *Global Competitiveness Report 2002–2003*, ed. P. Cornelius, K. Schwab, and M. E. Porter, New York: Oxford University Press.

PART II

THE SYSTEMIC NATURE OF INNOVATION

Introduction to Part II

A CENTRAL finding in innovation research is that firms seldom innovate in isolation. Interaction with customers, suppliers, competitors and various other private and public organizations is very important, and a "system perspective" is useful in understanding and analyzing such interaction. In Chapter 7, Edquist traces the development of such an "innovation system" perspective, particularly with respect to nations (so-called "national systems of innovation"), and discusses the achievements, shortcomings and potential of this approach. He also considers the role of some central activities of such systems, e.g., R&D and education. Following this, Mowery and Sampat (Chapter 8) examine one of the central organizations in national systems of innovations: universities. Chapter 9 by O'Sullivan focuses on another aspect of innovation systems, finance, discussing the varied approaches adopted on the relationship between finance and innovation. An essential feature in innovation systems is the conditions for appropriation of the economic returns to innovation. Granstrand (Chapter 10) provides a historical overview of intellectual property rights and surveys the extensive literature on this topic. The next two chapters deal with the boundaries of systems of innovation. As Edquist notes in Chapter 7 these boundaries do not have to be national, but may be regional, global or sectoral ("sectoral systems" are discussed in more detail in Part III of this volume). Regional systems of innovation are analyzed by Asheim and Gertler (Chapter 11) and Narula and Zanfei in Chapter 12 examine the "globalization" of innovation and the role of multinational enterprises in this process.

SYSTEMS OF INNOVATION

PERSPECTIVES AND CHALLENGES

CHARLES EDQUIST

7.1 INTRODUCTION[1]

THIS chapter presents an overview and assessment of the systems of innovation approach. I focus mainly on national systems of innovation, but in addition address sectoral and regional systems of innovation to a limited extent.[2] The chapter addresses the emergence and development of the systems of innovation (SI) approach, its strengths and weaknesses, the criticism that it is "undertheorized," the constituents of SIs, the main function and activities in SIs, the boundaries of SIs, and proposals for further research. I also discuss how the rigour and specificity of the SI approach could be increased.[3] The most central terms used in this chapter are specified in Box 7.1.

Box 7.1 Systems of innovation: main terms used

Innovations = product innovations as well as process innovations. Product innovations are new—or better—material goods as well as new intangible services. Process innovations are new ways of producing goods and services. They may be technological or organizational.

SI = *system of innovation* = the determinants of innovation processes = all important economic, social, political, organizational, institutional, and other factors that influence the development, diffusion, and use of innovations.

Constituents of SIs = *components* + *relations* among the components.

Main components in SIs = organizations and institutions.

Organizations = formal structures that are consciously created and have an explicit purpose. They are players or actors.

Institutions = sets of common habits, norms, routines, established practices, rules, or laws that regulate the relations and interactions between individuals, groups, and organizations. They are the rules of the game.

An SI has a *function*, i.e. it is performing or achieving something. The *main function* in SIs is to pursue innovation processes, i.e. to develop, diffuse and use innovations.

Activities in SIs are those factors that influence the development, diffusion, and use of innovations. The activities in SIs are the same as the determinants of the main function.

7.2 THE EMERGENCE AND DEVELOPMENT OF THE SI APPROACH

The chapter by Fagerberg in this volume highlights the systemic nature of innovation processes, noting that firms do not normally innovate in isolation, but in collaboration and interdependence with other organizations. These organizations may be other firms (suppliers, customers, competitors, etc.) or non-firm entities such as universities, schools, and government ministries. The behavior of organizations is also shaped by institutions—such as laws, rules, norms, and routines—that constitute incentives and obstacles for innovation. These organizations and institutions are components of systems for the creation and commercialization of knowledge. Innovations emerge in such "systems of innovation."

The innovation concept used in this chapter is wide and include product innovations as well as process innovations. Product innovations are new—or better—material goods as well as new intangible services. Process innovations are new ways of producing goods and services. They may be technological or organizational (Edquist, Hommen, and McKelvey 2001).

The expression "national system of innovation" (NSI) was, in published form, first used in Freeman (1987). He defined it as "the network of institutions in the public and private sectors whose activities and interactions initiate, import, and diffuse new technologies" (Freeman 1987: 1). Two major books on national systems of innovation (NSI) are Lundvall (1992) and Nelson (1993), which employ different approaches to the study of NSIs. Nelson (1993) emphasizes empirical case studies more heavily than theory development[4] and some of the studies focus narrowly on nations' R&D systems. By contrast, Lundvall (1992) is more theoretically oriented and seeks to develop an alternative to the neo-classical economics tradition by placing interactive learning, user–producer interaction and innovation at the center of the analysis (Lundvall 1992: 1).

Lundvall argues that "the structure of production" and "the institutional set-up" are the two most important dimensions that "jointly define a system of innovation" (Lundvall 1992: 10). In a similar way, Nelson and Rosenberg single out organizations supporting R&D, i.e. they emphasize those organizations that promote the creation and dissemination of knowledge as the main sources of innovation (Nelson and Rosenberg 1993: 5, 9–13).[5] Lundvall's broader approach recognizes that these "narrow" organizations are "embedded in a much wider socio-economic system in which political and cultural influences as well as economic policies help to determine the scale, direction and relative success of all innovative activities" (Freeman 2002: 195).

Both Nelson and Lundvall define national systems of innovation in terms of determinants of, or factors influencing, innovation processes.[6] However, they single out different determinants in their actual definitions of the concept, presumably reflecting what they believe to be the most important determinants of innovation. Hence, they propose different definitions of the concept, but use the same term. This reflects the lack of a generally accepted definition of a national system of innovation.

A more general definition of (national) systems of innovation includes "all important economic, social, political, organizational, institutional and other factors that influence the development, diffusion and use of innovations" (Edquist 1997b: 14). If all factors that influence innovation processes are not included in a definition, one has to argue which potential factors should be excluded—and why. This is quite difficult, since, at the present state of the art, we do not know the determinants of innovation systematically and in detail. It seems dangerous to exclude some potential determinants, since these might prove to be very important, once the state of the art has advanced. For example, twenty-five years ago it would have been natural to exclude the interactions between organizations as a determinant of innovation processes. Included in this general definition are the relationships among the factors listed and the actions of both firms and governments.

There are other specifications of systems of innovation than national ones. Carlsson and colleagues focus on "technological systems," arguing that these are unique to technology fields (Carlsson 1995). The sectoral approach of Breschi and

Malerba (1997) similarly focuses on a group of firms that develop and manufacture the products of a specific sector and that generate and utilize the technologies of that sector. The concept of "regional innovation system" has been developed and used by Cooke et al. (1997) and Braczyk et al. (1998), Cooke (2001), and Asheim and Isaksen (2002).

The three perspectives—national, sectoral and regional—may be clustered as variants of a single generic "systems of innovation" approach (Edquist 1997*b*: 3, 11–12). Much of the discussion in this chapter is relevant for the generic approach, and is based on the premise that the different variants of systems of innovation coexist and complement each other. Whether the most appropriate conception of the system of innovation, in a certain context, should be national, sectoral or regional, depends to a large extent on the questions one wants to ask.[7]

7.3 STRENGTHS AND WEAKNESSES OF THE SI APPROACH

7.3.1 The Diffusion of the SI Approach

The diffusion of the SI approach has been surprisingly rapid, and is now widely used in academic circles. The approach also finds broad applications in policy contexts— by regional authorities and national governments, as well as by international organizations such as the OECD, the European Union, UNCTAD and UNIDO. In Sweden, a public agency has even been named after the approach, i.e. the Swedish Agency for Innovation Systems (VINNOVA). The practice of VINNOVA is strongly influenced by the SI approach, an approach that appears to be especially attractive to policy makers who seek to understand differences among economies' innovative performance, and develop ways to support technological and other kinds of innovation. The next section briefly addresses some of the strengths of the generic SI approach.

7.3.2 The Strengths of the SI Approach

The SI approach places innovation and learning processes at the center of focus. This emphasis on learning acknowledges that innovation is a matter of producing new knowledge or combining existing (and sometimes new) elements of knowledge in new ways. This focus distinguishes the SI approach from other approaches that regard technological change and other innovations as exogenous.

The SI approach adopts a holistic and interdisciplinary perspective. It is "holistic" in the sense that it tries to encompass a wide array—or all—of the important determinants of innovation, and allows for the inclusion of organizational, social, and political factors, as well as economic ones. It is "interdisciplinary" in the sense that it absorbs perspectives from different (social science) disciplines, including economic history, economics, sociology, regional studies, and other fields.

The SI approach employs historical and evolutionary perspectives, which makes the notion of optimality irrelevant. Processes of innovation develop over time and involve the influence of many factors and feedback processes, and they can be characterized as evolutionary. Therefore, an optimal or ideal system of innovation cannot be specified. Comparisons can be made between different real systems (over time and space), and between real systems and target systems, but not between real systems and optimal ones. Although this is a complex view of the innovation process, it is far richer and more realistic than its alternatives.

The SI approach emphasizes interdependence and non-linearity. This is based on the understanding that firms normally do not innovate in isolation but interact with other organizations through complex relations that are often characterized by reciprocity and feedback mechanisms in several loops. Innovation processes are not only influenced by the components of the systems, but also by the relations between them. This captures the non-linear features of innovation processes and is one of the most important characteristics of the SI approach.

The SI approach can encompass both product and process innovations, as well as subcategories of these types of innovation. Traditionally, innovation studies have, to a large extent, focused upon technological process innovations and to some extent upon product innovations, but less on non-technological and intangible ones, i.e. service product innovations and organizational process innovations (as specified in Section 7.2). As argued in this Handbook, there are good reasons to use a comprehensive innovation concept,[8] and the systems of innovation approach is well suited to this comprehensive perspective, since all the categories of innovations specified in this chapter can be analyzed within it. That non-technological forms of innovation deserve more attention is also argued in OECD (2002a: 24.d).

The SI approach emphasizes the role of institutions. Practically all specifications of the SI concept highlight the role of institutions, rather than assuming them away from the list of determinants of innovation. This is important since institutions strongly influence innovation processes. There is, however, no agreement about what the term "institutions" means (see Section 7.3.3).

These six characteristics are often considered to be strengths of the SI approach by academic analysts, policy makers, and—increasingly—by firm strategists, and partly explain its rapid diffusion. However, the SI approach also has weaknesses, which represent challenges for future research on systems of innovation.

7.3.3 The Weaknesses of the SI Approach

The SI approach is still associated with conceptual diffuseness. One example is the term "institution," which is used in different senses by different authors: it is sometimes used to refer to organizational actors as well as to institutional rules. Sometimes the word means different kinds of organizations or "players" (according to the definitions in Section 7.4.2). At other times, the term means laws, rules, routines, and other "rules of the game." For Nelson and Rosenberg (1993), institutions are basically different kinds of organizations, while for Lundvall (1992) the term "institution" means primarily the rules of the game. Hence "institution" is used in several different senses in the literature (Lundvall 1992: 10; Nelson and Rosenberg 1993: 5, 9–13; Edquist 1997*b*: 26–8).

Another example of conceptual diffuseness is that the originators of the SI approach did not indicate what exactly should be included in a "(national) system of innovation"; they did not specify the boundaries of the systems (Edquist 1997*b*: 13–15). Nelson and Rosenberg provided "no sharp guide to just what should be included in the innovation system, and what can be left out" (Nelson and Rosenberg 1993: 5–6). Lundvall insisted that "a definition of the system of innovation must be kept open and flexible" (Lundvall 1992: 13).

With regard to the status of the SI approach, it is certainly not a formal theory, in the sense of providing specific propositions regarding causal relations among variables. It can be used to formulate conjectures for empirical testing, but this has been done only to a limited degree (see Section 7.5). Because of the relative absence of well-established empirical regularities, "systems of innovations" should be labeled an approach or a conceptual framework rather than a theory (Edquist 1997*b*: 28–9).

Scholars disagree on the seriousness of these weaknesses of the SI approach and on how they should be addressed. According to some, the approach should not be made too rigorous; the concept should not be "overtheorized" and it should remain an inductive one.[9] Another position argues that the SI approach is "undertheorized," that conceptual clarity should be increased and that the approach should be made more "theory-like."[10]

Hence, the international community within innovation studies is currently divided on this issue. In what follows, I try to increase the rigor and specificity of the SI approach. This effort is intended as a step towards developing the approach further. If it reveals weaknesses associated with the approach, this is a good thing. Acknowledging such weaknesses may lead to additional research and new insights into the operation of innovation systems.

7.4 The Constituents, Function, Activities and Boundaries of SIs

7.4.1 What is a System?

In an effort to develop the SI approach, it might be useful to relate it explicitly to "general systems theory," which has been used much more in the natural sciences than in the social sciences. In everyday language, as well as in large parts of the scientific literature, the term "system" is used generously and with limited demands for a precise definition. To the question "What is a system?" there is, however, a common answer in everyday language as well in scientific contexts (Ingelstam 2002: 19):

- A system consists of two kinds of constituents: There are, first, some kinds of components and, second, relations among them. The components and relations should form a coherent whole (which has properties different from the properties of the constituents).
- The system has a function, i.e. it is performing or achieving something.
- It must be possible to discriminate between the system and the rest of the world; i.e. it must be possible to identify the boundaries of the system. If we, for example, want to make empirical studies of specific systems, we must, of course, know their extent.[11]

Making the systems of innovations approach more theory-like does not require that all components and all relations among them must be specified. Such an ambition would certainly be unrealistic. At present, it is not a matter of transforming the SI approach into a "general theory of innovation," but rather we need to make it clearer and more consistent so it can better serve as a basis for generating hypotheses about relations between specific variables within SIs (which might be rejected or supported through empirical work). Even the much more modest objective of specifying the main function of SIs, the activities and components in them and some important relations among these, would represent a considerable advance in the field of innovation studies. Used in this way, the SI approach can be useful for the creation of theories about relations between specific variables within the approach.

7.4.2 The Main Components of SIs

Organizations and institutions are often considered to be the main components of SIs, although it is not always clear what is meant by these terms (as argued in Section 7.3.3). Let me therefore specify what organizations and institutions mean here.

Organizations are formal structures that are consciously created and have an explicit purpose (Edquist and Johnson 1997: 46–7). They are players or actors.[12] Some important organizations in SIs are firms, universities, venture capital organizations and public agencies responsible for innovation policy, competition policy or drug regulation.

Institutions are sets of common habits, norms, routines, established practices, rules, or laws that regulate the relations and interactions between individuals, groups, and organizations (Edquist and Johnson 1997: 46). They are the rules of the game. Examples of important institutions in SIs are patent laws, as well as rules and norms influencing the relations between universities and firms. Obviously, these definitions are of a "Northian" character (North 1990: 5), discriminating between "the rules of the game" and "the players" in the game.

SIs may differ from one another in many respects. For example, the set-ups of organizations and institutions, constituting components of empirically existing SIs, vary among them. Research institutes and company-based research departments may be important R&D performers in one country (e.g., Japan) while research universities may play a similar role in another (e.g., the United States). In some countries, such as Sweden, most research is carried out in universities, while the independent public research institutes are weak. In Germany, the latter are much more important. That the organizational set-up varies considerably among NSIs is shown in profiles of the national systems in Austria, Belgium, Finland, Germany, Spain, Sweden, Switzerland, and the United Kingdom, presented in OECD (1999a: Annex 3).

Institutions such as laws, rules, and norms also differ considerably among national SIs. For example, the patent laws are different between countries. In the USA, an inventor can publish before patenting, whereas this is not possible according to European laws. With regard to the patent rights of university teachers, individuals in this category own their patents outright in Sweden, thanks to the so-called "university teachers' privilege." However, this is not the case in the USA, where different laws apply. In Denmark and Germany, new laws have recently transferred ownership from the teachers to the universities, while in Italy, a transfer has occurred in the opposite direction. Many OECD governments are currently experimenting with changes in the ownership of knowledge created in universities, in the belief (based on little evidence—see the chapter by Mowery and Sampat) that such changes will influence the propensity to patent and accelerate the commercialization of economically useful knowledge.

In summary, there seems to be general agreement that the main components in SIs are organizations—among which firms are often considered to be the most important ones—and institutions. However, the specific set-ups of organizations and institutions vary among systems.

7.4.3 Functions and Activities in SIs

Although a system is normally considered to have a function, this was not addressed in a systematic manner in the early work on SIs. Somewhat later, some hints in this direction were made by Galli and Teubal (1997: 346–7). As we will see below, some recent contributions to the literature have started to address this theoretical gap.

7.4.3.1 *Functions and Activities in SIs and Determinants of Innovation Processes*

Xielin Liu and Steven White (2001) address what they call a fundamental weakness of national innovation system research, namely "the lack of system-level explanatory factors" (Liu and White 2001: 1092). To remedy this, they focus upon the "activities" in the systems, "activities" being related to "the creation, diffusion and exploitation of technological innovation within a system" (Liu and White 2001: 1093). On this basis, they compile a list of five fundamental activities in innovation systems.[13]

Johnson and Jacobsson (2003) emphasize that, for an innovation system "to support the growth of an industry, a number of functions have to be served within it, e.g. the supply of resources" (Johnson and Jacobsson 2003: 2). They suggest that "a technology or product specific innovation system may be described and analysed in terms of its 'functional pattern,' i.e. in terms of how these functions are served" (Johnson and Jacobsson 2003: 3). These authors present a list of five functions.[14] Rickne (2000: 175) provides a list of eleven functions that are important for new technology-based firms (i.e. not for innovations in an immediate sense).[15] Clearly, there is no consensus as to which functions or activities should be included in a system of innovation and this provides abundant opportunities for further research.

One way of addressing what happens in SIs is the following. At a general level, the main function—or the "overall function"—in SIs is to pursue innovation processes, i.e. to develop, diffuse and use innovations. What I, from now on, call activities in SIs are those factors that influence the development, diffusion, and use of innovation.[16] Examples of activities are R&D as a means of the development of economically relevant knowledge that can provide a basis for innovations, or the financing of the commercialization of such knowledge, i.e. its transformation into innovations.

A satisfactory explanation of innovation processes almost certainly will be multi-causal, and therefore should specify the relative importance of various determinants. These determinants cannot be expected to be independent of each other, but instead must be seen to support and reinforce—or offset—one another. Hence, it is important to also study the relations among various determinants of innovation processes. One way to try to do that would be to establish "a hierarchy" of causes à la E. H. Carr.

Carr argues that the study of history is a study of causes and that the historian continuously asks the question, "Why?" Further, the historian commonly assigns several causes to the same event (Carr 1986: 81, 83). He continues: "The true historian, confronted with this list of causes of his own compiling, would feel a professional compulsion to reduce it to order, to establish some hierarchy of causes which would fix their relation to one another, perhaps to decide which cause, or which category of causes, should be regarded 'in the last resort' or 'in the final analysis'... as the ultimate cause, the cause of all causes" (Carr 1986: 84). I do not believe that we will ever reach such an objective in a detailed and systematic manner or that we will be able to identify all determinants of innovation—because of the complexity of the task. However, there are good reasons to try to strive in this direction by developing theories about relations between specific variables within the approach in a pragmatic way (as proposed in Section 7.4.1.).

I believe that it is important to study the activities (causes, determinants) in SIs in a systematic manner. The hypothetical list of activities presented below is based upon the literature, e.g. the lists mentioned earlier, and on my own knowledge about innovation processes and their determinants. The activities listed are not ranked in order of importance, but start with knowledge inputs to the innovation process, continues with the demand side factors, the provision of constituent of SIs, and ends with support services for innovating firms.

The following activities can be expected to be important in most SIs:

(1) Provision of Research and Development (R&D), creating new knowledge, primarily in engineering, medicine, and the natural sciences.
(2) Competence building (provision of education and training, creation of human capital, production and reproduction of skills, individual learning) in the labor force to be used in innovation and R&D activities.
(3) Formation of new product markets.

(4) Articulation of quality requirements emanating from the demand side with regard to new products.

(5) Creating and changing organizations needed for the development of new fields of innovation, e.g. enhancing entrepreneurship to create new firms and intrapreneurship to diversify existing firms, creating new research organizations, policy agencies, etc.

(6) Networking through markets and other mechanisms, including interactive learning between different organizations (potentially) involved in the innovation processes. This implies integrating new knowledge elements developed in different spheres of the SI and coming from outside with elements already available in the innovating firms.

(7) Creating and changing institutions—e.g. IPR laws, tax laws, environment and safety regulations, R&D investment routines, etc—that influence innovating organizations and innovation processes by providing incentives or obstacles to innovation.

(8) Incubating activities, e.g. providing access to facilities, administrative support, etc. for new innovative efforts.

(9) Financing of innovation processes and other activities that can facilitate commercialization of knowledge and its adoption.

(10) Provision of consultancy services of relevance for innovation processes, e.g. technology transfer, commercial information, and legal advice.

This list is provisional and will be subject to revision as our knowledge about determinants of innovation processes increases. In addition to a set of activities that is likely to be important in most SIs, there are activities that are very important in some kinds of SIs and less important in others. For example, the creation of technical standards is critically important in some (sectoral) systems, such as mobile telecommunications.[17]

The systematic approach to SIs suggested here does not imply that they are or can be consciously designed or planned. On the contrary, just as innovation processes are evolutionary, SIs evolve over time in a largely unplanned manner. Even if we knew all the determinants of innovation processes in detail (which we certainly do not now, and perhaps never will), we would not be able to control them and design or "build" SIs on the basis of this knowledge. Centralized control over SIs is impossible and innovation policy can only influence the spontaneous development of SIs to a limited extent.

7.4.3.2 *Three Kinds of Learning in the SI Approach*

Regarding competence building as an important activity in SIs—and given that R&D has earlier been a central activity in SI studies—means that the SI approach focuses on three kinds of learning:

- *Innovation* (in new products as well as processes) takes place mainly in firms and leads to the creation of "structural capital," which is a knowledge-related asset controlled by firms (as opposed to "human capital"); it is a matter of organizational learning.
- *Research and Development* (R&D) is carried out in universities and public research organizations as well as in firms and leads to publicly available knowledge as well as knowledge owned by firms and other organizations, as well as by individuals.
- *Competence Building* (e.g. training and education) which occurs in schools and universities (schooling, education) as well as in firms, and leads to the creation of "human capital." Since human capital is controlled by individuals, it is a matter of individual learning.

An important area for further research in the SI tradition concerns the relationship among these three kinds of learning, which appear to be closely related to one another. One objective of such studies would be to address what types and levels of education and training are most important for specific kinds of innovations—e.g. process innovations and product innovations, or incremental and radical innovations.

As exemplification, I will now discuss two central learning activities—R&D and competence building—in some greater detail. This discussion constitutes the beginning of an examination of the relations among activities and constituents in SIs that is continued in section 7.4.4. It also serves as a guide to and synthesis of some of the recent work on these issues within the systems of innovation literature.

A. *Research and Development.* Considerable work on NSIs has been carried out within the OECD. However, although most of the OECD contributions mentioned here have "systems of innovation" in the title, many of them actually use this approach more as a label than as an analytical tool. The first phase of this work included the development of quantitative indicators, country case studies, and work within six focus groups on innovative firms, innovative firm networks, clusters, mobility of human resources, organizational mapping, and catching-up economies. Some of the empirical results were presented in (OECD 1998a) and a synthesis is found in (OECD 1999a).

The second phase provided a deepening of the analysis in three areas: innovative clusters (OECD 2001a and 1999b), cooperation in national innovation systems (OECD 2001b), and mobility of skilled personnel in NSIs (OECD 2001c). Yet another study (OECD 2002a) summarizes the findings of the second phase of the project and derives policy implications. In the studies mentioned, R&D as well as competence building is addressed to some extent.

In most countries, universities are the most important public organizations performing R&D (OECD 1998a: ch. 3). Governments fund university R&D activities in a number of ways. Traditionally, they have provided general support via block grants from the Ministries of Education, part of which was used by university staff to carry out R&D. Such funding is still very important in small, highly R&D-intensive

countries such as the Netherlands, Sweden, and Switzerland. Governments may also provide grants to encourage research "for the advancement of knowledge" or grants to obtain the knowledge needed for government missions such as defense or health care. In most countries, block grants have declined and direct support has grown in importance (OECD 1998*a*: ch. 3).

In certain countries, universities fall under the responsibility of the national government. In others, such as Germany, they are the responsibility of the regional governments. Whatever the form of organization, a growing regional influence can be observed in most countries. In Germany, the universities are financially very autonomous. In the UK, financial support is provided by research councils to individual projects selected on a competitive basis.

In many countries, the science system also includes public research institutes (or "national laboratories") which carry out the same type of R&D activities as universities, as well as more applied research and technical development work. Although the relative importance of universities in terms of performing R&D has increased in most countries (see Mowery and Sampat in this volume), public research organizations remain important. These organizations may be linked to the universities and included in the higher education sector, or they may be independent of them. The largest single case in the OECD area is the Centre National de la Recherche Scientifique (CNRS) in France, which receives the lion's share of direct funding of R&D in the higher education sector. The CNRS provides support for projects that are normally carried out in collaboration with university researchers. In this regard, the CNRS can be clearly distinguished from its counterparts in Germany (Max Planck Gesellschaft), Italy (CNR) and Spain (CSIC) (Laredo and Mustar 2001*c*: 502). In the United States, the higher education sector contains a large number of public research laboratories. (OECD 1998*a*: 83–4). Other countries with a large institute sector include Norway, Taiwan and Germany (e.g. Max Planck Gesellshaft and Fraunhofer). A number of national governments have tried to change these organizations and promote their links with the rest of the economy and society. This has, for example, been done in quite different ways in France and the UK (Laredo and Mustar 2001*c*: 503).

As this short discussion suggests, different kinds of public organizations (such as universities and public research institutes) can perform the same activity (R&D) in an NSI. NSIs differ significantly with regard to which organizations that perform public R&D and with regard to the institutional rules that govern or influence these organizations (Laredo and Mustar 2001*b*: 6–7).

In most NSIs, especially in low- and medium-income nations, only modest sums are invested in R&D and most of the R&D is performed by public organizations. The few countries that invest heavily in R&D are all rich, and much of their R&D is carried out by private organizations. This group includes some large countries, such as the USA and Japan, but also some small and medium-sized countries, such as Sweden, Switzerland, and South Korea. There are also some rich countries that do rather little R&D, e.g. Denmark and Norway. As mentioned, a considerable part of the R&D in many rich countries is carried out and financed by the private sector, primarily firms (although there are also public financial support schemes to stimulate firms to perform R&D). The proportion of all R&D performed in high-income OECD member states that is financed by firms ranged between 21 per cent (Portugal) and 72 per cent (Japan) in 1999 (OECD 2002*b*). Acknowledging such differences may help to distinguish between different types of NSIs.

Most of the R&D carried out by private organizations may be characterized as development work rather than research. Innovation certainly does not depend solely on R&D results, but requires also other actions, such as technical experimentation, technology adoption, market investigations, and entrepreneurial initiative. R&D and innovation activities are normally driven by different rationales and motivations—i.e. the advancement of knowledge and the quest for profits, respectively.

One implication of the complex interface between "research" and "innovation" is that links between universities/public research organizations and innovating firms are especially important to the performance of NSIs.[18] Innovating firms often need to collaborate with public research organizations or universities. Here, publicly created institutions are important. Governments may, for example, support collaborative centers and programs, remove barriers to cooperation, and facilitate the mobility of skilled personnel among different kinds of organizations. This might involve the creation of institutional rules, such as those in Sweden stating that university professors should perform a "third task" in addition to teaching and doing research—i.e. interact with the society surrounding the university, including firms. However, such "linkage activity" is carried out in different ways and to different extents in different NSIs.

B. *Competence building.* The early work within the SI approach largely neglected learning in the form of education and training.[19] However, competence building is increasingly considered to be an important activity in systems of innovation, reflecting the importance of skilled personnel for most innovative activities (Smith 2001: 8).[20] But no rigorous analyses of competence building have, to my knowledge, been conducted as part of the analysis of innovation systems.

Nevertheless, there is a large literature on various aspects of competence building outside the SI context. Competence building (e.g. training and education) is the same as enhancement of human capital and is carried out largely, though not exclusively, in schools and universities. Competence building also occurs in firms (in the form of training, learning-by-doing, learning-by-using, and individual learning) often throughout working life.

A recent OECD study analyzed vocational and technical education and training in some detail in Australia, Austria, Denmark, England (including Wales and Northern Ireland), France, Germany, Italy, the Netherlands, Quebec, and Switzerland (OECD 1998*b*). This study pointed out many differences across countries with regard to vocational and technical training. One difference concerns the stage prior to vocational and technical training, i.e. the structure of middle and lower secondary education. This structure is unified in most countries, but is divided into distinct programs in Germany, Austria, the Netherlands, and Switzerland. One of these programs is the beginning of the academic pathway, while the others lead essentially to vocational and technical training. Another difference concerns the relative numeric importance of vocational and technical training as opposed to academic pathways in upper secondary education. In the countries of the British Commonwealth—Australia, Canada, and the UK—the academic pathway is very much in the majority, while in the countries of continental Europe, vocational and technical training dominates (OECD 1998*b*).

The ways in which people access skilled jobs (and then climb the career "ladders" of enterprises) differ greatly among NSIs:

This may occur at a certain time after recruitment, when the young person has proved himself; after a fixed and codified period of service, according to a specific labour contract; or on recruitment, depending on the qualifications previously acquired. For vocational training, these three modes of access lead to three broad traditions: on-the-job training, formal apprenticeship, school training.   (OECD 1998*b*: 12)

These practices coexist in various countries, but their relative importance varies considerably; frequently, one of them dominates and determines training policy.

The models for transition from education to employment also differ across countries. Apprenticeship is important in some countries—e.g. in Germany, where it caters to about two-thirds of the age group (OECD 1996: 48). In other countries, school-based learning and productive work are combined in alternative ways—e.g. in Sweden, Australia, France, the United Kingdom, and Korea (OECD 1996: 146).

The organizational and institutional contexts of competence building thus vary considerably among NSIs. There are particularly significant differences between the systems in the English-speaking countries and continental Europe. However, scholars and policy makers lack good comparative measures on the scope and structure of such differences. There is little systematic knowledge about the ways in which organization of education and training influences the development, diffusion and use of innovations. Since labor, including skilled labor, is the least mobile production factor, domestic systems for competence building remain among the most enduringly "national" elements of NSIs.

7.4.4 The Relations Between Activities and Components and among Components

This chapter has placed greater emphasis on "activities" than much of the early work on SIs. Nonetheless, this emphasis does not mean that we can disregard or neglect the "components" of SIs and the relations among them. Organizations or individuals perform the activities and institutions provide incentives and obstacles influencing these activities. In order to understand and explain innovation processes, we need to address the relations between activities and components, as well as among different kinds of components.

What then are the relations between the components and the activities in SIs? As we saw in Section 7.4.3.2.A, the activity of research (the creation of new knowledge) can be carried out by research institutes, universities, or research-oriented firms. Most of the other activities mentioned earlier can also be performed by different organizations. Further, many categories of organizations can perform more than one activity. For example, universities provide new knowledge and educate people (human capital). Hence, there is not a one-to-one relation between activities and organizations.[21] However, there are limits to this flexibility—for instance, primary schools cannot carry out basic research. The relations between activities and institutions are less direct, since institutions influence whether or not, and how, certain organizations perform certain activities. It seems that the set-up of activities can be expected to vary less across NSIs than the set-up of organizations performing them and the set-up of institutions influencing those organizations. However, the "quantity" of each activity and the efficiency with which each activity is performed might vary considerably among NSIs.[22]

As we saw in section 7.3.2, the SI approach emphasizes the relations or interactions among the components in SIs. Interactions among different organizations may be of a market or non-market kind. That market, as well as non-market, relations should be addressed in SI research is stressed in a recent OECD report. There the concept of interaction is specified as including:

- Competition, which is an interactive process wherein the actors are rivals, and which creates or affects the incentives for innovation.
- Transaction, which is a process by which goods and services, including technology-embodied and tacit knowledge, are traded between economic actors.
- Networking, which is a process by which knowledge is transferred through collaboration, cooperation and long term network arrangements. (OECD 2002a: 15)

With regard to interactions among organizations in their pursuit of innovations, empirical work inspired by and designed on the basis of the SI approach has been carried out in many countries. One example is the Community Innovation Surveys (CIS) that have been coordinated by Eurostat of the European Union and carried out in all EU countries and in several additional countries (see Smith, Chapter 6 in this volume, for a detailed discussion of CIS). The CIS results include data on collaboration among innovating organizations, and indicate that such collaboration is very common and important. This result is supported by other surveys which have shown that between 62 per cent and 97 per cent of all product innovations were achieved in collaboration between the innovating firm and some other organizations in Austria, Norway, Spain, Denmark, and the region of East Gothia in Sweden (Christensen et al. 1999; Örstavik and Nås 1998; Edquist, Ericsson, and Sjögren 2000: 47).

These findings constitute empirical support for one of the main tenets of the SI approach, i.e. that interactive learning among organizations is crucial for innovation processes. This also illustrates the dynamics of this field of research over time. The emergence of the SI approach in its Danish version (Lundvall 1992) took inspiration from case studies indicating that user–producer interaction was very important for innovations, e.g. in the Danish dairy industry; the SI approach was formulated partly on this basis. One of its central elements—the importance of relations of interactive learning among organizations—has since been verified by systematic empirical research in Denmark and elsewhere. This is a good example of fruitful interaction between theoretical and empirical work.

Another example of empirical work partly based on the NSI approach is Furman, Porter, and Stern (2002), who introduce the concept of national innovative capacity, which is the ability of a country to produce and commercialize a flow of new-to-the-world technologies over the long term. This concept is explicitly based upon ideas-driven endogenous growth theory à la Romer (see Verspagen, Ch. 18 in this volume), the cluster-based approach à la Porter (1990), and the NSI approach. On this basis, they estimate the relationship between international patenting (patenting by foreign countries in the USA) and observable measures of national innovative capacity. Their results suggest that a small number of observable factors describe a country's national innovative capacity—i.e. they identify determinants of the production of new-to-the-world technologies. They find that a great deal of variation in patenting across countries is due to differences in the level of inputs devoted to innovation (R&D manpower and spending). They also find that an extremely important role is played by factors associated with differences in R&D productivity, e.g. policy choices such as the extent of protection of intellectual property and openness to international trade, the share of research performed by the academic sector and funded by the private sector, the degree of technological specialization, and each individual country's knowledge "stock" (Furman et al. 2002).

The relations between organizations and institutions are important for innovations and for the operation of SIs. Organizations are strongly influenced and shaped by institutions, so that organizations can be said to be "embedded" in an institutional environment or set of rules, which include the legal system, norms, routines, standards, etc. But institutions are also embedded in and develop within organizations. Examples are firm-specific rules with regard to bookkeeping or concerning norms with regard to the relations between managers and employees. Hence, there is a complicated relationship of mutual embeddedness between institutions and organizations (Edquist and Johnson 1997: 59–60).

Some organizations create institutions that influence other organizations. Examples are organizations that set standards and public organizations that formulate and implement those rules that we might call innovation policy. Examples are the NMT 450 and the GSM mobile telecom standards. The NMT 450 was created by the Nordic public telephone operators, which were state-owned monopolies at the time. The development and implementation of NMT 450 was an example of the importance of user–producer relations in innovation processes, which is stressed so strongly in the SI approach. The public organizations provided a technical framework for private equipment producers and thereby decreased uncertainty. The Nordic equipment producers, Ericsson and Nokia, greatly benefited from this, and it was an important factor contributing to their leading role in mobile telecommunications equipment production today. In essence, the NMT 450 provided the cradle for the development of mobile telecommunications in Europe (Edquist 2003: 21–3).

Institutions may also be the basis for the creation of organizations, as when a government makes a law that leads to the establishment of an organization. Examples of such organizations include patent offices or public innovation policy agencies.

There may also be important relations between different institutions, for example, between patent laws and informal rules concerning exchange of information between firms. Institutions of different kinds may support and reinforce each other, but may also contradict and be in conflict with each other, as discussed in some detail by Edquist and Johnson (1997). This work has been carried forward by Coriat and Weinstein (2002), who discuss different levels of institutions and focus on the principle of a hierarchy among rules themselves (Coriat and Weinstein 2002: 280).[23]

Our knowledge about the complex relations—characterized by reciprocity and feedback—between organizations and institutions is limited. Since the relations between two phenomena cannot be satisfactorily analyzed if they are not conceptually distinguished from each other, it is important to make a clear distinction between organizations and institutions when specifying the concepts.[24]

7.4.5 Boundaries of SIs: Spatial, Sectoral and in Terms of Activities

The distinction between what is inside and outside a system is crucial—i.e., the issue of the boundaries of SIs cannot be neglected (see Section 7.4.1). It is therefore necessary to specify the boundaries if empirical studies of specific SIs—of a national, regional, or sectoral kind—are to be carried out. As will be discussed later, one way to identify the boundaries of SIs is to identify the causes or determinants of innovations.

Although "national systems of innovation" is only one of several possible specifications of the generic SI concept, it certainly remains one of the most relevant.

One reason is the fact that the various case studies in Nelson (1993) show that there are sharp differences among various national systems in such attributes as institutional set-up, organizational set-up, investments in R&D, and performance. For example, the differences in these respects between Denmark and Sweden are remarkable—in spite of the fact that these two small countries in northern Europe are very similar in many other respects (Edquist and Lundvall 1993: 5–6).

Another reason to focus on national SIs is that most public policies influencing innovation processes or the economy as a whole are still designed and implemented at the national level. For very large countries, the national SI approach is less relevant than for smaller countries, but institutions such as laws and policies are still mainly national, even in a country such as the USA. In other words, the importance of national SIs has partly to do with the fact that they capture the importance of the policy aspects of innovation. It is not only a matter of geographical delimitation, as the state, and the power attached to it, is also important.

SIs may be supranational, national, or subnational (regional, local)—and at the same time they may be sectoral within any of these geographical demarcations.[25] All these approaches may be fruitful, but for different purposes or objects of study. Generally, the variants of the generic SI approach complement rather than exclude each other and it is useful to consider sectoral and regional SIs in relation to—and often as parts of—national ones.

There are three ways in which we can identify boundaries of SIs:

(1) spatially / geographically;
(2) sectorally; and
(3) in terms of activities.

1. To define the *spatial* boundaries is the easiest task, although it also has its problems. Such boundaries have to be defined for regional and national SIs, and sometimes also for sectoral ones. The problem of *geographical* boundaries is somewhat more complicated for a regional than for a national SI. One question is which criteria should be used to identify a "region."

For a regional SI, the specification of the boundaries should not only be a question of choosing or using administrative boundaries between regions in a mechanical manner (although this might be useful from the point of view of availability of data). It should also be a matter of choosing geographical areas for which the degree of "coherence" or "inward orientation" is high with regard to innovation processes. One possible operationalization of this criterion could be a sufficient level of localized learning spill-overs (among organizations), which is often associated with the importance of transfer of tacit knowledge among (individuals and) organizations. A second could be localized mobility of skilled workers as carriers of knowledge, i.e. an operationalization which shows that the local labor market is important. A third possibility could be that a minimum proportion of the innovation-related collaborations among organizations should be with partners within the region. This is a matter of localized networks, i.e. the extent to which learning processes among organizations are contained within regions.

For a national SI, the country's borders normally provide the boundaries. However, it could be argued that the criteria for regional SIs are as valid for national ones. In other words, if the degree of coherence or inward orientation is very low, the country might not reasonably be considered to have a national SI. It was also mentioned above that the national SI approach is less relevant for large than for smaller countries. In Germany, for example, the appropriate unit of analysis may be "Länder." The choice of approach may not only be a question of size of the country, but also whether it is federally organized or not.

2. Leaving the geographical dimension, we can also talk about "sectorally" delimited SIs, i.e., systems that include only a part of a regional, national or international system. These are delimited to specific technological fields (generic technologies) or product areas. The "technological systems" approach belongs to this category (although it did not initially use language associated with systems of innovation) (Carlsson and Stankiewicz 1995: 49).

According to Breschi and Malerba, "a Sectoral Innovation System (SIS) can be defined as that system (group) of firms active in developing and making a sector's products and in generating and utilising a sector's technologies" (Breschi and Malerba 1997: 131; see also Ch. 14 by Malerba in this volume). Specific technologies or product areas are used to define the boundaries of sectoral systems, but they must also normally be geographically delimited (if they are not global). However, it is not self-evident what a sector is, i.e., the *sectoral* boundaries are partly a theoretical—or social—construction, which may reflect the specific purpose of the study. It should also be noted here that the specification of sectoral boundaries is particularly difficult with regard to new sectors or sectors going through radical technological shifts.

3. Within a geographical area (and perhaps also limited to a technology field or product area), the whole socio-economic system cannot, of course, be considered to be included in the SI. The question is, then, which parts should be included? This is a matter of defining the boundaries of SIs in terms of activities. These have to be defined for all kinds of SIs: national, regional, and sectoral. This is more complicated than in the cases of spatial and sectoral boundaries.

Early work in the SI approach did not address the activities in SIs in a systematic way, and therefore failed to provide clear guidance as to what should be included in a system of innovation. Nor have the boundaries of the systems in terms of activities been defined in an operational way since then.

In Section 7.2, a system of innovation was defined as including "all important economic, social, political, organizational, institutional, and other factors that influence the development, diffusion, and use of innovations." If the concept of innovations has been specified (e.g., as in the beginning of Section 7.2), and if we know the determinants of their development, diffusion, and use, we will be able to define the boundaries of the SIs in terms of activities. This is one reason why it is so important to identify the activities in SIs. Admittedly, this is not as easy in practice as in principle, since we simply do not know in detail and systematically all the activities in SIs or determinants of innovation processes. As pointed out in Section 7.4.3.1, any list of activities in an SI must be treated as provisional and subject to change as our knowledge increases.

7.5 Research Gaps and Opportunities

In innovation studies, there has traditionally been a tendency to focus much more on technological process innovations and goods product innovations than on organizational process innovations and service product innovations. There are strong reasons to use a comprehensive innovation concept and give more attention to non-technological and intangible kinds of innovation (as proposed in Section 7.3.2). Such an orientation is implicit in the fact that we talk about systems of innovation and not systems of technological change.

More research should be done on the activities in SIs, i.e., on the determinants of the development, diffusion, and use of innovations. One particular task may be to revise and restructure the preliminary list of important activities in SIs presented in Section 7.4.3.1. Such a list can provide an important point of entry for empirical innovation studies.

A stronger focus on activities would increase our knowledge of, and capacity for, explaining innovation processes. Given our limited systematic knowledge about determinants of innovations, *case studies* of the determinants of specific innovations or specific (and narrow) categories of innovations would be very useful. In particular, I believe that *comparative* case studies have great potential, comparing innovations systems of various kinds as well as the determinants of innovation processes within them. Relevant questions to ask would include: Which activities of which organizations are important for the development, diffusion, or use of specific innovations? Is it possible to distinguish between important activities and less important ones? Which institutional rules influence the organizations in carrying out these activities? Such work could further develop the SI approach and contribute to the creation of partial theories about relations between variables within SIs. Such theories would also improve our ability to specify the boundaries of innovation systems.

In this chapter, I have accounted for many of the existing empirical studies that claim to have been carried out within a SI framework. The result has, on the whole, been rather disappointing in the sense that many of the studies cited have not been related to the SI approach in a profound way, although there are exceptions. The SI approach has often been used more as a label than as an analytical tool. It has not influenced the empirical studies in depth; for example, it has not been used to formulate hypotheses to be confronted to empirical observations. This has made a virtuous fertilization between conceptual and empirical work, that is so important to scholarly progress in this and other fields of research, difficult to achieve. The state of the art of the SI approach is partly responsible for this: it is often presented in too vague and unclear a way.

Clearly defined concepts are necessary in order to identify empirical correspondents to theoretical constructs and to identify the data that should be collected. Therefore, conceptual specifications are crucial for empirical studies and it is important to increase the rigor and specificity of the SI approach. This can be done by clarifying the meaning of key concepts such as innovation, function, activities, components, organizations, and institutions, as well as the relations among them. Moving in this direction does not mean transforming social science into something similar to natural science. For example, one cannot abstract from time and space, since there are no universal laws in the social sciences. It is also important to continue the work of specifying the boundaries of SIs of various kinds.

There are strong reasons to integrate conceptual and theoretical work with empirical studies in an effort to identify determinants of the development, diffusion, and use of innovations. Such integration can be expected to lead to cross-fertilization—just as in the case of work on interactive learning referred to in Section 7.4.4. The SI approach should be used as a conceptual framework in specific empirical analyses of concrete conditions. Testable statements or hypotheses should be formulated on the basis of the approach and these should be investigated empirically, by using qualitative as well as quantitative observations. Theoretically based empirical work is the best way to straighten up the SI approach conceptually and theoretically; the empirical work will, in this way, serve as a "disciplining" device in an effort to develop the conceptual and theoretical framework. Such work would increase our empirical knowledge about relations between the main function, activities, organizations, and institutions in SIs. This knowledge could then be a basis for further empirical generalizations to develop the framework—including theory elements. In other words, empirically based theoretical work is also very fruitful. Independently of where one starts, the important thing is that there should be a close relationship between theoretical and empirical work.

The array of determinants of innovations and the relations among them can be expected to vary over time and space, i.e. between innovation systems, as well as among different categories of innovation. For example, the determinants will probably vary between process and product innovations as well as between incremental and radical innovations (and between subcategories of these). It is therefore important to pursue the explanatory work at a meso- or micro-level of aggregation.

Taxonomies of different categories of innovations can therefore be expected to be an important basis for this work.

Innovation studies have traditionally included research on R&D and its signifi-cance for innovation processes. A well-educated labor force is necessary for both R&D and innovation, and competence building therefore should receive greater emphasis in innovation studies and in the SI approach. We should not only address those learning processes that lead directly to process and product innovations, but also address the knowledge infrastructure and learning in a more generic way.

This "widening" might eventually transcend the SI approach and move into thinking along the lines of "Systems of Learning" rather than "Systems of Innov-ation." Systems of Learning would include individual learning (leading to the creation of human capital) as well as organizational learning (leading to the creation of structural capital, e.g. innovations). It would include work on three kinds of learning: R&D, innovation, and competence building, and, above all, the relations between them. This also points out one direction in which the SI approach is currently developing.

Notes

1. For comments on previous versions of this chapter, I want to thank the editors of this Handbook, my discussant at the Lisbon workshop (John Cantwell), and my discussants at the Roermond workshop (Jan Fagerberg, Bill Lazonick, and Rikard Stankiewicz). I have also greatly profited from comments by other participants in the workshops and from Pierre Bitard, Susana Borras, David Doloreux, Leif Hommen, Björn Johnson, Rachel Parker, Lars Mjöset, and Annika Rickne. I also want to thank The Swedish Agency for Innovation Systems (VINNOVA) for supporting my work with this chapter. However, I remain responsible for the contents.

2. The regional and sectoral versions are dealt with in more detail in Ch. 11 by Asheim and Gertler and in Ch. 14 by Malerba in this volume.

3. In this sense, this chapter is a continuation along the same trajectory as earlier attempts, e.g. Edquist (1997*b*), Edquist and Johnson (1997) and Edquist (2001).

4. "[T]he orientation of this project has been to carefully describe and compare, and try to understand, rather than to theorise first and then attempt to prove or calibrate the theory" (Nelson and Rosenberg (1993: 4).

5. They mention organizations such as firms, industrial research laboratories, research universities, and government laboratories.

6. Their definitions of NSIs do not include, for example, consequences of innovation—which does not, of course, exclude the fact that innovations, emerging in innovation systems, have tremendously important consequences for socio-economic variables such as productivity growth and employment. To distinguish between determinants and consequences does not, of course, exclude feedback mechanisms.

7. It should also be mentioned that the publications mentioned in Section 7.2 by no means exhaust the stock of literature addressing or using the SI approach. Edquist and

McKelvey (2000) is a reference collection containing forty-two articles on SIs, some of which are reviewed in this chapter. Other contributions will be addressed in later sections of this chapter.

8. There are chapters in the Handbook on service product innovations (Ch. 16 by Ian Miles) and on organizational process innovations (Ch. 5 by Alice Lam).

9. See Lundvall et al. (2002: 221) and Lundvall (2003: 9), where it is argued that the pragmatic and flexible character of the concept may be seen to be a great advantage. However, Lundvall et al. (2002: 221) also argue that efforts should be made to give the concept a stronger theoretical foundation.

10. Such a view has, for example, been expressed by the OECD: "There are still concerns in the policy making community that the NIS approach has too little operational value and is difficult to implement" (OECD 2002a: 11). A similar position is taken by Fischer (2001: 213–14).

11. Only in exceptional cases is the system closed in the sense that it has nothing to do with the rest of the world (or because it encompasses the whole world). Like the SI approach, "general systems theory" might rather be considered to be an approach than a theory.

12. Although there are other kinds of actors than organizations—e.g. individuals—the terms "organizations" and "actors" are used interchangeably in this chapter.

13. The five activities are R&D, implementation, end-use, education, and linkage.

14. These are: to create new knowledge, to guide the direction of the search process, to supply resources, to create positive external economies, and to facilitate the formation of markets (Johnson and Jacobsson 2003: 3–4). Anna Johnson—now Anna Bergek—previously discussed these issues in Johnson (1998). There she identified functions mentioned or implicitly addressed in various previous contributions to the development of the SI approach. She also listed and stressed various benefits of using the concept of "function" in SI studies.

15. These functions are to create human capital, to create and diffuse technological opportunities, to create and diffuse products, to incubate (provide facilities, equipment, and administrative support), to facilitate regulation that may enlarge the market and facilitate market access, to legitimize technology and firms, to create markets and diffuse market knowledge, to enhance networking, to direct technology, market and partner search, to facilitate financing, and to create a labor market that the new technology-based firms can utilize.

16. The activities in SIs are the same as the determinants of the main function. An alternative term to "activities" could have been "subfunctions." I chose "activities" in order to avoid the connotation with "functionalism" or "functional analysis" as practiced in sociology, which focuses on the consequences of a phenomenon rather than on its causes, which are in focus here.

17. The activities in this sectoral system are discussed in Edquist (2003: 11).

18. Specific ways in which knowledge transfer takes place between universities and firms are analyzed in detail for the case of Austria in Schibany and Schartinger (2001).

19. When designing the anthology edited by Lundvall (1992), the Aalborg group planned to have a chapter on the education system. However, in the end it was not included (Lundvall and Christensen (1999: 3).

20. Competence building has also been addressed in some OECD publications, including a study on knowledge management in the learning society, managed by the Centre for Educational Research and Innovation (CERI) (OECD 2000). Another CERI study

includes a conceptual framework which tries to integrate "individual learning" (e.g. education) and "organizational learning" (e.g. innovation) into a generic conceptual framework on "learning." It also contains empirical studies of the respective roles of education and innovation for economic growth at a regional level (OECD 2001*d*). Another contribution is the DISKO project in Denmark as reported by Lundvall (2002).

21. In Rickne 2000: ch. 7, there is a more detailed discussion of the relations between activities and organizations.

22. As we saw in Section 7.4.3.1, there are also important relations between activities, i.e. relations between determinants of innovations processes.

23. Coriat and Weinstein address the relations between organizations and institutions as well, although they consider firms to be both institutions and organizations (Coriat and Weinstein 2002: 279).

24. The so-called "varieties of capitalism" literature has a wider perspective and focuses on a broader range of institutions and organizations. Examples are Hollingsworth and Boyer (1997), Whitley (1999), Hall and Soskice (2001), and Whitley (2002). Space limitations prevent me from going into this literature here. The same applies to "the social systems of innovation" perspective (e.g. Amable 2000) and the Triple-Helix perspective (e.g. Etzkowitz and Leydesdorff (2000).

25. An "industrial complex" or "cluster" as used by Porter (1990, 1998) can, if it is regionally delimited, be seen as a combination of a sectoral and a regional SI.

REFERENCES

AMABLE, B. (2000), "Institutional Complementarity and Diversity of Social Systems of Innovation and Production," *Review of International Political Economy* 7(4): 645–87.

ASHEIM, B., and ISAKSEN, A. (2002), "Regional Innovation Systems: The Integration of Local 'Sticky' and Global 'Ubiquitous' Knowledge," *Journal of Technology Transfer* 27: 77–86.

*BRACZYK, H.-J., COOKE, P., and HEIDENREICH, M. (eds.) (1998), *Regional Innovation Systems: The Role of Governance in a Globalised World*, London and Pennsylvania: UCL.

*BRESCHI, S., and MALERBA, F. (1997), "Sectoral Innovation Systems: Technological Regimes, Schumpeterian Dynamics, and Spatial Boundaries," in Edquist, 1997*a*: 130–56.

*CARLSSON, B. (ed.) (1995), *Technological Systems and Economic Performance: The Case of Factory Automation*, Dordrecht: Kluwer.

—— and STANKIEWICZ, R. (1995), "On the Nature, Function and Composition of Technological Systems," in Corlson 1995: 21–56.

CARR, E. H. (1986), *What is History?* Harmondsworth: Penguin.

CHRISTENSEN, J. L., ROGACZEWSKA, A. L., and VINDING, A. L., (1999) Summary Report of the Focus Group on Innovative Firm Networks, OECD home page.

COOKE, P. (2001), "Regional Innovation Systems, Clusters, and the Knowledge Economy," *Industrial and Corporate Change* 10(4): 945–74.

*—— GOMEZ URANGA, M., and ETXEBARRIA, G. (1997). "Regional Systems of Innovation: Institutional and Organisational Dimensions," *Research Policy* 26: 475–91.

* Asterisked items are suggestions for further reading.

CORIAT, B., and WEINSTEIN, O. (2002), "Organisations, Firms and Institutions in the Generation of Innovation," *Research Policy* 31(2): 273–90.

*EDQUIST, C. (ed.) (1997a), *Systems of Innovation: Technologies, Institutions and Organizations*, London: Pinter.

—— (1997b), "Systems of Innovation Approaches—their Emergence and Characteristics," in Edquist 1997a: 1–35. (The book is out of print, but this chapter has been republished in Edquist and McKelvey 2000.)

—— (2001), "The Systems of Innovation Approach and Innovation Policy: An Account of the State of the Art," Lead paper presented at the DRUID Conference, Aalborg, June 12–15 2001. Unpublished.

—— (2003), "The Fixed Internet and Mobile Telecommunications Sectoral System of Innovation: Equipment, Access and Content," in C. Edquist (ed.), *The Internet and Mobile Telecommunications System of Innovation: Developments in Equipment, Access and Content*, Cheltenham: Edward Elgar, 1–39.

—— ERICSSON, M.-L., and SJÖGREN, H. (2000), "Collaboration in Product Innovation in the East Gothia Regional System of Innovation," *Enterprise & Innovation Management Studies*, 1.

—— HOMMEN, L., and McKELVEY, M. (2001), *Innovation and Employment: Process versus Product Innovation*, Cheltenham: Edward Elgar.

—— and JOHNSON, B. (1997), "Institutions and Organisations in Systems of Innovation," in Edquist 1997a: 41–63. (The book is out of print, but this chapter has been republished in Edquist and McKelvey 2000.)

—— and LUNDVALL, B.-Å. (1993), "Comparing the Danish and Swedish systems of innovation," in Nelson 1993: 265–98.

—— and McKELVEY, M. (eds.) (2000), *Systems of Innovation: Growth Competitiveness and Employment*, Cheltenham: Edward Elgar.

ETZKOWITZ, H., and LEYDESDORFF, L. (2000), "The Dynamics of Innovation: From National Systems and 'Mode 2' to Triple Helix of University–Industry–Government Relations," *Research Policy* 29: 109–23.

FISCHER, M. F. (2001), "Innovation, Knowledge Creation and Systems of Innovation," *Regional Science* 35: 199–216.

FREEMAN, C. (1987), *Technology Policy and Economic Performance: Lessons from Japan*, London: Pinter.

*—— (2002), "Continental, National and Sub-national Innovation Systems—Complementarity and Economic Growth," *Research Policy* 31(2): 191–211.

FURMAN, J. L., PORTER, M. E., and STERN, S. (2002), "The Determinants of National Innovative Capacity," *Research Policy* 31: 899–933.

GALLI, R., and TEUBAL, M. (1997), "Paradigmatic Shifts in National Innovation Systems," in Edquist 1997a: 342–70.

HALL, P. A., and SOSKICE, D. (eds.) (2001), *Varieties of Capitalism. The Institutional Foundations of Comparative Advantage*, Oxford: Oxford University Press.

HOLLINGSWORTH, J. R., and BOYER, R., (eds.) (1997), *Contemporary Capitalism: The Embeddedness of Institutions*, Cambridge: Cambridge University Press.

INGELSTAM, L. (2002), *System—att tänka över samhälle och teknik* (Systems: To Reflect over Society and Technology—in Swedish), Energimyndighetens förlag.

JOHNSON, A. (1998), "Functions in Innovation System Approaches," Mimeo, Department of Industrial Dynamics, Chalmers University of Technology.

*——— and JACOBSSON, S. (2003), "The Emergence of a Growth Industry: A Comparative Analysis of the German, Dutch and Swedish Wind Turbine Industries," in S. Metcalfe and U. Cantner (eds.), *Transformation and Development: Schumpeterian Perspectives*, Heidelberg: Physica/Springer.

LAREDO, P., and MUSTAR, P. (eds.) (2001*a*), *Research and Innovation Policies in the New Global Economy: an International Comparative Analysis*, Cheltenham: Edward Elgar.

——— ——— (2001*b*), "General Introduction: A Focus on Research and Innovation Policies," in Laredo and Mustar 2001*a*: 1–13.

——— ——— (2001*c*), "General Conclusion: Three major Trends in Research and Innovation Policies," in Laredo and Mustar 2001a: 497–509.

*LIU, X. and WHITE, S. (2001), "Comparing Innovation Systems: A Framework and Application to China's Transitional Context," *Research Policy* 30: 1091–114.

*LUNDVALL, B.-Å. (ed.) (1992), *National Systems of Innovation: Towards a Theory of Innovation and Interactive Learning*, London: Pinter.

——— (2002), *Innovation, Growth and Cohesion: The Danish Model*, Cheltenham: Edward Elgar.

——— (2003), "National Innovation Systems: History and Theory," in *Elgar Companion to Neo-Schumpeterian Economics*, Cheltenham: Edward Elgar.

——— JOHNSON, B., ANDERSEN, E. S., and DALUM, B. (2002), "National Systems of Production, Innovation and Competence Building," *Research Policy* 31(2): 213–31.

——— and LINDGAARD CHRISTENSEN, J. (1999), "Extending and Deepening the Analysis of Innovation Systems—with Empirical Illustrations from the DISKO-project," Paper for the DRUID Conference on National Innovation Systems, Industrial Dynamics and Innovation Policy, Rebild, 9–12 June 1999.

*NELSON, R. R. (ed.) (1993), *National Systems of Innovation: A Comparative Study*, Oxford: Oxford University Press.

——— and ROSENBERG, N. (1993), "Technical Innovation and National Systems," in Nelson 1993: 3–21.

NORTH, D. C. (1990), *Institutions, Institutional Change and Economic Performance*, Cambridge: Cambridge University Press.

OECD (1996), *Lifelong Learning for All*, Paris: OECD.

——— (1998*a*), *Technology, Productivity and Job Creation: Best Policy Practices*, Paris: OECD.

——— (1998*b*), *Pathways and Participation in Vocational and Technical Education and Training*, Paris: OECD.

——— (1999*a*), *Managing National Innovation Systems*, Paris: OECD.

——— (1999*b*), *Boosting Innovation: The Cluster Approach*, Paris: OECD.

——— (2000), *Knowledge Management in the Learning Society*, Paris: OECD (CERI).

——— (2001*a*), *Innovative Clusters: Drivers of National Innovation Systems*, Paris: OECD.

——— (2001*b*), *Innovative Networks: Co-operation in National Innovation Systems*, Paris: OECD.

——— (2001*c*), *Innovative People: Mobility of Skilled Personnel in National Innovation Systems*, Paris: OECD.

——— (2001*d*), *Cities and Regions in the New Learning Economy*, written by C. Edquist, G. Rees, K. Larsen, M. Lorenzen, and S. Vincent-Lancrin, Paris: OECD (CERI).

——— (2002*a*), *Dynamising National Innovation Systems*, Paris: OECD.

——— (2002*b*), *Main Science and Technology Indicators*, vol. 2002/2, Paris: OECD.

ÖRSTAVIK, F., and NÅS, S.-O. (1998), "The Norwegian Innovation—Collaboration Survey, Oslo, the STEP Group," STEP Working Paper A-10.

PORTER, M. E. (1990), *The Competitive Advantage of Nations*, New York: Free Press.

——(1998), "Clusters and the New Economics of Competition," *Harvard Business Review* 77–90.

RICKNE, A. (2000), *New Technology-Based Firms and Industrial Dynamics: Evidence from the Technological Systems of Biomaterials in Sweden, Ohio and Massachusetts*, Department of Industrial Dynamics, Chalmers University of Technology.

SCHIBANY, A., and SCHARTINGER, D. (2001), "Interactions between Universities and Enterprises in Austria: An Empirical Analysis at the Micro and Sector Levels," in OECD 2001*b*: 235–52.

SMITH, K. (2001), "Human Resources, Mobility and the Systems Approach to Innovation," in OECD 2001*c*: ch. 1.

WHITLEY, R. (1999), *Divergent Capitalisms: The Social Structuring and Change of Business Systems*, Oxford: Oxford University Press.

——(ed.) (2002), *Competing Capitalisms: Institutions and Economics. An Elgar Reference Collection*, 2 vols., Cheltenham: Edward Elgar.

CHAPTER 8

UNIVERSITIES IN NATIONAL INNOVATION SYSTEMS

DAVID C. MOWERY

BHAVEN N. SAMPAT

8.1 INTRODUCTION

THE research university plays an important role as a source of fundamental knowledge and, occasionally, industrially relevant technology in modern knowledge-based economies. In recognition of this fact, governments throughout the industrialized world have launched numerous initiatives since the 1970s to link universities to industrial innovation more closely. Many of these initiatives seek to spur local economic development based on university research, e.g., by creating "science parks" located nearby research university campuses, support for "business incubators" and public "seed capital" funds, and the organization of other forms of "bridging institutions" that are believed to link universities to industrial innovation. Other efforts are modeled on a US law, the Bayh–Dole Act of 1980, that is widely (if perhaps incorrectly) credited with improving university–industry collaboration and technology transfer in the US national innovation system.

This chapter examines the roles of universities in industrial-economy national innovation systems, the complex institutional landscapes that influence the creation, development, and dissemination of innovations (for further discussion see Edquist, Ch. 7 in this volume). The inclusion of a chapter on university research in a volume on innovation is itself an innovation—it is likely that a similar handbook published two decades ago would have devoted far less attention to the role of universities in industrial innovation.[1] But scholarship on the role of universities in the innovation process, as opposed to their role in basic research, has grown rapidly since 1970. One important theme in this research is the reconceptualization of universities as important institutional actors in national and regional systems of innovation. Rather than "ivory towers" devoted to the pursuit of knowledge for its own sake, a growing number of industrial-economy and developing-economy governments seek to use universities as instruments for knowledge-based economic development and change.

Governments have sought to increase the rate of transfer of academic research advances to industry and to facilitate the application of these research advances by domestic firms since the 1970s as part of broader efforts to improve national economic performance. In the "knowledge-based economy," according to this view, national systems of higher education can be a strategic asset, if links with industry are strengthened and the transfer of technology enhanced and accelerated. Many if not most of these "technology-transfer" initiatives focus on the codification of property rights to individual inventions, and rarely address the broader matrix of industry–university relationships that span a broad range of activities and outputs.

Universities throughout the OECD also have been affected by tighter constraints on public funding since 1970. Growth in public funding for higher education has slowed in a number of OECD member states. In the United States, Cohen et al. (1998) note that federal research funding per full-time academic researcher declined by 9.4 per cent in real terms during 1979–91, in the face of significant upward pressure on the costs of conducting state-of-the-art research in many fields of the physical sciences and engineering. Financial support from state governments for US public universities' operating budgets (which obviously include more than research) declined from nearly 46 per cent of total revenues in 1980 to slightly more than 40 per cent in 1991 (Slaughter and Leslie 1997: Table 3.2), while the share of federal funds in US public university operating budgets declined from 12.8 to 10 per cent during the same period (the share of operating revenues derived from tuition and fees rose from 12.9 to 15 per cent). The UK government reduced its institutional funding of universities (as opposed to targeted, competitive programs for research) during the 1980s and 1990s, as did the government of Australia (Slaughter and Leslie 1997).

Faced with slower growth in overall public funding, increased competition for research funding, and continuing cost pressures within their operating budgets during the past two decades, at least some universities have become more aggressive and "entrepreneurial" in seeking new sources of funding. University presidents and vice-chancellors have promoted the regional and national economic benefits flowing from academic research and have sought closer links with industry as a means of expanding research support.

Both internal and external factors thus have led many nations' universities to promote stronger linkages with industry as a means of publicizing and/or strengthening their contributions to innovation and economic growth. In some cases, these initiatives build on long histories of collaboration between university and industry researchers that reflect unique structural features of national university systems and their industrial environment. In other cases, however, these initiatives are based on a misunderstanding of the roles played by universities in national innovation systems, as well as the factors that underpin their contributions to industrial innovation.

Although universities fulfill broadly similar functions in the innovation systems of most industrial and industrializing economies, the importance of their role varies considerably, and is influenced by the structure of domestic industry, the size and structure of other publicly funded research performers, and numerous other factors. Following a discussion of the (limited) evidence on the contrasting importance of universities within R&D performance and employment in national innovation systems, we examine other evidence on the contributions of universities to industrial innovation. Based on this discussion, we critically examine recent initiatives by governments in a number of OECD nations to enhance the contributions of universities to innovation and economic growth. We conclude with a discussion of the broad agenda for future research.

8.2 What Functions do Universities Perform within National Innovation Systems?

In varying degrees, universities throughout the OECD now combine the functions of education and research. This joint production of trained personnel and advanced research may be more effective than specialization in one or the other activity.[2] For example, the movement of trained personnel into industrial and other occupations can be a powerful mechanism for the diffusion of scientific research, and demands from students and their prospective employers for "relevance" in the curriculum can strengthen links between the academic research agenda and the needs of society.

The economically important "outputs" of university research have come in different forms, varying over time and across industries.[3] They include, among others: scientific and technological information[4] (which can increase the efficiency of applied R&D in industry by guiding research towards more fruitful departures), equipment and instrumentation[5] (used by firms in their production processes or their research), skills or human capital (embodied in students and faculty members), networks of scientific and technological capabilities (which facilitate the diffusion of new knowledge), and prototypes for new products and processes.[6]

Universities are widely cited as critical institutional actors in national innovation systems (see Nelson 1993; Edquist, Ch. 7 in this volume, and numerous other works). As Edquist notes in his chapter, the precise definition of "national innovation systems" remains somewhat hazy, but most of the large literature on the topic defines them as the institutions and actors that affect the creation, development, and diffusion of innovations. The literature on national innovation systems emphasizes the importance of strong linkages among these various institutions in improving national innovative and competitive performance, and this emphasis applies in particular to universities within national innovation systems.[7] The "national" innovation systems of the industrial economies appear more and more interdependent, reflecting rapid growth during the post-1945 period in cross-border flows of capital, goods, people, and knowledge. Yet the university systems of these economies retain strong "national" characteristics, reflecting significant contrasts among national university systems in structure, and the influence of historical evolution on contemporary structure and policy.

One influential conceptualization of the role of academic research within national innovation systems and economies was the so-called "linear model" of innovation widely associated with Vannevar Bush and his famous "blueprint" for the US post-1945 R&D system, *Science: The Endless Frontier*. Bush argued for expanded public funding for basic research within US universities as a critical contributor to economic growth, and argued that universities were the most appropriate institutional locus for basic research. This "linear model" of the innovation process asserted that funding of basic research was both necessary and sufficient to promote innovation. Bush's argument anticipated parts of the "market failure" rationale for the funding of basic academic research subsequently developed by Nelson (1959) and Arrow (1962). This portrayal of the innovation process has been widely criticized (see Kline and Rosenberg 1986, for one such rebuttal of the linear model). Many US policy makers during the 1970s and 1980s cited the Japanese economy as evidence that basic research may not be necessary or sufficient for a nation to improve its innovative performance.

Yet another view of the role of university research focuses on the contrasting "norms" of academic and industrial research. Merely contrasting the "fundamental" research activities of academics with the applied research of industrial scientists and engineers obscures as much as it illuminates—after all, there are abundant examples of university researchers making important contributions to technology development, as well as numerous cases of important basic research advances in industrial laboratories. Paul David and colleagues (Dasgupta and David 1994; David, Foray, and Steinmueller 1999) argue that the norms of academic research differ significantly from those observed within industry. For academic researchers, professional recognition and advancement depend crucially on being first to disclose and publish their result. Prompt disclosure of results and in most cases, the methods used to achieve them, therefore is central to academic research. Industrial innovation, by contrast, relies more heavily on secrecy and limitations to the disclosure of research results. The significance of these "cultural differences" for the conduct and dissemination of research may assume greater significance in the face of closer links between university and industrial researchers (see below).

But these contrasts also can be overstated, as David et al. (1999) acknowledge. The history of science is replete with examples of fierce competitions ("discovery races") between teams of researchers in a given field that systematically seek to mislead one another through the disclosure of false information. And recent research by Henderson and colleagues (Henderson, Orsenigo, and Pisano 1999; Henderson and Cockburn 1998) on pharmaceutical industry R&D highlights the increased emphasis by a number of large pharmaceutical firms on publication by industrial researchers as a means of improving their basic science capabilities. Nevertheless, the potential for clashes between the disclosure norms of academia and industry, and in particular, the potential risks posed by more restrictive disclosure norms for the educational functions and the broader pace of advance in scientific understanding, remains significant.

Still another conceptual framework that has been applied recently to descriptions of the role of academic research in "post-modern" industrial societies is the "Mode 2" concept of research identified with Michael Gibbons and colleagues (Gibbons et al. 1994). "Mode 2" research is associated with a more interdisciplinary, pluralistic, "networked" innovation system, in contrast to the previous system in which major corporate or academic research institutions were less closely linked with other institutions. Gibbons and other scholars argue that the growth of "Mode 2" research reflects the increased scale and diversity of knowledge inputs required for scientific research, a point echoed in the chapter by Pavitt in this volume. Increased diversity in inputs, in this view, is associated with greater interinstitutional collaboration and more interdisciplinary research. Because "Mode 2" involves the interaction of many more communities of researchers and other actors within any given research area, purely academic research norms may prove less influential even in such areas of fundamental research as biomedical research.

The "Mode 2" framework assuredly is consistent with some characteristics of modern innovation systems, notably the increased interinstitutional collaboration that has been remarked upon by numerous scholars. But this framework's claims that the sources of knowledge within modern innovation systems have become more diverse need not imply any decline in the role of universities as fundamental research centers. Several studies (Godin and Gingras 2000; Hicks and Hamilton 1999; see below for further discussion) support the "Mode 2" assertion that cross-institutional collaboration and diversification in knowledge sources have grown, but indicate no such decline.

Still another conceptual framework for analyzing the changing position of universities within national innovation systems is the "Triple Helix" popularized by Etzkowitz and Leytesdorff (1997). Like the "Mode 2" framework, the triple helix emphasizes the increased interaction among these institutional actors in industrial economies' innovation systems. Etzkowitz and co-authors (Etzkowitz et al. 1998) further assert that

In addition to linkages among institutional spheres, each sphere takes the role of the other. Thus, universities assume entrepreneurial tasks such as marketing knowledge and creating companies even as firms take on an academic dimension, sharing knowledge among each other and training at ever-higher skill levels. (p. 6)

The "triple helix" scholarship devotes little attention to the "transformations" in industry and government that are asserted to complement those in universities. The helix's emphasis on a more "industrial" role for universities may be valid, although it overstates the extent to which these activities are occurring throughout universities, rather than in a few fields of academic research. But the "triple helix" has yet to yield major empirical or research advances, and its value as a guide for future empirical research appears to be limited.

The "national systems," "Mode 2," and "triple helix" frameworks for conceptualizing the role of the research university within the innovation processes of knowledge-based economies emphasize the importance of strong links between universities and other institutional actors in these economies. And both "Mode 2" and the "Triple Helix" argue that interactions between universities and industry, in particular, have grown. According to the "Triple Helix" framework, increased interactions are associated with change in the internal culture and norms of universities (as noted, this framework has much less to say about the change in the characteristics of industrial and governmental research institutions). What is lacking in all of these frameworks, however, is a clear set of criteria by which to assess the strength of such linkages and a set of indicators to guide the collection of data.

8.3 The Role of Universities in National Innovation Systems: Cross-National Data

8.3.1 Comparative Data on the Structure of National Systems

The first universities appeared during the Middle Ages in Bologna and Paris, and were autonomous, self-governing institutions recognized by both church and local governmental authorities.[8] This situation persisted through much of the period prior to the eighteenth century. But the rise of the modern state was associated with the assertion by governments of greater control over public university systems in much of continental Europe, notably France and Germany, as well as Japan.[9] Such centralized control was lacking, however, in the British and especially, the US higher education systems throughout the nineteenth and twentieth centuries. Throughout the twentieth century, US universities retained great autonomy in their adminis-trative policies. Rosenberg (1999) and Ben-David (1968) argue that this lack of central control forced American universities to be more "entrepreneurial" and their research and curricula to be more responsive to changing socio-economic demands than their European counterparts. Data allowing for systematic cross-national comparisons of the structure of the higher educational systems of major industrial economies are surprisingly scarce.

This section summarizes and assesses the limited comparative data on the training and research roles of higher educational systems, as well as their relation-ships with industry. Enrollment data (summarized in Geiger 1986, and Graham and Diamond 1997) indicate that the US system enrolled a larger fraction of the 18–22-year-old population than those of any European nations throughout the 1900–1945 period. Not until the 1960s did European enrollment rates exceed 10 per cent of the relevant age cohorts, by which time US enrollment rates within this group were reaching 50 per cent (Burn et al. 1971). These contrasts in enrollment rates are reflected in enduring differences between the United States and European nations in the shares of their populations with university education. The share of the US population with university or "tertiary" educational degrees exceeded that of any other OECD economy as late as 1999.

These data also reveal that the US university degreeholder share is followed closely by that of Norway, at 25 per cent (OECD Education Database 2001). Surprisingly, Austria, with 6 per cent of the relevant population holding university or tertiary degrees, exhibits the lowest degreeholder share in this database. As Fagerberg and Godinho note in Chapter 19 of this volume, however, the large output of university degreeholders in the United States includes a significantly smaller share of natural science and engineering degreeholders than is true of such other nations as the

United Kingdom, Singapore, Finland, South Korea, and France. The share of 24-year-olds in the United States with "first degrees" from universities in natural sciences and engineering also lags well behind these and other nations.[10]

The limited data on the role of national higher education systems as R&D performers highlight other cross-national contrasts, including differences in their significance within the overall national R&D enterprise, their scale, their roles as employers of researchers, and their relationships with industry. As Figure 8.1 shows, the role of universities as R&D performers (measured in terms of the share of national R&D performed within higher education) is greatest in Italy, the Netherlands, and Canada, all of which show universities performing more than 25 per cent of total national R&D by 1998–2000 (Figure 8.1). The share of national R&D performance accounted for by US and Japanese universities, by contrast, was slightly more than 14 per cent during the same period.

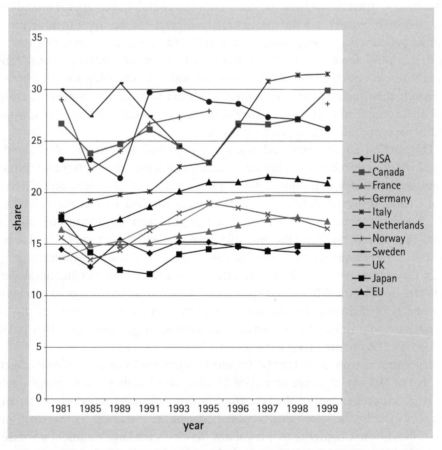

Fig. 8.1 Universities' performance share of total national R&D, 1981–99

Source: OECD, *Main Science and Technology Indicators, 2001.*

Cross-national data highlighting differences in the "division of labor" between universities and government laboratories in basic research indicate that the higher education sector's share of basic research performance is similar in most Western European economies and the United States, although higher than in most of the Eastern European and Asian countries for which data are available (OECD 2001*b*: Annex Table A.6.4.1). But a key difference between the United States and most European countries for which data are available is that a relatively low share of basic research outside the academic sector in the United States is performed by the government, and a relatively high share by industry. [11]

The data also reveal considerable variation among OECD member nations in the scale of the higher education research enterprise. Although the US higher education system is larger in absolute terms than those of other OECD member states, US universities' performance of R&D in fact accounts for a smaller share of GDP than is true of Sweden, France, Canada, the Netherlands, and Norway (Figure 8.2). Indeed, Figure 8.2 indicates that US universities' R&D as a share of GDP has in fact declined slightly during the 1989–99 period. At least a portion of this decline reflects the rapid growth in industrially funded R&D performed within US industry, especially during the 1995–9 period.

Comparison of the share of "employed researchers" in various nations' R&D systems that work in universities reveals that the United States and Japan rank very low, reflecting the fact that a much higher share of researchers in both nations

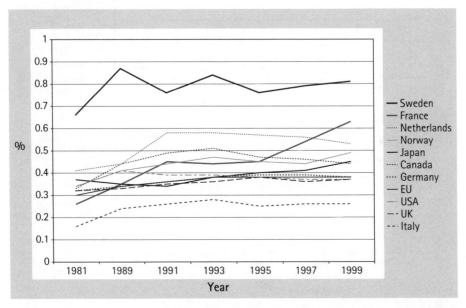

Fig. 8.2 R&D performed by the higher education sector as a percentage of GDP

Source: OECD, *Main Science and Technology Indicators, 2001.*

are employed by industry rather than higher education. In 1997, the last year for which reasonably complete data are available, 82.5 per cent of researchers were employed by industry in the United States (OECD 2001c: Table 39), significantly higher than in any other OECD nation. Korea ranks second (68.1 per cent) and Japan third (64.6 per cent), while the overall average for EU countries is much lower (48.4 per cent).

Figure 8.3 depicts the share of R&D funding within national higher education systems that is provided by industry. Despite the widely remarked closeness of US university–industry research ties and collaboration (see Rosenberg and Nelson 1994; Mowery et al. 2004), the share of R&D in higher education that is financed by industry is higher for Canada, Germany, and the United Kingdom than for the United States in the late twentieth century.

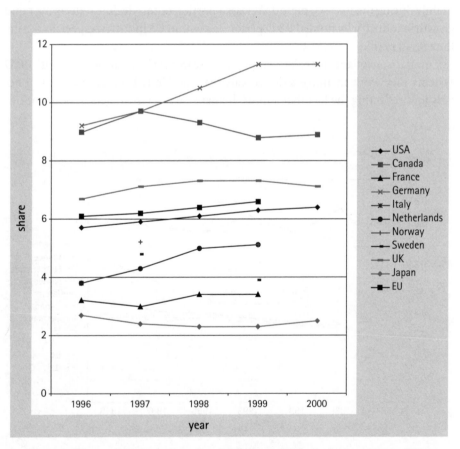

Fig. 8.3 Share of higher education R&D financed by industry, 1991–2000

Source: OECD, *Main Science and Technology Indicators, 2001.*

Other qualitative data from the OECD 2002 study of "science–industry relationships" (2002: 37) compare the labor mobility and other "network relationships" linking universities and industry for Austria, Belgium, Finland, Germany, Ireland, Italy, Sweden, the UK, the US, and Japan. "R&D consulting with firms by university researchers" is greater than the EU average (the basis for these characterizations is not provided by the OECD study) in Austria, Germany, the UK, US, and Japan; such consulting is rated as "low" in Belgium, Finland, Ireland, and Italy. The annual flow of university researchers to industrial employment, another potentially important channel for knowledge exchange, is significantly higher than the EU average in Belgium, Finland, Germany, Sweden, the UK, and the United States. Finally, the "significance of networks" linking universities and industry is rated as above the EU average for Finland, Germany, Sweden, the UK, the US, and Japan.

Surprisingly, in view of the frequency with which the United States is cited approvingly for the close links between university and industrial researchers, the evidence that university–industry relationships are "stronger" in the US than elsewhere is mixed: the qualitative data on labor mobility support this characterization, while the data on industrial support of academic research do not. An important gap in research on the role of universities in national innovation systems and a corresponding research opportunity is the development of better quantitative measures or indicators of the scope and importance of this role. If the stereotypical view of US universities as more closely linked with industrial research and innovation is indeed valid (and we believe that it is), it is striking that the available indicators shed so little light on the dimensions of these closer links.

Although universities serve similar functions in most industrial economies, these indicators suggest that their importance in training scientists and engineers and in research performance differs considerably among OECD member nations. These differences reflect cross-national differences in industry structure, especially the importance of such "high-technology" industries as electronics or information technology that are highly research-intensive and (at least since the end of the Cold War) rely heavily on private-sector sources for R&D finance. In addition, of course, the role of nonuniversity public research institutions differs among these economies, and is reflected in the contrasts in universities as performers of publicly funded R&D. These structural contrasts are the result of a lengthy, path-dependent process of historical development, in which institutional evolution interacts with industrial growth and change.

8.3.2 Recent Trends in University–Industry Linkages

Although comparative cross-sectional data reveal substantial differences in the sources of funding and other characteristics of the national systems of higher education among OECD member states, longitudinal data reveal an increase in co-authorship between university and industry researchers in many of these nations. Among other things, this evidence on increased co-authorship may indicate some growth, rather than decline, in the role of universities as centers for knowledge production within national innovation systems, the arguments of the "Mode 2" model notwithstanding. A recent paper by Calvert and Patel (2002) based on an examination of slightly more than 22,000 papers reveals a threefold increase in co-authorship between UK industry and university researchers during 1981–2000. Papers co-authored by industrial and university researchers expanded from approximately 20 to nearly 47 per cent of all UK scientific papers published by industrial researchers during the 1981–2000 period. The share of papers with UK university authors that were co-authored by industrial and university scientists also grew during this period, from 2.8 per cent in the early 1980s to 4.5 per cent in 2000.[12] Co-authored papers in computer science grew by more than eightfold, although the fields of chemistry, medicine, and biology accounted for the largest shares of co-authored papers (respectively, 20, 20, and 14 per cent).

Calvert and Patel found that the 1981/5–1986/90 period was characterized by the most rapid growth in such co-authorship. This finding is particularly interesting since the 1980s were characterized by cuts in UK central government spending on higher education, and the 1990s were a period of more aggressive governmental promotion of university–industry collaboration and technology transfer. In other words, the growth in co-authorship measured by these scholars appears to have occurred without any specific encouragement (beyond funding cuts) from government policy. The UK universities responsible for the majority of the co-authored papers were among the most distinguished research universities in Great Britain.

Another study of co-authorship between university and industry researchers is that by Hicks et al. (1995), which compares trends during the 1980–9 period in co-authorship in Japan and Western Europe. Overall co-authorship rates (covering all industrial sectors and including both domestic and foreign universities) were similar (roughly 20 per cent for European papers and slightly less for Japanese papers) for Western Europe and Japan in 1980. By 1989, however, co-authorship rates for Western Europe had risen to nearly 40 per cent of published papers, while Japanese co-authorship rates only slightly exceeded 20 per cent.

There is surprisingly little empirical work on co-authorship in the United States. A study by Hicks and Hamilton (1999) reports that between 1981 and 1994, the number of US papers co-authored by university and industry researchers more than doubled, considerably exceeding the 38 per cent increase in the total number of scientific papers published by US researchers during this period. The authors also suggest that these co-authored papers are less "basic" than academic articles without industrial co-authors.

Overall, these bibliometric studies present a rich descriptive and a relatively weak explanatory analysis of an important type of university–industry collaboration, inasmuch as they provide little explanation for trends or cross-national differences. Nonetheless, these data highlight a broad trend of growth in such co-authorship, and this area remains a very fruitful one for future research that spans more fields, nations, and types of publications. The results of the bibliometric work in this area provide some support for the "Mode 2" and "Triple Helix" frameworks' arguments that research collaboration between universities and industry is growing throughout the industrial economies, in university systems with very different structures (see Chapter 7 by Edquist in this volume, as well as the studies in Laredo and Mustar 2001).

8.4 How does University Research Affect Industrial Innovation? A Summary of Some US Studies

The quantitative indicators discussed in the previous section provide some information on the structure of universities within the OECD and their links with national innovation systems. But these data shed very little light on the characteristics of the knowledge flows between university research and the industrial innovation process. This issue is especially important in light of the numerous government policy initiatives that seek to enhance or exploit such knowledge flows (see below). Although their coverage is limited to US universities and industry, a number of recent studies based on interviews or surveys of senior industrial managers in industries ranging from pharmaceuticals to electrical equipment have examined the influence of university research on industrial innovation, and thereby provide additional insight into the role of universities within the US national innovation system.

All of these studies (GUIRR 1991; Mansfield 1991; Levin et al. 1987; Cohen, Nelson, and Walsh 2002) emphasize the significance of interindustry differences in the relationship between university and industrial innovation. The biomedical sector, especially biotechnology and pharmaceuticals, is unusual, in that university research advances affect industrial innovation more significantly and directly in this field than in other sectors. In these other technological and industrial fields, universities occasionally contributed relevant "inventions," but most commercially significant inventions came from nonacademic research. The incremental advances that were the primary focus of the R&D activities of firms in these sectors were almost exclusively the domain of industrial research, design, problem-solving, and development. University research contributed to technological advances by enhancing knowledge of the fundamental physics and chemistry underlying manufacturing processes and product innovation, an area in which training of scientists and engineers figured prominently, and experimental techniques.

The studies by Levin et al. (1987) and Cohen et al. (2002) summarize industrial R&D managers' views on the relevance to industrial innovation of various fields of university research (Table 8.1 summarizes the results discussed in Levin et al. 1987). Virtually all of the fields of university research that were rated as "important" or "very important" for their innovative activities by survey respondents in both studies were related to engineering or applied sciences. As we noted above, these fields of US university research frequently developed in close collaboration with industry. Interestingly, with the exception of chemistry, very few basic sciences appear on the list of university research fields deemed by industry respondents to be relevant to their innovative activities.

The absence of fields such as physics and mathematics in Table 8.1, however, should not interpreted as indicating that academic research in these fields does not contribute directly to technical advance in industry. Instead, these results reflect the fact that the effects on industrial innovation of basic research findings in such areas as physics, mathematics, and the physical sciences are realized only after a considerable lag. Moreover, application of academic research results may require that these advances be incorporated into the applied sciences, such as chemical engineering, electrical engineering and material sciences. The survey results summarized in Cohen et al. (2002) indicate that in most industries, university research results play little if any role in triggering new industrial R&D projects; instead, the stimuli originate with customers or from manufacturing operations. Here as elsewhere, pharmaceuticals is an exception, since university research results in this field often trigger industrial R&D projects.

Cohen et al. (2002) further report that the results of "public research" performed in government labs or universities were used more frequently by US industrial firms (on average, in 29.3 per cent of industrial R&D projects) than prototypes emerging from these external sources of research (used in an average of 8.3 per cent of industrial R&D projects). A similar portrait of the relative importance of different outputs of university and public laboratory research emerges from the responses to questions about the importance to industrial R&D of various information channels (Table 8.2). Although pharmaceuticals once again is unusual in its assignment of considerable importance to patents and license agreements involving universities and public laboratories, respondents from this industry still rated research publications and conferences as a more important source of information. For most industries, patents and licenses involving inventions from university or public laboratories were reported to be of very little importance, compared with publications, conferences, informal interaction with university researchers, and consulting.

Data on the use by industrial R&D managers of academic research results are needed for other industrial economies. Nonetheless, the results of these US studies

Table 8.1 The relevance of university science to industrial technology

Science	Number of Industries with "relevance" scores		Selected industries for which the reported "relevance" of university research was large
	≥ 5	≥ 6	(≥ 6)
Biology	12	3	Animal feed, drugs, processed fruits/vegetables
Chemistry	19	3	Animal feed, meat products, drugs
Geology	0	0	None
Mathematics	5	1	Optical instruments
Physics	4	2	Optical instruments, electronics
Agricultural science	17	7	Pesticides, animal feed, fertilizers, food products
Applied math/operations research	16	2	Meat products, logging/sawmills
Computer science	34	10	Optical instruments, logging/sawmills, paper machinery
Materials science	29	8	Synthetic rubber, nonferrous metals
Medical science	7	3	Surgical/medical instruments, drugs, coffee
Metallurgy	21	6	Nonferrous metals, fabricated metal products
Chemical engineering	19	6	Canned foods, fertilizers, malt beverages
Electrical engineering	22	2	Semiconductors, scientific instruments
Mechanical engineering	28	9	Hand tools, specialized industrial machinery

Source: Data from the Yale Survey on Appropriability and Technological Opportunity in Industry. For a description of the survey, see Levin et al. (1987).

consistently emphasize that the relationship between academic research and industrial innovation in the biomedical field differs from that in other knowledge-intensive sectors. In addition, these studies suggest that academic research rarely

Table 8.2 Importance to industrial R&D of sources of information on public R&D (including university research)

Information source	Rating it as "very important" for industrial R&D (%)
Publications & reports	41.2
Informal Interaction	35.6
Meetings & conferences	35.1
Consulting	31.8
Contract research	20.9
Recent hires	19.6
Cooperative R&D projects	17.9
Patents	17.5
Licenses	9.5
Personnel exchange	5.8

Source: Cohen et al. (2002).

produces "prototypes" of inventions for development and commercialization by industry—instead, academic research informs the methods and disciplines employed by firms in their R&D facilities. Finally, the channels rated by industrial R&D managers as most important in this complex interaction between academic and industrial innovation rarely include patents and licenses. Perhaps the most striking aspect of these survey and interview results is the fact that they have not informed the design of recent policy initiatives to enhance the contributions of university research to industrial innovation.

8.5 FROM "SCIENCE PUSH" TO "TECHNOLOGY COMMERCIALIZATION"

As we suggested in Section 8.1, since 1980 a number of industrialized countries have implemented or considered policies to strengthen "linkages" between universities (and public research organizations) and industry, in order to enhance the contributions of university-based research to innovation and economic performance.

These initiatives all share the premise that universities support innovation in industry primarily through the production by universities of "deliverables" for commercialization (e.g., patented discoveries), despite the modest support for this premise in the research discussed above. We illustrate these points in this section with case studies of two types of policies: (1) policies encouraging the formation of regional economic "clusters" and spin-offs based on university research, and (2) policies attempting to stimulate university patenting and licensing activities.

The global diffusion of these "technology commercialization" policies illustrates a phenomenon that has received too little attention in the literature on innovation policy—the efforts by policy makers to "borrow" policy instruments from other economies and apply these instruments in a very different institutional context. As Lundvall and Borrás point out in their chapter, history, path dependence, and institutional "embeddedness" all make this type of "emulation" very difficult. Nonetheless, such emulation has been especially widespread in the field of technology policy. International policy emulation of this sort is characterized by two key features: (1) the "learning" that underpins the emulation is highly selective; and (2) the implementation of program designs based on even this selective learning is affected by the different institutional landscape of the emulator.

8.5.1 Universities and Regional Economic Development

In many OECD countries, efforts to increase the national economic returns from public investments in university research have attempted to stimulate the creation of "regional clusters" of innovative firms around universities. These undertakings seek to stimulate regional economic development and agglomeration via facilitating the creation of "spin-off" firms to commercialize university technologies (OECD 2002).[13]

These policy initiatives are motivated by the high-technology regional clusters in the United States, notably Silicon Valley in California and Route 128 in the Boston area. Both of these high-technology clusters have a spawned a large number of new firms and have major research universities in their midst (in California, the University of California at Berkeley, Stanford University, and the University of California at San Francisco; in Boston, Harvard University and MIT). At least some of the successful new firms in these regions have been involved in commercializing technologies developed at regional universities.

Other evidence (notably, Trajtenberg, Jaffe, and Henderson 1997) suggests that the "knowledge spillovers" from university research within the United States, measured by the location of inventors citing university patents, tend to be localized at the regional level. Recent work by Hicks et al. (2001) similarly indicates that patents filed by US inventors disproportionately cite scientific papers from research institutions located in the same state as these inventors.

But little evidence supports the argument that the presence of universities some-how "causes" the development of regional high-technology agglomerations. And even less evidence supports the argument that the regional or innovation policies of governments are effective in creating these agglomerations. One can point to high-technology clusters with highly productive research universities in a number of areas in the United States and other industrial economies; but there are also a number of research universities that have not spawned such agglomerations. Moreover, efforts to replicate the "Silicon Valley model" in other economies have proven difficult and the results of these efforts have been mixed (a fascinating historical account of the efforts by Frederick Terman of Stanford University to promote such "exports" may be found in Leslie and Kargon 1996).

National and local governments in many OECD countries have attempted to stimulate the formation of these clusters via funding for "science parks" (occasion-ally also called incubators, technology centers, or centers of excellence). Interest-ingly, there is considerable disagreement about exactly what a "science park" is and what they do; the International Association of Science Parks characterizes them as follows:

A Science Park is an organisation managed by specialised professionals, whose main aim is to increase the wealth of its community by promoting the culture of innovation and the competitiveness of its associated businesses and knowledge-based institutions . . . To enable these goals to be met, a Science Park stimulates and manages the flow of knowledge and technology amongst universities, R&D institutions, companies and markets; it facilitates the creation and growth of innovation-based companies through incubation and spin-off processes; and provides other value-added services together with high quality space and facilities. (http://www.iaspworld.org/information/definitions.php)

Despite the widespread interest in science parks, there is little evidence that they positively affect universities' contributions to innovation or spur regional economic development. Using data on US science parks, Felsenstein (1994) finds no evidence that firms located on university-based science parks are more innovative than other local firms, and Wallsten (2001) finds that science parks have little effect on regional economic development and rates of innovation.

The research on "science parks" in other industrial economies is also limited. One examination of "science parks" in the UK (Massey et al. 1992) is dated, but presents interesting evidence on the characteristics of nearly 200 firms in twenty UK science parks. The study found that startup firms represented 25–30 per cent of the tenants in the science parks surveyed; in the absence of some kind of "control population," it is difficult to reach conclusions about whether startup firms are over-represented or under-represented in these UK science parks. Perhaps more surprising was the study's finding that

formal research links between academic institutions and establishments on science parks were no more evident than similar links with firms located off-park . . . Formal research links such as "employment of academics," "sponsoring trials or research," "testing and analysis," "student project" work and "graduate employment" were fairly similar for park firms and off-park firms. However, significantly more park firms than off-park firms mentioned "informal contacts with academics" and the use of academic facilities such as computers, libraries or dining facilities as being important.   (Massey et al. 1992: 38)

This and other evidence on the results of government policies to promote university-based regional agglomerations suggests that such policies have a mixed record of success. And even successful regional agglomerations may require considerable time to emerge. Recent work by Sturgeon (2000) argues that Silicon Valley's history as a center for new-firm formation and innovation dates back to the early decades of the twentieth century, suggesting that much of the region's innovative "culture" developed over a much longer period of time and predates the ascent to global research eminence of Stanford University. Similarly, the North Carolina "Research Triangle," which was promoted much more aggressively by the state government, was established in the late 1950s and became a center for new-firm formation and innovation only in the late 1980s.

Still other work on the development of Silicon Valley by Leslie (1993, 2000) and Saxenian (1988) emphasizes the massive increase in federal defense spending after 1945 as a catalyst for the formation of new high-technology firms in the region. In this view, the presence of leading research universities may have been necessary, but was by no means sufficient, to create Silicon Valley during the 1950s and 1960s. Saxenian in particular emphasizes the very different structure of British defense procurement policies in explaining the lack of similar dynamism in the Cambridge region.

The links between university research and the emergence of regional high-technology agglomerations thus are more complex than is implied by the correlation between the presence of high-technology firms and research universities in a number of locales. The US experience suggests that the emergence of such agglomerations is a matter of contingency, path-dependence, and (most importantly) the presence of other supporting policies (intentional or otherwise) that may have little to do with university research or the encouragement of university–industry linkages.

The policy initiatives in the United States and other OECD economies that seek to use university research and "science parks" to stimulate regional economic development suffer from a deficiency that is common to many of the other recent efforts to stimulate university–industry linkages in OECD countries, i.e. a lack of attention to supporting institutions, a focus on "success stories" with little attention to systematic evidence on the casual effects of the policies, and a narrow focus on commercialization of university technologies, rather than other more economically important outputs of university research. These characteristics are also seen in recent efforts elsewhere within the OECD to emulate the Bayh–Dole Act.

8.5.2 Patenting the Results of Publicly Funded
Academic Research

As we noted above, this increased interest by governments in "Bayh–Dole type" policies is rooted in motives similar to those underpinning policy initiatives that seek to create "high-technology" regional clusters. But the "emulation" of Bayh–Dole in other industrial economies overlooks the importance and effects on university–industry collaboration and technology transfer of the many other institutions that support these interactions and the commercialization of university technologies in the United States. In addition, these "emulation" initiatives are based on a misreading of the empirical evidence on the importance of intellectual property rights in facilitating the "transfer" and commercialization of university inventions, as well as a misreading of the evidence on the effects of the Bayh–Dole Act.

8.5.2.1 *Origins of the Bayh–Dole Act*

Although some US universities were patenting faculty inventions as early as the 1920s, few institutions had developed formal patent policies prior to the late 1940s, and many of these policies embodied considerable ambivalence toward patenting. Public universities were more heavily represented in patenting than private universities during the 1925–45 period, both within the top research universities and more generally.

These characteristics of university patenting began to change after 1970, as private universities expanded their share of US university patenting, universities generally expanded their direct role in managing patenting and licensing, and the share of biomedical patents within overall university patenting increased. Lobbying by US research universities active in patenting was one of several factors behind the passage of the Bayh–Dole Act in 1980.

The Bayh–Dole Patent and Trademark Amendments Act of 1980 provided blanket permission for performers of federally funded research to file for patents on the results of such research and to grant licenses for these patents, including exclusive licenses, to other parties. The Act facilitated university patenting and licensing in at least two ways. First, it replaced a web of Institutional Patent Agreements (IPAs) that had been negotiated between individual universities and federal agencies with a uniform policy. Second, the Act's provisions expressed Congressional support for the negotiation of exclusive licenses between universities and industrial firms for the results of federally funded research.

The passage of the Bayh–Dole Act was one part of a broader shift in US policy toward stronger intellectual property rights.[14] Among the most important of these policy initiatives was the establishment of the Court of Appeals for the Federal Circuit (CAFC) in 1982. Established to serve as the court of final appeal for patent cases throughout the federal judiciary, the CAFC soon emerged as a strong champion of patentholder rights. But even before the establishment of the CAFC, the 1980 US Supreme Court decision in Diamond v. Chakrabarty upheld the validity of a broad patent in the new industry of biotechnology, facilitating the patenting and licensing of inventions in this sector.

Rather than emphasizing public funding and relatively liberal disclosure and dissemination, the Bayh–Dole Act assumes that restrictions on dissemination of the results of many R&D projects will enhance economic efficiency by supporting their commercialization. In many respects, the Bayh–Dole Act is the ultimate expression of faith in the "linear model" of innovation—if basic research results can be purchased by would-be developers, commercial innovation will be accelerated.

8.5.2.2 *The Effects of Bayh–Dole*

How did the Bayh–Dole Act affect technology transfer by US universities? Figure 8.4 depicts US research university patenting as a share of domestically assigned US patents during 1963–99, in order to remove the effects of increased patenting in the United States by foreign firms and inventors during the late twentieth century. Universities increased their share of patenting from less than 0.3 per cent in 1963 to nearly 4 per cent by 1999, but the rate of growth in this share begins to accelerate before rather than after 1980. The growth rate of the ratio of research university patents to academic research spending remains surprisingly constant through the 1963–93 period, suggesting no structural break in trends in universities' "patent propensity" after passage of the Bayh–Dole Act in 1980.

Figure 8.5 displays trends during 1960–99 in the distribution among technology classes of US research university patents, highlighting the growing importance of biomedical patents in the patenting activities of the leading US universities during the period. Non-biomedical university patents increased by 90 per cent from the 1968–70 period to the 1978–80 period, but biomedical university patents increased by 295 per cent. The increased share of the biomedical disciplines within overall federal academic R&D funding, the dramatic advances in biomedical science that occurred during the 1960s and 1970s, and the strong industrial interest in the results of this biomedical research during this period all contributed to this shift in the composition of university patent portfolios.

During the late 1990s and early twenty-first century, many commentators and policy makers portrayed the Bayh–Dole Act as a critical catalyst to growth in US universities' innovative and economic contributions. Indeed, the OECD went so far as to argue that the Bayh–Dole Act was an important factor in the remarkable growth of incomes, employment, and productivity in the US economy of the late 1990s.[15] Remarkably, virtually none of these characterizations of the positive effects of the Bayh–Dole Act cite any evidence in support of their claims beyond the clear growth in patenting and licensing by universities. Nor does evidence of increased patenting

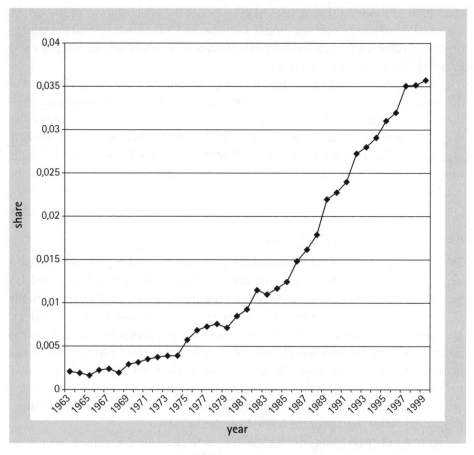

Fig. 8.4 US research university patents as a percentage of all domestic–assignee US patents, 1963–99

and licensing by universities by itself indicate that university research discoveries are being transferred to industry more efficiently or commercialized more rapidly, as Colyvas et al. (2002) and Mowery et al. (2001) point out.

These "assessments" of the effects of the Bayh–Dole Act also fail to consider any potentially negative effects of the Act on US university research or innovation in the broader economy. Some scholars have suggested that the "commercialization motives" created by Bayh–Dole could shift the orientation of university research away from "basic" and towards "applied" research (Henderson et al. 1998), but thus far there is little evidence of substantial shifts since Bayh–Dole in the content of academic research.

A second potentially negative effect of increased university patenting and licensing is the potential weakening of academic researchers' commitments to "open science," leading to publication delays, secrecy, and withholding of data and

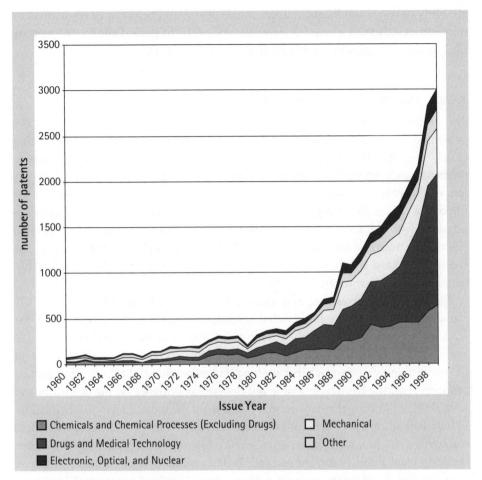

Fig. 8.5 Technology field of US "research university" patents, 1960–99

materials (Dasgupta and David 1994; Liebeskind 2001). In view of the importance assigned by industrial researchers to the "nonpatent/licensing" channels of inter-action with universities in most industrial sectors, it is crucially important that these channels not be constricted or impeded by the intensive focus on patenting and licensing in many universities. The effects of any increased assertion by institutional and individual inventors of property rights over inputs to scientific research have only begun to receive serious scholarly attention. Patenting and restrictive licensing of inputs into future research ("research tools") could hinder downstream research and product development (Heller and Eisenberg 1998; Merges and Nelson 1994).

Although there is little evidence as yet that the Bayh–Dole Act has had significant, negative consequences for academic research, technology transfer, and industrial innovation in the United States, the data available to monitor any such effects are very limited. Moreover, such data are necessarily retrospective, and in their nature are likely to reveal significant changes in the norms and behavior of researchers or universities only with a long lag. Any negative effects of Bayh–Dole accordingly are likely to reveal themselves only well after they first appear.

8.5.2.3 *International "Emulation" of the Bayh–Dole Act*

The limited evidence on the Act's effects (both positive and negative) has not prevented a number of other OECD governments from pursuing policies that closely resemble the Bayh–Dole Act. Like the Bayh–Dole Act, these initiatives focus narrowly on the "deliverable" outputs of university research, and typically ignore the effects of patenting and licensing on the other, more economically important, channels through which universities contribute to innovation and economic growth. Moreover, such emulation is based on a misreading of the limited evidence concerning the effects of Bayh–Dole and on a misunderstanding of the factors that have encouraged the long-standing and relatively close relationship between US universities and industrial innovation.

The policy initiatives that have been debated or implemented in most OECD economies have sought to shift ownership of the intellectual property rights (IPR) for academic inventions to either the academic institution or the researcher (see OECD 2002, for an excellent summary). In some university systems, such as those of Germany or Sweden, researchers have long had ownership rights for the intellectual property resulting from their work, and debate has centered on the feasibility and advisability of shifting these ownership rights from the individual to the institution. In Italy, legislation adopted in 2001 shifted ownership from universities to individual researchers. In Japanese universities, ownership of intellectual property rights resulting from publicly funded research is determined by a committee, which on occasion awards title to the researcher. No single national policy governs IPR ownership within the British or Canadian university systems, although efforts are underway in both nations to grant ownership to the academic institution rather than the individual researcher or the funding agency. In addition, the Swedish, German, and Japanese governments have encouraged the formation of external "technology licensing organizations," which may or may not be affiliated with a given university.

These policy proposals and initiatives display the classic signs of international emulation described above—selective "borrowing" from another nation's policies for implementation in an institutional context that differs significantly from that of the nation being emulated. Inasmuch as patenting and licensing are rated by industrial R&D managers as relatively unimportant for technology transfer in most fields, emulation of the Bayh–Dole Act is insufficient and perhaps even unnecessary to stimulate higher levels of university–industry interaction and technology transfer. Instead, reforms to enhance interinstitutional competition and autonomy within national university systems, as well as support for the external institutional contributors to new-firm formation and technology commercialization, appear to be more important. Indeed, emulation of Bayh–Dole could be counterproductive in other industrial economies, precisely because of the importance of other channels for technology transfer and exploitation by industry.

8.6 Conclusion

Universities play important roles in the "knowledge-based" economies of modern industrial and industrializing states as sources of trained "knowledge workers" and ideas flowing from both basic and more applied research activities. But conventional (and, perhaps, evolutionary) economic approaches to the analysis of institutions are very difficult to apply to universities, for several reasons. First, with the exceptions of the US and British university systems, inter-university "competition" has been limited in most national systems of higher education. Inter-university competition was a very important historical influence on the evolution of US universities and their links with industry; but this aspect of the "selection environment" is lacking in most other national systems of higher education.

Second, analyzing universities as economic institutions requires some definition of the objectives pursued by individual universities. Partly because universities perform multiple roles in many national systems and partly because the internal structure of most research universities more closely resembles that of a cooperative organization rather than the hierarchical structure associated with industrial firms, characterizing "the objectives of the university" is difficult if not meaningless. The modern university has its roots in the Middle Ages, rather than the Industrial Revolution, and its medieval origins continue to influence its organization and operation. If universities are to be conceptualized as economic institutions for purposes of analyzing their evolution, the current analytic frameworks available in neo-classical or evolutionary economics are insufficient.

The development of such an analytic framework is important, not least for understanding the consequences for academic research of government policies that seek to accelerate the transfer of research results to industrial firms. The intensified demands from governments to raise the (measurable) economic returns to their substantial investments in academic research and education makes the development of better tools for understanding and measuring the operations and outputs of universities all the more important. As we argued above, many of the current initiatives in the United States and other industrial economies to enhance the economic returns from university research are based on a poor understanding of the full spectrum of roles fulfilled by research universities in industrial economies, as well as a tendency to emphasize the outputs of university research that can be easily quantified.

Although the analytic frameworks provided by the "national innovation systems," "Mode 2," and "Triple Helix" models of scientific research and innovation shed some light on the roles of universities and largely agree in their assessment of these roles, these frameworks provide limited guidance for policy or evaluation. Moreover, these frameworks tend to downplay the very real tensions among the different roles of research universities within knowledge-based economies. Such tensions are likely to intensify in the face of pressure from policy makers and others on universities to accelerate their production and transfer to commercial interests of tangible, measurable research outputs.

The development of useful theoretical or conceptual tools or models for analyzing universities as economic or other institutions within knowledge-based economies is seriously hampered by the lack of data on the roles of universities that enable comparisons across time or across national innovation systems. Indicators that enable longitudinal analysis of the roles of universities in training scientists and engineers, contributing to "public knowledge," or transferring inventions to industrial firms are scarce if not entirely lacking for most national systems of higher education. Few of these indicators incorporate information on the geographic dimensions of university–industry interactions, despite the importance of agglomeration economies in the current policy approaches of many governments in this area. Moreover, such indicators as do exist rarely are comparable across national systems of higher education.

The absence of broader longitudinal and cross-nationally comparable indicators of university–industry interaction thus impedes both the formulation and the evaluation of policies. And the lack of better indicators reflects the lack of a stronger analytic framework for understanding the roles of universities within national innovation systems. Such a framework must adopt a more evolutionary, historically grounded approach to the understanding of the roles of universities, especially the influence of the structure of national higher education systems on these roles. As we have argued in this chapter, many of the efforts by OECD governments to encourage technology transfer and to increase the economic payoffs to investments in university research are hampered by a lack of such understanding. More comparative institutional work on the evolution and roles of research universities, including the contrasting "division of labor" among universities and other publicly supported research institutions in both industrial and industrializing economies, is an indispensable starting point for analysis of the current and likely future position of the research university within national innovation systems.

The development of better indicators of the full array of channels through which industries and universities interact within knowledge-based economies represents another important research opportunity. In addition, more information is needed on measures of firm-level "absorptive capacity" and investments in its creation and maintenance—how do existing firms develop these various channels of interaction? How and why are new firms formed to exploit university research advances, and how does this "spinoff" process vary across time, geographic space, and national innovation systems? The extensive discussion of all of these important economic phenomena still lacks a strong evidentiary basis for making comparisons among the higher education systems of the industrialized economies. The current emphasis on the countable rather than the important aspects of university–industry interactions could have unfortunate consequences for innovation policy in the industrial and industrializing world.

Notes

1. Godin and Gingras (2000: 273) note that "After having been left out of major government policies centered on industrial innovation, universities seem, over the past 5 years, to have become the object of a renewed interest among students of the system of knowledge production."
2. Mowery and Rosenberg (1989: 154) note that the conduct of scientific research and education within many research universities "exploits a great complementarity between research and teaching. Under the appropriate set of circumstances, each may be performed better when they are done together."
3. This list draws from Rosenberg (1999), Cohen et al. (1998), and other sources.
4. David, Mowery, and Steinmueller (1992) and Nelson (1982) discuss the economic importance of the "informational" outputs of university research.
5. See Rosenberg's (1994) discussion of universities as a source of innovation in scientific instruments.
6. See Rosenberg (1999).
7. Thus, Nelson's concluding chapter in his 1993 collection of studies of national innovation systems argues that "One important feature distinguishing countries that were sustaining competitive and innovative firms was education and training systems that provide these firms with a flow of people with the requisite knowledge and skills. For industries in which university-trained engineers and scientists were needed, this does not simply mean that the universities provide training in these fields, but also that they consciously train their students with an eye to industry needs" (1993: 511).
8. According to Ben-David (1971: 48), "To create order among the turbulent crowds of scholars and to regulate their relationships with the environing society, corporations were established. Students and scholars were formed into corporations authorized by the church and recognized by the secular ruler. The relationships of their corporation with that of the townspeople, with the local ecclesiastical officials, and with the king were carefully laid down and safeguarded by solemn oaths. . . . The important result of this corporate device—which was not entirely unique to Europe but which attained a much greater importance there than elsewhere—was that advanced studies ceased to be conducted in isolated circles of masters and students. Masters and/or students came to form a collective body. The European student of the thirteenth century no longer went to study with a particular master but at a particular university."

9. The Japanese higher education system has a large number of private universities, although the bulk of these are devoted primarily to undergraduate education.

10. US "science and engineering" degreeholders also account for a smaller share of all advanced degrees awarded in the United States in 1999 than is true of France, Taiwan, and the United Kingdom, although this share in the United States exceeded those for Finland and South Korea (National Science Foundation 2002).

11. These data must be interpreted with caution, since the definitions used by the national statistical agencies whence they are drawn often differ. For example, in France CNRS is classified as part of the Higher Education Sector, whereas in Italy similar organizations are treated as part of the government sector. See OECD 2001*b*: Annex 2.

12. By comparison, the share of US university publications co-authored by industrial and university researchers grew from 4.9 per cent in 1989 to 7.3 per cent in 1999, while this share in Canadian university publications grew from 1.4 per cent in 1980 to 3.5 per cent in 1998.

13. A recent OECD report notes that "Spinning off is the entrepreneurial route to commercializing knowledge developed by public research and as such is attracting a great deal of attention, given the 'start-up' fever in many countries" and that governments "have a special interest in this specific type of industry–science linkage because it may be one of the factors that explain differences in performance in newfast-growing science based industries" (OECD 2002: 41).

14. According to Katz and Ordover (1990), at least fourteen Congressional bills passed during the 1980s focused on strengthening domestic and international protection for intellectual property rights, and the Court of Appeals for the Federal Circuit created in 1982 has upheld patent rights in roughly 80 per cent of the cases argued before it, a considerable increase from the pre-1982 rate of 30 per cent for the Federal bench.

15. "Regulatory reform in the United States in the early 1980s, such as the Bayh–Dole Act, have [sic] significantly increased the contribution of scientific institutions to innovation. There is evidence that this is one of the factors contributing to the pick-up of US growth performance" (OECD 2000: 77).

References

Arrow, K. (1962), "Economic Welfare and the Allocation of Resources for Invention," in R. R. Nelson (ed.), *The Rate and Direction of Inventive Activity*, Princeton: Princeton University Press.

*Ben-David, J. (1968), *Fundamental Research and the Universities*, Paris: OECD.

—— (1971), *The Scientist's Role in Society*, New York: Prentice-Hall.

Burn, B. B, Altbach, P. G., Kerr, C., and Perkins, J. A. (1971), *Higher Education in Nine Countries*, New York: McGraw-Hill.

Bush, V. (1945), *Science: The Endless Frontier*, Washington, DC: US Government Printing Office.

* Asterisked items are suggestions for further reading.

CALVERT, J., and PATEL, P. (2002), "University–Industry Research Collaborations in the UK," unpublished working paper, Science Policy Research Unit, University of Sussex, Brighton, UK.

COHEN, W., FLORIDA, R., RANDAZZESE, L., and WALSH, J. (1998), "Industry and the Academy: Uneasy Partners in the Cause of Technological Advance," in R. Noll (ed.), *Challenges to the Research University*, Washington, DC: Brookings Institution.

*COHEN, W. M., NELSON, R. R., and WALSH, J. P. (2002), "Links and Impacts: The Influence of Public Research on Industrial R&D," *Management Science* 48(1): 1–23.

*COLYVAS, J., CROW, M., GELIJNS, A., MAZZOLENI, R., NELSON, R. R., ROSENBERG, N., and SAMPAT, B. N. (2002), "How Do University Inventions Get into Practice?" *Management Science* 48(1): 61–72.

DASGUPTA, P., and DAVID, P. (1994), "Towards a New Economics of Science," *Research Policy* 23(5): 487–521.

DAVID, P., FORAY, D., and STEINMUELLER, W. E. (1999), "The Research Network and the New Economics of Science: From Metaphors to Organizational Behaviours," in F. Malerba and A. Gambardella (eds.), *The Organization of Economic Innovation in Europe*, Cambridge: Cambridge University Press.

—— MOWERY, D. C., and STEINMUELLER, W. E. (1992), "Analyzing the Economic Payoffs from Basic Research," *Economics of Innovation and New Technology*: 73–90.

ETZKOWITZ, H., and LEYTESDORFF, L. (1997), *Universities in the Global Economy: A Triple Helix of Academic–Industry–Government Relation*. London: Croom Helm.

—— WEBSTER, A., and HEALEY, P. (1998), "Introduction," in H. Etzkowitz, A. Webster, and P. Healey (eds.), *Capitalizing Knowledge*, Albany: State University of New York Press.

FELSENSTEIN, D. (1994), "University-Related Science Parks—'Seedbeds' or 'Enclaves' of Innovation?" *Technovation* 14: 93–100.

GEIGER, R. (1986), *To Advance Knowledge: The Growth of American Research Universities, 1900–1940*, New York: Oxford University Press.

GIBBONS, M. et al. (1994), *The New Production of Knowledge*, London: Sage.

GODIN, B., and GINGRAS, Y. (2000), "The Place of Universities in the System of Knowledge Production," *Research Policy* 29: 273–8.

GOVERNMENT UNIVERSITY INDUSTRY RESEARCH ROUNDTABLE (GUIRR) (1991), *Industrial Perspectives on Innovation and Interactions with Universities*, Washington, DC: National Academy Press.

GRAHAM, H. D., and DIAMOND, N. (1997), *The Rise of American Research Universities*. Baltimore: Johns Hopkins University Press.

*HELLER, M. A., and EISENBERG, R. S. (1998), "Can Patents Deter Innovation? The Anti-commons in Biomedical Research," *Science* 280: 298.

HENDERSON, R., and COCKBURN, I. (1998), "Absorptive Capacity, Coauthoring Behavior, and the Organization of Research in Drug Discovery," *Journal of Industrial Economics* 46(2): 157–82.

—— JAFFE, A. B., and TRAJTENBERG, M. (1998), "Universities as a Source of Commercial Technology: A Detailed Analysis of University Patenting, 1965–88," *Review of Economics & Statistics* 80: 119–27.

*—— ORSENIGO, L., and PISANO, G. (1999), "The Pharmaceutical Industry and the Revolution in Molecular Biology: Interactions among Scientific, Institutional and Organizational Change," in D. C. Mowery and R. R. Nelson (eds.), *Sources of Industrial Leadership*, New York: Cambridge University Press, 267–311.

HICKS, D., BREITZMAN, T., OLIVASTRO, D., and HAMILTON, K. (2001), "The Changing Composition of Innovative Activity in the US: A Portrait Based on Patent Analysis," *Research Policy* 30: 681–703.

—— and HAMILTON, K. (1999), "Does University–Industry Collaboration Adversely Affect University Research?" *Issues in Science and Technology Online*, http://www.nap.edu/issues/15.4/realnumbers.htm

—— ISARD, P. A., and MARTIN, B. R. (1995), "A Morphology of Japanese and European Corporate Research Networks," *Research Policy* 25: 359–78.

KATZ, M. L., and ORDOVER, J. A. (1990), "R&D Competition and Cooperation," *Brookings Papers on Economic Activity: Microeconomics*, 137–92.

*KLINE, S., and ROSENBERG, N. (1986), "An Overview of Innovation," in R. Landau and N. Rosenberg (eds.), *The Positive Sum Strategy : Harnessing Technology for Economic Growth*, Washington, DC: National Academy Press, xiv, 640.

LAREDO, P., and MUSTAR, P. (2001), *Research and Innovation Policies in the New Global Economy: An International Comparison*, Cheltenham: Elgar.

LESLIE, S. W. (1993), *The Cold War and American Science: The Military–Industrial–Academic Complex at MIT and Stanford*, New York: Columbia University Press.

—— (2000), "The Biggest Angel of Them All: The Military and the Making of Silicon Valley," in M. Kenney (ed.), *Understanding Silicon Valley: The Anatomy of an Entrepreneurial Region*, Stanford, Calif.: Stanford University Press.

*—— and KARGON, R. H., (1996), "Selling Silicon Valley : Frederick Terman's Model for Regional Advantage," *Business History Review* 70: 435–72.

*LEVIN, R. C., KLEVORICK, A., NELSON, R. R., and WINTER, S. (1987), "Appropriating the Returns from Industrial Research and Development," *Brookings Papers on Economic Activity* 3: 783–820.

LIEBESKIND, J. (2001), "Risky Business: Universities and Intellectual Property," *Academe* 87(5): http://www.aaup.org/publications/Academe/01SO/s001lie.htm

MANSFIELD, E. (1991), "Academic Research and Industrial Innovations," *Research Policy* 20: 1–12.

MASSEY, D. B., QUINTAS, P., and WIELD, D. (1992), *High-Tech Fantasies: Science Parks in Society, Science, and Space*. London and New York: Routledge.

MERGES, R., and NELSON, R. R. (1994), "On Limiting or Encouraging Rivalry in Technical Progress: The Effect of Patent Scope Decisions," *Journal of Economic Behavior and Organization* 25: 1–24.

MOWERY, D. C., NELSON, R. R. SAMPAT, B. N., and ZIEDONIS, A. A. (2001), "The Growth of Patenting and Licensing by US Universities: An Assessment of the Effects of the Bayh-Dole Act of 1980," *Research Policy* 30: 99–119.

*———————— (2004), *The Ivory Tower and Industrial Innovation: University–Industry Technology Transfer Before and After the Bayh–Dole Act*, Stanford, Calif: Stanford University Press.

—— and ROSENBERG, N. (1989), *Technology and the Pursuit of Economic Growth*, New York: Cambridge University Press.

—— —— (1993), "The US National Innovation System," in Nelson 1993:

NATIONAL SCIENCE BOARD (2002), *Science and Engineering Indicators: 2002*, Washington, DC: US Government Printing Office.

NELSON, R. R. (1959), "The Simple Economics of Basic Scientific Research," *Journal of Political Economy* 67: 297–306.

—— (1982), "The Role of Knowledge in R and D Efficiency," *Quarterly Journal of Economics* 97(3): 453–70.

—— (1993), *National Innovation Systems: A Comparative Analysis.* New York: Oxford University Press.

OECD (2000), *A New Economy,* Paris: OECD.

—— (2001*a*), "Education at a Glance. OECD Indicators 2001 Edition," Paris: OECD.

—— (2001*b*), *Main Science and Technology Indicators,* Paris: OECD.

—— (2001*c*), *Science, Technology, and Industry Scoreboard,* Paris: OECD.

—— (2001*d*), *Basic Science and Technology Statistics,* Paris: OECD.

—— (2002), *Benchmarking Science–Industry Relationships,* Paris: OECD.

ROSENBERG, N. (1992), "Scientific Instrumentation and University Research," *Research Policy* 21: 381–90.

—— (1999), "American Universities as Economic Institutions," Mimeo.

*—— and NELSON, R. R. (1994), "American Universities and Technical Advance in Industry," *Research Policy* 23: 323–48.

SAXENIAN, A. (1988), "The Cheshire Cat's Grin: Innovation and Regional Development in England," *Technology Review* 91: 67–75.

SLAUGHTER, S., and LESLIE, L. L. (1997), *Academic Capitalism: Politics, Policies, and the Entrepreneurial University,* Baltimore: Johns Hopkins University Press.

STURGEON, T. J. (2000), "How Silicon Valley Came to Be," in M. Kenney (ed.), *Understanding Silicon Valley: The Anatomy of an Entrepreneurial Region,* Stanford, Calif.: Stanford University Press, xvi, 285.

TRAJTENBERG, M., HENDERSON, R., and JAFFE, A. B. (1997), "University Versus Corporate Patents: A Window on the Basicness of Inventions," *Economics of Innovation and New Technology* 5: 19–50.

TROW, M. (1991), "American Higher Education: 'Exceptional' or Just Different," in B. E. Shafer (ed.), *Is America Different? A New Look at American Exceptionalism,* Oxford: Oxford University Press.

WALLSTEN, S. (2001), "The Role of Government in Regional Technology Development: The Effects of Public Venture Capital and Science Parks," Stanford University: SIEPR Working Paper.

CHAPTER 9

FINANCE AND INNOVATION

MARY O'SULLIVAN

9.1 INTRODUCTION[1]

INNOVATION is an expensive process; significant resources must be expended to initiate, direct and sustain it. It is a process that takes time which means that the resources that support it must be committed until the process is complete. Finally, its outcomes are uncertain so the returns to innovative investments are not assured. The importance of resource allocation to innovation, as well as the complexity of its relationship to that process, makes its systematic analysis crucial to a comprehensive economic theory of innovation.

It is not surprising, therefore, that Joseph Schumpeter, widely regarded as the pioneer in the economic analysis of innovation, made the study of resource allocation, especially the allocation of financial resources, central to his study of innovation. In contrast, as I show in Section 9.2, contemporary economists of innovation have largely neglected the relationship between finance and innovation. Though there are a few exceptions to this rule, they are too recent and too few to suggest that we are on the brink of any systematic change in this regard.

However, some financial economists working in the fields of enterprise finance[2] and finance and growth have begun to explore concerns that are closely related to those that animated Schumpeter's research. In Section 9.3 I discuss research in these fields that is relevant to the study of finance and innovation. I note that empirical research has not kept pace with theoretical developments and the evidence that does exist, even on basic propositions, is often ambiguous. Furthermore, in Section 9.4, I point to serious limitations associated with the dominant analytical approaches employed in micro- and macroeconomic research on finance for analyzing the dynamics of economic change.

Intellectual exchange between evolutionary economist and financial economists seems a better route to an improved understanding of the relationship between finance and innovation. However, methodological differences, especially with respect to the importance of history in economic analysis, are major obstacles to integration between the fields. Therefore, it may be more fruitful for economists of innovation to collaborate with economic and business historians who research the origins and evolution of financial systems. In closing, I draw attention to some crucial questions that need to be addressed in new research on finance and innovation.

9.2 THE ROLE OF FINANCE IN THE ECONOMICS OF INNOVATION

Schumpeter's analysis of the relationship between innovation and resource allocation, especially the allocation of financial resources, was central to his study of the economics of innovation. Many of the details of Schumpeter's economics of innovation were controversial and/or incomplete and contemporary research on the subject has made considerable progress in refining, and expanding on, his analysis. In particular, questions about the appropriate characterization of the innovation process have received considerable scrutiny. However, with a few notable exceptions, the implications of characteristics of innovative activity for resource allocation, especially the allocation of financial resources, have been largely overlooked.

9.2.1 The Pioneering Work of Schumpeter

The role of innovation as the primary stimulus to the process of economic development was the central preoccupation in Joseph Schumpeter's work as an economist. Schumpeter focused on two different, but related, units of analysis in conceptualizing the links between innovation and resource allocation. On the one hand, he was concerned with the implications of the microeconomic characteristics of innovative activity, notably features of entrepreneurial behavior, for resource allocation. In addition, at a more aggregate level of economic analysis, he studied the interaction between structural economic change and resource allocation. In both his micro- and macroeconomic analyses, he paid particular attention to the role of finance in facilitating economic change though, especially in his microeconomic analysis, his understanding of that role shifted over time.

9.2.1.1 *The Microeconomics of Innovation*

In Schumpeter's early writings on the microeconomics of innovation, especially the *The Theory of Economic Development* (*TED*) and *Business Cycles: A Theoretical, Historical and Statistical Analysis of the Capitalist Process* (*BC*), the process of credit creation featured prominently. Later in his life, however, especially in *Capitalism, Socialism and Democracy* (*CSD*), Schumpeter downplayed the role of credit creation in facilitating innovation and economic development. Instead, he emphasized the self-financing of innovative investment by dominant enterprises. The shift in Schumpeter's thinking about the relationship between finance and economic change reflected the well-known transformation in his characterization of innovation from a process driven by new, entrepreneurial ventures to one dominated by large, persistent enterprises.

The Entrepreneur, the Financier, and Innovation. Beginning with an analysis of the allocation of economic resources in the absence of innovation, *The Theory of Economic Development* is Schumpeter's initial treatise on the economics of innovation. Schumpeter's central claim was that the competitive capitalist economy would settle into a routine that might be in motion but would tend towards a stationary or equilibrium state. Resources would flow around the economy, along routinized paths, in what Schumpeter described as the "circular flow of economic life." They would be fully utilized in that perpetual motion and as such could not accumulate into stocks (Schumpeter 1996: 45).

Schumpeter then posed the question of how innovation could occur under these conditions. He defined innovation as the commercial or industrial application of something new, such as a new product or process, a new type of organization, a new source of supply or product market (Schumpeter 1996: 66). He emphasized three characteristics of the innovation process that had crucial implications for resource allocation.

First, innovation depended on the investment of resources: "major innovations and also many minor ones entail construction of New Plant (and equipment)—or the rebuilding of old plant—requiring nonnegligible time and outlay" (Schumpeter 1939: 68). Second, innovation, as a general rule, was embodied in new firms that are founded to undertake the new combination (Schumpeter 1964: 69; see also Schumpeter 1996: 66). Finally, innovation was usually driven by entrepreneurs who were "new" men, that is, who were not already prominent in business circles (Schumpeter 1996: 78; 1964: 70).

Through the process of innovation, the economy's existing productive resources were to be combined in new ways but how were new men, operating in new firms, to get the control over the resources that they needed in order to innovate (Schumpeter 1996: 68)? Since existing resources were already fully utilized in the circular flow, somehow they had to be detached from their current uses and made available to entrepreneurs so that they could combine them in new ways. What was the mechanism through which resources would be reallocated from existing to new uses in the economy?

The solution that Schumpeter provided to the puzzle put the financial system at center stage. Specifically, he argued that innovation was financed through the creation of credit. The purchasing power required by entrepreneurs to detach resources from the circular flow to undertake new combinations was generated *ex nihilo*. Credit creation did not need to be backed by an existing stock of money or goods (Schumpeter 1996: 106).

Schumpeter believed that credit could be created in a variety of ways but he gave prominence to the role of the commercial bank in generating new purchasing power and making it available to entrepreneurs. In entrusting an entrepreneur with the productive resources of society, "[the banker] makes possible the carrying out of new combinations, authorises people, in the name of society as it were, to form them. He is the ephor of the exchange economy" (Schumpeter 1996: 74).

The Large-Scale Enterprise, Innovation, and the Role of Finance. Over the course of his scholarly life, Schumpeter dramatically altered his characterization of innovation from a process dominated by new men and new firms to one driven by the activities of large-scale industrial enterprises. During the period of almost forty years between the writing of *TED* and *CSD*, Schumpeter moved to the United States. He was struck by the major changes underway in the US economy in the first decades of the twentieth century and he believed that they represented a shift from nineteenth-century "competitive capitalism" to twentieth-century "trustified capitalism."

Whereas previously he had argued that the life and purpose of new firms faded over time, in *CSD* he claimed that the "perfectly bureaucratized giant industrial unit" had succeeded in rationalizing and routinizing the process of innovation to such an extent that the large-scale enterprise had become "the most powerful engine" of economic progress (Schumpeter 1942: 106). As a result, technological progress had increasingly become the business not of individual entrepreneurs but of "teams of trained specialists who turn out what is required and make it work in predictable ways" (Schumpeter 1942: 132; see also Schumpeter 1949: 71).

Schumpeter's revised characterization of the innovation process led to important changes in his analysis of resource allocation in a dynamic economy and, in particular, in his analysis of finance. Specifically, it led him to downplay the role of external finance and the banking system in favor of an emphasis on internal finance for facilitating innovative investment. Indeed, for a scholar who once described the process of credit creation as "the monetary complement of innovation" (Schumpeter 1964: 85), the absence of any reference in *CSD* to the process of credit creation, or the specific functions of banks, represents a dramatic change in perspective.

9.2.1.2 *Finance, Innovation, and Structural Economic Change*

Although *TED* is dominated by microeconomic analysis, towards the end of the book Schumpeter sketched the outlines of a macroeconomic analysis of innovation and economic development which he articulated in greater detail in *Business Cycles*. His main concern was with the question of why economic development "does not proceed evenly as a tree grows, but as it were jerkily," of why there was a business cycle, "the wave-like movement of alternating periods of prosperity and depression," which he claimed had characterized the economic system throughout the capitalist period (Schumpeter 1996: 223).

Schumpeter's analysis of the business cycle built on the microeconomic analysis of innovation that he had laid out in *TED*, notably in assuming that innovation was embodied in new firms established by new men. It went beyond it, however, in suggesting that the process of venture creation had important characteristics that could only be identified at the aggregate level. Specifically, he claimed that innovative activity was unevenly distributed over time and across industries.

As a result, economic development was an uneven and jerky process (Schumpeter 1964: 76). The effect of booms in entrepreneurial activity was to fundamentally alter the economic system that existed beforehand, disturbing its equilibrium and starting "an apparently irregular movement in the economic system, which we conceive as a struggle towards a new equilibrium position" (Schumpeter 1996: 235). Therefore, he rejected the notion of economic development as a mere quantitative change in the level of aggregate economic activity and argued that it was a process of qualitative change that could only be understood with reference to the structural composition and evolution of the economy.

Having characterized the process of economic development in this way, the challenge for Schumpeter was to explain how such qualitative economic change might occur. Once again, the financial system or, more precisely, the banking system and the credit expansion and contraction that it facilitated, featured as the crucial mechanism that facilitated the reallocation of resources necessary to induce dramatic changes in the structure of economic activity. For Schumpeter, therefore, the evolution of a country's financial system was of crucial importance for facilitating the waves of innovation that he regarded as the motive force behind its economic development.

9.2.2 The Neglect of Finance in the Contemporary Economics of Innovation

Given Schumpeter's expansive vision of the economics of innovation, it is hardly a surprise that the details of his arguments have been subject to significant critical scrutiny. Most of the attention has focused on his characterization of the innovation process. For example, his assertion of a general historical transformation from an innovation process dominated by new ventures (often described as Type 1 innovation) to one driven by large industrial firms (Type 2) has been rejected. Contemporary scholars have emphasized that both of these patterns of innovation coexist in the economy with some industries characterized by Type 1 and others by Type 2 innovation (Winter 1984: 295).

However, hardly any attention has been devoted to the relationship between innovation and resource allocation even though Schumpeter's analysis of it is as controversial and incomplete as his characterization of innovation. Nevertheless, questions about resource allocation are latent in much of the work that has been done on characteristics of the innovation at the level of the enterprise, the industry and the economy. Like Schumpeter, economists of innovation must develop an explicit analysis of the implications of these characteristics for the allocation of resources and, in particular, financial resources. Otherwise, we might well ask whether contemporary research on innovation as yet amounts to an economics of innovation at least if we define economics, as most people do, as centrally concerned with resource allocation.

9.2.2.1 *Finance, Innovation, and the Enterprise*

One way of reading the implications of Schumpeter's intellectual evolution from *TED* to *CSD* is as a call for a theory of the innovative enterprise. Though Schumpeter was confident that such an important research agenda would not be ignored, the subject attracted almost no attention for more than three decades after his demise. One major exception was a book entitled *The Theory of the Growth of the Firm* published in 1959 by Edith Penrose. Penrose's main argument was that the basic foundation for the growth of firms was a dynamic process of organizational learning that occurred within enterprises.[3] In recent decades, there has been a veritable explosion of interest in conceptualizing innovation as it occurs at the level of the firm. Whether derived directly from Penrose or not, the notion that firm-level innovation is based on a process of organizational learning is an idea that is common, even pervasive, in the literature on the innovative enterprise (for a similar conclusion, see Nelson 1991; Teece, Pisano, and Shuen 1997; see also Lazonick, Ch. 2, and Lam, Ch. 5, this volume).

Most of the research on the subject of organizational learning in firms makes some reference to the process through which resources are allocated to allow them to initiate, direct and sustain organizational learning. However, it is rare to find system-atic discussions of the implications of organizational learning for the allocation of financial resources to investments and returns. One basic question that needs to be addressed is whether, and to what extent, the process of organizational learning affects the scale of investment that must be made by an innovative firm. Resources have to be invested in the learning process itself, as well as the implementation of the results of such learning within the organization. In addition, financial resources are required to develop or acquire the necessary complementary assets (Teece 1986) to commercialize innovations based on this organizational learning. However, very little research has been conducted on the implications of the characteristics of organizational learning for the resource requirements of different types of enterprises.

If the organizational character of the innovation process has implications for the scale of enterprise investment, it will also affect the organization and governance of the process of resource allocation within the firm. Who are the key decision makers in allocating resources to innovative investments? On what basis do they make their investment decisions? How do they coordinate investment decisions across the firm? What capabilities do they have for making these decisions? How do their incentives relate to those of participants in the learning process and to organiza-tional objectives?

These questions only hint at the plethora of issues that are raised by a consider-ation of the organizational dimension of firm-level innovation. Economists have also highlighted other important characteristics of the innovation process, such as its cumulative character and inherent uncertainty; their implications for resource allocation also require consideration in any comprehensive economics of innovative enterprise. Scholarly analysis of these issues is, as yet, uncommon. Indeed, studies of the internal processes that companies use to allocate financial resources to any type of investment, not just innovative ones, are rare (for exceptions, see Bower 1970; Burgelman and Sayles 1986).

9.2.2.2 Finance, Innovation, and the Industry

Empirical studies have confirmed the importance of cross-industry variation in certain basic characteristics of innovative activity. Industries vary not only in the distribution of innovative activity between entrants and incumbents at a point in time but also over time as measured by the stability or turbulence of these popula-tions' shares of innovative inputs and outputs (for a summary, see Malerba, Ch. 14, this volume). Sectoral differences in innovative activity have important implications for resource allocation.

To the extent that entrants dominate innovative activity, for example, the ques-tion of how they get access to the resources that they require to innovate is crucial. While the financing of *de novo* entry is relevant here, empirical studies have shown that entrants are not necessarily new ventures; entry also occurs through spin-offs from established firms as well as diversification by incumbent firms in other industries. As a result, understanding how the resources of existing firms are reallocated, by decision makers in those firms or the employees that leave them to establish new ventures, is also important.

If the heterogeneity of entrants' origins generates a range of questions about resource allocation, so too do other types of observed variation in sectoral patterns of innovation. Industries display important differences in competitive interactions among firms. If we focus on the relationship between entrants and incumbents, for example, we find that some entrants compete directly with incumbent firms for customers whereas others build an innovative strategy based on licensing arrangements or joint ventures. Gans, Hsu, and Stern contrast biotechnology, where joint ventures and alliances between entrants and incumbents are common, with hard disk drives, where entrants confronted incumbents in head-to-head competition in product markets (Gans, Hsu, and Stern 2000: 3). These differences have implications for the type and amount of investment that needs to be made by innovative firms. They affect the extent to which firms are involved in different business activities, such as production or marketing, and, more generally, the costs for different players of organizing themselves to innovate.

Industries differ not only in the competitive interactions among firms but also in terms of the relationships between firms and other industry actors. Competitive firms may have relationships with their suppliers and customers as well as universities and governments (local, regional, national, and international) that exert an important influence on their allocation of resources (Malerba, Ch. 14, this volume). In the early stages of the development of the US software industry, for example, the federal government played a crucial role in supporting the education and training of large numbers of software engineers (Mowery and Langlois 1996). As a result, many start-up companies were able to enter the industry as viable competitors without having access to the resources that would have been necessary if they had undertaken this effort on their own. Similarly, in the US biotechnology industry the citation patterns for patents show a major reliance by private companies on public science in their innovative activities with similar implications for their capacity to economize on the resources that they needed to invest to become players (McMillan, Narin, and Deeds 2000).

In addition to intersectoral variation, the relationship between finance and innovation may vary over time within a given industry. Industries in which entrants drive innovation in one era may change over time to a structure in which innovation becomes more incremental in character and is dominated by incumbent firms, as in the automobile industry. Cases of the opposite trend also exist, as incumbent-dominated industries face waves of innovation-led entry—the development of biotechnology within pharmaceuticals is a good example. The characteristics of the enterprises that require finance, the sources of finance on which they rely, and the implications of these financial arrangements for innovation, are likely to differ significantly as a result of these evolutionary changes.

We must also consider the possibility that the direction of causation goes in the other direction, from finance to innovation. As far as the distribution of innovative activity between entrants and incumbents is concerned, for example, we could ask whether incumbent firms dominate because they are more innovative or because entrants are too financially constrained to compete with them. Conversely, when entrants dominate is it because they are more innovative than incumbents or because a ready, and even excessive, availability of financial resources allows them to do so?

The latter question is the subject of two related studies of the hard disk drive industry. In an article called "Capital Market Myopia," William Sahlman and Howard Stevenson (1985) argue that venture capitalists and the stock market massively over-invested in the industry during the period from 1977 to 1984 with negative consequences for financial returns and for innovation. However, another study of the disk drive industry claims that that the diagnosis of capital market myopia is suspect in long-term perspective. Bygrave, Lange, Roedel, and Wu accept that many players in the disk drive industry failed but they argue that the survivors ultimately enjoyed sufficient commercial success to justify the financial bets that were made on the industry (Bygrave, Lange, Roedel, and Wu 2000: 17).

These two studies hint at the potential of analyses of the joint influence of finance and innovation on industry evolution but they are exceptions (see also Carpenter, Lazonick, and O'Sullivan 2003). The extent to which labour is reallocated from incumbent firms to entrants through spin-off activity has attracted some recent interest (see e.g. Klepper 2001). To date, however, the relationship between the allocation of financial resources and industrial patterns of innovative activity has been largely ignored.

9.2.2.3 Finance, Innovation, and the Economy

Contemporary economists have developed several approaches to thinking about the economics of innovation at the level of the economy, that is, to the relationship between innovative activity and economic development. One perspective, that of techno-economic paradigms, builds directly on Schumpeter's work on business cycles (for a summary, see Freeman and Louçã, 2001). A second approach—the national systems of innovation framework—has been developed over the last 25 years to explain comparative–historical patterns in economic development (Nelson 1993).

Both approaches share a commitment to the Schumpeterian emphasis on the structural composition and evolution of the economy in analyses of economic development. The two approaches differ, however, in their attention to resource allocation. In the national systems of innovation literature, the process through which labor and capital are allocated to innovation processes is rarely discussed even when its significance is recognized (Nelson 1993: 13). In contrast, the literature on techno-economic paradigms has long been concerned with the interaction between innovative activity and the allocation of labour (Freeman 1977). With the recent publication of *Technological Revolutions and Financial Capital: The Dynamics of Bubbles and Golden Ages* by Carlota Perez (2002), the relationship between finance and techno-economic paradigms has also received systematic treatment.

Perez follows Schumpeter in placing technological revolutions, that is, "clusters of radical innovation forming successive and distinct revolutions that modernize the whole productive structure," at the heart of her theory. However, her concept of a revolution places much greater emphasis on the diffusion of innovation than is found in Schumpeter's work.[4] She also emphasizes, to a much greater degree than Schumpeter did in his work on business cycles, that the effects of technological revolutions go far beyond their economic impact to include "a transformation of the institutions of governance, of society and even of ideologies and culture" (Perez 2002: 24–5).[5]

Perez's analysis of the interaction between the financial and productive system is also more comprehensive than that of Schumpeter. While the latter focused primarily on the role of the financial system in funding an initial burst of innovative investment, Perez characterizes the ways in which the financial system may be involved in the productive system throughout the life cycle of technological revolutions. She argues that the relationship between the financial and productive sectors alters as the economy moves from one stage of the life cycle to another.

Perez's book represents an important contribution to our understanding of finance and innovation. Not surprisingly, given its originality, it suffers from certain limitations that will need to be addressed in future work. In particular, systematic empirical support for some of the key arguments that she makes about the role of finance in funding technological revolutions is not provided.[6]

However, a similar Schumpeterian analysis of finance and technology is found in several recent empirical articles by Boyan Jovanovic and his co-authors on the relationship between the development of the stock market and technological revolutions in the US economy from the late nineteenth century until the present (Jovanovic and Greenwood 1999; Jovanovic and Rousseau 2001; Hobijn and Jovanovic 2001). For example, Jovanovic and Rousseau (2001) compare and contrast the implications of the IT revolution with that of the "electricity-era" technological revolution for the US stock market. They show that the transition periods from founding to listing on the stock market, from incorporation to listing and from first product or process innovation to listing were shorter at the beginning and end of the twentieth century than in the intervening years, and argue that the explanation for this pattern can be found in the characteristics of technological change (Jovanovic and Rousseau 2001: 336).

9.3 INNOVATION AND THE ECONOMICS OF FINANCE

In the previous section, I emphasized that economists of innovation, as a general rule, have not been greatly inspired by Schumpeter's concern with finance and innovation. For a long time, there was a general neglect of the interaction between finance and the real economy in other branches of economics. However, in recent years that has begun to change.

Theoretical developments in the field of corporate finance have stimulated interest in the interaction between enterprise finance and investment. In the literature on economic growth, the influence of endogenous growth theory has led economists to consider how and to what extent the financial system might affect the rate and process of economic growth. As a result, some of the issues that are now being explored by economists, such as the characteristics and importance of venture capital as a source of finance, as well as the financing of R&D-intensive firms and new ventures, are of direct interest to those concerned with the economics of innovation. However, empirical research has lagged theoretical development and the evidence that does exist is often ambiguous even on some of the basic theoretical propositions that have been advanced. Moreover, there are serious limits to the dominant conceptual approaches employed by financial economists for understanding the relationship between finance and innovation.

9.3.1 The Microeconomics of Enterprise Finance

From the late 1970s, there was a major shift in theoretical research on corporate finance based on the economics of information. A whole new set of theories of corporate finance emerged that took as their starting assumption the importance of "information asymmetry"[7] as a determinant of enterprise finance. One impact of the growing influence of information economics on corporate finance was a transformation in the way financial economists thought about alternative sources of finance. There was a move away from the rather simple cost–benefit analyses of different sources of finance that had been inspired by the work of Modigliani and Miller (1958). Comparisons of alternative sources of finance increasingly considered their implications for the information, incentives and control of different economic actors.

These developments stimulated fresh treatments of old distinctions among alternative sources of finance; traditional comparisons between debt and equity and bank- and market-based finance, for example, were conceptualized in new ways based on the economics of information (Allen and Gale 2000). Moreover, sources of finance that had previously received little attention from financial economists attracted interest given what appeared to be their distinctive characteristics from an information-economics perspective; the best example is the burgeoning literature on venture capital.

An article by William Sahlman (1990) pioneered in using the logic of asymmetric information as the foundation for a theory of venture finance. Sahlman identified several different mechanisms, commonly employed in the process of venture capital investing, that he claimed venture capitalists used to overcome the problems of financing ventures in the presence of asymmetric information. In subsequent work, other scholars delved further into the roles played by these mechanisms, such as staging, compensation, monitoring and control of investee companies, and exit, in the relationship between venture capitalists and their investee companies (for a summary of this research, see Gompers and Lerner 1999).

Important theoretical advances have been made in analyzing the economics of alternative sources of enterprise finance empirical analysis has lagged behind. Even basic evidence on the relative importance of different sources of enterprise finance, such as internal and external sources of funds, as well as alternative sources of external funds like banks, stock and bond markets, and venture capitalists is modest (see Box 9.1 for a summary history of the US venture capital industry). More

---

Box 9.1 Venture capital in comparative–historical perspective

The United States is the country most closely associated with a professional venture capital industry. Although individuals and families in the United States, as in most industrial economies, long have used their private fortunes to fund new ventures, the US venture capital "industry" began with the founding in 1946 of American Research and Development (ARD) for the finance of new ventures. ARD was founded in Boston by members of the local investment community, as well as professors and administrators from MIT. With the expansion of federally funded research at MIT during World War II, the principal sources of ARD's deal were the federally funded laboratories at Harvard and MIT. By far its most successful deal was its 1957 investment in a new firm founded by Kenneth Olsen of MIT's Lincoln Laboratory, the Digital Equipment Corporation (DEC) (Rosegrant and Lampe 1992: 72, 110–14).

The enormous success of ARD's investment in DEC—when the computer firm went public in 1966 ARD's original investment of $70,000 was valued at $37 million—generated huge interest in venture capital in the financial community. By that time, a new breed of financiers had emerged out of Silicon Valley's high-technology enterprises. Successful entrepreneurs like Fairchild's Eugene Kleiner and Don Valentine reinvested the capital that they accumulated in promising local start-ups and brought cash and technical skill, operating experience, and networks of industry contacts to the ventures they funded. Silicon Valley's venture capitalists were heavily involved with their ventures, advising entrepreneurs on business plans and strategies, helping find co-investors, recruiting key managers, and serving on boards of directors (Saxenian 1994: 39).

Box 9.1 (continued)

Several regulatory changes contributed to the further expansion of the US venture capital industry in the 1980s and 1990s. Particularly important were a series of legislative initiatives in the late 1970s that made venture capital a much more attractive investment option: the capital gains tax rate was reduced from 49.5 to 20 per cent and it became much easier for pension funds to invest in venture capital partnerships. There was a major increase in the flow of money into the US venture capital from then on (see fig. 1.1 in Gompers and Lerner 1999: 7). Historical trends in investments and disbursements by the industry displayed considerable volatility, with major boom–bust cycles recorded in the 1960s, the 1980s and again in the late 1990s/early 2000s.

Investments by the US venture capital industry were highly concentrated by sector. For the period from 1965 to 1992, four industries—office and computing machinery, communications and electronics, pharmaceuticals, and scientific instruments—accounted for 81 per cent of all investments by the US venture capital industry. The relative importance of these sectors changed considerably over time; office and computing machines became less important over time whereas the drugs sector increased its share of venture capital investment. In the 1990s, especially in the second half of the decade, Internet-related investments were estimated to account for 70 per cent and 75 per cent of all venture capital investments in 1999 and 2000 respectively (Venture Economics).

Measured as a share of gross economic output, the US venture capital industry is the largest in the world. The venture capital industries of other countries enjoyed considerable growth in the 1990s; for the EU as a whole, for example, venture capital investment increased from 0.04 to 0.12 per cent of GDP for the period from 1989 to 1999 with Belgium, Sweden and the Netherlands recording particularly strong expansion (Table 9.1). Nevertheless, even more rapid growth in the US venture capital industry during this period meant that it actually increased its lead over other countries, particularly in early-stage investment activity by venture capitalists.

challenging questions, such as whether enterprises choose alternative sources of finance for the reasons posited by financial economists or whether their use of these sources has the implications for performance that theories of enterprise finance suggest, remain largely unexplored (Rajan and Zingales 1995: 1421).

The growing influence of information economics stimulated another body of theoretical analysis of enterprise finance, encouraging interest in the long-neglected interaction between the financing and investment behavior of enterprises. Since information asymmetries between enterprises and financiers made external finance more expensive for companies than internal finance, *ceteris paribus*, enterprises with access to substantial liquidity from internal sources would invest more than enterprises that have to resort to external finance.

There is now a substantial body of empirical evidence that suggests that "liquidity constraints" matter to capital investment (for a summary see Hubbard 1998: 199). However, serious methodological concerns have been raised about these studies.

Table 9.1 Venture capital and early-stage investment as a percentage of GDP

Country	VC Investment as percentage of GDP		Early-Stage Investment as percentage of GDP	
	1989	1999	1989	1999
Austria	0.01	0.03	0.006	0.007
Belgium	0.05	0.26	0.015	0.093
Denmark	0.01	0.05	0.009	0.019
Finland	0.01	0.11	0.003	0.057
France	0.05	0.12	0.009	0.039
Germany	0.01	0.13	0.004	0.051
Greece	n.a.	0.06	n.a.	0.017
Ireland	0.05	0.09	0.002	0.048
Italy	0.02	0.05	0.002	0.014
The Netherlands	0.05	0.25	0.006	0.096
Portugal	0.02	0.05	0.004	0.008
Spain	0.02	0.09	0.009	0.018
Sweden	0.02	0.19	0.004	0.113
United Kingdom	0.13	0.20	0.023	0.020
EU	0.04	0.12	0.008	0.036
US	0.11	0.59	0.027	0.056

Source: Christofidis and Debande, 2001, p. 20, p. 23.

Particularly controversial has been the common practice of imputing the importance of liquidity constraints from the sensitivity of a company's investment to its cash flow. Some scholars claim that "investment-cash flow sensitivities provide no evidence of the presence of financing constraints" thus calling into question the validity of the empirical results reported in this literature (Kaplan and Zingales 2000; Fazzari, Hubbard, and Petersen 2000).

Theories that posited that the influence might go in the other direction, that the characteristics of enterprise investment might influence enterprise finance, were also developed. One stream of literature posited a relationship between the types of activities in which enterprises invested and their financial behavior. Another suggested that characteristics of the investing enterprise, for example, its stage of development, mattered to the way it was financed.

In analyses of the influence of enterprises' activities on their financing, R&D investments attracted particular attention since they were deemed to create acute information asymmetries between corporate managers and financiers (Bah and Dumontier 2001: 675; Himmelberg and Petersen 1994; Hall 2002). The main implication that has been drawn from such analyses is that the gap between the costs of financing R&D investment from internal and external sources should be greater than for other forms of investment. Therefore, R&D-intensive firms should be more inclined than other firms to rely on internal funds to finance their investments. Moreover, financing constraints arising from imperfections in capital markets should have a much greater impact on R&D, than other, investments (Hall 2002; Carpenter and Petersen 2002: F55). Recently, some scholars have suggested that these arguments apply not just to R&D investments but to all investments in high-technology industries (see Bank of England 2001: Annex, 81–5).

Empirical analyses of the relationship between finance and R&D are primarily analyses of the links between cash flow and R&D expenditures. The common finding of these studies is that R&D investment is indeed positively correlated with cash flow (for a summary see Hall 2002). However, these studies are subject to the same methodological criticisms as the empirical work on liquidity constraints and capital investment to which we have already referred, and until these issues are resolved we cannot be confident of their findings.

Financial economists have also begun to analyse whether the characteristics of investing firms might matter to their financing behavior. Particular attention has been paid to a company's stage of development and the concept of a "financing growth cycle" is now widely used to characterize the challenges for firms as they evolve from new venture to going concern highlighting once again the extent of informational asymmetry involved (Berger and Udell 1998: 622). Perhaps the most straightforward implication of this type of analysis is that firms at earlier stages of the cycle, such as start-up companies, are likely to have difficulties raising external finance. As a result, they should be more heavily dependent on insider finance than firms at later stages of development.

Once again, however, empirical research has not kept pace with theoretical developments. The evidence that has been compiled, moreover, does not provide clear support for some of the most basic propositions advanced in the theoretical literature. In this regard, Berger and Udell highlight two findings from their empirical analysis which seem particularly surprising. The first is that the funds provided by the principal owner are more important as the firm gets older than at early stages. Second, their evidence suggests that finance from insiders never outweighs that provided by outsiders even for the youngest firms (Berger and Udell 1998: 625).

The two streams of literature on the influence of investment characteristics on enterprise finance—the one on characteristics of the investments being made and the other on the characteristics of investing firms—have recently been brought together in research on the financing of small firms making R&D, high-technology, or technology-based investments, often referred to as technology-based small firms (TBSFs). Theoretical models from the "new" corporate finance predict that these firms will be much more tightly constrained by their own internal resources in financing their investment than other firms. As yet, however, the jury is out on whether even this basic proposition is borne out by empirical evidence. As a recent report by the Bank of England concluded: "the evidence from such studies is conflicting on the key issue of whether TBSFs face greater difficulties in accessing finance than SMEs [small and medium enterprises] in general" (Bank of England 2001: 83).

9.3.2 Financial Systems and Economic Growth

Contemporary growth theory, in recognizing the importance of technological change for economic growth, echoes Schumpeter's emphasis on the importance of innovation as the primary impetus for the process of economic development. Moreover, whereas Schumpeter emphasized the importance of financial systems in fueling innovation and, therefore, economic development, the idea that the development and structure of a country's financial system might have implications for its economic growth was treated with skepticism or indifference by most macroeconomists in the second half of the twentieth century. In recent years, however, there has been a major increase in interest in the relationship between financial development and economic growth, a trend that is usually attributed to the influence of endogenous growth theory in macroeconomics. A variety of theoretical articles model the mechanisms through which the financial system might affect long-run growth. Empirical studies have also been undertaken, primarily based on large-scale cross-country regressions, to analyze the relationship between financial systems and economic growth.

Most of these studies of finance and growth seek to relate the level of development and structural characteristics of the financial system to aggregate economic activity. Consistent with the dominant approach taken in neoclassical growth theory, economic development is understood as an undifferentiated quantity generated by an aggregate production function. As a result, contemporary analyses of finance and growth make no reference to the structural composition and evolution of the economy and, therefore, ignore what Schumpeter regarded as the essential characteristic of the process of economic development, that is, its lumpiness over time and across sector.

However, there are a few exceptions to the general rule. Rajan and Zingales (1998) differentiated among industries in terms of their investment and financing behavior, arguing that their financial requirements are technologically determined:

there is a technological reason why some industries depend more on external finance than others. To the extent that the initial project scale, the gestation period, the cash harvest period, and the requirement for continuing investment differ substantially between industries, this is indeed plausible. (Rajan and Zingales 1998: 563)

Their main hypothesis is that industries that are more dependent on external finance should grow faster in countries with more developed financial markets. Several other studies have followed Rajan and Zingales in discriminating among industries in terms of their demand for finance. Some of them go further to consider whether financial structure—notably whether a financial system is market- or bank-based—matters to the development of different industries (Beck and Levine 2002; Demirgüç-Kunt and Maksimovic 2002; Carlin and Mayer 2003).

The central implication of this disaggregated approach to the relationship between finance and growth is that the economic impact of financial systems may be reflected not only in aggregate rates of economic growth but also in the differential development of particular industries. Differences in the growth trajectories of particular industries will in turn be reflected in variations across country in the composition of economic growth; sectors favoured by a nation's financial system will become more and more prominent in the economy over time while other, less-favored sectors will languish or fail to develop. However, as yet, the task of analyzing the relationship between financial systems and the structural evolution of the economy has not been treated in any detailed way in these studies.

9.4 A NEW AGENDA FOR RESEARCH ON FINANCE AND INNOVATION

Contemporary research in economics has a long way to go before it can help us to understand the relationship between finance and innovation. In the economics of innovation, that relationship has been largely neglected though questions about resource allocation are latent in existing research. In financial economics, some scholars have begun to explore concerns that are directly relevant to the study of finance and innovation, but financial economists have been more effective at generating new theoretical arguments than adducing empirical evidence to support them.

In principle, there is a good case for integration of the two fields as a route to new insights on the relationship between finance and innovation. However, there are serious limitations of the dominant theoretical approaches that financial economists employ for analyzing the process of economic change. Moreover, barriers of mutual ignorance and, more fundamentally, methodological difference, make such integration unlikely. Collaboration between economists of innovation and historians of finance is a much more promising path to a better understanding of the relationship between finance and innovation. These scholars share a commitment to history as a technique of economic analysis which is crucial given that the essential processes that need to be understood in analyzing the interaction between finance and innovation are historical processes.

9.4.1 Finance Theory and the Dynamics of Economic Change

Important weaknesses of the research of financial economists become evident when we bring the literature on the economics of finance into contact with what we know about the dynamics of innovation. As a result, it is doubtful that, left to their own devices, financial economists will ever develop a satisfying understanding of the role of finance in the process of innovation. In the microeconomic literature, the main problem is the centrality of the concept of asymmetric information in theories of enterprise finance. It is true that analytical space for the "real" economy was created by incorporating this concept; so long as one could argue that a real phenomenon led to important asymmetries of information, financial economists were willing to take it seriously in theories of enterprise finance. However, the versatility of the concept hints at its limitations.

One problem is that it has been invoked in a rather casual manner without much in the way of proof of the extent of asymmetries of information or their importance in influencing economic relationships. For the case of R&D investments, for example, one typically finds an assertion that they are subject to acute problems of asymmetric information. Yet, why should we assume that investors have less information about the likely success of pharmaceutical companies' R&D efforts than the factors that determine the productivity of a new automobile plant?

More fundamentally, it is not clear that privileged access to information by some economic actors is the major determinant of the challenges for enterprises in financing such investments. When making innovative investments, a more important challenge than asymmetric information is the fundamental uncertainty that characterizes the relationship between investments and their outcomes. In an environment characterized by fundamental uncertainty, the crucial problem is not that one person knows something whereas another does not; rather the challenge to decision making is ignorance, the fact that nobody really knows anything.

Uncertainty in this sense is different from the concept as it is used in neoclassical economics (Arrow and Debreu 1954; Arrow 1974) where the main concern is with parametric uncertainty.[8] The environment in which economic decisions are made is characterized as a set of mutually exclusive but collectively exhaustive possible states of the world. In such a world, one that is closed and deterministic, rational decisions based on probabilistic estimates are a reasonable basis for action.

When uncertainty is fundamental, as it is when innovation occurs, economic agents are uncertain not just about which possible state will obtain but about which ones are even possible. In making innovative investments, therefore, there are really no objective guidelines for making decisions or for resolving disputes. So how does anyone act under these circumstances? How do innovative investments ever get made given the uncertainty that surrounds them?

Questions about decision making under fundamental uncertainty are further complicated by the fact that the process of innovation reveals new, possible states of the world (Kline and Rosenberg 1986: 297–8). In other words, the uncertainty inherent in innovation unfolds as the process evolves. As a result, the future state of the world cannot be defined until it is discovered through the process of innovation (Rosenberg 1994: 53–4). Through involvement in that process, or close familiarity with it, decision makers learn and, as they do so, their perceptions of the possibilities and problems of innovative investment change. There is no reason to assume that, in altering their perceptions based on new information, decision makers necessarily move closer to some kind of truth. To the contrary, they are likely to make many mistakes in interpreting what the information that they have implies for the challenges and opportunities of innovation.

Economists who take seriously the fundamental uncertainty that surrounds innovative investment have tended to emphasize the importance of subjective judgements, based on perceptions and belief systems, for decision making. They have also suggested that, in reacting to the unfolding of the uncertainty inherent in innovation, decision making is experiential as well as interpretative. Financial economists tend to overlook the basic cognitive characteristics—ignorance, learning and error—of resource allocation and the mechanisms that allow enterprises to make commitments of resources to innovative activity notwithstanding the challenges of doing so. To date, in analyzing enterprise finance, financial economists have emphasized rationality, indeed a rather limited and static concept of rationality, to the exclusion of the subjective and experiential dimensions of decision making. While rational analysis may well feed into the process of innovative resource allocation, it cannot dominate it.

If financial theory is of limited use for understanding the cognitive challenges associated with innovative investment, it provides no help at all in dealing with other important features of firm-level innovation such as its organizational character. Notwithstanding the important progress that has been made in microeconomics in conceptualizing the economics of the firm, theorists of corporate finance have failed to embed their analyses in a substantive theory of the firm (Zingales 2000). In fact, many financial economists remain wedded to the idea of the firm as a nexus of contracts and reject the notion of firms as organizations that have a logic that is distinct from that of markets.

As far as the macroeconomic literature is concerned, its main problem for understanding the relationship between finance and growth is that it conceptualizes it in a rather mechanical way. Most of the models used in this literature treat economic growth as if it was generic across the economy and over time. Typically, they make no allowance for variation and change in the organizational and institutional contexts in which financial resources are allocated and employed to facilitate economic growth. Instead, the relationship between finance and growth is understood in terms of the influence of the quantity of finance provided and, to a lesser extent, the price at which it is supplied, on the "amount" of economic growth.

In contrast, research on innovation and technological change suggests the central importance of the context in which these processes occur to their impact on productivity and economic development. It is for this reason that Gavin Wright, a prominent economic historian, has recently exhorted economists to embrace "[a] conception of technology that is historically-contingent and institutionally-specific" (Wright 1997: 1562). From this perspective, research on finance and innovation must devote greater attention to the role that contextual factors play in shaping financial relationships. Who gets financial resources, when they get them, how they use them and other factors that can only be identified by making qualitative distinctions among enterprises, time periods, and investments, are likely to be more important than the overall quantity of financial resources that is invested in an industry.

Recent studies that incorporate sectoral differences in their analyses of finance and growth tend to overlook the importance of these contextual factors. They assume that given characteristics of technologies matter to the scale and timing of enterprise investment and, as result, to enterprise demand for finance. However, even if we take technological characteristics as given, a range of variables, such as the balance between entrants and incumbents and, more generally, the competitive structure of the industry, defines the context that shapes the relationship between technology, investment requirements and financing needs. Moreover, the relationship between technology and finance is likely to change within industries over time, in part because technological characteristics of industries evolve, but also since the intervening variables that determine their relationship to finance also change.

9.4.2 The Possibilities and Problems of Integration between Fields

Some of the weaknesses of financial economics for dealing with the relationship between finance and the dynamics of economic change could be overcome, at least in principle, through closer integration with research on innovation. There is some recognition of this fact within financial economics. One prominent scholar in the field, Luigi Zingales, has been quite explicit in his call for "new foundations," notably a more sophisticated analysis of the firm, for the study of enterprise finance. Particularly interesting from my perspective is the fact that he emphasizes the importance of "a theory of entrepreneurship" for understanding how firms innovate (Zingales 2000).

Economists of innovation, in turn, can learn from research in finance. In part, their having greater contact with financial economists might stimulate a general interest in finance among them but the specific details of existing research can also be a source of useful insights. For example, the recent attention by financial economists to the relationship between stages of an enterprise's development and its financing requirements could be fruitfully incorporated by economists of innovation.

If the only barrier to intellectual integration was lack of familiarity with each other's work then it would be relatively easy to overcome. However, ignorance of another's field of inquiry typically conceals deeper barriers. In addition to theoretical differences described above, methodological differences pose a formidable barrier to the integration of research on finance and innovation.

Perhaps the most important methodological obstacle to cross-fertilization between the fields of innovation and finance is the different priorities that they assign to history in economic analysis. In recognition of the importance of change over time in the process that they study, economists of innovation often describe themselves as "evolutionary economists" and they typically assign considerable importance to historical research in their economic analysis. In their empirical work, they tend to emphasize the importance of qualitative variations across the economy and over time in innovation and technological change. They are often skeptical of theories of innovation and economic development that are abstracted from the historical contexts and emphasize the importance of "reasoned history" and "history-friendly models" (Freeman and Louçã 2001; Lazonick 1991; Malerba et al. 1999).

In contrast, most financial economists neglect the fact that capitalism is an evolutionary process. Equilibrium analysis overwhelmingly dominates theoretical research to the neglect of the historical origins and evolution of economic behavior. In empirical research, quantitative analysis is the norm and attention to qualitative variations within the economy and over time is modest. It is hardly surprising, therefore, that limited attention is paid in theoretical or empirical research to the historical development of the financial system, the structural evolution of the economy or change over time in the relationship between the financial system and the real economy.

It seems unlikely that such methodological differences can be overcome in the foreseeable future. A more promising avenue to intellectual progress on the relationship between innovation and finance, therefore, is for economists of innovation to collaborate with financial historians. Evolutionary economists have already had direct experience of the value of a shared methodological commitment to historical analysis in facilitating intellectual exchange through their collaboration in research on technological change with historians of technology, labor, and business. Similarly fruitful exchanges are likely to result from interaction with economic and business historians who study the history of financial systems. Indeed, there are already some promising signs of research in this direction (see e.g. Lamoreaux and Sokoloff 2004).

9.5 CONCLUSION

What are the priorities in new research on finance and innovation? It is widely recognized that the paucity of, and flaws in, existing empirical research on the financial system are major barriers to economic analyses of its role. The dearth of evidence is a problem even if we focus on contemporary patterns of financial demand and supply. When we look to historical data, the holes in our understanding of these patterns are even more gaping (Levine 2003; O'Sullivan 2004a, b; Zingales 2003).

It is true that patterns of financial demand and supply by enterprises, firms, and economies serve only as background evidence for more specific empirical questions about the relationship between finance and innovation. In an ideal world, scholars who are primarily interested in researching that relationship could compile these data from readily available sources. However, the current, rather dire, state of empirical research on patterns of financial demand and supply mean that this option is not available.

It is hard to see, therefore, how substantial progress in our understanding of finance and innovation can be made unless scholars who are interested in the subject are willing to contribute to the compilation of these data so that basic empirical questions can be addressed. At a minimum, we need evidence that allows us to identify variation and change in the historical patterns of financing by firms, industries, and nations. We also need to understand long-term trends in the supply of resources by different financial institutions. For example, how important has the stock market been as a source of finance for corporate investment in different economies at various times? What types of industries and firms did it finance during different periods? Are there marked differences in the characteristics of firms and industries funded by bond markets and banks? (For the weaknesses of existing empirical research on these questions, see O'Sullivan 2004a, b.)

Only when we have identified these basic patterns in financial demand and supply can we hope to understand how they are related to the dynamics of economic change. For example, in the case of a particular industry we might ask how patterns of financial demand and supply are related to the balance between incumbents and entrants in innovative activity. For an economy, the relationship between financial patterns and changes in the structure of the economy, such as the shift from manufacturing to services, are clearly relevant. For understanding these types of interactions between finance and the dynamics of economic change, empirical studies that treat the demand for, and supply of, finance as characterized primarily by quantity and price can only take us so far. Detailed case studies are needed to generate insights on the qualitative dimensions of these interactions that could then be used to illustrate, enrich and perhaps confront some of the implications that can be drawn from quantitative studies.

Of course, progress in the analysis of finance and innovation cannot be made based on empirical research alone. At a minimum, empirical studies need to be theoretically informed in terms of the questions that they pose. Therefore, a certain amount of conceptual development on the relationship between finance and innovation will be necessary to facilitate empirical studies on the subject. One promising route to new ideas is the further development of the implications for resource allocation of observed characteristics of innovation at the level of the firm, the industry and the economy along the lines discussed above. However, theoretical research needs to go farther than an analysis of the implications of the characteristics of innovation for finance to consider how the structural characteristics and evolution of the financial system influence innovative activity in the real economy.

Notes

1. The author gratefully acknowledges the helpful critiques and suggestions that she received from the participants in the various workshops organized to produce this book. She would like to thank Jan Fagerberg, Bronwyn Hall, Bill Lazonick, David Mowery, and Richard Nelson for the particularly detailed comments that they provided on earlier drafts of this chapter as well as Ron Adner and Bruce Kogut, her colleagues at INSEAD, for their helpful suggestions.
2. The term "enterprise finance" is used herein to refer to research on corporate and venture finance.
3. The other major exception is Alfred D. Chandler. From a theoretical perspective, much of his work can be seen as contributing to a similar concept of the firm as that of Penrose. Indeed, in recent years, in reflecting on the general implications of his research, Chandler has tended to use the language of organizational learning (see e.g. Chandler 1992).
4. This is true of most of the contemporary literature on techno-economic paradigms (Freeman and Louçã 2001: 149)
5. In *CSD*, however, Schumpeter was centrally preoccupied with the social and political, as well as the economic, implications of changes in the characteristics of the innovation process.

6. Indeed, Perez explicitly portrays her book as "a 'think-piece,' the spelling out of an interpretation, with enough illustrations to strengthen the case and stimulate discussion" (Perez 2002: xix).
7. In economics, the term "information asymmetry" is used to refer to a situation in which one economic agent has more information than another about a task that affects both agents' welfare.
8. For general discussions of the difference between structural or radical uncertainty and parametric uncertainty, see Loasby 1976; O'Driscoll and Rizzo 1985; Langlois 1986; Shackle 1992.

References

ALLEN, F., and GALE, D. (2000), *Comparing Financial Systems*, Cambridge, Mass.: MIT Press.

ARROW, K. (1974), *The Limits of Organization*, New York: Norton.

—— and DEBREU, G. (1954), "Existence of an Equilibrium for a Competitive Economy," *Econometrica*, 22: 265–90.

BAH, R., and DUMONTIER, P. (2001), "R&D Intensity and Corporate Financial Policy: Some International Evidence," *Journal of Business Finance and Accounting* 28(5/6): 671–92.

Bank of England (2001), *Financing of Technology-Based Small Firms*, London: Domestic Finance Division, Bank of England, February.

BECK, T., and LEVINE, R. (2002), "Industry Growth and Capital Allocation: Does Having a Market- or Bank-Based System Matter?" *Journal of Financial Economics* 64(2): 147–80.

BERGER, A., and UDELL, G. (1998), "The Economics of Small Business Finance: The Roles of Private Equity and Debt Markets in the Financial Growth Cycle," *Journal of Banking & Finance* 22(6–8): 613–73.

BOWER, J. (1970), *Managing the Resource Allocation Process: A Study of Corporate Planning and Investment*, Cambridge, Mass.: Division of Research, Graduate School of Business, Harvard University.

BURGELMAN, R., and SAYLES, L. (1986), *Inside Corporate Innovation: Strategy, Structure, and Managerial Skills*, New York: Free Press.

*BYGRAVE, W., LANGE, J., ROEDEL, J. R., and WU, G. (2000), "Capital Market Excesses and Competitive Strength: The Case of the Hard Disk Drive Industry, 1984–2000," *Journal of Applied Corporate Finance*, 13(2): 8–19.

CARLIN, W., and MAYER, C. (2003), "Finance, Investment and Growth," *Journal of Financial Economics* 69: 191–226.

*CARPENTER, M., LAZONICK, W., and O'SULLIVAN, M. (2003), "The Stock Market and Innovative Capability in the New Economy: The Optical Networking Industry," *Industrial and Corporate Change* 12(5): 963–1034.

*CARPENTER, R., and PETERSEN, B. (2002), Capital Market Imperfections, High-Tech Investment, and New Equity Financing," *Economic Journal* 112: F54–F72.

CHANDLER, A. (1992), "Organisational Capabilities and the Economic History of the Industrial Enterprise," *Journal of Economic Perspectives* 6(3): 79–100.

CHRISTOFIDIS, C., and DEBANDE, O. (2001), "Financing Innovative Firms Through Venture Capital," EIB Sector Papers.

DEMIRGÜÇ-KUNT, A., and MAKSIMOVIC, V. (2002), "Funding Growth in Bank-Based and Market-Based Financial Systems: Evidence from Firm-Level Data," *Journal of Financial Economics*, 65: 337–63.

FAZZARI, S., HUBBARD, G., and PETERSEN, B. (2000), "Investment-Cash Flow Sensitivities are Useful: A Comment on Kaplan and Zingales," *Quarterly Journal of Economics* 115: 695–705.

FREEMAN, C. (1977), "The Kondratiev Long Waves, Technical Change and Unemployment," in *Structural Determinants of Employment*, vol. 2, Paris: OECD, 181–96.

—— and LOUÇÃ, F. (2001), *As Time Goes By: From the Industrial Revolutions to the Information Revolution*, Oxford: Oxford University Press.

GOMPERS, P., and LERNER, J. (1999), *The Venture Capital Cycle*, Cambridge, Mass.: MIT Press.

* Asterisked items are suggestions for further reading.

GANS, J., HSU, D., and STERN, S. (2000), "When Does Start-Up Innovation Spur the Gale of Creative Destruction?" NBER Working Paper 7851.

*HALL, B. (2002), "The Financing of Research and Development," NBER, Working Paper 8773.

HIMMELBERG, C., and PETERSEN, B. (1994), "R&D Internal Finance: A Panel Study of Small Firms in High-Tech Industries," *Review of Economics and Statistics* 76(1): 38–51.

HOBIJN, B., and JOVANOVIC, B. (2001), "The Information-Technology Revolution and the Stock Market: Evidence," *American Economic Review* 91(5): 1203–20.

HUBBARD, G. (1998), "Capital-Market Imperfections and Investment," *Journal of Economic Literature* 36(1): 193–225.

JOVANOVIC, B., and GREENWOOD, J. (1999), "The Information-Technology Revolution and the Stock Market," *American Economic Review* 89(2): 116–22.

*——and ROUSSEAU, P. (2001), "Why Wait? A Century of Life before IPO," *American Economic Review* 91(2): 336–41.

KAPLAN, S., and ZINGALES, L. (2000), "Investment-Cash Flow Sensitivities are not Valid Measures of Financing Constraints," *Quarterly Journal of Economics* 115: 707–12.

KLEPPER, S. (2001), "Employee Startups in High-Tech Industries," *Industrial and Corporate Change* 10: 639–74.

KLINE, S., and ROSENBERG, N. (1986), "An Overview of Innovation," in R. Landau and N. Rosenberg (eds.), *The Positive Sum Strategy: Harnessing Technology for Economic Growth*, Washington, DC: National Academic Press, 275–305.

* LAMOREAUX, N., and SOKOLOFF, K. (eds.) (2004), *The Financing of Innovation in Historical Perspective*, Cambridge, Mass.: MIT Press.

LANGLOIS, R. (1986), "Rationality, Institutions and Explanation," in id., *Economics as a Process: Essays in the New Institutional Economics*, Cambridge and New York: Cambridge University Press, 225–55.

LAZONICK, W. (1991), *Business Organization and the Myth of the Market Economy*, New York: Cambridge University Press.

LEVINE, R. (2003), "More on Finance and Growth: More Finance, More Growth?" Federal Reserve Bank of St. Louis, July–August, 31–46.

LOASBY, B. (1976), *Choice, Complexity, and Ignorance: An Inquiry into Economic Theory*, Cambridge and New York: Cambridge University Press.

MALERBA, F., NELSON, R., ORSENIGO, L., and WINTER, S. (1999), "'History-friendly' Models of Industry Evolution: The Computer Industry," *Industrial and Corporate Change* 8: 3–40.

McMILLAN, G. S., NARIN, F., and DEEDS, D. (2000), "An Analysis of the Critical Role of Public Science in Innovation: The Case of Biotechnology," *Research Policy* 29(1): 1–8.

MODIGLIANI, F. and MILLER, M. H. (1958) "The Cost of Capital, Corporation Finance and the Theory of Investment," *American Economic Review* 48(3): 261–97.

MOWERY, D., and LANGLOIS, R. (1996), "Spinning off and Spinning on (?): The Federal Government Role in the Development of the US Computer Software Industry," *Research Policy* 25(6): 947–66.

NELSON, R. (1991), "Why do Firms Differ and how does it Matter?" *Strategic Management Journal* 12: 61–74.

——(1993), *National Innovation Systems: A Comparative Analysis*, Oxford: Oxford University Press.

O'DRISCOLL, G., and RIZZO, M. (1985), *The Economics of Time and Ignorance*, Oxford and New York: Basil Blackwell.

O'SULLIVAN, M. (2004*a*), "The Financing Role of the US Stock Market in the 20th Century," Working Paper, INSEAD.

—— (2004*b*), "Historical Patterns of Corporate Finance at General Electric and Westinghouse Electric," Working Paper, INSEAD.

* PEREZ, C. (2002), *Technological Revolutions and Financial Capital: The Dynamics of Bubbles and Golden Ages*, Cheltenham, UK and Northampton, Mass.: Edward Elgar.

RAJAN, R., and ZINGALES, L. (1995), "What do we know about Capital Structure: Some Evidence from International Data," *Journal of Finance* 50(5): 1421–60.

*—— —— (1998), "Financial Dependence and Growth," *American Economic Review* 88(3): 559–86.

ROSEGRANT, S., and LAMPE, D. (1992), *Route 128: Lessons from Boston's High-Tech Community*, New York: Basic Books.

ROSENBERG, N. (1994), *Exploring the Black Box: Technology, Economics, and History*, Cambridge and New York: Cambridge University Press.

SAHLMAN, W. (1990), "The Structure and Governance of Venture-Capital Organizations," *Journal of Financial Economics* 27(2): 473–521.

*—— and STEVENSON, H. (1985), "Capital Market Myopia," *Journal of Business Venturing* 1(1): 7–30.

SAXENIAN, A. (1994), *Regional Advantage: Culture and Competition in Silicon Valley and Route 128*, Cambridge, Mass. and London: Harvard University Press.

SCHUMPETER, J. (1939), *Business Cycles*, vol. 1, New York: McGraw Hill.

—— (1942), *Capitalism, Socialism, and Democracy*, New York and London: Harper & Brothers.

—— (1949), "Economic Theory and Entrepreneurial History," in Research Center in Entrepreneurial History, Harvard University, *Change and the Entrepreneur*, Cambridge, Mass.: Harvard University Press.

—— (1954), *History of Economic Analysis*, Oxford: Oxford University Press.

—— (1964), *Business Cycles: A Theoretical, Historical, and Statistical Analysis of the Capitalist Process*, abridged, with an introd., by Rendigs Fels, New York: McGraw-Hill.

—— (1975), *Capitalism, Socialism and Democracy*, New York: Harper Torchbooks.

—— (1996), *The Theory of Economic Development*, New Brunswick: Transaction Publishers.

SHACKLE, G. L. S. (1992), *Epistemics and Economics: A Critique of Economic Doctrines*, New Brunswick, Transaction Publishers.

TEECE, D. (1986), "Profiting from Technological Innovation," *Research Policy* 15(6): 285–305.

—— PISANO, G., and SHUEN, A. (1997), "Dynamic Capabilities and Strategic Management," *Strategic Management Journal* 18(7): 524–6.

WINTER, S. (1984), "Schumpeterian Competition in Alternative Technological Regimes," *Journal of Economic Behavior and Organization* 5: 287–320.

WRIGHT, G. (1997), "Towards a More Historical Approach to Technological Change," *The Economic Journal* 107(444): 1560–6.

* ZINGALES, L. (2000), "In Search of New Foundations," *Journal of Finance* 55(4): 1623–53.

—— (2003), "Commentary," Federal Reserve Bank of St. Louis, July–August, 47–52.

CHAPTER 10

INNOVATION AND INTELLECTUAL PROPERTY RIGHTS

OVE GRANSTRAND

10.1 INTRODUCTION[1]

THE use of property-like rights to induce innovations of various kinds is perhaps the oldest institutional arrangement that is particular to innovation as a social phenomenon. It is now customary to refer to these rights as intellectual property rights (IPRs), comprising old types of rights such as patents for inventions (judged as sufficiently novel, non-obvious and useful), trade secrets, copyrights, trademarks, and design rights, together with newer ones such as breeding rights and database rights.[2] The various IPRs usually have long legal and economic histories, often with concomitant controversies. Nonetheless, despite their long history, until recently IPRs did not occupy a central place in debates over economic policy, national competitiveness, or social welfare. In the last quarter of the twentieth century, however, a new era—dubbed the pro-patent or pro-IP era—emerged, first in the US and then diffused globally. This change was embedded in a deeper, more broad-based and much slower flow of events towards a more information- (knowledge-) intensive and innovation-based economy. (This type of economy has in recent years been dubbed the "new economy" somewhat misleadingly, as if the entire economy has suddenly changed into something new, replacing the old.) These changes provided policy makers in both developed and developing countries with new challenges.

10.2 HISTORY OF THE IPR SYSTEM

The brief historical account below will focus primarily on patents, being in general the most important and representative IPR, and will be divided into eras, summarized in Table 10.1.

10.2.1 The Non-Patent and Pre-Patent Era

Ancient cultures, as in Babylonia, Egypt, Greece, and the Roman Empire are not known to have had any patent-like institutions for technical inventions, but there are clear indications of other forms of IP in these cultures. It was not until late medieval times that patent-like institutions started to appear, mostly in the form of privileges granted by rulers to special individuals or professions.

10.2.2 The National Patent Era

Concepts of IP became more elaborate and closely linked to political institutions as trade and technology developed in the Middle Ages. In 1474 Venice promulgated the first formal patent code. Inventions shown (at least by a model) to be workable and useful received ten years of protection from imitation, subject to certain compulsory licensing provisions. The 1474 patent code constituted a policy for Venice to attract engineers from the outside and stimulate orderly technical progress. These laws signified the emergence of a new era, which we refer to as the "national patent era," since patent systems typical of this period were national (or local) phenomena pertaining only to single city-states or nations.

The granting of patent-like privileges by governments or rulers was not confined to Venice and the practice spread within Europe. As nation states with more absolutist governments emerged, controversies also emerged between governments and rulers regarding the conditions for granting patents and monopoly privileges.[3]

The practice of granting patents also spread in England and France during the sixteenth century as part of national mercantilist policies. Thus patents became

Table 10.1 Eras in the history of patents and IP[1]

Era	Characteristics
1. Non-patent era (Ancient cultures: Egypt, Greece, etc.)	Emergence of science separated from technology Emergence of cultural and industrial arts Secrecy and symbols emerging as recognized IP No patent-like rights or institutions for technical inventions
2. Pre-patent era (Middle Ages to Renaissance)	Emergence of universities Secrecy, copyright and symbols (artisan/trade marks/names) as dominant IP, also collectively organized Emerging schemes to grant privileges and remunerate disclosure Extensions of mining laws to inventions
3. National patent era (Late 15th–late 18th cent.)	Breakthrough of natural sciences Local codifications of laws for patents (Venice 1474, England 1623, etc.), copyrights (Venice 1544, England 1709, etc.), etc. Regulation of privileges Conscious stimulation of technical progress at national level, linked to economic policies (e.g. mercantilistic)
4. Multinational patent era (Late 18th–late 19th cent.)	Emergence of modern nation states Industrialization Continued international diffusion of the patent system Local anti-patent movements Emerging international patent relations (e.g. disputes)
5. International patent era (Late 19th–late 20th cent.)	Emerging industrial and military R&D International coordination of the patent system (Paris Convention 1883, WIPO, PCT, EPO, etc.) Separate IP regimes in socialist countries and LDCs
6. The pro-patent/pro-IP era (Late 20th cent.–?)	Intellectual capital surpasses physical capital for many entities Intensified international competition Global activism for IP from industrialized countries, especially from the US (leading to TRIPS and the WTO) Almost worldwide adoption of the patent system Increased international patenting

Note:
[1] Discerning eras, epochs or stages in a historical stream of events may be a useful sorting device but it always involves some arbitrariness, even if good criteria are used. (Here the degrees of codification and geographical diffusion of the patent system are used as primary criteria for distinguishing different eras.) Also, beneath the events that surface in an era is often an undercurrent of events that lead up to a later era.

linked to trade policies, a link that has been important as well as controversial ever since.

An important event in the early diffusion of the patent system was the passage in 1623 of the Statute of Monopolies by the English Parliament, which gave a clear recognition of the underlying ideas and specific form of a patent system.[4] This later came to serve as a model, for example, for British colonies in North America, which started to adopt similar patent laws in the seventeenth century. An interesting feature of the statute was that although the patent granted monopoly privileges to the true and first inventor, the invention had to be new in England. This provision was intended to stimulate domestic technical progress (e.g. by attracting foreign engineers and entrepreneurs to England) and reflected concern by England's political leadership that the nation had fallen back in some technical areas and needed to catch up. The statute established a 14-year lifetime for a patent (twice the time needed for a master to train a generation of apprentices). A third interesting feature of the statute was its explicit shift of the granting authority from a royal ruler or sovereign to a government or its bureaucracy. The government was considered the source of patent rights, in contrast to the views that patent rights derived from sovereigns or were natural rights of the individual. The latter view underlined the French patent law at the time of the French Revolution in 1791 and lived on in nineteenth-century France.

Another important event was the US enactment of a federal patent law in 1790. The importance attached to patents and individual IPRs in the newly created USA is clear from the fact that the American Constitution stated that Congress had the power "to promote the progress of science and useful arts, by securing for limited times to authors and inventors the exclusive right to their respective writings and discoveries".

Thomas Jefferson played a key role in the early days of the US patent system. As Secretary of State he was responsible for administering the patent laws, and as a head of a newly created "Patent Board", he personally examined patent applications. He was noted for his opposition to monopolies but believed in the value of limited monopolies for authors and inventors. The new US patent system had a slow start, just as it had had in Venice three centuries earlier and would have in Japan a century later. The Act of 1793 made substantial changes, omitting the requirement that a patentable invention had to be "sufficiently useful and important". The examination of applications for novelty and usefulness was replaced by mere registration, making the issuing of patents more or less a clerical matter, and the Patent Board was abolished. The Patent Act of 1836 in essence reestablished the examination system that had been in place until 1793 and created an executive Patent Office as a separate bureau within the Department of State. The US Patent Office created by the 1836 Act was administered by a Commissioner of Patents, appointed by the President upon approval by the Senate. The present US system for reviewing and administering the patents is largely based on the principles set forth in the Act of 1836.

Box 10.1 International IP conventions

The Paris International Convention for the Protection of Industrial Property of 1883 (the "Paris Convention") covering patents, trademarks, and designs and the accompanying Berne Convention for the Protection of Literary and Artistic Works of 1886 covering copyright and some related rights were the results of complex interplay between different interests during preceding decades. Switzerland without a patent system at the time but with an increasingly pro-patent watch industry was one of the countries active in pulling through the Paris Convention (plus housing the Berne Convention) and was entrusted with secretariats to administer and supervise these conventions. The secretariats shortly thereafter merged into a bureau ("BIRPI"). In 1967 it was reorganized as the World Intellectual Property Organization (WIPO). Largely through the diplomatic efforts of the former US representative to the Paris Union and the Berne Union (respectively consisting of the signatories to the Paris Convention and the Berne Convention), WIPO later became a United Nations agency in 1974.

The Paris Convention is based on two major principles: (*a*) Foreigners and foreign patent applications should receive the same treatment in a member state as domestic applicants and applications (non-discrimination); (*b*) a priority claim established in one member state should be recognized by all others, i.e. once an application for a patent is filed in a member state, the applicant can within twelve months file a patent application for the same invention in any other member state, which must regard the latter applications as being filed on the date of the original first application.

10.2.3 The Multinational Patent Era

The period from the late eighteenth to the late nineteenth century was characterized by the diffusion of the patent system throughout the industrial and industrializing economies, though at an uneven rate and not without setbacks. An anti-patent movement emerged in Germany and somewhat later in Holland, where patent laws were repealed in 1869, and Switzerland rejected several patent law proposals. Even England considered a proposal to weaken significantly her patent laws, and France had already weakened patent protection at the time of the French Revolution.

The anti-patent movement was a consequence of free trade and anti-monopoly movements which considered patents to be associated with mercantilist policies and monopoly privileges. However, interest groups in emerging industries and in some strong-patent nations created pro-patent lobbying groups that gradually gained influence. Finally, the worldwide depression in the 1870s revived protectionism and the anti-patent era by and large ended in the 1870s.

The case of Switzerland provides an interesting illustration of the forces affecting the international diffusion of the patent system. After popular referendums in 1866 and 1882 had rejected proposals to introduce patent laws, Switzerland finally approved such laws in a referendum in 1887, mainly because its important watch industry was under pressure from foreign imitations. However, the 1887 law limited patent protection within Swiss borders to mechanical inventions, since firms in the emerging Swiss chemical industry wanted to imitate and catch up with the more advanced German chemical industry. After Germany threatened Swiss chemical firms with retaliatory tariffs, Switzerland extended its patent coverage to include chemical process (but not product) inventions in 1907 (see Penrose 1951 and Kaufer 1989).

10.2.4 The International Patent Era

Eventually the patent system was widely adopted, concomitant with the growth of international trade and competition in industrial goods. Nation states adopted various policies for promoting their industries, policies that often discriminated against foreign individuals or firms, in turn creating a need for international cooperation in patent matters. The Paris Convention of 1883 was the first milestone in this respect, followed by several other treaties and agreements, such as the Berne Convention for copyrights in 1886, covering a wide range of IPRs (see Box 10.1).

The emergence of industrial R&D in the twentieth century transformed the modes and settings for innovative work. The individual inventor, who was the original target for patent laws, gradually became less important. Inventions increasingly required large resources, and industrial firms became the prime movers of technology in both the East and the West. Economic and industrial differences between various categories of nations increased and became alarmingly large, creating tension among institutions, including national IP regimes in developed and developing nations. Science and technology progressed and accumulated tremendously at an increasing pace. Nonetheless, the IP system and its essential ideas survived and continued to spread internationally, not least after the downfall of the Soviet Union and other planned economies.

International harmonization of the world's patent systems received a new impetus after the end of World War II, as part of a broader set of efforts to establish or strengthen international organizations. A convention establishing the World Intellectual Property Organization—WIPO—was promulgated in 1967 by fifty-one governments, mostly from developed nations. WIPO joined the UN system in 1974 and thereby came under much stronger influence from developing countries. Although WIPO was established to administer and supervise various international IP treaties such as the Paris Convention, the organization also became involved in teaching, arbitration, and consultancy, and the processing of patent applications within the framework of the Patent Cooperation Treaty (PCT), signed in 1970 but not effective until 1978. This treaty was an important step in the process of international harmonization, since it established an international clearing house enabling a patent application to take effect in some or all of the PCT member states (103 in 1999) at the applicant's choosing.

In Europe the European Patent Convention (EPC), which was signed in 1973 and became effective in 1978, began a process for the adjustment of national patent laws of signatory nations (thirteen in 1986) to a European standard. The European Patent Office (EPO) was established in 1977 in Munich to process patent applications for the protection of an invention in some or all signatory nations. However, a patent issued by EPO is only a bundle of national patent rights that are enforceable according to the local law and court system in each national jurisdiction. A European Community Patent Convention was signed in 1975 to establish a unified European patent that would be valid in all member states, but this goal has not been achieved as of 2003. A key issue in the full harmonization of the European IPR system is the design of a court system and court procedures for enforcement, potentially including the creation of a single European Court of Appeal for IPR disputes similar to the US Court of Appeals for the Federal Circuit (CAFC), discussed below.

The case of Japan provides an interesting illustration of the creation of a patent system for the purpose of catching up with advanced nations. The visit by US Commodore Perry to Japan in 1852 demonstrated to Japanese leaders the power of modern military innovations and forced Japan to reopen the country to foreigners. The Meiji Restoration and its broad program of industrial modernization and "catch-up" led to Japan's first patent law in 1871. In the following decades, new laws were enacted for various IPRs (patents, trademarks, utility models, and designs), each of them modeled on various European and US laws. A Japanese Patent Office was established in 1885, with K. Takahashi, who subsequently served as Japan's Prime Minister, as its first Director General. The Japanese patent system evolved over the years into an important vehicle for catching up and promoting national interests. In the beginning, foreigners were barred altogether from obtaining patent rights but became eligible when Japan in 1899 became a member of the Paris Convention.

The postwar IP system in Japan was but one component of a broader complex of policies for trade, industry, and technology that focused on reconstruction and "catch-up" with the West, especially the United States. Laws were passed in 1950 for regulating foreign investment, exchange, and trade, inaugurating a period of substantial technology imports from the US and Europe. Japanese government agencies and firms collected and analyzed technical information, including information disclosed in domestic and foreign patent documents, to evaluate technological developments abroad and within Japan. The Japanese requirement for publication of patent applications within eighteen months of the filing of an application (a policy similar to that of many European patent systems that was adopted by the United States only in 1999) supported domestic as well as international diffusion of technical information.

The Japanese patent system limited both the number and scope of patent claims. Many Japanese firms acquired large portfolios of relatively narrow domestic patents and participated in dense patent networks (also in their foreign patenting, see Granstrand 1999). IP disputes were avoided and cross-licensing and diffusion of technical information were promoted by special features of Japanese patent laws and practices (see Ordover 1991). Use of patents (both foreign and domestic) by Japanese firms for technological "catch-up" purposes was facilitated by the often lax enforcement by Western firms of their IPRs, as well as the limited attention paid to dynamic competition and IP matters by Western nations until the 1980s.

Nonetheless, Japan supported international harmonization efforts. In 1978 Japan acceded to the PCT. The Japanese Patent Office (JPO), together with a small number (around ten) of other patent offices (in other PCT member states), was entrusted with authority to perform international searches for prior art to assess whether the novelty criterion for patentability was met. Japan subsequently became active in trilateral patent office cooperation among the EPO, JPO, and USPTO (United States Patent and Trademark Office), another vehicle for international coordination and harmonization among the industrial nations.

By 1999, 155 nations had adopted the Paris Convention, which in 1883 had been signed by ten. In 1994, another large step towards international harmonization was taken with the signing of the US-inspired TRIPS agreement (see Box 10.2), considered by most experts to be the most important international IPR agreement since the Paris Convention. The TRIPS agreement has been criticized for favoring developed nations and impeding economic development in developing ones. The least developed nations in particular, many of which lack the capabilities to enter a virtuous catch-up development circle by themselves, may be hindered by the TRIPS agreement, although most such nations have a considerable period of time to adhere to all of the provisions of TRIPS (see e.g. the collection of articles in Mansfield and Mansfield 2000).

Despite long-standing efforts to coordinate and harmonize the national patent laws, many important differences remain, and a global patent system, with international or global patents, seems far away.

10.2.5 The Pro-Patent Era

Towards the end of the twentieth century a new era—the pro-patent era—emerged, characterized by stronger enforcement of a broader array of IPR-holder rights and by additional efforts at international coordination and harmonization. The downfall of the Soviet empire and US diplomatic pressure contributed to a higher rate of convergence of IP regimes in the world, exemplified by the TRIPS agreement and the creation of the WTO (see below).

Four developments in the United States led to the "pro-patent era" (see Table 10.2 for an overview). The first concerned the creation in 1982 of the Court of Appeals for the Federal Circuit (CAFC) to hear patent appeals in lieu of the other circuit courts of appeal.[5] This type of specialized court had been discussed for a long time in patent circles.[6] As the complexity of patent disputes grew, the pressures within pro-patent circles in law and industry for a specialized court of appeals mounted and finally resulted in the creation of the CAFC. As many of its proponents had hoped, the CAFC began to act in a pro-patent manner, in contrast to what US courts had done previously. The validity of patents was upheld far more often (as if they were "born valid"), and patent damages were increased. The effect of the CAFC's creation and its decisions was to increase the economic value of patentholder rights.

A second factor behind the emergence of the pro-patent era was linked to a change of attitude within the Antitrust Division of the US Department of Justice in the early 1980s under Assistant Attorney General William Baxter. Since the late 1930s, the Antitrust Division had been hostile to IP legislation and IP licensing, interpreting patents as monopolies harming competition. Baxter was instrumental in shifting the Justice Department's enforcement policy to emphasize the role of patents in promoting innovation, emphasizing the dynamic benefits rather than the static costs. This change in attitude could be traced back to ideas and perspectives emerging in the 1960s among economists, especially within the emerging field of law and economics.[7] The shift in antitrust policy in the early 1980s in the USA is a good (albeit rare) example of how changes in scholarly thinking have had a direct impact on policies.

The third stream of events contributing to the rise of the pro-patent era came from large US corporations that pressed for stronger IP protection and enforcement against infringers and counterfeiters domestically and abroad. US industry also pressed for a "trade-based approach" to improve IP protection by including IP matters in US trade negotiations and in the GATT framework of international trade negotiations, resulting in a number of "trade related aspects of IPRs" (TRIPS) subjected to negotiations (see Box 10.2). These initiatives, which were spearheaded by US pharmaceuticals, entertainment, and electronics firms, were part of a larger upsurge in political concern over the competitiveness of US industry and a growing belief that technology was a key asset that had to be protected. Individual US corporations such as Texas Instruments and Motorola became aggressive litigators against both domestic and foreign, especially Japanese, infringers in the mid-1980s. Most of the largest awards of damages in patent infringement cases, however, occurred in litigation among US firms. A landmark case in this connection was *Polaroid Corp. v. Eastman Kodak*, which in 1991 resulted in a damages award to Polaroid of almost $US900 million.[8] Cases like these and the financial success of the litigation strategy of Texas Instruments were widely publicized and drew the attention of top corporate management to IP matters and the economic value of strong patent portfolios and well-conceived IP strategies.[9]

Box 10.2 Trade-Related aspects of Intellectual Property Rights—TRIPS

The idea of linking IP policies to trade policies can be traced far back in history (for instance, in the national patent era IPRs were often used in a mercantilistic way). The acronym TRIPS refers to a US initiative of the 1980s that sought to link more stringent, internationally harmonized IP policies to international trade policy. The US strategy was to move IPR issues from the auspices of WIPO (seen by US as too weak and narrowly focused) into the GATT Uruguay Round of multilateral trade negotiations in which the US had more influence. The outcome was a success for the US and its allies, but developing nations were frustrated. When the World Trade Organization (WTO) came into being in 1995 as a successor to GATT, the TRIPS agreement was one of its founding components. The agreement consisted of seven parts and seventy-three articles covering all aspects of IPRs, their enforcement and institutional arrangements. It provided general obligations regarding national nondiscrimination and transparency, it stipulated substantive minimum standards in almost all IPR areas (patents, copyrights, trademarks etc.) plus standards for effective enforcement of IPRs (also involving dispute settlement mechanisms at the WTO). It further set up a TRIPS Council for monitoring the operations of the agreement. Finally, transition periods were stipulated, giving one year for developed nations from entry into the WTO to comply with all TRIPS requirements and eleven years for least-developed nations (i.e. until 1 January 2006) with an option to request extensions. The TRIPS agreement implied particularly significant changes in the coverage of patents (forcing many nations to extend patent protection to chemical, pharmaceutical, and biotechnological inventions), requirements for protection of plant varieties, protection of computer software and effective measures to protect trademarks and trade secrets (see e.g. Maskus 2000 for details).

The TRIPS agreement of 1994 has been characterized as the most significant international harmonization effort of IPRs in history, certainly on par with the Paris Patent Convention of 1883. It also appears to become the most controversial one, perhaps creating an anti-IP movement of much larger international proportions than the anti-patent movement in Europe in the 1850s to 1870s. Particularly controversial issues concern developing nations' access to new technologies, especially drugs, and the effects of stronger IPRs on the efforts of these nations to catch-up economically (see e.g. Scherer 2004, Scherer and Watal 2002, and the chapters by Anawalt, Barton, and Verspagen in Granstrand 2003).

A fourth force behind the emergence of the pro-patent era was the US Government, especially the Reagan administration. This "political stream" was also related to the growing domestic concern of the 1980s for US industrial competitiveness, which included the widespread perception that a number of Asian economies were "free-riding" on US technology as they made significant inroads into US markets. In addition, US industrially funded R&D spending was growing slowly during the early 1980s with little or no increase in patenting. Meanwhile foreign corporations, especially Japanese firms, increased their patenting in the USA.[10] One component of a broader policy response to the perceived decline in US competitiveness was legislative action to strengthen IPRs and other incentives to invest in R&D (such as R&D tax credits and favorable conditions for the creation of R&D consortia), and to encourage patenting of the results of federally funded R&D to facilitate interinstitutional R&D collaboration and technology transfer. The Bayh–Dole Act of 1980 simplified the procedures under which US universities could patent and license the results of federally funded R&D (see Ch. 8 by Mowery and Sampat in this volume).

Table 10.2 Chronological overview of major events in US post-war IPR development (through 2000)

Year	Event
1949	Patents so frequently declared invalid when litigated that Supreme Court Justice Jackson remarks, "the only patent that is valid is one which this Court has not been able to get its hands on". (*Jungerson v. Ostby & Barton Co.*)
1952	The present (as of 2003) US Patent Law is passed. Revisions have occurred continually.
1976	US Copyright Act enacted.
1979	US Senate and President Carter desire to strengthen domestic patent enforcement.
1980	US Supreme Court declares man-made microorganisms to be patentable and states in a dictum that "anything under the sun that is made by man" can be patented. Bayh–Dole Act enacted, facilitating for universities to patent inventions from federally funded research.
1981	The US Justice Department revises its antitrust enforcement activity to make it easier for patents not to violate antitrust statutes. US Supreme Court decision in the Diehr case leads through its USPTO interpretation to patentability of certain computer software.
1982	CAFC is established. In quick order, the court changes the validity of litigated patents from 30% to 89%, thus initiating an era in which patents are of much greater interest to industry.
1983	Patent Commissioners' trilateral conference started.
1985	WIPO Harmonization conference. USITC litigations increased. The Young Report delivered to President Reagan by the Commission on Industrial Competitiveness (headed by Hewlett-Packard's John Young).
1986	TI semiconductor patent litigation initiated at USITC. GATT TRIPS negotiations started.
1988	US Trade Act (Special 301). US Tariff Act 337 amended.
1989	The Structural Impediments Initiative (SII) talks initiated between the USA and Japan remove structural impediments to trade between the two nations, and include intellectual property protection. Japan on Watch List of Special 301.
1992	US Patent Law reform report. Honeywell won patent litigation against Minolta.

1993	GATT TRIPS negotiations completed.
1994	World's industrialized nations agree to harmonize aspects of their intellectual property protection under the auspices of GATT, known as the TRIPS agreement. US–Japan Patent Commissioners' Understanding signed. After years of favorable court decisions, all software is now clearly patentable.
1995	GATT-related TRIPS agreement causes USA (and other nations) to amend its patent laws to expand the patent term to 20 years from filing date (from previous 17 years from issuing date, thus giving mixed effects depending upon the application processing time at the USPTO), allow inventive activity abroad to be considered by the patent office, and permit the filing of provisional patent applications.
1998	The CAFC declares inventions of so-called business methods to be patentable (which include e.g. financial inventions, teaching methods, and e-commercial methods) in *State Street Bank and Trust v. Signature Financial Group* by stating that "since the 1952 Patent Act, business methods have been, and should have been, subject to the same legal requirements for patentability as applied to any other process or method". The Digital Millennium Copyright Act enacted.

The CAFC and the change in antitrust policies paved the way for effective domestic enforcement of existing US IP laws. The trade-based approach to IP legislation, however, focused primarily on international standards and enforcement of intellectual property protection. This effort was largely successful (from the US point of view), in part because the US Congress created leverage for US trade negotiators through a number of changes in US trade laws.[11] However, the pro-patent era, set in motion by the actions of US corporations and policy makers, gained ground internationally for other reasons as well. Technology-based MNCs, not only in the USA but also in Europe and especially in Japan, shared an interest in stronger international protection for intellectual property.

There is an ongoing debate, fueled by the bursting of the "IT bubble," as to whether in fact a new type of economy has emerged and what characterizes such a "New Economy". Although much of the "New Economy" rhetoric is now discredited, many scholars believe a new type of economy has emerged, albeit gradually, in which intellectual capital has surpassed physical capital in importance. "Intellectual capitalism", then, refers to a capitalist economic system with a dominance of intellectual capital (see Granstrand 1999). What role did the IP system and the pro-patent era play in the emergence of "Intellectual capitalism"? A definitive answer to this question is difficult at this stage, but a few observations may still be in order.

ICTs are generally recognized as a key technological contributor to the emergence of "intellectual capitalism" as well as the "New Economy." It is natural to ask, therefore, how important the IP system was for the emergence of ICTs; let us consider some well-known cases. The transistor was patented at Bell Labs but licensed liberally (in part because of antitrust litigation and pressure from the US Justice Department). The subsequent emergence of the semiconductor industry was significantly spurred by public procurement and a lax IP regime (Mowery 1996). The same could be said about the emergence of Internet under the Defense Advanced Research Projects Agency (DARPA). The software industry also emerged under a lax IP regime (Samuelson 1993). The telecom industry was largely operated by national monopolies until the 1980s and 1990s, and IPRs played little role in the rapid advance of technology in this industry. Mobile telephony also emerged until the late 1980s under a lax IP regime (Granstrand 1999). The conclusion seems to be that the IP system has not been of major importance for the emergence of ICTs (at least not in the early stages). In fact it may even be argued that lax IP regimes were instrumental for the emergence of several ICT industries.

The strengthening of the IP regime may have reinforced some features of intellectual capitalism, but it appears that the pro-patent era was as much a consequence of intellectual capitalism as a cause of it, and that it was not a necessary condition for the emergence of the industries and technologies that fostered it.

10.3 ROLE OF IPRs IN INNOVATION SYSTEMS

10.3.1 Perspectives on IPRs

The IP system has over the years spawned a series of legal and economic controversies. Among the legal controversies is the nature of IPRs: are they rights in the first place? Couldn't a liability approach do better? Do they have to be exclusive and/or temporary rights? And if a right, what kind of right? Is it an individual natural (or moral) right or a right conferred to the individual by society, justified on the grounds that its consequences are beneficial to the society? These types of questions are primarily addressed by legal scholars in jurisprudence and, although important, will not be discussed further here.

As for economic issues, there has been a continuing discussion (with varying intensity) over the centuries about the pros and cons of the patent system. One key question is whether the system can correct for (or lead to) over- or underinvestment in R&D and innovation (from a societal point of view). Another issue is whether the system distorts, redirects, or blocks technological progress. Still another (but related) topic concerns how the patent system affects static and dynamic efficiency through its impact on competition and trade. For a classic review of these issues, see Machlup (1958); for a more recent contribution, see Mazzoleni and Nelson (1998).

Surprisingly, for much of the twentieth century, economists devoted little attention to the patent system, and even less to other IPRs. From the 1960s onwards, however, the literature on the economics of IPRs, and patents in particular, has grown significantly. A seminal work in this category is Arrow (1962) who argued that, from society's point of view, private firms will underinvest in R&D because of their inability to appropriate sufficient returns of their R&D investments.[12] Following this view, patent protection can be justified as one of several alternative means, such as contracts, prizes, subsidies, and research consortia, to deal with this market failure. Alternative means to address market failures in the innovation process are also discussed by, among others, Wright (1983) and David (1993) emphasizing conditions and factors (such as uncertainty and elasticity of research supply) determining their relative advantages.

The opposite view, that capitalist economies may overinvest in R&D and innovation, also has its adherents. A number of recent theoretical contributions in industrial economics, growth theory, and behavioral finance examine this possibility, highlighting the ways in which competitive races affect incentives for innovation and can result in overinvestment in R&D. The intuition behind such models is that patent races may result in substantial duplication of R&D investments by the competing firms and lead to industry-wide levels of R&D investment for which the social returns may be less than the cost of the overall investment.[13] Such overinvestment is most likely in races in which rewards are skewed heavily to the early finishers. In this case agents face strong incentives to accelerate both the start and the completion of R&D projects (at least until rents are dissipated).

Thus appropriability problems may lead to underinvestment in innovation, e.g. in "waiting games", on the one hand, or prospects of a quick success may lead to "patent races" that result in overinvestment. Which type of game is actually observed will depend on many circumstances. Most of this literature is theoretical, and few empirical studies have addressed these theories and models.

Another important issue that attracts considerable interest concerns the length of the time for which an inventor should be awarded a patent. For instance, Nordhaus (1969) argued that increasing the length of patent protection increases the incentives for investment in process innovation (and hence "dynamic efficiency") but at the expense of "static efficiency" (since increased protection means less competition, higher prices, and slower diffusion). An "optimal patent length", Nordhaus pointed out, involves a trade-off between these two effects, and will depend on the nature of competition, the price elasticity of demand and the R&D elasticity of process cost reduction.

More recent research focuses on the optimal breadth or scope of a patent (see Jaffe 2000 for an overview) as well as optimal combinations of length and breadth (Klemperer 1990). The scope of a patent defines the range of its industrial applications by delineating the set of technological designs that the claims in the issued patent give protection to (i.e., exclusion of imitators). In most industrial-economy patent systems, patent scope is initially determined by negotiations between the applicant and a patent examiner; ultimately, however, the scope for "important" patents (those deemed by patentholder or competitors as especially valuable) is likely to be determined in the courts as a result of private lawsuits initiated by the patentholder or others.

The scope of a patent is far more difficult to parametrize than its duration. Thus, the "optimal" scope for patents is a very complex issue, as shown by Merges and Nelson (1990). The scope of a patent affects the private as well as the social rates of return from patented industrial innovations (just as the time duration of a patent does), and these returns will vary among industries and technologies. Accordingly, it is difficult for legislators to design an overall "optimal" patent system that fits equally well everywhere. For example, patent offices often issue broad patents to applicants pioneering in a new technological field characterized by little patent-based "prior art" and a great deal of uncertainty. (Such patents are often regarded as being too broad by ex post observers.)

In order to obtain a patent for an invention, the inventor has to disclose information. The resulting disclosure of patent information accelerates the diffusion of patented technical information, and may reduce duplicate R&D, induce substitute technologies (through "inventing around" an important patent), stimulate new ideas, direct R&D efforts to opportunity-rich areas or bottleneck problems, provide a basis for bench-marking and competitive intelligence, and stimulate technology exchange and cooperation.[14] Thus, the disclosure requirement is one of the key features of the patent system and a rationale behind it (see Ordover 1991). It should be kept in mind, however, that there are several other channels for dissemination than just disclosure of patent information, and that, in general, information about new technologies leaks out fairly quickly (see Mansfield 1985).

In summary, IPRs, particularly patents, play several important roles in innovation systems—to encourage innovation and investment in innovation, and to encourage dissemination (diffusion) of information about the principles and sources of innovation throughout the economy. However, as we shall see, the importance of these roles varies across sectors (industries) and countries, and over time.

10.3.2 Evidence on IPRs

Mansfield (1986), in an empirical study of US firms, shed light on the impact of a possible abolition of the patent system on the rate of invention and innovation, and concluded that the effects of abolition would be small in most industries. The exceptions were pharmaceuticals and chemicals, for which the patent system was shown to be essential. In spite of this evidence on the relatively modest effects of patents on innovation, however, the firms in Mansfield's study patented extensively.

The propensity to utilize the patent system (instead of alternative means) is the subject of a large literature: see e.g. Scherer (1983) and Arundel and Kabla (1998). The Yale study by Levin, Nelson and others (Levin et al. 1987) investigated, through a survey of hundreds of R&D managers in the US in more than a hundred industries, sector-specific variations in appropriability conditions and the role of patents. Their study also concluded that innovations would continue to appear in the absence of patent protection, and that in general patents were not sufficient to appropriate or capture all benefits from innovation (once again, a significant exception to this general finding was pharmaceuticals, where patent protection was deemed to be especially valuable). The Yale study was followed up by an expanded international study (the Carnegie–Mellon study), which revealed substantial nation-and sector-specific differences in the use of patents, secrecy, lead times and other means for appropriation of the returns from innovation (see e.g. Cohen et al. 2003). For instance, the latter study indicated that lead time and patents were the most important appropriation mechanisms for Japanese firms, while lead time and secrecy were most important for US firms.

A comparative study of Japanese and Swedish corporations, accounting for more than 50 per cent of industrial R&D in Japan and over 90 per cent in Sweden (Granstrand 1999), also shed light on this issue. Table 10.3 compares the results from this study with the results in Levin et al. (1987). As in the Carnegie–Mellon study, Japanese firms stand out by assigning greater importance to patent protection than

Table 10.3 Means for commercializing new product technologies
(Scale: No importance = 0,1,2,3,4 = Major importance)

Means	Japan[1]	Sweden[1]	US[2]
Taking out patents to deter imitators (or to collect royalties)	3.3	1.9	2.0
Exercising secrecy	2.4	2.0	1.7
Creating market lead times	2.7	2.4	2.9
Creating production cost reductions	2.9	2.7	2.7
Creating superior marketing	2.7	3.0	3.1
Creating switching costs at user end	1.9	1.7	n.a.

Notes:
[1] Sample of 24 large corporations. Perceptions for 1992.
[2] As reported in Levin et al. (1987). Perceptions for mid-1980s, rescaled to the scale used in Granstrand (1999).
Source: Granstrand (1999).

firms from other countries. A 1988 study of Swiss firms, however, concluded that lead time was the most important mechanism for appropriating the returns from innovation, while patents ranked lowest (Harabi 1995). One has to interpret these results with care, though. First, the different mechanisms are to some extent complementary (both patents and secrecy may be seen as means to create lead time by increasing speed to market relative to competitors). Second, the attitudes of firms towards IPRs have changed recently, following the emergence of the pro-patent era. Particularly European industry in general was slow to adapt to this new environment.

10.3.3 Differences in IPRs across Sectors

As shown in most if not all studies of the role of IPRs, especially patents, differences across industries or sectors are strikingly large (see e.g. the studies by Mansfield, Scherer, Levin et al., and Cohen et al. cited above and also Malerba, Ch. 14 in this volume).[15] Table 10.4 provides an illustration of this for the sample of large Japanese companies mentioned above. Similar findings have also been reported for US and UK firms (Mansfield 1986, Taylor and Silberston 1973). Several explanations have been set forth for these inter-industry differences, including industry and market structure (competitive conditions, size and diversification of firms, barriers to entry, market growth, R&D intensity etc.), the nature of the technology (technological opportunities, codifiability, capital intensity etc.) and the nature of IPRs (patents for technology, copyright for software and creative industries, trademarks in mass consumer markets, etc.). But these largely static cross-industry comparisons rarely have incorporated considerations of the stage of industries' evolution.[16]

Table 10.4 Sensitivity of the R&D investments of large Japanese corporations to length of term (1992)

What would the effect be on your company's total R&D budget (as a rough percentage), if the maximum length of patent protection was:	Chemical (n=9)	Electrical (n=10)	Mechanical (n=5)	Total (n=24)
(a) Increased by 3 years	+8.5	+2.8	+0.3	+4.8
(b) Decreased to 10 years	−21.2	−3.7	−0.3	−10.7
(c) Decreased to 0 years (i.e. patent protection ceases)	−59.2	−40.0	−5.5	−38.2

Source: Granstrand (1999).

As mentioned above, the role of a strong IP regime in emerging industries is ambiguous. There is some evidence that several leading-edge US industries based on ICTs developed after World War II under a fairly lax IP regime (see above). On the other hand, there also exist cases in which a strong IP regime has fostered the emergence of new leading-edge industries—pharmaceuticals and chemicals are among the best-known examples of such industries.

In general, patents are most likely to support the growth of knowledge-intensive industries in fields characterized by low ratios of imitation to innovation costs. Such low ratios are likely in areas with large-scale R&D projects, especially if the R&D results in highly codified knowledge, as in chemicals (see Table 10.4), and reverse engineering is cheap.[17] In such industries other institutional means to induce innovation than patents are also commonly employed, e.g. procurement contracts, consortia, or natural monopolies. However, many emerging industries are characterized by relatively low innovation costs and strong "first-mover advantages", making firms in such sectors less sensitive to free-rider problems and "waiting games" and, hence, reducing the importance of patents.

In the later stages of industry evolution, the R&D scale is often high, and barriers to entry tend to be built up by incumbents, especially against small firms (see e.g. Granstrand and Sjölander 1990, and Arora et al. 2001). The use of various patent portfolio strategies such as blanketing or "evergreening" together with litigation threats by large firms (both incumbents and diversifying entrants) may serve this purpose (see Granstrand 1999, 2004).[18] This may result in a division of R&D labor, in which small firms specialize in early-stage R&D, and license their new technologies to established firms specializing in later stages of the innovation process, and/or seek to be acquired by established firms (rather than investing in production and marketing).

These intersectoral differences in the importance of IPRs have led several scholars to criticize the "one-size-fits-all" design of the patent system. Reform proposals have suggested a more differentiated industry tailored system, e.g. regarding patent duration or special new (*sui generis*) types of IPRs for certain industries such as software (see e.g. Thurow 1997 and Reichman 1994). Counterarguments to industry-specific schemes for IPRs emphasize the resulting high transaction costs, including IP administration costs, and the fact that a certain amount of industry tailoring already exists in the laws and practices of patent offices and courts. Perhaps because of these complexities, the debate over "industry-specific" patent systems has raged for many years with little or no action by industrial-economy governments.

10.3.4 Differences in IPRs across Nations

There is widespread consensus today that technical progress, the promotion of which is one of the purposes of the patent system, is the major determinant behind economic progress.[19] It is therefore logical to turn to economic history for evidence on the role of IPRs. However, a strong patent system has not been necessary for countries' industrialization and economic growth. Although many countries, including Japan, successfully industrialized in the presence of a patent system (see e.g. Dutton 1984), other countries such as Germany, Holland, and Switzerland did not (see e.g. Kaufer 1989). Schiff (1971), studying Holland and Switzerland, found no evidence that industrialization in these countries was hampered by the absence of a patent system. Moser (2003), studying two nineteenth-century world fairs (in London 1851 and Philadelphia 1876), found no evidence that strong patent laws increased national levels of innovative activities but concluded that they influenced the intersectoral distribution of innovative activities. Lerner (2000), examining 177 policy changes across sixty countries over the past 150 years, found that changes increasing patent protection had much stronger effect on inward patenting by foreigners than on patenting by domestic entities.

Thus the IPR system in general, and the patent system in particular, have been neither necessary nor sufficient for historically significant technical and/or economic progress at national and company level. Although hardly surprising, this is an important conclusion. There seems to be some consensus in the scholarly literature that the patent system has made positive contributions to technical progress, but these contributions are secondary and complementary to other factors, particularly other institutional developments such as a general property rights system (see North 1981 and Nelson 1993).

Moreover, current research provides little guidance on the potential contributions of an internationally strong patent system to the prospects for "catch-up" by the less developed countries in the contemporary world. Indeed, a certain amount of "free-riding" under a weak IP regime, elements of which are apparent in the nineteenth-century United States or the Japanese economy of the 1950–80 period, may aid in successful catch-up. In this context TRIPS may be seen as an attempt by leading countries and companies to increase the economic payoffs of their R&D, making it more costly for developing countries to catch up.

10.4 SUMMARY AND CONCLUSIONS

Instruments, strategies, and policies for protection of IP, ranging from trade secrets and trademarks to copyrights and patents, have a long history. From this perspective, TRIPS is only the latest expression of the links between trade and intellectual property protection. Despite this long history, surprisingly little scholarly attention has been devoted to the study of intellectual property rights and innovation. One impetus for research on IPRs has been a series of controversies in various national jurisdictions over their effects. These controversies reveal many basic similarities across nations and centuries, although they often concern different types of IPRs (patent, copyrights, trademarks, design rights, etc.).

It seems fair to say that until recently the role of IPRs in national, sectoral and corporate innovation systems has been relatively modest (although not without exceptions). However, a US-originated pro-patent or pro-IP era has emerged since the 1980s, perhaps as a consequence rather than a cause of a much broader and gradual transition into a new type of capitalist economy that is based more heavily on knowledge (and information), innovation and intellectual capital. The consequences of the pro-IP era have been far-reaching and seem likely to broaden, since there are no signs of a reversal of these global trends. IPRs have been applied to more forms of intellectual property through extensions of old and creation of new types of rights; they have become more economically valuable and have assumed a more strategic role in national, sectoral and corporate innovation systems.

More research on the complex relations between IP and innovation is clearly needed, especially since the emergence of the pro-patent era has led to significant change and created new challenges (e.g. the so-called "anticommons" problem that stems from a proliferation of interdependent IPRs: see Heller and Eisenberg 1998 for a pioneering work). Future research in this area needs to focus on the interplay between economic, technical, and legal dynamics and, in a better way than before, link theoretical work with empirical longitudinal studies.

Notes

1. Helpful comments from the editors and participants in the TEARI-project and the assistance of Thomas Ewing are gratefully acknowledged.
2. Breeding rights refer to exclusive time limited rights to commercialize certain cultivated plant or animal varieties. Database rights refer to exclusive time limited rights to commercialize certain collections of data or content material, including literary, artistic, and musical works or collections of material such as texts, sound, images, numbers, and facts. However, the term "database" should in this context not be taken to extend to computer programs used in the making or operation of a database. Database rights exist in Europe since the mid-1990s (although with some older precedents) but not in the US, and has so far created a great deal of controversies.
3. The history of the term "patent" is interesting in this context. In English the term was short for "letters patent," in turn deriving from "litterae patentes," which in medieval Europe referred to sealed but open royal letters, granting the holders certain rights, privileges, titles, or offices. The term derives from Latin *patere*, meaning "to be open."
4. In fact, the patent monopoly rights became an exemption in the Statute of Monopolies, which generally limited monopoly privileges. The handing out of such privileges by the

royalty had degenerated, and the English Parliament wanted to put an end to it, apparently recognizing the exceptional importance of technical progress.

5. See further the Federal Courts Improvement Act of 1982 and Dreyfuss (1989).
6. A proposal can e.g. be found in recommendations of the US Senate Committee TNEC from the 1940s, see Folk (1942: 281–95).
7. Prof. William Baxter, personal communications.
8. The total cost to Kodak was much more, however, since Kodak paid damages to customers and legal fees, had to shut down a plant and fire about 700 people, and lost investments and goodwill (see Granstrand 1999 and Rivette and Klein 1999). The original Polaroid damages claim was over $US 5 billion, which if awarded could bankrupt even a large corporation such as Eastman Kodak.
9. A number of law suits against (alleged) Japanese infringers were brought by US firms, and a number of out-of-court settlements also were reached among these parties. Royalty rates for licenses were also raised. In general, these events signified the outbreak of the so-called patent war between USA and Japan. (See Warchofsky (1994) and Granstrand (1999).) The increased management attention paid to patenting also contributed to the surge of patenting in the US (see Kortum and Lerner 1999), as well as in several other developed nations.
10. In fact, the share of foreigners' patenting in the USA rose from 22 per cent in 1967 to 40 per cent in 1980 (Evenson in Griliches 1984, p. 92).
11. Important new trade legislation included the Trade and Tariff Act of 1984, which included Section 301 (authorizing US government to take retaliatory action against countries judged to give an inadequate IP protection) and Section 501 (authorizing the President to judge the adequacy of IP protection in granting tariff preferences to a country), combining to a stick and carrot approach. The Omnibus Trade and Competitiveness Act of 1988 moved further along these lines with a "Special 301" that required the US Trade Representative to watch, identify and investigate foreign states denying adequate IP protection to US firms.
12. Empirical studies have followed this up by collecting data on imitation costs and times in relation to innovation costs and times for a number of sectors. Patents do increase imitation costs, especially in pharmaceuticals, but apart from that industry, patents have not been essential for the rate of innovations, at least not before the pro-patent era. (See in particular Mansfield et al. 1981. See also below.)
13. Central references in this literature include Scherer (1966, 1967), Barzel (1968), Dasgupta and Stiglitz (1980), Fudenberg et al. (1983), and Aghion and Howitt (1992). For overviews, see Baldwin and Scott (1987), Tirole (1988), Romer (1996) and Aghion and Howitt (1998).
14. There is in fact a whole service industry emerging around the processing of patent information for various purposes. Japan is a nation that has significantly fostered and benefitted from patent information analysis through methods subsumed under "patent mapping". Disclosure of patent information imposes a disadvantage upon patentees but its perceived advantages to patentees were found to be significantly higher in Japanese firms (see Granstrand 1999).
15. Cross-national differences might be larger than industry differences, however, as was the case between Japan and Sweden in Granstrand (1999). One has also to keep in mind that firms are diversified and more so regarding their patent portfolio than their product portfolio (see Pavitt 1999).

16. An exception to this statement is the longitudinal study of patenting behaviour in an industry by Hall and Ziedonis (2001).

17. Note that both secrecy building and secrecy breaking in itself is technology dependent and thus results in cost changes as new technologies appear (e.g. in cryptography and chemical analysis).

18. "Evergreening" refers to a patent strategy aimed at prolonging the effective duration of patent protection in a business area through continually patenting related inventions, often incremental ones covering product and process improvements and new applications, as well as through patenting new, emerging technologies for technology transitions into new product generations in the area.

19. Note that a patent is granted to a technical invention based on whether it is novel to the world and non-obvious (i.e. its technical advance has a certain size, i.e. it fulfills a minimum inventive step requirement) but not based on its economic merits (apart from a general requirement of industrial applicability or usefulness of the invention), although the underlying assumption is that by so doing, economic progress will be stimulated.

REFERENCES

AGHION, P., and HOWITT, P. (1992), "A Model of Growth through Creative Destruction," *Econometrica*, 60: 323–51.

——— (1998), *Endogenous Growth Theory*, Cambridge, Mass., London, England: The MIT Press.

ANAWALT, H. (2003), "Intellectural Property Scope: International Intellectual Property, Progress, and the Rule of Law," in Granstrand (2003: 55–76).

*ARORA, A., FOSFURI, A., and GAMBARDELLA, A. (2001), *Markets for Technology: The Economics of Innovation and Corporate Strategy*, Cambridge, Mass.: MIT Press.

*ARROW, K. J. (1962), "Economic Welfare and the Allocation of Resources for Invention," in R. R. Nelson (ed.), *The Rate and Direction of Inventive Activity: Economic and Social Factors*, Princeton: Princeton University Press for the National Bureau of Economic Research, 609–26.

ARUNDEL, A., and KABLA, I. (1998), "What percentage of innovations are patented? Empirical estimates for European firms," *Research Policy* 27(2): 127–41.

*BALDWIN, W. L., and SCOTT, J. T. (1987), *Market Structure and Technological Change*, New York: Harwood.

BARTON, J. H. (2003), "New International Arrangements in Intellectual Property and Competition Law", in Granstrand (2003: 105–22).

BARZEL, Y. (1968), "Optimal Timing of Innovations," *Review of Economics and Statistics* 50: 248–355.

BRANDI-DOHRN, M. (1994), "The Unduly Broad Claim," *International Review of Industrial Property and Copyright Law*, 25(5): 648–57.

CHANDLER, Jr. A. D. (1990), *Scale and Scope—The Dynamics of Industrial Capitalism*, Cambridge, Mass., and London: Belknap Press of Harvard University Press.

* Asterisked items are suggestions for further reading.

COHEN, W. M., GOTO, A., AKIYA, A., NELSON, R. R., and WALSH, J. P. (2003), "R&D Information Flows and Patenting in Japan and the United States," in Granstrand (2003: 123–54).

DASGUPTA, P., and STIGLITZ, J. (1980), "Industrial Structure and the Nature of Innovative Activity," *The Economic Journal* 90: 265–93.

DAVID, P. A. (1993), "Intellectual Property Institutions and the Panda's Thumb: Patents, Copyrights, and Trade Secrets in Economic Theory and History," in M. B. Wallerstein, M. E. Mogee, and R. A. Schoen, (eds.), *Global Dimensions of Intellectual Property Rights in Science and Technology*, Washington, DC: National Academy Press, 19–61.

DREYFUSS, R. C. (1989), "The Federal Circuit: A Case Study in Specialized Courts," *New York University Law Review* 64(1): 1–77.

DUTTON, H. I. (1984), *The Patent System and Inventive Activity during the Industrial Revolution 1750–1852*, Manchester: Manchester University Press.

EVENSON, R. E. (1984), "International Invention: Implications for Technology Market Analysis," in Griliches (1984: 89–126).

FOLK, G. E. (1942), *Patents and Industrial Progress: A Summary, Analysis and Evaluation of the Record on Patents of the Temporary National Economic Committee*, New York: Harper & Brothers.

FUDENBERG, D., GILBERT, R., STIGLITZ, J., and TIROLE, J. (1983), "Preemption, Leapfrogging, and Competition in Patent Races," *European Economic Review* 22: 3–31.

GRANSTRAND, O. (1982), *Technology, Management and Markets: An Investigation of R & D and Innovation in Industrial Organizations*, London: Pinter.

*—— (1999), *The Economics and Management of Intellectual Property*, Cheltenham: Edward Elgar.

*—— (ed.) (2003), *Economics, Law and Intellectual Property*, Dordrecht: Kluwer Academic Publishers.

—— (2004), "The Economics and Management of Technology Trade—Towards a Pro-licensing Era?" *International Journal of Technology Management* 27(2/3): 209–40.

—— and SJÖLANDER, S. (1990), "The Acquisition of Technology and Small Firms by Large Firms," *Journal of Economic Behavior and Organization* 13: 367–86.

GRILICHES, Z. (1984), *R&D, Patents, and Productivity*, Chicago: University of Chicago Press.

HALL, B. H., and ZIEDONIS, R. H. (2001), "The Patent Paradox Revisited: An Empirical Study of Patenting in the U.S. Semiconductor Industry, 1979–1995," *RAND Journal of Economics* 32(1): 101–28.

HARABI, N. (1995), "Appropriability of Technical Innovations: An Empirical Analysis," *Research Policy* 24: 981–92.

JAFFE, A. (2000), "The U.S. Patent System in Transition: Policy Innovation and the Innovation Process," *Research Policy* 29: 531–57.

KAUFER, E. (1989), *The Economics of the Patent System*, New York: Harwood Academic Publishers.

KLEMPERER, P. (1990), "How Broad should the Scope of Patent Protection be?" *RAND Journal of Economics* 21(1): 113–30.

KORTUM, S., and LERNER, J. (1999), "What is behind the Recent Surge in Patenting?" *Research Policy* 28: 1–22.

LERNER, J. (2000), "150 Years of Patent Protection," *NBER*, Working paper No. 7478.

LEVIN, R. C., KLEVORICK, A. K., NELSON, R. R., and WINTER, S. G. (1987), "Appropriating the Returns from Industrial Research and Development," *Brookings Papers on Economic Activity* 3: 783–831.

MACHLUP, F. (1958), *An Economic Review of the Patent System*, Study No 15 of the Subcommittee on Patents, Trademarks, and Copyrights of the Committee on the Judiciary, US Senate, Washington, DC: US Government Printing Office.

MANSFIELD, E. (1985), 'How Rapidly Does New Industrial Technology Leak Out?', *The Journal of Industrial Economics*, XXXIV, No. 2. 217–23

—— (1986), "Patents and Innovation: An Empirical Study", *Management Science* 32(2): 173–81.

—— RAPOPORT, J., ROMEO, A., WAGNER, S., and BEARDSLEY, G. (1977), "Social and private rate of return from industrial innovations," *Quarterly Journal of Economics* 71 (May): 221–40.

—— SCHWARTZ, M., and WAGNER, S. (1981), "Imitation Costs and Patents: An Empirical Study," *Economic Journal* 91 (Dec.): 907–18.

*MANSFIELD, E. D., and MANSFIELD, E. (guest eds.) (2000), "Intellectual Property Protection and Economic Development," *International Journal of Technology Management, Special issue*, 19 (1/2).

*MASKUS, K. E. (2000), *Intellectual Property Rights in the Global Economy*, Washington, DC: Institute for International Economics.

*MAZZOLENI, R., and NELSON, R. R. (1998), "The Benefits and Costs of Strong Patent Protection: A Contribution to the Current Debate," *Research Policy* 27(3): 273–84.

MERGES, R. P., and NELSON, R. R. (1990), "On the Complex Economics of Patent Scope," *Columbia Law Review* 90(4): 839–916.

MOSER, P. (2003), "How Do Patent Laws Influence Innovation? Evidence from 19th-Century World Fairs," National Bureau of Economic Research, Working Paper No. 9909.

MOWERY, D. (1996), *The International Computer Software Industry*. New York: Oxford University Press.

NELSON, R. R. (ed.) (1993), *National Innovation Systems: A Comparative Analysis*, New York and Oxford: Oxford University Press.

NORDHAUS, W. D. (1969), *Invention, Growth and Welfare*, Cambridge, Mass.: MIT Press.

NORTH, D. C. (1981), *Structure and Change in Economic History*, New York: W.W. Norton.

ORDOVER, J. A. (1991), "A Patent System for Both Diffusion and Exclusion," *Journal of Economic Perspectives* 5(1): 43–60.

PAVITT, K. (1999), *Technology, Management and Systems of Innovation*, Cheltenham: Edward Elgar.

PENROSE, E. T. (1951), *The Economics of the International Patent System*, Baltimore: Johns Hopkins University Press.

REICHMAN, J. H. (1994), "Legal Hybrids between the Patent and Copyright Paradigms," *Columbia Law Review* 94(8): 2432–558.

RIVETTE, K. G., and KLINE, D. (1999), *Rembrandts in the Attic: Unlocking the Hidden Value of Patents*, Boston: Harvard Business School Press.

ROMER, D. (1996), *Advanced Macroeconomics*, Berkeley, University of California: McGraw-Hill.

SAMUELSON, P. (1993), "A Case Study on Computer Programs," in Wallerstein et al. (1993: 284–318).

SCHERER, F. M. (1966), "Time-Cost Tradeoffs in Uncertain Empirical Research Projects," *Naval Research Logistics Quarterly* 13: 71–82.

—— (1967), "Research and Development Resource Allocation under Rivalry," *Quarterly Journal of Economics* 81: 359–94.

SCHERER, F. M. (1983), "The Propensity to Patent," *International Journal of Industrial Organization* 1(1): 107–28.

—— (2004), "Global Welfare in Pharmaceutical Patenting," *The World Economy* (forthcoming).

*—— and ROSS, D. (1990), *Industrial Market Structure and Economic Performance*, 3rd edn., Boston: Houghton Mifflin.

—— and WATAL, J. (2002), "Post-trips Options for Access to Patented Medicines in Developing Nations", *Journal of International Economic Law* 5(4): 913–39.

SCHIFF, E. (1971), *Industrialization without Patents*, Princeton: Princeton University Press.

TAYLOR, C. T., and SILBERSTON, Z. A. (1973), *The Economic Impact of the Patent System. A Study of the British Experience*, Cambridge: Cambridge University Press.

THUROW, L. C. (1997), "Needed: A New System of Intellectual Property Rights," *Harvard Business Review* (Sept.–Oct.): 95–103.

TIROLE, J. (1988), *The Theory of Industrial Organization*, Cambridge, Mass.: MIT Press.

VERSPAGEN, B. (2003), "Intellectual Property Rights in the World Economy," in Granstrand (2003: 489–518).

*WALLERSTEIN, M. B., MOGEE, M. E., and SCHOEN, R. A (eds.) (1993), *Global Dimensions of Intellectual Property Rights in Science and Technology*, Washington DC: National Academy Press.

WARSHOFSKY, F. (1994), *Patent Wars*, Chichester: John Wiley & Sons.

WRIGHT, B. D. (1983), "The Economics of Invention Incentives: Patents, Prizes, and Research Contracts," *American Economic Review* 73(4): 691–707.

CHAPTER 11

THE GEOGRAPHY OF INNOVATION

REGIONAL INNOVATION SYSTEMS

BJØRN T. ASHEIM

MERIC S. GERTLER

11.1 INTRODUCTION

THERE are two paradoxical characteristics of the contemporary global economy. First, innovative activity is not uniformly or randomly distributed across the geographical landscape. Indeed, the more knowledge-intensive the economic activity, the more geographically clustered it tends to be. The best examples include industries such as biotechnology or financial services, which have become ever more tightly clustered in a small number of major centers, despite the attempts of many other places to attract or generate their own activities in these sectors. Second, this tendency toward spatial concentration has become more marked over time, not less (Leyshon and Thrift 1997; Feldman 2001; Cortright and Mayer 2002). This reality contradicts longstanding predictions that the increasing use of information and communication technologies would lead to the dispersal of innovative activity over time. Given these rather striking stylized facts, it would appear that the process of knowledge production exhibits a very distinctive geography.

We argue in this chapter that this geography is fundamental, not incidental, to the innovation process itself: that one simply cannot understand innovation properly if one does not appreciate the central role of spatial proximity and concentration in this process. Our goal is to demonstrate why this is true, and to examine how innovation systems at the subnational scale play a key part in producing and reproducing this uneven geography over time.

This chapter addresses four key issues. First, why does location "matter" when it comes to innovative activity? If one considers the production and circulation of new knowledge to be the core of innovation, then it is important to have a sound understanding of the nature of the different types of knowledge involved and their geographical tendencies. Second, what are regional innovation systems, and what role do they play in generating and circulating new knowledge leading to innovation? Third, what is the relationship between regional systems of innovation and institutional frameworks at the national level? Finally, what is the relationship between local and global knowledge flows, and is there any evidence that the global nature of today's economy has weakened or altered the influence of proximity on the geography of innovation?

11.2 TYPES OF KNOWLEDGE AND THEIR GEOGRAPHIES

A growing body of thought argues that in a competitive era in which success depends increasingly upon the ability to produce new or improved products and processes, tacit knowledge constitutes the most important basis for innovation-based value creation (Pavitt 2002). As Maskell and Malmberg (1999: 172) have put it, when everyone has relatively easy access to explicit/codified knowledge, the creation of unique capabilities and products depends on the production and use of tacit knowledge:

Though often overlooked, a logical and interesting consequence of the present development towards a global economy is that the more easily codifiable (tradable) knowledge can be accessed, the more crucial does tacit knowledge become for sustaining or enhancing the competitive position of the firm. . . . In other words, one effect of the ongoing globalisation is that many previously localised capabilities and production factors become ubiquities. What is not ubiquified, however, is the non-tradable/non-codified result of knowledge creation— the embedded tacit knowledge that at a given time can only be produced in practice. The fundamental exchange inability of this type of knowledge increases its importance as the internationalisation of markets proceeds.

Implicit in the above quote is a fundamentally spatial argument: tacit knowledge is a key determinant of the *geography* of innovative activity. There are two closely related elements to this argument. First, because it defies easy articulation or codification (Polanyi 1958, 1966), tacit knowledge is difficult to exchange over long distances. It is heavily imbued with meaning arising from the social and institutional context in which it is produced, and this context-specific nature makes it spatially sticky (Gertler 2003). The second relates to the changing nature of the innovation process itself and, in particular, the growing importance of socially organized learning processes. The argument here is that innovation has come to be based increasingly on the interactions and knowledge flows between economic entities such as firms (customers, suppliers, competitors), research organizations (universities, other public and private research institutions), and public agencies (technology transfer centers, development agencies). This is fundamental to Lundvall and Johnson's (1994) learning economy thesis, and is especially well reflected in their concept of "learning through interacting." When one combines these two features of the innovation process—the centrality of "sticky," context-laden tacit knowledge and the growing importance of social interaction—it becomes apparent why geography now "matters" so much.

The recent literature on learning regions further explores the character and geographical consequences of tacit knowledge (see Lundvall and Johnson 1994; Florida 1995; Asheim 1996, 2001; Morgan 1997; Cooke and Morgan 1998; Lundvall and Maskell 2000). It argues that tacit knowledge does not "travel" easily because its transmission is best shared through face-to-face interaction between partners who already share some basic commonalities: the same language; common "codes" of communication and shared conventions and norms that have been fostered by a shared institutional environment; and personal knowledge of each other based on a past history of successful collaboration or informal interaction. These commonalities are said to serve the vital purpose of building trust between partners, which in turn facilitates the local flow of tacit (and codified) knowledge between partners.

This approach adopts the learning-by-interacting model as the cornerstone of its conceptual framework, and argues that the production of tacit knowledge occurs simultaneously with the act of transmission—primarily through the mechanism of user–producer interaction (Lundvall 1988; Gertler 1995). According to this perspective, knowledge does not flow unidirectionally from technology producers to users. Instead, users provide tacit and proprietary, codifiable knowledge to producers in order to enable the latter to devise innovative solutions to users' practical problems. But at the same time, by supplying users with innovative technologies, producers are also sharing their tacit and other proprietary knowledge with their customers. The end product arising from this close interaction benefits both users and producers, and embodies within it new knowledge that could not have been produced by either party working in isolation. This, in effect, describes a social process of joint innovation and knowledge production.

Lam (1998, 2000) points out that the skills required for effective knowledge transfer within collective learning processes are highly time- and space-specific. Interactive, collective learning is based on compatible intra- or interorganizational routines, tacit norms and conventions regulating collective action as well as tacit mechanisms for the absorption of codified knowledge. This requires that the actors in question have a shared understanding of "local codes," on which collective tacit as well as disembodied codified knowledge is based (Asheim 1999; Lundvall 1996). Thus, the ability to interpret local codes in consistent ways will be critical for the integration of the operations of a firm within a local interfirm learning network.

Since spatial proximity is key to the effective production and transmission/sharing of tacit knowledge, this reinforces the importance of innovative clusters, districts, and regions. Moreover, as Maskell and Malmberg (1999) point out, these regions also benefit from the presence of localized capabilities and intangible assets that further strengthen their centripetal pull (Dosi 1988; Storper 1997). Many of these are social assets—i.e. they exist between rather than within firms. Although they are therefore not fully appropriable by individual firms, only local firms can enjoy their benefits. These assets include the region's unique institutional endowment, which can act to support and reinforce local advantage. Because such assets evolve slowly over time, exhibiting strong tendencies of path-dependent development (David 1994; Zysman 1994), they may prove to be very difficult to emulate by would-be imitators in other regions, thereby preserving the initial advantage of "first mover" regions. Maskell and Malmberg argue (1999: 181):

It is the region's distinct institutional endowment that embeds knowledge and allows for knowledge creation which—through interaction with available physical and human resources—constitutes its capabilities and enhances or abates the competitiveness of the firms in the region. The path-dependent nature of such localised capabilities makes them difficult to imitate and they thereby establish the basis of sustainable competitive advantage.

We discuss the precise nature of this "distinct institutional endowment" in the following section of this chapter. Before doing so, however, it is important to explore further the different types of knowledge base in the economy, since the precise roles of tacit and codified (or codifiable) knowledge tend to differ accordingly.

11.2.1 Industrial Knowledge Bases

When one considers the actual knowledge base of various industries and sectors of the economy, it is clear that knowledge and innovation have become increasingly complex in recent years. There is a larger variety of knowledge sources and inputs to be used by organizations and firms, and there is more interdependence and a finer division of labour among actors: individuals, companies, and other organizations (Cowan et al. 2000). Nonaka and Takeuchi (1995) and Lundvall and Borrás (1999) have pointed out that the process of knowledge generation and exploitation requires a dynamic interplay between, and transformation of, tacit and codified forms of knowledge as well as a strong interaction of people within organizations and between them. Thus, these knowledge processes have become increasingly inserted into various forms of networks and innovation systems—at regional, national and international levels (see Ch. 3 by Powell and Grodal in this volume for a discussion of the role of networks in innovation).

Despite the general trend towards increased diversity and interdependence in the knowledge process, Pavitt (1984) and others have argued that the innovation process of firms is also strongly shaped by their *specific* knowledge base, which tends to vary systematically by industrial sector (see also Ch. 1 by Fagerberg and Ch. 15 by von Tunzelmann and Acha, in this volume). For the purposes of this chapter, we distinguish between two types of knowledge base: "analytical" and "synthetic" (Laestadius 1998). These types entail different mixes of tacit and codified knowledge, as well as different codification possibilities and limits. They also imply different qualifications and skills, reliance on different organizations and institutions, as well as contrasting innovation challenges and pressures.[1]

A synthetic knowledge base prevails in industrial settings where innovation takes place mainly through the application or novel combination of existing knowledge. Often this occurs in response to the need to solve specific problems arising in the interaction with clients and suppliers. Industry examples include specialized industrial machinery, plant engineering, and shipbuilding. R&D is in general less important than in other sectors of the economy. When it occurs, it tends to take the form of applied research, but more often it involves incremental product or process development related to the solution of specific problems presented by customers (von Hippel 1988). University–industry links are relevant, but they are clearly more significant in the realm of applied research and development than in basic research. Knowledge is created less in a deductive process or through abstraction than through an inductive process of testing, experimentation, computer-based simulation, or practical work. Knowledge embodied in the respective technical solution or engineering work is at least partially codified. However, tacit knowledge seems to be more important than in other types of activity, due to the fact that knowledge often results from experience gained at the workplace, and through learning by doing, using, and interacting. Compared to the second knowledge type ("analytical") described below, more concrete know-how, craft and practical skill is required in the knowledge production and circulation process. These forms of knowledge are often provided by professional and polytechnical schools, or by on-the-job training.

The innovation process for industries with a synthetic knowledge base tends to be oriented towards the efficiency and reliability of new solutions, or the practical utility and user-friendliness of products from the perspective of the customers. Innovation-related activities are dominated by the modification of existing products and processes. Since these types of innovation are less disruptive to existing routines and organizations, most of them take place in existing firms, making spin-offs and new firm formation for the development and exploitation of new synthetic knowledge relatively infrequent.

In contrast, an analytical knowledge base dominates economic activities where scientific knowledge is highly important, and where knowledge creation is often based on formal models, codified science and rational processes. Prime examples are biotechnology and information technology. Both basic and applied research, as well as the systematic development of products and processes, are central activities in this form of knowledge production. Companies typically have their own in-house R&D departments but they also rely on the research output of universities and other research organizations in their innovation process. University–industry links and networks are thus important, and this type of interaction is more frequent than in the synthetic type of knowledge base. Knowledge inputs and outputs in this type of knowledge base are more often codified (or readily codifiable) than in the case of synthetic knowledge. This does not imply that tacit knowledge is irrelevant, since both kinds of knowledge are always involved in the process of knowledge creation and innovation (Nonaka et al. 2000, Johnson, Lorenz, and Lundvall 2002).

The importance of codification in analytic knowledge reflects several factors: knowledge inputs are often based on reviews of existing (codified) studies, knowledge generation is based on the application of widely shared and understood scientific principles and methods, knowledge processes are more formally organized (e.g. in R&D departments), and outcomes tend to be documented in reports, electronic files, or patent descriptions. Knowledge application takes the form of new products or processes, which are more likely to constitute radical innovations than in those industries for which synthetic knowledge constitutes the principal knowledge base. New firms and spin-off companies (i.e. new market entrants rather than existing firms) are an important conduit for the application of knowledge embodied in these radically new inventions or products.[2]

How is the importance of tacit, as opposed to codified, knowledge, as well as the geography of innovation, affected by this differential importance of synthetic and analytical knowledge bases across industries and technologies? Clearly, the "learning through interacting" scenario at the core of the learning economy and learning regions thesis seems to be based implicitly on activities for which synthetic forms of knowledge are central. For instance, many of Lundvall's (1988) original examples come from the realm of mechanical engineering and specialized industrial machinery, where non-linear, iterative interaction between users and producers represents the primary mode of innovation. For such economic activities, the spatial concentration of interacting firms sharing a common social and institutional context is an obvious prerequisite to socially organized, interactive learning processes (Gertler 2004). But what about those sectors for which analytical knowledge is pre-eminent? Given the greater prominence of codified and codifiable knowledge in the innovation process, might we not expect innovation processes within analytically based industries to be more widely distributed spatially?

Apparently not. For starters, economists have produced much striking evidence about the highly uneven geography of innovation in analytically based activities. One important approach proceeds by measuring knowledge spillovers through the use of indicators such as patent citations.[3] For example, in their classic study, Jaffe et al. (1993) find evidence that patent applicants in analytically based industries cite other patents originating in the same city more frequently than they cite patents originating non-locally. Furthermore, they find that patent citations are more likely to be localized in the first year following the establishment of the patent, with the effect fading over time, as the knowledge diffuses more widely.

A related approach tracks knowledge spillovers in analytically based industries such as biotechnology and pharmaceuticals through the analysis of "star scientists." Zucker, Darby, and colleagues have tracked the location of these highly productive scientists and their impact on innovation in the local economy, demonstrating that the rates of start-up of new biotech firms are significantly higher in those regions in which these key scientists live and work (Zucker and Darby 1996; Zucker, Darby, and Armstrong 1998; Zucker, Darby, and Brewer 1998). Moreover, firms that have established working relationships with star scientists outperform firms that do not enjoy this kind of access, in terms of productivity growth, new product development, and employment growth.

Both of these sets of findings strongly suggest that in fact the innovation process in industries based on analytical forms of knowledge is no less spatially concentrated than those forms of innovative economic activity based on synthetic types of knowledge. Indeed, if anything, there is compelling evidence to suggest that the former may exhibit an even higher degree of geographical concentration than the latter (Cortright and Mayer 2002).[4]

How can one explain this counterintuitive finding? What are the processes underlying innovation in analytically based industries that explain their distinctive and highly uneven geography? There are three principal forces at work here. First, it is clear that, despite the importance of codifiable knowledge in analytically oriented sectors, the circulation of new knowledge remains highly localized, as the economic literature on knowledge spillovers (reviewed above) attests. This is because these spillovers occur first, fastest and most readily within established local social networks of scientists—often by word of mouth, well before formal results are published in widely accessible outlets. Some forms of valuable knowledge are almost never transmitted non-locally. For example, knowledge concerning failures in scientific experiments is rarely, if ever, published. Yet, the knowledge that a particular research strategy failed to yield expected results can save research teams considerable time and expense if it prevents them from pursuing unproductive lines of inquiry

(Enright 2003). The existence of this type of localized knowledge circulation—underpinned by commonly shared frames of experience and understandings—has been highlighted in the recent work of Storper and Venables (2003), who have coined the term "buzz" to capture this phenomenon.

Second, the central importance of highly educated (and potentially footloose) workers in the production of innovations in analytically based industries means that those places that offer the most attractive employment opportunities will be favored over others. Only a relatively small number of places offer a local labor market that is sufficiently rich and deep to promise not just one but a series of challenging employment opportunities in which these people can work at the cutting edge for well-known firms or research institutes (Florida 2002a). In other words, these workers are attracted to those places that offer this kind of career-based "buzz," and where they can also find a critical mass of people working in the same or similar occupational categories. Once a particular place becomes recognized by such workers for its portfolio of attractive employment prospects, as well as by employers for its deep pool of highly skilled labor, increasing-returns dynamics will generate a powerful virtuous circle of long-term growth and dynamism for analytically based sectors.

Third, those locations that offer a high quality of life in addition to attractive career opportunities will have an even more marked advantage in the "battle for talent." These highly talented workers can live in many places, but they tend to choose to live in those cities that offer a high quality of place, defined by a particular social character. According to Florida (2002b), such places are imbued with a critical mass of creative activity and workers, strong social diversity (measured in terms of ethnic or national origin) and tolerance (best indicated by, for example, a large gay population). Florida argues that such places have low barriers to entry for talented newcomers from diverse social backgrounds, making it easy for them to gain entry to local social networks and labor markets. They are also likely to offer colorful, attractive neighborhoods and cultural amenities that further enhance the attractive power of such places. The more highly educated (and creative) the worker, the stronger this effect will be. Hence, in those industries with the most knowledge-intensive workforce, we ought to find the strongest degrees of geographical concentration.

11.3 REGIONAL INNOVATION SYSTEMS AND LOCALIZED LEARNING

Having presented the most important arguments to explain the consistent tendency towards the geographical concentration of innovative activities, we turn now to consider the role of innovation systems at the subnational level in fostering and promoting this process.

The concept of a regional innovation system (RIS) is a relatively new one, having first appeared in the early 1990s (Asheim 1995, Asheim and Isaksen 1997; Cooke 1992, 1998, 2001), following Freeman's use of the innovation system concept in his analysis of Japan's economy (Freeman 1987), and at approximately the same time that the idea of the national innovation system was examined in books by Lundvall (1992) and Nelson (1993). As this chronology suggests, the regional innovation system concept was inspired by the national innovation system concept, and it is based on a similar rationale that emphasizes territorially based innovation systems.[5]

One such rationale stems from the existence of technological trajectories that are based on "sticky" knowledge and localized learning within the region. These can become more innovative and competitive by promoting stronger systemic relationships between firms and the region's knowledge infrastructure. A second rationale stems from the presence of knowledge creation organizations whose output can be exploited for economically useful purposes by supporting newly emerging economic activity. The emergence of the concept of a regional innovation system coincides with the success of regional clusters and industrial districts in the post-Fordist era (Asheim 2000; Asheim and Cooke 1999; Piore and Sabel 1984; Porter 1990, 1998), and the elaboration of the concept represents an attempt by students of the geographical economy to understand better the central role of institutions and organizations in promoting innovation-based regional growth (Asheim et al. 2003; Gertler and Wolfe 2004).[6]

The regional innovation system can be thought of as the institutional infrastructure supporting innovation within the production structure of a region. Taking each element of the term in turn (Asheim and Cooke 1999), the concept of region highlights an important level of governance of economic processes between the national level and the level of the individual cluster or firm. Regions are important bases of economic coordination at the meso-level: "the region is increasingly the level at which innovation is produced through regional networks of innovators, local clusters and the cross-fertilizing effects of research institutions" (Lundvall and Borrás 1999: 39). In varying degrees, regional governance is expressed in both private representative organizations such as branches of industry associations and chambers of commerce, and public organizations such as regional agencies with powers devolved from the national (or, within the European Union, supranational) level to promote enterprise and innovation support (Asheim et al. 2003; Cooke et al. 2000).

The systemic dimension of the RIS derives in part from the team-like character associated with innovation in networks. Although an innovation system is a set of relationships between entities or nodal points involved in innovation (see Lundvall 1992 for more discussion), it is much more than this. Such relationships, to be systemic, must involve some degree of interdependence, though to varying degrees. Likewise, not all such systemic relations need be regionally contained, but many are.[7]

As the interactive mode of innovation grows in importance, these relations are more likely to become regionally contained, especially in the case of specialized suppliers with a specific technology or knowledge base. Such suppliers often depend on tacit knowledge, face-to-face interaction and trust-based relations and, thus, benefit from cooperation with customers in regional clusters, while capacity subcontractors are increasingly sourced globally.[8] Further reinforcing the systemic character of the RIS is the prevalence of a set of attitudes, values, norms, routines, and expectations—described by some as a distinctive "regional culture"—that influences the practices of firms in the region. As noted earlier, it is this common regional culture—itself the product of commonly experienced institutional forces—that shapes the way that firms interact with one another in the regional economy.

11.3.1 Varieties of Regional Innovation Systems

The "innovation system" concept can be understood in both a narrow as well as a broad sense (see Ch. 7 by Edquist, in this volume). A narrow definition of the innovation system primarily incorporates the R&D functions of universities, public and private research institutes and corporations, reflecting a top-down, linear model of innovation as exemplified by the triple helix approach (Etzkowitz and Leydesdorff 2000). A broader conception of the innovation systems includes "all parts and aspects of the economic structure and the institutional set-up affecting learning as well as searching and exploring" (Lundvall 1992: 12). This broad definition incorporates the elements of a bottom-up, interactive innovation model of the sort described in our earlier discussion of the "learning regions" concept.

In order to reflect the conceptual variety and empirical richness of the relationships linking the production structure to the "institutional set-up" in a region, Asheim (1998) distinguishes among three types of RISs (see also Cooke 1998; Asheim and Isaksen 2002). The first type may be denoted as *territorially embedded regional innovation systems*, where firms (primarily those employing synthetic knowledge) base their innovation activity mainly on localized learning processes stimulated by geographical, social and cultural proximity, without much direct interaction with knowledge organizations. This type is similar to what Cooke (1998) calls "grassroots RIS," and implies the broader definition of innovation systems described by Lundvall (1992) above.

The best examples of *territorially embedded regional innovation systems* are networks of SMEs in industrial districts. Thus in Italy's Emilia-Romagna, for example, the innovation system can be described as territorially embedded in spatial structures of social relations within that particular region (Granovetter 1985). These territorially embedded systems provide bottom-up, network-based support through, for example, technology centers, innovation networks, or industry centers providing market research and intelligence services, to promote the "adaptive technological and organizational learning in territorial context" (Storper and Scott 1995: 513).

Another type of RIS is the *regionally networked innovation system*. The firms and organizations are still embedded in a specific region and characterized by localized, interactive learning. However, policy interventions lend these systems a more planned character through the intentional strengthening of the region's institutional infrastructure—for example, through a stronger, more developed role for regionally based R&D institutes, vocational training organizations, and other local organizations involved in firms' innovation processes. The networked system is commonly regarded as the ideal type of RIS: a regional cluster of firms surrounded by a regional "supporting" institutional infrastructure. Cooke (1998) also calls this type "network RIS." The network approach is most typical of Germany, Austria, and the Nordic countries.

Box 11.1 Baden-Württemberg's regionally networked innovation system

The German state of Baden-Württemberg is one of the country's most prosperous regions. It is home to some of Germany's most important mechanical engineering firms, including Daimler-Chrysler, Porsche, and Robert Bosch. These firms are well supported by a highly developed network of small and medium-sized enterprise specializing in the development, production, and supply of components, machinery, and systems, within a finely articulated social division of labor. Their most important local competence is their ability to solve the complex technological problems of their customers, resulting in custom-designed solutions or improvements to existing products and processes. While analytical knowledge is not irrelevant, synthetic knowledge predominates throughout this set of industries.

Given the importance of these supplier firms to the competitiveness of the region's large, flagship firms (and hence, to the overall performance of the regional economy), the regional innovation system has evolved to produce and diffuse these competencies in incremental mechanical engineering innovation. The most important elements of this regionally networked innovation system are:

- A strong vocational education, apprenticeship, and training system that produces a highly skilled and versatile work force.
- A well-developed infrastructure for technology transfer, incorporating both basic research facilities and market-oriented development, with special focus on the needs of SMEs. The Steinbeis Foundation operates a region-wide network of tech-transfer offices to help SMEs solve technical problems.
- A well-organized Chamber of Commerce (IHK), in which membership is mandatory. Among other responsibilities, the Chamber plays a leading role in co-ordinating the design of training programs tailored to local industry's needs.
- Highly developed and specialized, regionally organized producer associations. These organizations conduct research on market trends, economic forecasting and emerging, market-ready technologies on behalf of producer firms.

In addition to these regionally based elements, the entire regional innovation system is embedded within a national regulatory framework that reinforces innovative activity at the regional scale. The most important features of this system are:

- Labor market structures that foster stable employment relations, facilitating learning by doing and strengthening employers' incentives to train.
- An industrial relations system that formalizes worker participation in day-to-day and longer-term strategic decision making, enabling employers to harness workers' tacit knowledge acquired through learning by doing and using.
- Centralized collective bargaining systems that minimize interfirm variation in wage and benefit levels, inducing firms to compete on the basis of quality and innovativeness.
- Capital market structures that encourage longer-term time horizons in firm-level decision making, thus further reinforcing stability in the workplace.

Sources: Morgan (1999); Gertler (2004).

The regionally networked innovation system is a result of policy intervention to increase innovation capacity and collaboration. SMEs, for example, may have to supplement their informal knowledge (characterized by a high tacit component) with competence arising from more systematic research and development in order to carry out more radical innovations. In the long run, most firms cannot rely exclusively on informal localized learning, but must also gain access to wider pools of both analytical and synthetic knowledge on a national and global basis. The creation of regionally networked innovation systems through increased cooperation with local universities and R&D institutes, or through the establishment of technology transfer agencies and service centers, may provide access to information and competence that supplements firms' locally derived competence. This not only increases their collective innovative capacity, but may also serve to counteract technological "lock-in" (the inability to deviate from an established but outmoded technological trajectory) within regional clusters of firms.

The third main type of RIS, the *regionalized national innovation system,* differs from the two preceding types in several ways. First, parts of industry and the institutional infrastructure are more functionally integrated into national or international innovation systems—i.e. innovation activity takes place primarily in cooperation with actors outside the region. Thus, this represents a development model in which exogenous actors and relationships play a larger role. Cooke (1998) describes this type as "dirigiste RIS," reflecting a narrower definition of an innovation system incorporating mainly the R&D functions of universities, research institutes, and corporations. Second, the collaboration between organizations within this type of RIS conforms more closely to the linear model, as the cooperation primarily involves specific projects to develop more radical innovations based on formal analytical–scientific knowledge. Within such systems, cooperation is most likely to arise between people with the same occupational or educational back-ground (e.g. among engineers). This functional similarity facilitates the circulation and sharing of knowledge through "communities of practice," whose membership may cross inter-regional and even international boundaries (Amin and Cohendet 2004).

One special example of a *regionalized national innovation system* is the clustering of R&D laboratories of large firms and/or governmental research institutes in planned "science parks." These may be located in close proximity to universities and technical colleges, but the evidence suggests that science park tenants typically have limited linkages to local industry (Asheim 1995). Science parks are, thus, a typical example of a planned innovative milieu comprised of firms with a high level of internal resources and competence, situated within weak local cooperative envir-onments. These parks have generally failed to develop innovative networks based on interfirm cooperation and interactive learning within the science parks themselves (Asheim and Cooke 1998; Henry et al. 1995). Technopoles, as developed in countries such as France, Japan, and Taiwan, are also characterized by a limited degree of innovative interaction between firms within the pole, and by vertical subcontracting relationships with non-local external firms. In those rare cases where local innova-tive networks arise, they have normally been orchestrated by deliberate public sector intervention at the national level. These characteristics imply a lack of local and regional embeddedness, and lead us to question the capability of science parks and technopoles to promote innovativeness and competitiveness more widely within local industries (especially SMEs) as a prerequisite for endogenous regional devel-opment[9] (Asheim and Cooke 1998; and Longhi and Quére 1993).

11.4 THE RELATIONSHIP BETWEEN REGIONAL AND NATIONAL INNOVATION SYSTEMS

Recent work applying the RIS concept has begun to explore the linkage between the larger institutional frameworks of the national innovation system and national business system, and the character of regional innovation systems. This question has recently been addressed by Cooke (2001) in studies of biotechnology in the UK, the USA, and Germany. Cooke has introduced a distinction between the traditional regional innovation system (which he refers to as the institutional regional innov-ation system—IRIS) and the new economy innovation system (NEIS), which he refers to as an entrepreneurial regional innovation system (ERIS) (Cooke 2003). The traditional IRIS is more typical of German regions such as Baden-Württemberg or regions in the Nordic countries, whose leading industries draw primarily from synthetic knowledge bases. Its effectiveness flows from the positive effects of systemic relationships between the production structure and the knowledge infrastructure embedded networking governance structures and supporting regulatory and institutional frameworks. According to Cooke, the IRIS form "works well where technology and innovation tends to be path dependent rather than disruptive (the latter being more typical of the ERIS set-up), where institutions have grown incrementally to meet needs in an evolving but well-understood sectoral innovation system" (Cooke 2003: 57).

In contrast, the NEIS or ERIS (found in the US, the UK, and other Anglo-American economies) lacks the strong systemic elements of the IRIS form discussed above, and instead gets its dynamism from local venture capital, entrepreneurs, scientists, market demand, and incubators to support innovation that draws primarily from an analytical knowledge base. Thus, Cooke calls this a "venture capital driven" system. Such a system is more flexible and adjustable and, thus does not run the same risk of ending up in "lock-in" situations. On the other hand, new economy innovation systems do not have the same long-term stability and provide less systemic support for historical technological trajectories, raising important questions about their long-term economic sustainability.

Box 11.2 US biotechnology clusters: entrepreneurial regional innovation systems

---

A recent study (Cortright and Mayer 2002) of biotechnology in the United States concludes that innovation in this industry is dominated by just a handful of metropolitan centers. Boston and San Francisco are the two largest and best-established centers, followed by recent entrants San Diego, Seattle, and Raleigh-Durham. Philadelphia, New York, Washington-Baltimore, and Los Angeles also have significant concentrations of biotech activity. The authors of this study conclude that the two most important overarching factors supporting the emergence of strong biotech clusters are: (i) the presence of first-class pre-commercial medical research in a local university or government laboratory, and (ii) local systems to support and encourage entrepreneurial activity leading to successful translation of research into commercially viable outputs.

In emphasizing these two factors, this study emphasizes the key roles played by both public and private sectors actors in such innovation systems. Government support, through key granting councils such as the National Institutes of Health and state-level programs to invest in university systems and research, is at least a necessary condition for the emergence of a local biotechnology cluster. It also plays a role in the active-recruitment of "star scientists." But this process also requires the local presence of venture capital and managerial expertise in the development of technology-based companies. For the reasons outlined earlier in this chapter, a high local quality of life is also a crucial determinant of a region's ability to attract and retain highly educated scientific workers.

Despite these important local processes, non-local forces and relationships also play a key role. Three of the most important include: organizations such as the National Institutes of Health mentioned above; research alliances with large, global pharmaceutical firms with the financial resources to bankroll expensive research and clinical trials; and non-local venture capitalists with money and expertise to identify and support promising local firms and commercially viable research.

Once such local centers become established and attain critical mass, they begin to attract inward investment from multinational firms, who set-up their own research facilities in these locations in order to tap into distinctive local research competencies. This further reinforces the technological dynamism, entrepreneurial capabilities and commercialization potential of the region, setting in motion a virtuous circle of increasing returns.

Additional sources: Cooke (2001; 2003); Feldman (2001); Zucker et al. (1998).

In making these arguments about a general correspondence between the macro-institutional characteristics of the economy and the dominant form and character of its regional innovation systems, Cooke provides a link to another useful literature on "varieties of capitalism" and national business systems (Lam 1998, 2000; Whitley 1999; Hall and Soskice 2001). Soskice (1999) argues that different national institutional frameworks evolve to support particular forms of economic activity—i.e. that coordinated market economies such as Germany and the Nordic countries base their competitive advantage in "diversified quality production" (Streeck 1992), while liberal market economies such as the US and UK are most competitive in industries characterized by science-based innovative activities. Within the coordinated market economies, the driving force is the non-market coordination and cooperation that exists inside the business sphere and between private and public actors, as well as the degree to which labor is meaningfully "incorporated" within the production process and the financial system is able to supply long-term finance (Soskice 1999). In a comparison between coordinated market economies such as Sweden, Germany, and Switzerland on the one hand, and liberal market economies such as the US and UK on the other, he found that the coordinated economies performed best in the production of "relatively complex products, involving complex production processes and after-sales service in well-established industries" (e.g. synthetically based sectors such as the machine tool industry). By contrast, the US performed best in industries producing complex systemic products such as IT and defense technology, where the importance of analytical, scientific-based knowledge—often with the major support of the state—is significant (Soskice 1999: 113–14).

Thus, Soskice argues that competitive strength in markets for diversified quality production is based on problem-solving knowledge developed through interactive learning and accumulated collectively in the workforce (Soskice 1999). This type of production system is incompatible with an employment relation in which work processes are controlled exclusively by management—a preference generated by certain finance and governance systems found in liberal market economies. Competitive strength in other markets—e.g. markets characterized by a high rate of change through radical innovations—is based on the institutional freedom as well as financial incentives to continuously restructure production systems in light of new market opportunities (Gilpin 1996). While coordinated market economies on the macro level support cooperative, long-term, and consensus-based relations between private as well as public actors, liberal market economies inhibit the development of these relations but instead offer the opportunity to quickly adjust formal structures to new requirements. Such institutional specificities both contribute to the formation of divergent national business systems, and constitute the context within which different organizational forms with different mechanisms for learning, knowledge accumulation and knowledge appropriation have evolved (Asheim and Herstad 2003).

Christopherson (2002) has argued that the kinds of organizational features and labor market characteristics of interest to Lam (1998, 2000) are shaped by the structure of capital markets and "investment regimes" determined at the national level. Moreover, these different investment regimes produce the societal conditions for divergent forms of competitive advantage in global markets. An American-style "market governance model" dominated by the drive to maximize short-term investment returns has promoted the emergence of US strengths in analytically based sectors such as biotechnology and ICT, as well as in a set of "project-oriented" industries such as electronic media and entertainment, advertising, management consulting, public relations, engineering and industrial design, and computer services.

The essence of Christopherson's argument is that, under divergent sets of national institutions governing capital and labor markets and corporate governance, the kinds of social relationships that are likely to develop between economic actors locally—and hence the social organization of local innovation and production systems—will vary dramatically. Clearly, there is considerably more emphasis in the US system (than in, say, the German system) on the role of individual workers as mobile agents of knowledge circulation and local social learning, since they are the principal actors responsible for the sharing of knowledge between firms. Grabher's (2001, 2002) recent work on the project-based nature of production organization in the London advertising industry documents many structural similarities with Christopherson's description of US-style, project-based economic activity, strongly suggesting the continuing viability of a distinctive "Anglo-American" model of regionally based production and innovation systems.

Thus, liberal market economies as represented by the US and the UK seem to have advantages in industries characterized by an analytical knowledge base, as well as in those sectors that depend on a high degree of mobility in labor markets. Concerning the former, the elite universities and education institutions, often privately organized, provide strengths in R&D, the generation of formalized knowledge, inventions, and radical innovations. Other institutional features such as close university–industry links, academic spin-offs and an active scientific labor market all operate to promote the transfer and application of scientific knowledge.

Placed in this context the classic "traditional" institutional regional innovation system typified by a region such as Germany's Baden-Württemberg is most compatible with the institutional framework of a coordinated market economy, while the new economy innovation system (London advertising, Silicon Valley, or New York City's new media "Silicon Alley") reflects the institutional framework of a liberal market economy.

This raises an important issue that has been the subject of some debate in the literature, concerning the extent to which markedly different regional innovation systems can emerge within the same national institutional space. Saxenian's (1994) landmark study comparing the electronics and computing industries in two dominant regions of the United States—California's Silicon Valley and Route 128 in Massachusetts—has reinforced the view that widely divergent regional innovation systems can and do emerge within a single national institutional framework. She argues that Silicon Valley outperformed Route 128 in terms of employment growth and new firm formation because of its more open, flexible, high-mobility system compared to Route 128's more closed, rigid, hierarchical, loyalty-based system. While both regions were home to world-renowned institutions of higher learning and research, the Silicon Valley system proved more effective in generating successful innovations in response to profound competitive challenges from abroad.

It is important to note, however, that this analysis was based on the evolution of these two regional systems during the 1980s and early 1990s. In fact, the Massachusetts innovation system experienced a profound transformation over the 1990s. Best (2001) documents this transition, describing the emergence of new industries in biotechnology, medical devices, nanotechnology, and related fields. He argues that this transformation was underpinned by a more fundamental shift in the social organization of the leading sectors of the regional economy towards open systems architecture—in other words, through its evolution towards a structure that much more closely resembles the new economy innovation system of Silicon Valley. Best's analysis suggests that Saxenian's earlier case studies captured two regional innovation systems at a time when one of them (Route 128) was exhausting the innovative capabilities of an older, already outmoded system, but before a coherent, fully-formed alternative had emerged (Kenney and von Burg 1999; Saxenian 1999). Now that the organizational contours of this new system are clear, they suggest that the character of different *regional* innovation systems within the same *national* institutional space may vary within a considerably narrower range than was previously thought. This variation is likely to depend primarily upon regionally specific technological trajectories and knowledge bases.

11.5 ALTERNATIVE ORGANIZATIONAL FORMS AND EMERGING RELATIONSHIPS BETWEEN LOCAL AND GLOBAL KNOWLEDGE

Questions have lately been raised over whether the spatial embeddedness of learning and knowledge creation might be challenged by alternative organizational forms—in particular, temporary organizations—which some see as becoming more prevalent in the global economy (Asheim 2002; Grabher 2002). For example, Gann and Salter (2000) suggest that firms in the construction and engineering sector now rely on projects to organize the production of knowledge-intensive and complex products and systems. What impact might the adoption of temporary forms of organization have on the spatial embeddedness of learning and innovation? Grabher's (2002) work on projects in London advertising (discussed above) shows how co-location facilitates the continuous and rapid reconfiguration of project teams as well as the circulation of knowledge concerning the competencies and experience of potential project partners.

In contrast, Alderman (2004) argues that "there are ... important a priori or theoretical reasons why a project-based model does not fit comfortably with ideas about clustering, localized learning and local innovation networks." His argument relies on a recent literature that sees "communities of practice" as key entities driving the firm's knowledge-processing activities. This literature argues that routines and established practices shaped by organizations (or subset communities within organizations) promote the production and sharing of tacit and codifiable knowledge (Brown and Duguid 1996, 2000; Wenger 1998). Communities of practice are defined as groups of workers informally bound together by shared experience, expertise and commitment to a joint enterprise. These communities normally self-organize for the purpose of solving practical problems facing the larger organization, and in the process they produce innovations (both product and process). The commonalities shared by members of the community facilitate the identification, joint production and sharing of tacit knowledge through collaborative problem-solving assisted by story-telling and other narrative devices for circulating tacit knowledge.

According to this view, organizational or relational proximity and occupational similarity are more important than geographical proximity in supporting the production, identification, appropriation and flow of tacit knowledge (Allen 2000; Amin 2000; Amin and Cohendet 2004). The resulting geography of innovation differs from that envisioned by adherents to the learning region approach. In this view, the joint production and diffusion/transmission of tacit and codifiable knowledge across intra-organizational boundaries is possible, so long as it is mediated within these communities. Moreover, because communities of practice may extend outside the single firm to include customers or suppliers, knowledge can also flow across the boundaries of individual organizations.

Furthermore, the communities of practice literature asserts that tacit knowledge may also flow across regional and national boundaries if organizational or "virtual community" proximity is strong enough. In other words, learning (and the sharing of tacit knowledge) need not be spatially constrained if relational proximity is present. For large, multinational firms with "distributed" knowledge bases and multiple sites of innovation, the use of communities of practice, aided and supported by ever-cheaper and more powerful ICTs and air travel, is seen as an effective strategy for overcoming geographical separation.

These arguments are useful reminders of the importance of relationships and the strength of underlying similarities rather than geographical proximity *per se* in determining the effectiveness of knowledge-sharing between economic actors. However, they fail to answer a key question: what forces shape or defines this "relational proximity," enabling it to transcend physical, cultural, and institutional divides? How are shared understandings produced? Much of the communities of practice literature is largely silent on this question. A notable exception is the work of Brown and Duguid (2000: 143) who stake out a very different position on the spatial reach of communities of practice:

They are usually face-to-face communities that continually negotiate with, communicate with, and coordinate with each other directly in the course of work. And this negotiation, communication, and coordination is highly implicit, part of work practice ... In these groups, the demands of direct coordination inevitably limit reach. You can only work closely with so many people.

11.6 CONCLUSIONS AND ISSUES FOR FUTURE RESEARCH

In this chapter we have argued that the geographical configuration of economic actors—firms, workers, associations, organizations, and government agencies—is fundamentally important in shaping the innovative capabilities of firms and industries. We have distinguished between two types of knowledge base—synthetic and analytical—and have demonstrated that, while the nature of innovation processes may differ in each case, innovative activity tends to be spatially clustered in both cases, though for somewhat different reasons.

We then introduced the concept of regional innovation systems, describing the elements, relationships, and systemic character that comprise a key part of a region's distinct institutional endowment. We also explored the variety of different types of regional innovation systems that can be identified, noting how particular regional systems may be more strongly associated with particular regimes of business systems and institutional frameworks at the national level. Although there is significant variation in economic performance across different regions within the same national system, the characteristics of *successful* regional innovation systems under the same national regime will exhibit certain consistencies from case to case.

Despite the emergence of a strong consensus around the above issues, there remain a number of contentious or unresolved questions that are likely to provide the focus for future research. First, in the ongoing discourse on the nature and impact of globalization, some authors have argued that tacit knowledge has become increasingly codified and hence ubiquitous, ultimately eroding the competitive advantage of high-cost regions and nations (Maskell et al. 1998; Maskell 1999). Others maintain that much strategic knowledge remains "sticky," and that important parts of the learning process continue to be localized as a result of the enabling role of geographical proximity and local institutions in stimulating interactive learning (Asheim 1999; Markusen 1996).

Nevertheless, global knowledge networks and flows are important sources of innovative ideas for a growing number of economic activities (Mackinnon et al. 2002). If so, then how should we understand their impact on the geographical distribution of innovative activity, and on the future importance of regionally based innovation systems? In a recent conceptual paper, Bathelt et al. (2004) seek to reconcile these divergent views. They argue that firms clustered in particular locales require access to non-local sources of knowledge as an essential complement to the knowledge they generate and share locally. The metaphor they have adopted to capture the dual nature of emerging geographies of innovation is "local buzz and global pipelines." They view these global knowledge pipelines as extending between different nodal geographical concentrations of firms and other knowledge-producing organizations around the world. On the one hand, no firm—especially in analytical, science-based sectors such as biotechnology—can afford to cut themselves off from non-local knowledge sources. To do so would be to court potential disaster, as regional innovation systems would be prone to encouraging technological stagnation and "lock-in" tendencies. On the other hand, the abilities of firms to make the most effective use of this knowledge—that is, to convert it most effectively into economic value—still depends on their access to important place-based assets, both tangible and intangible, and the close interaction with other organizations around them that such locations foster (see also Asheim and Herstad 2003; Cooke et al. 2000; Freeman 2002).

This work is a welcome conceptual contribution, but the empirical basis for its framework remains underdeveloped. Clearly, there is a need for future research—both case studies and aggregate statistical analyses—to investigate the prevalence of this "dual geography" of innovation more systematically. At the same time, this approach raises a further question that remains unanswered: is this metaphor of local buzz and global pipelines appropriate only for those science-intensive industries whose innovation rests on an analytical knowledge base? To what extent are non-local knowledge flows and learning relationships extending between localized centers of innovation becoming important for those industries that rely more heavily on a synthetic knowledge base?

A problematic aspect of the learning economy and learning regions literature has been its focus on learning by doing and using based largely on local synthetic knowledge with a high tacit content and incremental innovations. We continue to agree with Freeman's useful insight concerning "the tremendous importance of incremental innovation, learning by doing, by using and by interacting in the process of technical change and diffusion of innovations" (Freeman 1993: 9–10). Yet in a highly competitive, globalizing economy, it may be increasingly difficult for the reproduction and growth of a learning economy to rely primarily on incremental improvements to products and processes, and not on new products (i.e. radical innovations). Crevoisier (1994: 259) argues that the exclusive reliance on incremental innovations "would mean that these areas will very quickly exhaust the technical paradigm on which they are founded."

In future studies it will be important to follow these tendencies, which undoubtedly will be reinforced by globalization processes (see Ch. 12 by Narula and Zanfei in this volume). The basic rationale of regional innovation systems is that the systemic promotion of localized learning processes can improve the innovativeness and competitive advantage of regional economies. What remains to be seen is how the capacity of regional innovation systems to upgrade the knowledge bases of firms in regional clusters will develop over time.

NOTES

1. Pavitt (1984: 353–65) offers a three-way taxonomy of industries based on the predominant nature and sources of technical change. Supplier-dominated industries include agriculture and traditional manufacturing sectors such as textiles. Production-intensive industries can be further subdivided into scale-intensive sectors such as steel, consumer durables, and automobiles, and specialized supplier sectors such as machinery and instruments. Science-based industries include electronics and chemicals (including pharmaceuticals). In the discussion that follows, our observations about industries with

synthetic knowledge bases correspond closely to those sectors encompassed by the first two of Pavitt's categories (supplier-dominated and production-intensive). Similarly, the analytical category corresponds directly to Pavitt's science-based industries.

2. We should acknowledge that many industries draw significantly upon both synthetic and analytical forms of knowledge. A clear example would be medical devices and technologies, whose development rests upon knowledge drawn from fields as diverse as bioscience, ICT, software, advanced materials, nanotechnology, and mechanical engineering. For this reason, it makes sense to conceive of individual industrial sectors being arrayed along a continuum between purely analytical and synthetic industries, with many—such as the automotive industry—occupying an intermediate position along this spectrum.

3. See Feldman (2000) for a useful overview of this literature.

4. Cortright and Mayer produce evidence to show that the degree of geographical concentration in the US life sciences industries is considerably higher than the population as a whole. They also demonstrate—using indicators such as venture capital, funded research conducted through inter-firm alliances, and new firm formation rates—that this concentration has increased dramatically during the past two decades.

5. This conceptualization of regional innovation systems corresponds with the one found in Cooke et al. (2000). In their words any functioning regional innovation system consists of two subsystems: (i) the knowledge application and exploitation subsystem, principally occupied by firms within vertical supply-chain networks; and (ii) the knowledge generation and diffusion subsystem, consisting mainly of public organizations.

6. There is a strong historical correspondence between these concepts and approaches and agglomeration theories within regional science and economic geography, such as Perroux's (1970) growth pole theory.

7. In a recent study Carlsson (2003) shows that the majority of theoretical as well as empirical analyses of innovation systems have a regional focus (see also Bathelt 2003 for a critical discussion of RIS).

8. A recent comparative study of European clusters shows that firms increasingly find relevant research activities and other supporting services inside the cluster boundaries (Isaksen 2004).

9. See Ch. 8 in this volume by Mowery and Sampat for a similarly critical assessment of science parks.

REFERENCES

ALDERMAN, N. (2004), "Mobility versus Embeddedness: The Role of Proximity in Major Capital Projects," in A. Lagendijk and P. Oinas (eds.), *Proximity, Distance and Diversity: Issues on Economic Interaction and Local Development*, Aldershot: Ashgate.

ALLEN, J. (2000), "Power/Economic Knowledge: Symbolic and Spatial formations," in J. R. Bryson, P. W. Daniels, N. Henry, and J. Pollard (eds.), *Knowledge, Space, Economy*, London: Routledge, 15–33.

AMIN, A. (2000), "Organisational Learning through Communities of Practice," Paper presented at the workshop on *The Firm in Economic Geography*, University of Portsmouth, UK, 9–11 March.

—— and COHENDET, P. (2004), *Architectures of Knowledge*, Oxford: Oxford University Press.

ASHEIM, B. T. (1995), "Regionale innovasjonssystem—en sosialt og territorielt forankret teknologipolitikk," *Nordisk Samhällsgeografisk Tidskrift* 20: 17–34.

—— (1996), "Industrial Districts as 'Learning Regions': A Condition for Prosperity?" *European Planning Studies* 4(4): 379–400.

—— (1998), "Territoriality and Economics: On the Substantial Contribution of Economic Geography," in O. Jonsson and L.-O. Olander, (eds.), *Economic Geography in Transition, The Swedish Geographical Yearbook*, vol. 74, Lund, 98–109.

—— (1999), "Interactive Learning and Localised Knowledge in Globalising Learning Economies," *GeoJournal* 49(4): 345–52.

—— (2000), "Industrial Districts: The Contributions of Marshall and Beyond," in G. Clark, M. Feldman, and M. Gertler (eds.), *The Oxford Handbook of Economic Geography*, Oxford: Oxford University Press, 413–31.

—— (2001), "Learning Regions as Development Coalitions: Partnership as Governance in European Workfare States?" *Concepts and Transformation. International Journal of Action Research and Organizational Renewal* 6(1): 73–101.

—— (2002), "Temporary Organisations and Spatial Embeddedness of Learning and Knowledge Creation," *Geografiska Annaler, Series B, Human Geography*, 84B(2): 111–24.

—— and COOKE, P. (1998), "Localised Innovation Networks in a Global Economy: A Comparative Analysis of Endogenous and Exogenous Regional Development Approaches," *Comparative Social Research* 17, Stamford, Conn: JAI Press, 199–240.

—— —— (1999), "Local Learning and Interactive Innovation Networks in a Global Economy," in E. Malecki and P. Oinas (eds.), *Making Connections: Technological Learning and Regional Economic Change*, Aldershot: Ashgate, 145–78.

—— and HERSTAD, S. J. (2003), "Regional Innovation Systems, Varieties of Capitalism and Non-Local Relations: Challenges from the Globalising Economy," in B. T. Asheim and Å. Mariussen (eds.), *Innovations, Regions and Projects*, Stockholm: Nordregio, R 2003:3, 241–74.

—— and ISAKSEN, A. (1997), "Location, Agglomeration and Innovation: Towards Regional Innovation Systems in Norway?" *European Planning Studies* 5(3): 299–330.

—— —— (2002), "Regional Innovation Systems: The Integration of Local 'Sticky' and Global 'Ubiquitous' Knowledge," *Journal of Technology Transfer* 27: 77–86.

—— et al. (eds.) (2003), *Regional Innovation Policy for Small-Medium Enterprises*, Cheltenham: Edward Elgar.

BATHELT, H. (2003), "Geographies of Production: Growth Regimes in Spatial Perspective 1: Innovation, Institutions and Social Systems, *Progress in Human Geography* 27(6): 789–804.

*—— MALMBERG, A., and MASKELL, P. (2004), "Clusters and Knowledge: Local Buzz, Global Pipelines and the Process of Knowledge Creation," *Progress in Human Geography*, 28(1): 31–56.

BEST, M. (2001), *The New Competitive Advantage*, Oxford: Oxford University Press.

BROWN, J. S., and DUGUID, P. (1996), "Organisational Learning and Communities-of-Practice: Towards a Unified Theory of Working, Learning and Innovation," in M. Cohen and L. Sproul (eds.), *Organisational Learning*, New York: Sage, 58–82.

*—— —— (2000), *The Social Life of Information*, Boston: Harvard Business School Press.

* Asterisked items are suggestions for further reading.

CARLSSON, B. (2003), "Innovation Systems: A Survey of the Literature from a Schumpeterian Perspective," in A. Pyka (ed.), *The Companion to Neo-Schumpeterian Economics*, Cheltenham: Edward Elgar.

CHRISTOPHERSON, S. (2002), "Why do National Labor Market Practices Continue to Diverge in the Global Economy? The "Missing Link" of Investment Rules," *Economic Geography* 78(1): 1–20.

COOKE, P. (1992), "Regional Innovation Systems: Competitive Regulation in the New Europe," *Geoforum* 23: 365–82.

—— (1998), "Introduction: Origins of the Concept," in H. Braczyk, P. Cooke, and M. Heidenreich (eds.), *Regional Innovation Systems*, London: UCL Press, 2–25.

*—— (2001), "Regional Innovation Systems, Clusters, and the Knowledge Economy," *Industrial and Corporate Change* 10(4): 945–74.

—— (2003), "Integrating Global Knowledge Flows for Generative Growth in Scotland: Life Sciences as a Knowledge Economy Exemplar," in *Inward Investment, Entrepreneurship and Knowledge Flows in Scotland—International Comparisons*, Paris: OECD.

—— and MORGAN, K. (1998), *The Associational Economy: Firms, Regions and Innovation*, Oxford: Oxford University Press.

—— BOEKHOLT, P., and TÖDTLING, F. (2000), *The Governance of Innovation in Europe. Regional Perspectives on Global Competitiveness*, London: Pinter.

CORTRIGHT, J., and MAYER, H. (2002), *Signs of Life: The Growth of Biotechnology Centers in the U.S.*, Washington, DC: Center on Urban and Metropolitan Policy, The Brookings Institution.

COWAN, R., DAVID, P. A., and FORAY, D. (2000), "The Explicit Economics of Knowledge Codification and Tacitness," *Industrial and Corporate Change* 9: 211–53.

CREVOISIER, O. (1994), Book review of G. Benko and A. Lipietz (eds.), *Les Régions qui gagnent*, Paris, 1992, *European Planning Studies* 2: 258–60.

DAVID, P. A. (1994), "Why are Institutions the 'Carriers of History'? Path Dependence and the Evolution of Conventions, Organizations and Institutions," *Structural Change and Economic Dynamics* 5: 205–20.

DOSI, G. (1988), "The Nature of the Innovative Process," in G. Dosi et al. (eds.), *Technical Change and Economic Theory*, London: Pinter, 221–38.

ENRIGHT, M. (2003), "Competitiveness, Innovative Clusters and Positive Externalities", paper presented at the Sixth Annual Conference of the Competitiveness Institute, Gothenburg, Sweden, 17–19 September.

ETZKOWITZ, H., and LEYDESDORFF, L. (2000), "The Dynamics of Innovation: From National Systems and 'Mode 2' to a Triple Helix of University–Industry–Government Relations," *Research Policy* 29: 109–23.

*FELDMAN, M. P. (2000), "Location and Innovation: The New Economic Geography of Innovation, Spillovers, and Agglomeration," in G. L. Clark, M. P. Feldman, and M. S. Gertler (eds.), *The Oxford Handbook of Economic Geography*, Oxford: Oxford University Press, 373–94.

—— (2001), "Where Science Comes to Life: University Bioscience, Commercial Spin-offs, and Regional Economic Development," *Journal of Comparative Policy Analysis: Research and Practice* 2: 345–61.

FLORIDA, R. (1995), "Toward the Learning Region," *Futures* 27: 527–36.

*—— (2002a), "The Economic Geography of Talent," *Annals of the Association of American Geographers* 92: 743–55.

—— (2002b), *The Rise of the Creative Class*, New York: Basic Books.

FREEMAN, C. (1987), *Technology Policy and Economic Performance: Lessons from Japan*, London: Pinter.

—— (1993), "The Political Economy of the Long Wave," Paper presented at EAPE 1993 conference on "The economy of the future: Ecology, technology, institutions". Barcelona, October 1993.

—— (1995), "The 'National System of Innovation' in Historical Perspective," *Cambridge Journal of Economics* 19: 5–24.

*—— (2002), "Continental, National and Sub-National Innovation Systems—Complementarity and Economic Growth," *Research Policy* 31: 191–211.

GANN, D. M. and SALTER, A. J. (2000), "Innovation in Project-Based, Service Enhanced Firms: The Construction of Complex Products and Systems," *Research Policy* 29: 955–72.

GERTLER, M. S. (1995), " 'Being there': Proximity, Organization, and Culture in the Development and Adoption of Advanced Manufacturing Technologies," *Economic Geography* 71: 1–26.

—— (2003), "Tacit Knowledge and the Economic Geography of Context, or the Undefinable Tacitness of Being (there)," *Journal of Economic Geography* 3: 75–99.

—— (2004), *Manufacturing Culture: The Institutional Geography of Industrial Practice*, Oxford: Oxford University Press.

—— and WOLFE, D. A. (2004), "Ontario's Regional Innovation System: The Evolution of Knowledge-Based Institutional Assets," in P. Cooke, and M. Heidenreich, H. Braczyk (eds.), *Regional Innovation Systems*, 2nd edn., London: Taylor and Francis.

GILPIN, R. (1996), "Economic Evolution of National Systems," *International Studies Quarterly* 40: 411–43.

GRABHER, G. (2001), "Ecologies of Creativity. The Group, the Village and the Heterarchic Organisation of the British Advertising Industry," *Environment and Planning A* 33: 351–74.

—— (2002), "Cool Projects, Boring Institutions: Temporary Collaboration in Social Context," *Regional Studies* 36: 205–14.

GRANOVETTER, M. (1985), "Economic Action and Social Structure: The Problem of Embeddedness," *American Journal of Sociology* 91: 481–510.

HALL, P., and SOSKICE, D. (2001), "An Introduction to Varieties of Capitalism," in Hall and Soskice (eds.), *Varieties of Capitalism: The Institutional Foundations of Comparative Advantage*, Oxford: Oxford University Press, 1–68.

HENRY, N. et al. (1995), "Along the Road: R&D, Society and Space," *Research Policy* 24: 707–26.

ISAKSEN, A. (2004), "Regional Clusters between Local and Non-Local Relations: A Comparative European Study," in A. Lagendijk and P. Oinas (eds.), *Proximity, Distance and Diversity: Issues on Economic Interaction and Local Development*, Aldershot: Ashgate.

JAFFE, A., TRAJTENBERG, M., and HENDERSON, R. (1993), "Geographical Localization of Knowledge Spillovers as Evidenced by Patent Citations," *Quarterly Journal of Economics* 108: 577–98.

JOHNSON, B., LORENZ, E., and LUNDVALL, B.-Å. (2002), "Why All this Fuss about Codified and Tacit Knowledge?" *Industrial and Corporate Change* 11: 245–62.

KENNEY, M., and VON BURG, U. (1999), "Technology, Entrepreneurship and Path Dependence: Industrial Clustering in Silicon Valley and Route 128," *Industrial and Corporate Change* 8(1): 67–103.

LAESTADIUS, S. (1998), "Technology Level, Knowledge Formation and Industrial Competence in Paper Manufacturing," in G. Eliasson et al. (eds.), *Microfoundations of Economic Growth. A Schumpeterian Perspective*, Ann Arbor: University of Michigan Press, 212–26.

LAM, A. (1998), "The Social Embeddedness of Knowledge: Problems of Knowledge Sharing and Organisational Learning in International High-Technology Ventures," *DRUID Working Paper* 98-7, Aalborg.

—— (2000), "Tacit Knowledge, Organizational Learning and Societal Institutions: An Integrated Framework," *Organization Studies* 21(3): 487–513.

LEYSHON, A., and THRIFT, N. J. (1997). *Money/Space: Geographies of Monetary Transformation*, London: Routledge.

LONGHI, C., and QUÉRE, M. (1993), "Innovative Networks and the Technopolis Phenomenon: The Case of Sophie-Antipolis," *Environment and Planning C: Government & Policy* 11: 317–30.

LUNDVALL, B-Å. (1988), "Innovation as an Interactive Process: From User–Producer Interaction to the National System of Innovation," in G. Dosi, C. Freeman, G. Silverberg, and L. Soete (eds.), *Technical Change and Economic Theory*, London: Pinter.

—— (ed.) (1992), *National Innovation Systems: Towards a Theory of Innovation and Interactive Learning*, London: Pinter.

—— (1996), "The Social Dimension of the Learning Economy," *DRUID Working Papers* 96-1, Aalborg.

—— and JOHNSON, B. (1994), "The Learning Economy," *Journal of Industry Studies* 1: 23–42.

—— and BORRÁS, S. (1999), *The Globalising Learning Economy: Implications for Innovation Policy*, DGXII-TSER, The European Commission.

—— and MASKELL, P. (2000), "Nation States and Economic Development—from National Systems of Production to National Systems of Knowledge Creation and Learning," in G. L. Clark, M. P. Feldman, and M. S. Gertler (eds.), *The Oxford Handbook of Economic Geography*, Oxford: Oxford University Press, 353–72.

MACKINNON, D., CUMBERS, A., and CHAPMAN, K. (2002), "Learning, Innovation and Regional Development: A Critical Appraisal of Recent Debates," *Progress in Human Geography* 26(3): 293–311.

MARKUSEN, A. (1996), "Sticky Places in Slippery Space: A Typology of Industrial Districts," *Economic Geography* 72(3): 293–313.

MASKELL, P. (1999), "Globalisation and Industrial Competitiveness: The Process and Consequences of Ubiquitification," in E. Malecki and P. Oinas (eds.), *Making Connections: Technological Learning and Regional Economic Change*, Aldershot: Ashgate, 35–59.

—— et al. (1998), *Competitiveness, Localised Learning and Regional Development*, London: Routledge.

*—— and MALMBERG, A. (1999), "Localised Learning and Industrial Competitiveness," *Cambridge Journal of Economics* 23: 167–86.

*MORGAN, K. (1997), "The Learning Region: Institutions, Innovation and Regional Renewal," *Regional Studies* 31: 491–504.

—— (1999), "Reversing Attrition? The Auto Cluster in Baden-Württemberg," in T. J. Barnes and M. S. Gertler (eds.), *The New Industrial Geography: Regions, Regulation and Institutions*, London: Routledge, 74–97.

NELSON, R. (ed.) (1993), *National Innovation Systems: A Comparative Analysis*, Oxford: Oxford University Press.

NONAKA, I., and TAKEUCHI, H. (1995), *The Knowledge Creating Company*, Oxford and New York: Oxford University Press.

—— et al. (2000), "SECI, Ba and Leadership: A Unified Model of Dynamic Knowledge Creation," *Long Range Planning* 33: 5–34.

PAVITT, K. (1984), "Sectoral Patterns of Technical Change: Towards a Taxonomy and a Theory," *Research Policy* 13: 343–73.

—— (2002), "Knowledge about Knowledge since Nelson and Winter: A Mixed Record," Electronic Working Paper Series Paper No. 83, SPRU, University of Sussex, June.

PERROUX, F. (1970), "Note on the Concept of Growth Poles," in D. McKee et al. (eds.), *Regional Economics: Theory and Practice*, New York: The Free Press, 93–103.

PIORE, M., and SABEL, C. (1984), *The Second Industrial Divide: Possibilities for Prosperity*, New York: Basic Books.

POLANYI, M. (1958), *Personal Knowledge: Towards a Post-Critical Philosophy*, London: Routledge & Kegan Paul.

—— (1966), *The Tacit Dimension*, New York: Doubleday.

PORTER, M. (1990), *The Competitive Advantage of Nations*, New York: Basic Books.

PORTER, M. E. (1998), "Clusters and the New Economics of Competition," *Harvard Business Review* (Nov.–Dec.): 77–90.

*SAXENIAN, A. (1994), *Regional Advantage: Culture and Competition in Silicon Valley and Route 128*, Cambridge, Mass.: Harvard University Press.

—— (1999), "Comment on Kenny and von Burg, 'Technology, Entrepreneurship and Path Dependence: Industrial Clustering in Silicon Valley and Route 128'," *Industrial and Corporate Change* 8(1): 105–10.

SCHOENBERGER, E. (1997), *The Cultural Crisis of the Firm*, Oxford: Blackwell.

SOSKICE, D. (1999), "Divergent Production Regimes: Coordinated and Uncoordinated Market Economies in the 1980s and 1990s," in H. Kitschelt et al. (eds.), *Continuity and Change in Contemporary Capitalism*, Cambridge: Cambridge University Press, 101–34.

STORPER, M. (1997), *The Regional World: Territorial Development in a Global Economy*, New York: Guilford Press.

—— and SCOTT, A. (1995), "The Wealth of Regions," *Futures*, 27(5): 505–26.

*—— and VENABLES, A. J. (2003), "Buzz: The Economic Force of the City," Paper presented at the DRUID Summer Conference 2003 on "Creating, sharing and transferring knowledge: the role of geography, institutions and organizations," Elsinore, Denmark.

STREECK, W. (1992), *Social Institutions and Economic Performance—Studies of Industrial Relations in Advanced Capitalist Economies*, New York: Sage.

VON HIPPEL, E. (1988), *The Sources of Innovation*, Oxford: Oxford University Press.

WENGER, E. (1998), *Communities of Practice: Learning, Meaning and Identity*, Cambridge: Cambridge University Press.

WHITLEY, R. (1999), *Divergent Capitalism: The Social Structuring and Change of Business Systems*, Oxford: Oxford University Press.

ZUCKER, L. G., and DARBY, M. R. (1996), "Star Scientists and Institutional Transformation: Patterns of Invention and Innovation in the Formation of the Biotechnology Industry," *Proceedings of the National Academy of Science* 93: 12709–16.

—— —— and ARMSTRONG, J. (1998), "Geographically Localized Knowledge: Spillovers or Markets?" *Economic Inquiry* 36: 65–86.

—— —— and BREWER, M. B. (1998), "Intellectual Human Capital and the Birth of U.S. Biotechnology Enterprises," *American Economic Review* 88: 290–306.

ZYSMAN, J. (1994), "How Institutions Create Historically Rooted Trajectories of Growth," *Industrial and Corporate Change* 3: 243–83.

CHAPTER 12

GLOBALIZATION OF INNOVATION

THE ROLE OF MULTINATIONAL ENTERPRISES

RAJNEESH NARULA

ANTONELLO ZANFEI

12.1 INTRODUCTION

ECONOMIC globalization implies a growing interdependence of locations and economic units across countries and regions. Technological change and multinational enterprises (MNEs) are among the primary driving forces of this process. In this chapter we attempt to evaluate the changing extent and importance of MNEs as conduits for cross-border knowledge flows.

MNEs affect the development and diffusion of innovations across national borders through a number of mechanisms, among which FDI (through which MNEs acquire existing assets abroad or set-up new wholly or majority owned activities in foreign markets) is only one. International knowledge flows also move through trade, licensing, cross-patenting activities, and international technological and scientific collaborations. These other modalities involve a wide variety of economic actors, but the MNE occupies a central role among these actors. This chapter emphasizes the MNE's multifaceted role in the more general process of globalization of innovation.

12.2 TRENDS IN THE INTERNATIONALIZATION OF INNOVATIVE ACTIVITIES

A useful taxonomy proposed by Archibugi and Michie (1995) identifies three main categories of the globalization of innovation (Table 12.1). Although a variety of economic actors undertake innovation and are engaged in its internationalization, the MNE is the only institution which by definition can carry out and control the global generation of innovation within its boundaries. We briefly discuss each of the three categories below.

12.2.1 The Cross-Border Commercialization of National Technology

The first category involves national and multinational firms as well as individuals engaged in the international commercialization of technology developed at "home." Key indicators of these activities are international trade flows and cross border patenting, both of which are responsible for growing levels of global transfer of technology.

The share of high-tech products (including electrical and electronic equipment, aerospace products, precision instruments, fine chemicals and pharmaceuticals) in world exports rose from 8 per cent in 1976 to 23 per cent in 2000. Exports of information and communications technology products showed the highest annual growth rate among all products in 1985–2000 (UNCTAD 2002: 146–7). The rise in the share of world trade represented by R&D-intensive sectors suggests that the globalization of technology flows is increasing.[1]

Table 12.2 reveals a growth in the "internationalization" of patenting: the share of non-resident patenting in virtually all OECD economies has grown during the 1980s

Table 12.1 A taxonomy of the globalization of innovation

Categories	Actors	Forms
International Exploitation of Nationally Produced Innovations	Profit-seeking (national and multinational) firms and individuals	Exports of innovative goods. Cession of licenses and patents. Foreign production of innovative goods internally designed and developed.
Global Generation of Innovations	MNEs	R&D and innovative activities both in the home and the host countries. Acquisitions of existing R&D laboratories or green-field R&D investment in host countries.
Global Techno-Scientific Collaborations	Universities and Public Research Centers	Joint scientific projects. Scientific exchanges, sabbatical years. International flows of students.
	National and Multinational Firms	Joint ventures for specific innovative projects. Productive agreements with exchange of technical information and/or equipment.

Source: elaboration on Archibugi and Michie 1995.

and early 1990s, and external patenting (i.e. patent applications of national inventors abroad) has also rapidly increased.

12.2.2 Technological and Scientific Collaborations

Domestic and international technical and scientific collaborations involve both private and public institutions, including national and multinational firms, universities and research centers. Since the 1970s, the use by industrial firms of "non-internal" options that include cooperation with competitors, suppliers, customers, and other external institutions (e.g. universities), which we denote as strategic technology partnering (STP), has grown. Available indicators of international STP

Table 12.2 Rates of growth of industrial R&D and patenting in the OECD countries

| Countries | Average annual rates of change (per cent) | | | | | | | |
| | Industrial R&D (1) | | Resident patents (2) | | Non-resident patents (3) | | External patents (4) | |
	1970–80	1985–95	1970–80	1984–94	1970–80	1984–94	1970–80	1985–95
United States	2.0	1.3	-2.0	5.7	5.0	6.6	-0.6	15.6
Japan	6.1	5.4[e]	5.1	2.2	-0.8	5.1	5.5	8.3
Germany	4.9[a]	1.1	-0.7	1.4	0.8	4.6	1.7	8.0
France	3.7	3.2	-2.4	1.0	0.2	5.3	3.0	8.4
United Kingdom	3.0[b]	0.3[e]	-2.4	-0.4	0.8	4.8	-1.7	16.2
Italy	3.6	-0.5	n.a.	2.5[l]	n.a.	3.8	1.8	10.3
Netherlands	1.4	3.3[e]	-2.1	-1.5	1.5	6.8	0.1	14.1
Belgium	6.7[c]	1.7[f]	-3.0	-1.6	-0.1	7.7	0.5	13.4
Denmark	3.8	7.4[g]	1.7	3.0	-0.3	19.9	1.0	22.5
Spain	12.7	1.8[e]	-4.5	2.0	0.2	19.2	1.3	16.0
Ireland	5.2[c]	15.4	6.8	2.3	4.9	31.1	6.7	24.3
Portugal	4.6[d]	2.2[h]	-6.4	0.9	-0.5	37.2	-24.2	52.4
Greece	n.a.	-1.4[i]	-0.8	-13.4[m]	2.4	37.0	n.a.	21.5
Sweden	5.9[c]	0.2[g]	-0.5	0.0	2.5	7.1	3.0	14.2
Austria	9.8[a]	5.1[g]	0.3	-1.6	3.4	9.0	1.4	10.1
Finland	6.8[c]	5.1	4.7	2.7	0.7	13.4	5.7	23.1
Switzerland	0.8[a]	-0.5[l]	-3.1	-1.5	2.2	7.8	-1.3	5.5
Norway	7.3	1.3[g]	-2.7	0.9	-0.1	11.1	0.8	21.1
Australia	n.a.	8.9[e]	5.2	1.5	-2.0	7.5	6.7	21.7
Canada	5.5	4.9	-1.1	2.2	-2.1	4.5	-0.5	21.5
OECD weighted average	n.a.	n.a.	1.3	2.7	0.9	9.3	0.9	13.3

Notes: n.a. = not available
[a] 1970–81   [b] 1972–81   [c] 1971–81   [d] 1971–80   [e] 1985–94   [f] 1985–91   [g] 1985–93   [h] 1986–92   [i] 1986–93
[l] 1992–94   [m] 1984–93

(1) Million $US at 1995 PPP
(2) Resident patents: inventors in their home country
(3) Non-resident patents: foreign inventors in the country
(4) External patents: national inventors patenting abroad

Source: Archibugi and Iammarino (2002) based on OECD, MSTI, various years.

have a number of well-known drawbacks due to the quality of data available.[2] In spite of these drawbacks, there is a general agreement in the literature that global inter-firm alliances have become increasingly popular over the past two decades (Hagedoorn 2002; see also Ch. 3 by Powell and Grodal, this volume).[3] International STP has grown considerably in absolute terms, although its share of all STP has remained steady in the 1970s and 1980s, oscillating around 60 per cent of all agreements, while the share has declined in the 1990s to about 50 per cent (Hagedoorn 2002). There has also been a gradual shift in the types of agreements favored by firms over time, according to the MERIT-CATI database. The percentage of equity agreements in the total has declined from about 70 per cent to less than 10 per cent between the mid-1970s and the end of the 1990s. The increasing share of non-equity alliances may indicate growing use by MNEs of STPs as relatively rapid, short-term vehicles to gain access to non-domestic knowledge sources (see Section 12.5 for further discussion).

STP agreements appear to be most common in the domain of new materials, biotechnology and information technology, and largely involve Triad economies rather than developing economies. Developed-country firms participate in 99 per cent of the STP agreements in the MERIT-CATI dataset (Hagedoorn 2002). Although R&D and manufacturing outsourcing agreements with developing-country firms have expanded in number during the last two decades, the share of these firms in STP has remained around 5–6 per cent since the 1990s (Narula and Sadowski 2002). Seventy per cent of all STP since the 1960s have had at least one US partner, with collaborations between European and North American firms increasing from 18.5 to 25.2 per cent of overall technological alliances between the 1970s and the 1990s (Hagedoorn 2002).

12.2.3 The Role of MNEs in the Cross-Border Generation of Innovation

The globalization of innovation has been associated with growth in MNE activity and FDI since World War II. FDI stocks as a percentage of GDP[4] stood at 21.46 per cent in 2001, up from just 6.79 per cent in 1982 (Table 12.3). Furthermore, MNEs engage in considerable intra- and inter-firm trade (Table 12.3). The primary source of outbound FDI—almost 90 per cent of the total in 2001—continues to be the industrialized countries. The EU accounted for the largest share of outward FDI, with Netherlands, UK, France, and Germany accounting for fully 41.3 per cent of all outward FDI stock from the developed world. Around 68 per cent of inward FDI is also directed towards Triad countries. The developing economies' increase in the share of inward FDI during the period 1982–2001 is almost entirely due to a small group of developing countries, primarily the Asian NICs and China.

Table 12.3 Selected indicators of FDI and international production,
1982–2001 ($US Billion at current prices and percentage values)

	1982	2001
FDI inflows	59	735
FDI outflows	28	621
FDI inward stock	734	6846
FDI outward stock	552	6582
Sales of foreign affiliates	2541	18517
Gross product of foreign affiliates	594	3495
Total assets of foreign affiliates	1959	24952
Exports of foreign affiliates	670	2600
Employment of foreign affiliates (thousands)	17987	53581
Inward FDI stocks to GDP ratio	6,79%	21,46%
Foreign affiliates' export to total exports	32,20%	34,99%

Source: UNCTAD, based on its FDI/TNC database and UNCTAD estimates.

The figures for R&D activity reflect similar patterns, since many of the largest firms engaged in FDI are key actors in the generation and diffusion of innovation. More than one-third of the top 100 MNEs are active in the most R&D-intensive industries, such as electronics and electrical equipment, pharmaceuticals, chemicals (UNCTAD 2002). Furthermore, large MNEs play a dominant role in the innovative activities of their home countries. For instance, Siemens, Bayer and Hoechst performed 18 per cent of the total manufacturing R&D expenditures in Germany in 1994 (Kumar, 1998). In 1997 three MNEs accounted for more than the 30 per cent of the overall UK R&D investment in manufacturing. These same MNEs also undertake a growing share of their total R&D activities outside their home countries.

Significant cross-national differences are also apparent in indicators of international R&D. The share of national R&D expenditures accounted for by non-domestic sources varies substantially within the industrialized and developing areas (see Table 12.4 for some details). The origins of international R&D investment flows also differ considerably among industrialized economies (Table 12.5). Cantwell (1995) suggests that countries such as Switzerland, UK, and the Netherlands, which have historically been home to large MNEs and that were long-time international investors in R&D, have greatly expanded their offshore R&D investments since World War II. Another group of countries (such as France and Germany) has relatively few large MNEs, and their outward R&D investments have grown more

Table 12.4 R&D expenditure of foreign affiliates as a percentage of total R&D expenditures by all firms in selected host economies, 1998 or latest year

Country	Percentage of R & D
Canada	34.2
Finland (1999)	14.9
France	16.4
Japan	1.7
Netherlands	21.8
Spain (1999)	32.8
UK (1999)	31.2
US	14.9
Czech Republic (1999)	6.4
Hungary	78.5
India (1994)	1.6
Turkey	10.1

Source: UNCTAD (2002), table I.10.

Table 12.5 Shares of US patenting of largest nationally owned industrial firms due to research located abroad, 1920–1990 (%)

	1920–1939	1940–1968	1969–1990
US	6.81	3.57	6.82
Europe	12.03	26.65	27.13
UK	27.71	41.95	43.17
Germany	4.03	8.68	13.72
Italy	29.03	24.76	14.24
France	3.35	8.19	9.55
Sweden	31.04	13.18	25.51
Netherlands	15.57	29.51	52.97

Source: Cantwell (1995).

gradually during the last eighty years. A third group includes countries that were major investors in offshore R&D during the first fourteen years of the twentieth century. Offshore investment by these economies actually declined after 1914 and returned to pre-World War I levels only recently. This group includes the United

States, home of a number of MNEs which have a relatively low proportion of their R&D and patenting activity abroad.[5]

On average, firms from EU countries obtain a larger share of patents from their foreign subsidiaries than is true of US or Japanese companies (Table 12.6). During the 1969–95 period, the share of total patents of EU firms attributable to foreign affiliates grew from 26.3 to 32.5 per cent. European firms tend to concentrate a considerable share of their international R&D activities in the US (over 50 per cent of their foreign R&D investment on average, with German, British, and Swiss firms showing the highest concentration of their foreign activities in the United States). The foreign patenting activity of US firms also increased during this period, but remained below 10 per cent.[6] Although US foreign R&D activities are relatively low compared to EU firms, they are much larger than Japanese companies, whose offshore patenting declined from 2.1 per cent in 1969–77 to approximately 1 per cent of their total patenting activity in 1987–95.

Overall, MNEs have increasingly internationalized their innovative activities, with a few relevant exceptions (most notably, Japanese MNEs). The importance of R&D activities of foreign affiliates has grown in most host economies over the 1990s. R&D by foreign firms is especially high in the UK, Ireland, Spain, Hungary, and Canada, and lowest in Japan, with other countries (including the US, France, and Sweden) in intermediate positions. Nevertheless, most R&D and patenting activities are still largely concentrated in the MNEs' home countries, and in a few host countries. Well over 90 per cent of the R&D expenditures of most MNEs is located

Table 12.6 Share of US patents of the world's largest firms attributable to research in foreign locations by main area of origin of parent firms, 1969–1995 (%)

Nationality of parent firm	1969–77	1978–86	1987–95
US	5.4	6.9	8.3
Japan	2.1	1.2	1.0
European countries[a]	26.3	25.6	32.5
Total all countries[b]	10.3	10.7	11.3
Total all countries excluding Japan	11.1	13.0	16.2

Notes:
[a] Germany, UK, Italy, France, Netherlands, Belgium, Luxembourg, Switzerland, Sweden, Denmark, Ireland, Spain, Portugal, Greece, Austria, Norway, Finland.
[b] Total includes all the 784 world's largest firms recorded by the University of Reading database, base year 1984.
Source: Cantwell and Janne (2000).

within the Triad.[7] While there are significant differences in the international dispersion of innovative activity across industries, firms have generally not internationalized their innovative activity at the same rate as their production activities.
Exceptions to this rule are MNEs originating from small economies, such as Belgium, the Netherlands, and Switzerland. A large proportion of even the most
internationalized MNEs concentrate at home their more "strategic" activities,
such as R&D and headquarters functions (Benito et al. 2003).

This relatively low—but increasing—degree of internationalization is associated
inter alia with the complex nature of systems of innovation, and the embeddedness
of the MNE's activities in the home environment (see e.g. Narula 2002*a*), the need
for internal cohesion within the MNE (Blanc and Sierra 1999, Zanfei 2000), and the
high quality of local infrastructures and appropriability regimes that R&D activities
tend to require. These factors, together with the difficulties of managing complex
technological portfolios, imply that the internationalization of innovation occurs at
a slower pace than the internationalization of production.

12.3 Overseas Innovative Activities of MNEs: Theoretical and Empirical Issues

The extensive literature on international R&D investment highlights two broad
firm-level motives. First, firms internationalize their R&D to improve the way in
which existing assets are utilized. That is, firms seek to promote the use of their
technological assets in conjunction with, or in response to, specific foreign locational conditions. This has been dubbed as asset-exploiting R&D (Dunning and
Narula 1995) or home-base exploiting (HBE) activity[8] (Kuemmerle 1996). For
example, some modification in these firms' products or processes may be necessary
to make them competitive in the relevant foreign market. This type of offshore R&D
investment typically is based on the technological advantages of the source firm,
which in turn reflect those of its home country.

Asset-exploiting strategies correspond to traditional views of the organization of
innovative activities and foreign direct investment, many of which were rooted in the
"product life cycle" theory of such investment. Referring mainly to US-based
multinationals, Vernon (1966), Kindleberger (1969), and Stopford and Wells (1972)
suggested that an MNE's foreign subsidiary replicated the parent's non-strategic

activities abroad, with strategic decisions—including R&D and innovation— being rigidly centralized in the home country. Vernon emphasized that coordinating international innovative activities was too costly, due to the difficulties of collecting and controlling relevant information across national borders. The R&D activities of foreign subsidiaries were limited largely to the adoption and diffusion of centrally created technology.

The second broad motive for offshore R&D investment is strategic asset-augmenting activity (Dunning and Narula 1995), also known as home-base aug-menting (HBA) activity (Kuemmerle 1996). Firms use these types of R&D invest-ments to improve existing assets or to acquire (and internalize) or create completely new technological assets through foreign-located R&D. The assumption in such cases is that the foreign location provides access to *complementary* location-specific advantages that are less available in its primary or "home" base (Ietto-Gillies 2001). In many cases, the strategic assets sought by the investing firm are associated with the presence of other firms. A location which is home to a major competitor may attract asset-augmenting investments by other firms in the same or in other related indus-tries (see Cantwell in this volume on the implications of these patterns of FDI for the competitiveness of host countries). Asset-augmenting motives and technology sourcing have been partially incorporated in formal models of the FDI decision.[9]

The asset-augmenting perspective, which considers local contexts more as sources of competencies and of technological opportunities, and less as constraints to the action of MNEs, marks a fundamental departure from the conventional wisdom. In a seminal contribution, Hedlund (1986: 20–1) caught the essence of this new way of conceptualizing the role of local contexts: "The main idea is that the foundations of competitive advantage no longer reside in any one country, but in many. New ideas and products may come up in many different countries and later be exploited on a global scale." (See Kogut 1989 for a similar view.)

There are several reasons why such asset-augmenting R&D activities are hard to achieve through means other than FDI. Some of these reasons are associated with the nature of technology. When the knowledge relevant for innovative activities is clustered in a certain geographical area and is "sticky," foreign affiliates engage in asset-augmenting activities in these areas in order to benefit from the external economies and knowledge spillovers generated by the concentration of production and innovation activities in the relevant clusters. The tacit nature of technology implies that even where knowledge is available through markets, it may still require modification to be efficiently integrated within the acquiring firm's portfolio of technologies. The tacit nature of knowledge associated with production and innov-ation activity in these sectors also means that "physical" or geographical proximity may be important for accessing and absorbing it (Blanc and Sierra 1999). The marginal cost of transmitting codified knowledge across geographic space does not depend on distance, but the marginal cost of transmitting, accessing, and absorbing tacit knowledge increases with distance. This leads to the clustering of innovation activities, especially in the early stages of an industry life cycle where tacit knowledge plays an important role (Audretsch and Feldman 1996).

In general, asset-exploiting activities are primarily associated with demand-driven innovative activities (e.g. localization of the parent-firm products for a specific offshore market). Asset-augmenting activities, on the other hand, are primarily undertaken with the intention to acquire and internalize technological spillovers that are host location-specific. Asset-exploiting activity, broadly speaking, represents an extension of R&D work undertaken at home, while asset-augmenting activity represents a diversification into new scientific problems, issues or areas.

An extensive literature has suggested that asset-augmenting internationalization of R&D has become more significant during the past two decades as a result of several factors that include: (*a*) the increasing costs and complexity of technological development, leading to a growing need to expand technology sourcing and inter-action with different and geographically dispersed actors endowed with comple-mentary bits of knowledge; (*b*) the faster pace of innovative activities in a number of industries, spurring firms to search for application opportunities which are mainly location-specific; (*c*) growing pressures from host governments, which have led MNEs to increase the interaction with local partners as key conditions to gain access to foreign markets.

Although the conceptual differences between these two motives for offshore R&D investment are clear, indicators of the importance of these two motives are scarce. Until recently, most empirical studies of international R&D investment (Mansfield et al. 1979; Lall 1979; Warrant 1991) reflected the view that the role played by foreign R&D units was determined by market or demand-side factors, i.e., asset-exploiting motives were assumed. More recent empirical work, however, has focused on asset-augmenting motives for R&D investments. Detailed analyses carried out by Miller (1994), Odagiri and Yasuda (1996), and Florida (1997) argue that technology sourcing strategies play an important role in a number of manufacturing industries in North America, Europe and Asia.[10] Some studies find that "market-oriented" R&D units established for asset-exploiting motives have evolved into asset-aug-menting ones (Rondstadt (1978). But other foreign R&D units experience no major shift in their characters (Kuemmerle 1999).

Several studies have used multivariate techniques to identify the relative import-ance of asset-augmenting vs. asset-exploiting motives for offshore R&D investment. Using patent citations Almeida (1996) found that foreign firms in the semiconductor industry not only learnt more from local sources, but they did so to a greater extent than their domestic counterparts. This study also found that, with the significant exception of subsidiaries of Japanese MNEs, foreign firms locate their technological activities overseas in areas where these firms exhibited a home country disadvantage (measured in terms of 'Revealed Technological Advantages' (RTA)). Using a similar methodology, Cantwell and Noonan (2002) showed that MNE subsidiaries located in Germany between 1975 and 1995 sourced a relatively high proportion of knowledge (especially new, cutting-edge technology) from this host country.

Data such as this lend support to the idea that foreign owned technological activities undertaken in Germany are often asset-augmenting. However, Patel and Vega (1999) obtained different results from their study of US patenting activities in high technology fields. By comparing the RTA of the MNE at home and the host location, they showed that a majority of firms undertook foreign innovative activities in the technological fields in which they were strong at home. They interpreted this as evidence that asset exploiting motives, i.e. adapting products and processes for foreign markets and providing technical support to offshore manufacturing plants, remained dominant in MNEs' foreign innovative activities. Their findings were supported by an extensive interview-based survey carried out by Pearce (1999). Employing a methodology similar to that of Patel and Vega, Le Bas and Sierra (2002) confirmed that MNEs rarely internationalize R&D to compensate for technological weaknesses at home. However, their research also showed that the lion's share of these investments went to technologically advanced locations, indicating that asset augmenting is very important and can coexist with asset exploiting in many cases. This may be interpreted as signalling the formation of global "centers of excellence" in specific technological fields (see Box 12.1 for details on the methodology used to measure alternative international R&D strategies).

Box 12.1 Asset exploiting, asset augmenting or both?

In one of the most extensive empirical exercises to date, Le Bas and Sierra (2002) studied the R&D investment strategies of the 345 MNEs with the greatest patenting activity in Europe between 1988 and1996. These companies, which accounted for about one half of total patenting through the European Patent Office (EPO) over this period, were predominantly of US, European, or Japanese origin.

To measure the technological strength of companies and locations, the authors used a patent-based indicator ("Relative Technolological Advantage," RTA). For a company, HomeRTA is defined as the firm's share of total European patents in a particular technological field relative to its overall share of all European patents. Patents from foreign affiliates of the firm (filed from outside the country in question) were excluded from the calculation. For a location (country) in which a given firm has invested, HostRTA is defined as the host country's share of all European patenting in that field, divided by its share of all European patents in all fields. In all cases an RTA> 1 signals a relative advantage of the country (firm). Based on these definitions, four different R&D strategies may be identified:

Corporate technological activities in the home country	Technological activities in the host country	
	Weak	Strong
Weak	Type 1: market-seeking HomeRTA < 1 HostRTA < 1 (Technology is not a driver of FDI) (10%)	Type 2: technology-seeking HomeRTA < 1 HostRTA > 1 (13%)
Strong	Type 3: asset-exploiting HomeRTA > 1 HostRTA < 1 (Efficiency-oriented FDI in R&D) (30%)	Type 4: asset-augmenting> HomeRTA > 1 HostRTA > 1 (Learning-oriented FDI in R&D) (47%)

Source: adapted from Patel and Vega (1999, p. 152) and from Le Bas and Sierra (2002 p. 606).

The numbers in brackets indicate the frequency of the strategy in question for the sample of firms considered. As is evident from the table, Le Bas and Sierra found that the great majority of MNEs located their activities abroad in technological areas or fields for which they were strong at home (strategies 3 and 4). However, the most frequent strategy is clearly number 4, in which case not only the firm but also the host country has a relative technological advantage (HostRTA > 1). This may indicate the formation of "centers of excellence" in which strong domestic research environments function as global attractors.

12.4 FORCES SUPPORTING CONCENTRATION AND DISPERSION OF R&D

The literature on the location of R&D activities views the location of MNEs' innovative activities as affected by centrifugal and centripetal forces that determine whether the MNE centralizes (in the home location) or internationalizes to create additional centers abroad. But all too often, this dichotomy—while substantially correct—presumes that the MNE has a single center in the first place. In order to allow for the possibility that the MNE may have multiple home bases or several locations of R&D concentration rather than a single "hub", this section uses the terms "concentration" and "dispersion."

We can single out at least four broad sets of factors affecting the concentration *and* dispersion of innovative activities. These forces are active at both the macro-level of countries, regions and systems of firms involved in the globalization of innovation; and the micro-level of individual firms and of their internal networks of innovative activities across national borders.

12.4.1 The Costs of Integrating Activities in Local Contexts

When firms engage in R&D in a foreign location to avail themselves of complementary assets that are location-specific (including those that are specific to local firms or institutions), they seek to internalize several aspects of the systems of innovation of the host location. Developing and maintaining strong linkages with external networks of local counterparts is expensive and time consuming. Networks of government-funding institutions, suppliers, university professors, private research teams, informal networks of like-minded researchers take considerable effort to create, but once developed, links with these entities or networks are less costly to maintain. Even where the host location is potentially superior to the home location, the high costs of becoming familiar with and integrating into a new location may be prohibitive. Firms are constrained by resource limitations, and some minimum threshold size of R&D activities exists in every distinct location. As such, maintaining more than one facility with a "critical mass" of researchers requires that the new (host) location offer significantly superior spillover opportunities, or provide access to complementary resources that are unavailable elsewhere and cannot be acquired by less costly means more efficiently.[11]

12.4.2 Local Technological Opportunities and Constraints

The high costs of integration into a host location's systems of innovation—in contrast to the low marginal cost of maintaining its embeddedness in its home location's innovation system—may increase the fixed costs firms have to overcome in order to expand internationally (Narula 2003). However, these costs must be tempered by other supply-side considerations. For example, development of the technologies in question may benefit from diversity and heterogeneity in the knowledge base, which might come from competitors, from interaction with customers and from other complementary technologies in the offshore site. A single national innovation system, especially in a small country, may be unable to offer the full range of interrelated technological assets required for this diversification strategy (see Box 12.2 on the interactions between innovation systems and R&D internationalization strategies).

Box 12.2 How innovation systems affect the internationalization of R&D

Innovation systems are built upon a relationship of trust, iteration, and interaction between firms and the knowledge infrastructure, within the framework of institutions based on experience of and familiarity with each other over relatively long periods of time. In engaging in foreign operations in new locations, firms which already face opportunities and constraints created by their home innovations systems gradually become embedded in the host environment. The self-reinforcing interaction between firms and infrastructure perpetuates the use of a specific technology or technologies, or production of specific products, and/or through specific processes. Increased special-ization often results in a systemic lock-in. Institutions develop that support and reinforce the interwoven relationship between firms and the knowledge infrastructure through positive feedback, resulting in positive lock-in. When innovation systems cannot respond to a technological discontinuity, or a radical innovation that has occurred elsewhere, there is a mismatch between what home locations can provide and what firms require, this is known as sub-optimal lock-in (Narula 2002*a*).

In general, national innovation systems and industrial and technological specializa-tion of countries change only very gradually, and—especially in newer, rapidly evolv-ing sectors—much more slowly than the technological needs of firms. In other words, there may be systemic inertia. Firms have three options open to them (Narula 2002*a*). Firms may seek either to import and acquire the technology they need from abroad, or venture abroad and seek to internalize aspects of other countries' innovation systems, thereby utilizing an "exit" strategy. Of course, firms rarely exit completely, preferring often to maintain both domestic and foreign presence simultaneously. There are costs associated with an exit strategy. On the one hand, they would weaken their contact with their home market and by so doing they might reduce their ability to absorb external knowledge. On the other hand, it must suffer the costs of entry in another location (in terms of effort, capital, and time), and firms may minimize this through a cooperative strategy with a local firm. Developing alternative linkages and becoming embedded in a non-domestic innovation system takes considerable time and effort.

They can also use a "voice" strategy which is to seek to modify the home-country innovation system. For instance, establishing a collective R&D facility, or by political lobbying. Firms are inclined towards voice strategies, because it may have lower costs, especially where demand forces are not powerful, or where the weakness of the innovation system is only a small part of their overall portfolio. But voice strategies have costs, and may not be realistic for SMEs, which have limited resources and political clout. Such firms usually cannot afford an "exit" strategy either, and end up utilizing a "loyalty" strategy, relying instead on institutions to evolve, or seeking to free-ride on the voice strategy of industry collectives, or larger firms.

Where local technological opportunities are sufficiently high, asset-augmenting activities are likely. Capturing foreign opportunities may require that a firm develop proximity to local "technology leaders" (see Ch. 20 by Cantwell, this volume) whose competences are rooted in the offshore system of innovation.[12] Whenever products are multi-technology-based, one firm may be marginally ahead in one technology, and its competitor in another; but on a macro-level, both may be associated with "powerful" innovation systems (Criscuolo et al. 2005). Thus, technology leadership can change rapidly. This is another reason why firms often engage in both asset-augmenting and asset-exploiting activities simultaneously.[13]

12.4.3 Firm Size and Market Structure

An important factor affecting internationalization is the size of the firm. The expansion of R&D activities—both at home and in overseas locations—requires considerable resources of capital and management expertise that smaller firms often lack. *Ceteris paribus*, large firms have more money and resources to use in overseas activity. As they have higher R&D budgets at home, they are also more likely to have the absorptive capacity to set up linkages with both foreign and domestic science bases. R&D is a costly and slow affair, and overseas R&D facilities are an expensive and risky option that is hard to justify for SMEs. Indeed, Belderbos (2001) finds that there is a non-linear relationship between firm size and overseas R&D, with medium-sized Japanese firms showing a higher propensity (in relative terms) to internationalize R&D than small- or large-sized firms. Many small firms operate as part of a domestic supplier network for larger firms, and are thus also bound to their home location (or the location of their main customers) (Narula 2002*b*). Internationalization of supplier firms often occurs in tandem with the internationalization of their primary customer, especially where the customer dominates their market. This motive was apparent in the investment by Japanese automobile firms' supplier firms in US and European production facilities during the 1980s and early 1990s (Florida 1997).

Industry-specific factors also encourage or discourage the locational concentration of innovative activities. The industrial structure of countries is path dependent, and technological specialization changes only gradually over time (Cantwell 1989; Zander 1995). At one extreme, mature technologies evolve slowly and demonstrate minor but consistent innovations over time. The technology is to a great extent codifiable, widely disseminated, and the property rights well defined. Under these circumstances, constant and close interaction with customers is not an important determinant of R&D: profits of firms depend on the costs of inputs, and proximity to the source of these inputs is often more significant than that of customers. At the other extreme, rapid technological change in "newer" technologies or engineering industries may require closer interaction between production and R&D (Lall 1979), or between users and producers of technology. In some circumstances both new technology and applications environments have a high tacit, uncodified element, requiring extensive interaction during new product development, design, and testing. This factor may account for the frequent establishment of both manufacturing and R&D plants close to applications abilities in foreign telecommunications markets (Ernst 1997). In other industries, however, a large variety of international linkages are required for R&D and innovation, as appears to be particularly the case of biotechnology (Arora and Gambardella 1990).

12.4.4 Organizational Issues

Another micro-level determinant is associated with the difficulties of managing cross-border R&D activities. It is not sufficient for the foreign affiliate to internalize spillovers if it cannot make these available to the rest of the MNE (Blanc and Sierra 1999). A dispersion of R&D activities across the globe requires extensive coordination between them, and particularly with headquarters, if they are to function efficiently. This acts as a centripetal force on R&D, and accounts for a tendency of firms to locate R&D (or at least the most strategically significant elements) closer to headquarters.

Complex linkages, both within the firm, and between external networks and internal networks, require complex coordination if they are to provide optimal benefits (Zanfei 2000). Such coordination may require expertise, managerial and financial resources that are most likely available to larger firms with more experience in transnational activity (Castellani and Zanfei 2004). Large firms tend to engage in both asset-augmenting and asset-exploiting activities. Indeed, large MNEs may have several semi-autonomous sister affiliates in the same location that operate in similar technological areas. Lastly, MNEs tend to engage in production activities (whether in the same or another physical facility) in the host location, and this prompts a certain level of asset-exploiting activity. Thus, an MNE in a given location may seek to internalize spillovers from non-related firms and to exploit intrafirm knowledge transfers within the same multinational group (Criscuolo et al. 2005).

12.5 INNOVATION THROUGH INTERNATIONAL STRATEGIC TECHNOLOGY PARTNERING

The previous sections have discussed the growing international dimension of R&D, concentrating on the intra-MNE aspect of this development. However, not all innovatory activity is undertaken within hierarchies; during the last two decades, "non-internal" R&D activities that rely on interfirm cooperative agreements have grown rapidly in number.

Fully examining the role of (international) networks in the generation and diffusion of innovation is beyond the scope of this chapter (see Powell and Grodal, Ch. 3 in this volume, for a more comprehensive discussion). A key issue for this discussion is whether and to what extent there is substitution or complementarity between internal innovative activities and technological collaborations on a global scale.

In some circumstances, international STP may substitute for internal innovative activities. One such circumstance is that of R&D alliances aiming to enter foreign markets protected by non-tariff barriers, as is the case for environmental regulations in the chemical industry. Nonetheless, there are limits to how much a firm can substitute STP for in-house R&D, and by extension, international STP for overseas R&D facilities.[14] STP tends to develop in areas in which partner companies share complementary capabilities, and these alliances create a greater degree of interaction between the partners' respective paths of learning and innovation (Mowery et al. 1998; Cantwell and Colombo 2000; Santangelo 2000).

One way to look at this issue is to tackle the problem of firm size, technological capabilities, and collaborations. Participation in STPs tends to be correlated with firm size in technology-intensive sectors. In these sectors, cooperation is a way to keep up with the technological frontier: by associating complementary resources and competencies, it makes it possible to explore and exploit new technological opportunities. But smaller technology-based MNEs also are involved in such agreements, and their growing significance raises numerous conundrums (Narula 2002b; see also Ch. 5 by Lam, and Ch. 3 by Powell and Grodal, this volume). Firms—regardless of size—must maintain a growing breadth of technological competences, and this may require participation in international internal and external networks. SMEs need to rely on non-internal sources, as they often experience wider gaps in terms of competencies and development abilities than their larger counterparts (Zanfei 1994) but must be more skilful at managing their portfolio of technological assets, because they have limited resources (Narula 2002b). Indeed, the costs of managing a web of cross-border agreements highlight the importance of transaction-type ownership advantages for the MNE. This complementarity between firm size, technological capabilities and the development of innovation networks is consistent with some of the trends highlighted in Section 12.2.2. In particular, the geographical concentration of STP activity within the Triad reflects *inter alia* the fact that firms from these areas tend to be larger and account for a major share of R&D activity.

The issue of complementarity or substitution between the internal and non-internal innovative activities of MNEs can also be examined by looking at the interdependencies between multinational expansion and international STP. Drawing on the transaction-cost literature, several works on international market entry strategies argue that multinational experience may lower the risks faced by an MNE in entering a new foreign market. In the absence of multinational experience, cooperative ventures may be more effective market entry tools than hierarchical control strategies. As MNEs accumulate greater experience in foreign markets, the information-gathering and risk-sharing advantages of collaboration will decline. As a result, the organizational costs of cooperation, in terms of shirking and conflicts of interest between partners, will exceed the benefits of this strategy for experienced MNEs (e.g. Gomez-Casseres 1989; Hennart and Larimo 1998). In summary, multinational experience is supposed to impact negatively on collaborative ventures and positively on equity-based, commitment-intensive linkages. This view is largely—but not exclusively—consistent with the argument that multinational experience helps facilitate the exploitation of MNEs' assets in foreign markets. That is, MNEs respond to uncertainty in host economies by utilizing their own assets as a means to penetrate these markets. Such a view regards STP as a second-best option.

A second body of literature focuses mainly on the evolution of high technology industries, and highlights an important motive for interfirm linkages, i.e. the need to explore and rapidly exploit new opportunities, either new businesses or new technological developments. From this perspective, strategic alliances provide "an attractive organisational form for an environment characterised by rapid innovation and geographical dispersion in the sources of know how" (Teece 1992: 20). As the relevant knowledge sources are dispersed globally in a number of industries, this perspective explains the formation of some types of international STP agreements. From this perspective, multinational experience—which is associated with the establishment and activity of foreign subsidiaries over time—may increase a firm's capacity to search for and absorb external knowledge (Cantwell 1995; Castellani and Zanfei 2004). This view is consistent with a number of studies on high technology industries which highlight the mutually reinforcing nature of intra- and interfirm networks. Multinational experience thus may expand a firm's exploration potential and hence expand its use of international STP.[15]

Some of the trends in the development of STP highlighted in section 12.2.2 seem to be consistent with the view that firms with multinational experience are more likely to use alliances as an exploratory strategy. As we have shown, the fraction of *non equity* STPs is growing, particularly in high technology industries. This trend may constitute evidence of the fact that low commitment intensive agreements are more effective as a mechanism to gain timely and extensive access to rapidly evolving technology across borders. From this perspective, STP may represent a "first-best" option to MNEs (Narula 2003), especially where innovative activities are concerned. In other words, firms do not necessarily resort to these strategies because they cannot have access to more effective and more profitable channels of technology transfer (as uncertainty is too high or institutional barriers constrain "internal" strategies). Instead, STPs, especially non equity agreements, are more flexible and more apt for knowledge development and learning.

12.6 CONCLUSIONS AND POLICY ISSUES

This chapter has discussed the internationalization of innovative activities, and highlighted that it has been driven by a myriad of factors. One of the most recurrent among these factors is the need to respond to different demand and market conditions across locations, and the need for the firms to respond effectively to these by adapting their existing product and process technologies through foreign-located R&D.

Nevertheless, supply factors and the need to gain access to local competencies have become an increasingly important motivation to engage in asset-augmenting R&D abroad. This is due, *inter alia*, to the growing tendency for multi-technology products, and to the fact that patterns of technological specialization are distinct across countries, despite the economic and technological convergence associated with economic globalization.

As a result, there is a growing mismatch between what home locations can provide and what firms require. In general, innovation systems and the industrial and technological specialization of countries change only very gradually, and—especially in newer, rapidly evolving sectors— much more slowly than the technological needs of firms. Firms must seek either to import and acquire the technology they need from abroad, or venture abroad and seek to internalize aspects of other countries' innovation systems. A third option, lobbying for modification of the home-country innovation system, is expensive and difficult (Narula 2002a). Thus, in addition to proximity to markets and production units, firms venture abroad to seek new sources of knowledge, which are associated with the innovation system of the host region. The interdependence of markets and the cross-fertilization of technologies—whether through arm's-length means, cooperative agreements, or equity based affiliates—means that that few countries have truly "national" systems. Of course, some innovation systems are more "national" than others, and the term is indicative rather than definitive (see also Ch. 7 by Edquist and Ch. 14 by Malerba in this volume for a discussion). Furthermore, firms need a broader portfolio of technological competences than they did in the past.

The internationalization of R&D raises crucial welfare issues, since it provides opportunities for spillovers between the MNE and its host economy, and in certain circumstances between the MNE affiliate and its home country. There has been some concern in the US with the potential loss of competitiveness of domestic firms and with the impoverishment of the "national knowledge base" which would be associated with the increasing local R&D presence of foreign-owned MNEs (e.g., Dalton et al. 1999). In other countries and areas of the world, the perception is very different, as a local presence of foreign R&D and value-added activities is expected to contribute to the upgrading of national technology systems. A few empirical studies seem to provide sound evidence on the existence of positive spillovers of multinational presence in some emerging economies such as Korea, Taiwan, and Singapore (Hobday 2000; Lim 1999), and some EU member states (Barry and Strobl 2002; Castellani and Zanfei 2003). However the evidence in the case of most developing countries does not point to significant spillovers (see Harrison 1999). Indeed, according to a recent survey on econometric studies of productivity spillovers from FDI, the number of studies in which negative or non-significant results are obtained is approximately as high as cases where positive spillovers were observed (Gorg and Strobl 2001). This suggests a cautious approach to this issue, and calls for a refinement of analytical tools (see Box 12.3). There is a need to develop more appropriate measures of technological spillovers, which are not properly captured by performance indicators like productivity. The channels through which spillovers occur also need to be examined more carefully, if FDI-related spillovers are to be explicitly used as means for technological upgrading.

Another position in this policy debate argues that the internationalization of R&D may lead to a "hollowing out" of the home country's innovatory capacity when the domestic innovation system does not meet the needs of firms in certain industries.

Box 12.3 Host country effects: technology gaps, technological upgrading, and absorptive capacity

One of the strongest and most popular arguments in favor of inward investment as a vehicle for local technological upgrading is that foreign firms usually outperform domestic ones (see Bellak 2002 for a review on empirical evidence on this aspect). The underlying policy issue is whether or not foreign presence can generate technological opportunities for the local economy. There is a clear connection here to the literature on technology gaps and catching up (Godinho and Fagerberg, Ch. 19 in this volume). On the one hand, some works suggest that the larger the productivity gap between host country firms and foreign-owned firms, the larger the potential for technology transfer and for productivity spillovers to the former. This assumption, can be derived from the original idea put forward by Findlay (1978), who formalized technological progress in relatively "backward" regions as an increasing function of the distance between their own level of technology and that of the "advanced regions," and of the degree to which they are open to direct foreign investment.

On the other hand, scholars have argued that the lower the technological gap between domestic and foreign firms, the higher the absorptive capacity of the former, and thus the higher the expected benefits in terms of technology transfer to domestic firms. It is worth noting that the role of absorptive capacity is also implicitly recognized in the catching up tradition, when it is acknowledged that a sort of lower bound of local technological capabilities exists, below which foreign investment cannot be expected to have any positive effects on host economies.[16] The "technological accumulation hypothesis" goes beyond this simplistic view of absorptive capacity and places a new emphasis on the ability to absorb and utilize foreign technology as a necessary condition for spillovers to take place.

Although there is currently little evidence to support or refute the hollowing out hypothesis, this has been raised by policy makers in several countries, and represents an important area for future research. The consequences of a potential hollowing out may be especially significant in small open economies that are specialized around a few products, and/or concentrated around a few large firms. Another related and potentially important area for future research is the need to distinguish between hollowing-out as a symptom of sub-optimal lock-in and the internationalization of innovation to supplement domestic supply limitations (Narula 2003). After all, no country can provide world-class competences in all technological fields. Even the largest, most technologically advanced countries cannot provide strong innovation systems to all their industries, and world-class competences in all technological fields. Some countries regard imported technologies as a sign of national weakness, and have sought to maintain and develop in-country competences, often regardless of the cost (Narula 2002a). Relying largely on in-country competences may however lead to a sub-optimal strategy, especially in this age of multi-technology products. In fact, the cross-border flow of ideas is fundamental to firms, and this imperative has increased with growing cross-border competition, and international production.

Notes

1. Both changes in the composition of world trade, and sectoral correlations between R&D intensity and internationalization should be considered with caution since definitions of industries change over time (see Von Tunzelman and Acha, Ch. 15 in this volume).

2. For instance, press releases are often used to construct data-sets, and these are not always factual, sometimes reflecting the public relations objectives of the firms; the coverage of large firms is higher than for smaller firms; STP failures are not reported as accurately (or as often) as STP formation; large databases are hard to update and are frequently subject to changes in the methodology of data collection over time.

3. STP refers to interfirm cooperative agreements where R&D is at least part of the collaborative effort, and which are intended to affect the long-term product-market positioning of at least one partner.

4. Strictly speaking, the two numbers are not comparable, because GDP is a flow figure. Nonetheless, it is generally accepted that FDI stock is a monotonic function of value added, so the change in this ratio gives a general idea of how the significance of FDI activities has changed.

5. Paradoxically, perhaps, this group also includes Swedish MNEs, whose much higher shares of offshore R&D and patenting throughout the twentieth century, nevertheless, display a sharp drop after 1940 and a recovery by 1969–90 to a share that is lower than that of 1920–39.

6. Although the degree of R&D internationalization of US firms is below average, it more than doubled between the mid-1960s and the end of the 1980s (Creamer 1976; Pearce 1990).

7. Even where MNEs do engage in R&D in developing countries (e.g. industries where demand considerations and regional variations are especially significant, such as food products and consumer goods), these tend to agglomerate in just a few locations such as China, India, Malaysia, Brazil, South Africa, and the Asian NICs.

8. Although "home-base exploiting" (HBE), and "home-base augmenting" (HBA) (which we define later) have become dominant in the literature, this terminology is less accurate than "asset exploiting" and "asset augmenting". HBA and HBE hold to a very traditional view of the MNE as centered in a dominant home base. In fact, by emphasizing the role of home bases, the HBA–HBE jargon cannot be easily made consistent with the possibility that firms are evolving towards network structures, hence reducing the importance of a single home and, by the same token, expanding the number of countries wherein the firm ends up being based. This chapter takes the view that being accurate is more important than being fashionable, and avoids using the HBE–HBA terminology except where necessary for historic accuracy.

9. Fosfuri and Motta (1999) and Siotis (1999) show that a technological laggard may choose to enter a foreign market by FDI because there are positive spillover effects associated with locational proximity to a technological leader in the foreign country. Where the beneficial knowledge spillover effect is sufficiently strong, Fosfuri and Motta show that it may even pay the laggard firm to run its foreign subsidiary at a loss to incorporate the benefits of advanced technology in all the markets in which it operates.

10. Miller (1994: 37) studied the factors affecting the location of R&D facilities of twenty automobile firms in North America, Europe, and Asia, and found that an important motivation is to establish "surveillance outposts" to follow competitors' engineering and styling activities. In their study of 254 Japanese manufacturing firms, Odagiri and Yasuda (1996: 1074) note that R&D units are often set up in Europe and in the US to be kept informed of the latest technological developments. Similar results are obtained by Florida (1997: 90) analyzing 186 foreign affiliated laboratories in the US.

11. With few exceptions (e.g. Narula 2002a), the costs and inertia of offshore R&D networks is a topic which has not as yet been properly explored and represents an important area for further research.

12. Technology leaders are not always synonymous with industry leaders: firms—particularly in technology intensive sectors—increasingly need to have multiple technological competences (see e.g. Granstrand 1998; Granstrand et al. 1997).

13. This is another area which has not as yet been fully studied (for an exception, see Zander 1999) and represents an important area for further research.

14. The attempt to understand the reasons behind a firm's choice between non-internal and internal technological development is not new. The work of Teece (1986) presents a pioneering analysis of this issue, which builds on Abernathy and Utterback (1978), Dosi (1982) among others. See also further developments by Pisano (1990), Henderson and Clark (1990), Nagarajan and Mitchell (1998), Veugelers and Cassiman (1999), Gambardella and Torrisi (1998), Nooteboom (1999), Narula (2001) and Brusoni et al. (2001).

15. Castellani and Zanfei (2004) have tried to provide some empirical basis to this view with reference to the electronics industry. They measure what they call "specific experience" in terms of the number of subsidiaries a MNE has established in a given country, which in

their view would reduce uncertainty about the foreign market. Controlling for a number of sources of heterogeneity, they show that this factor is positively correlated with the creation of new subsidiaries and of equity agreements. By contrast, what they call "variety experience," reflecting the heterogeneity and geographical dispersion of markets where a MNE is active, should increase the firm's exploratory capacity. They find that, in the examined industry, variety experience has a positive and significant impact on non equity technical alliances.

16. As Findlay (1978: 2–3) notes: "Stone age communities suddenly confronted with modern industrial civilisation can only disintegrate or produce irrational responses... Where the difference is less than some critical minimum, admittedly difficult to define operationally, the hypothesis does seem attractive and worth consideration." Findlay also observes that the educational level of the domestic labour force, which is a good proxy for what is currently named country's "absorptive capacity," might also affect, *inter alia*, the rate at which the backward region improves its technological efficiency (Findlay 1978: 5–6).

REFERENCES

ABERNATHY, W., and UTTERBACK, J. (1978), "Patterns of Industrial Innovation," *Technology Review* 80: 97–107.

ALMEIDA, P. (1996), "Knowledge Sourcing by Foreign Multinationals: Patent Citation Analysis in the Semiconductor Industry," *Strategic Management Journal* 17: 155–65.

ARCHIBUGI, D., and IAMMARINO, S. (2002), "The Globalisation of Technological Innovation: Definition and Evidence," *Review of International Political Economy* 9(1): 98–122.

————(2000), "Innovation and Globalisation: Evidence and Implications," in F. Chesnais, G. Ietto-Gillies, and R. Simonetti (eds.), *European Integration and Global Corporate Strategies*, London: Routledge, 95–120.

*——and MICHIE, J. (1995), "The Globalisation of Technology: A New Taxonomy," *Cambridge Journal of Economics* 19: 121–40.

ARORA, A., and GAMBARDELLA, A. (1990), "Complementarity and External Linkages: the Strategies of the Large Firms in Biotechnology," *The Journal of Industrial Economics* 38(4): 361–79.

AUDRETSCH, D., and FELDMAN, M. (1996), "R&D Spillovers and the Geography of Innovation and Production," *American Economic Review* 86: 253–73.

BARRY, F., and STROBL, E. (2002), "FDI and the Changing International Structure of Employment in the EU Periphery," CEPR-LdA workshop on *Labour Market Effects of European Foreign Investments*, Turin, 10–11 May.

BELDERBOS, R. (2001), "Overseas Innovations by Japanese Firms: An Analysis of Patent and Subsidiary Data," *Research Policy*, 30(2): 313–32.

BELLAK, C. (2002), "How Performance Gaps between Domestic and Foreign Firms Matter for Policy," *EIBA Annual Conference*, Athens, December.

BENITO, G., GROGAARD, B., and NARULA, R. (2003), "Environmental Influences on MNE Subsidiary Roles: Economic Integration and the Nordic Countries," *Journal of International Business Studies* 34: 443–56.

* Asterisked items are suggestions for further reading.

BLANC, H., and SIERRA, C. (1999), "The Internationalisation of R&D by Multinationals: A Trade-Off between External and Internal Proximity," *Cambridge Journal of Economics* 23: 187–206.

BRUSONI, S., PRENCIPE, A., and PAVITT, K. (2001), "Knowledge Specialization and the Boundaries of the Firm: Why do Firms Know More Than They Do?" *Administrative Science Quarterly* 46: 597–621.

BURETH, A., WOLFF, S., and ZANFEI, A. (1999), "Cooperative Learning and the Evolution of European Electronics Industry," in A. Gambardella and F. Malerba (eds.), *The Organisation of Inventive Activities in Europe*, Cambridge: Cambridge University Press, 202–38.

* CANTWELL, J. (1989), *Technological Innovation and Multinational Corporations*, Oxford: Basil Blackwell.

—— (1995), "The Gobalisation of Technology: What Remains of the Product Cycle Model," *Cambridge Journal of Economics*, 19, 155–174.

—— and COLOMBO, M. G. (2000), "Technological and Output Complementarities, and Inter-firm Cooperation in Information Technology Ventures," *Journal of Management and Governance* 4: 117–47.

—— and JANNE, O. (2000), "The Role of Multinational Corporations and National States in the Globalisation of Innovatory Capacity: The European Perspective," *Technology Analysis and Strategic Management* 12(2): 243–62.

—— and NOONAN, C. A. (2002), "Technology Sourcing by Foreign-Owned MNEs in Germany: An Analysis Using Patent Citations," *EIBA Annual Conference*, Athens, December.

CASTELLANI, D., and ZANFEI, A. (2003), "Technology Gaps, Absorptive Capacity and the Impact of Inward Investments on Productivity of European firms," *Economics of Innovation and New Technology* 12: 555–76.

—— —— (2004), "Choosing International Linkage Strategies in Electronics Industry: The Role of Multinational Experience," *Journal of Economic Behaviour and Organisation* 53: 447–75.

CREAMER, D. (1976), *Overseas Research and Development by United States Multinationals 1966–1975*, New York: The Conference Board Inc.

CRISCUOLO, P., NARULA, R., and VERSPAGEN, B. (2005), "The Relative Importance of Home and Host Innovation Systems in the Internationalisation of MNE R&D: A Patent Citation Analysis," *Economics of Innovation and New Technologies* (forthcoming).

DALTON, D., SERAPIO, M., and YOSHIDA, P. (1999), *Globalizing Industrial R&D*, U.S. Department of Commerce, Technology Administration, Office of Technology Policy.

DOSI, G. (1982), "Technological Paradigms and Technological Trajectories: A Suggested Interpretation of the Determinants and Directions of Technical Change," *Research Policy* 11: 147–62.

DUNNING, J. H. (1994), "Multinational Enterprises and the Globalization of Innovatory Capacity," *Research Policy* 23: 67–88.

*—— and NARULA, R. (1995), "The R&D Activities of Foreign Firms in the United States," *International Studies of Management & Organization* 25(1–2): 39–73.

ERNST, D. (1997), "From Partial to Systemic Globalisation: International Production Networks in the Electronics Industry, *BRIE Working Paper* 98, Berkeley Roundtable on the International Economy, University of California at Berkeley.

ETAN (1998), *Technology Policy in the Context of Internationalisation of R&D and Innovation. How to Strengthen Europe's Competitive Advantage in Technology*. Brussels: European Commission, Directorate-General Science, Research and Development.

FINDLAY, R. (1978), "Relative Backwardness, Direct Foreign Investment and the Transfer of Technology: A Simple Dynamic Model," *Quarterly Journal of Economics* 92: 1–16.

FLORIDA, R. (1997), "The Globalisation of R&D: Results of a Survey of Foreign-Affiliated R&D Laboratories in the USA," *Research Policy* 26: 85–103.

FOSFURI, A., and MOTTA, M. (1999), "Multinationals without Advantages," *Scandinavian Journal of Economics* 101(4): 617–30.

GAMBARDELLA, A., and TORRISI, S. (1998), "Does Technological Convergence Imply Convergence in Markets? Evidence from the Electronics Industry," *Research Policy,* 27: 445–63.

GOMEZ-CASSERES, B. (1989), "Ownership Structures of Foreign Subsidiaries," *Journal of Economic Behavior and Organization* 11: 1–25.

GORG H., and STROBL, E. (2001), "Multinational Companies and Productivity Spillovers: A Meta-analysis," *The Economic Journal* 11 (Nov): F723–F739.

GRANSTRAND, O. (1998), "Towards a Theory of the Technology Based Firm," *Research Policy* 27: 465–90.

—— PATEL, P., and PAVITT, K. (1997), "Multi-Technology Corporations: Why they have 'Distributed' rather than 'Distinctive Core' Competencies," *California Management Review* 39(4): 8–25.

HAGEDOORN, J. (2002), "Inter-Firm R&D Partnerships: An Overview of Patterns and Trends since 1960," *Research Policy* 31: 477–92.

HANNAN, M., and FREEMAN, J. (1984), "Structural Inertia and Organisational Change," *American Sociological Review* 49: 149–64.

* HEDLUND, G. (1986), "The Hypermodern MNC—a Heterarchy," *Human Resource Management* 25: 9–35.

HENDERSON, R., and CLARK, K. (1990), "Architectural Innovation: The Reconfiguration of Existing Product Technologies and the Failure of Established Firms," *Administrative Sciences Quarterly* 35: 9–30.

HENNART, J. F., and LARIMO, J. (1998), "The Impact of Culture on Strategy of Multinational Enterprises: Does National Origin Affect Ownership Decisions?" *Journal of International Business Studies* 29(3): 515–38.

HOBDAY, M. M. (2000), "East vs. South East Asian Innovation Systems: Comparing OEM- and MNE-led Growth in Electronics," in L. Kim and R. Nelson (eds.), *Technology, Learning and Innovation,* Cambridge: Cambridge University Press, 129–69.

IETTO-GILLIES, G. (2001), *Transnational Corporations: Fragmentation amidst Integration,* London: Routledge.

KINDLEBERGER, C. P. (1969), *American Business Abroad. Six Lectures on Direct Investment,* New Haven: Yale University Press.

KOGUT, B. (1989), "A Note on Global Strategies," *Strategic Management Journal* 10: 383–9.

KOKKO, A. (1994), "Technology, Market Characteristics and Spillovers," *Journal of Development Economics* 43(2): 279–93.

KUEMMERLE, W. (1996), "Home Base and Foreign Direct Investment in R&D," Unpublished Ph.D. dissertation, Boston: Harvard Business School.

—— (1999), "Foreign Direct Investment in Industrial Research in the Pharmaceutical and Electronic Industries: Results from a Survey of Multinational Firms," *Research Policy* 28: 179–93.

KUMAR, N. (1998), *Globalization, Foreign Direct Investment, and Technology Transfer,* London: Routledge.

LALL, S. (1979), "The International Allocation of Research Activity by U.S. Multinationals," *Oxford Bulletin of Economics and Statistics* 41: 313–31.

LE BAS, C., and SIERRA, C. (2002), "Location versus Country Advantages in R&D Activities: Some Further Results on Multinationals' Locational Strategies," *Research Policy* 31: 589–609.

LIM, Y. (1999), *Technology and Productivity: The Korean Way of Learning and Catching Up*, Cambridge, Mass.: MIT Press.

MANSFFELD, E., TEECE, D., and ROMEO, A. (1979), "Overseas Research and Development by US-Based Firms," *Economica* 46 (May): 187–96.

MILLER, R. (1994), "Global R&D Networks and Large Scale Innovations: The Case of Automobile Industry," *Research Policy* 23(1): 27–46.

MITCHELL, W., and SINGH, K. (1992), "'Incumbents' Use of Pre-Entry Alliances before Expansion into New Technical Subfields of an Industry," *Journal of Economic Behaviour and Organisation* 18: 347–72.

MOWERY, D. C., OXLEY, J. E., and SILVERMAN, B. S. (1998), "Technological Overlap and Interfirm Cooperation: Implications for the Resource-Based View of the Firm," *Research Policy* 27(5): 507–24.

NAGARAJAN, A., and MITCHELL, W. (1998), "Evolutionary Diffusion: Internal and External Methods used to acquire Encompassing, Complementary, and Incremental Technological Changes in the Lithotripsy Industry," *Strategic Management Journal* 19: 1063–77.

NARULA, R. (2001), "Choosing between Internal and Non-internal R&D Activities: Some Technological and Economic Factors," *Technology Analysis & Strategic Management* 13: 365–88.

—— (2002a), "Innovation Systems and 'Inertia' in R&D Location: Norwegian Firms and the Role of Systemic Lock-in," *Research Policy* 31: 795–816.

—— (2002b), "R&D Collaboration by SMEs: Some Analytical Issues and Evidence," in F. Contractor and P. Lorange (eds.), *Cooperative Strategies and Alliances*, Kidlington: Pergamon Press, 543–68.

*—— (2003), *Globalisation and Technology*, Cambridge: Polity Press.

—— and HAGEDOORN, J. (1999), "Innovating through Strategic Alliances: Moving towards International Partnerships and Contractual Agreements," *Technovation* 19: 283–94.

—— and SADOWSKI, B. (2002), "Technological Catch-up and Strategic Technology Partnering in Developing Countries," *International Journal of Technology Management* 23: 599–617.

NOOTEBOOM, B. (1999), "Inter-firm Alliances: Analysis and Design," London: Routledge.

ODAGIRI, H., and YASUDA, H. (1996), "The Determinants of Overseas R&D by Japanese Firms: An Empirical Study at the Industry and Company Levels," *Research Policy* 25(7): 1059–79.

PATEL, P. (1996), "Are Large Firms Internationalising the Generation of Technology? Some New Evidence," *IEEE Transactions on Engineering Management* 43: 41–7.

*—— and PAVITT, K. (2000), "National Systems of Innovation Under Strain: The Internationalisation of Corporate R&D," in R. Barrell, G. Mason, and M. O'Mahoney (eds.), *Productivity, Innovation and Economic Performance*, Cambridge: Cambridge University Press, 135–60.

—— and VEGA, M. (1999), "Patterns of Internationalisation and Corporate Technology: Location versus Home Country Advantages," *Research Policy* 28: 145–55.

* Pearce, R. (1990), *The Internationalisation of Research and Development*, London: Macmillan.

—— (1999), "Decentralised R&D and Strategic Competitiveness: Globalized Approaches to Generation and use of Technology in Multinational Enterprises (MNEs)," *Research Policy* 28(2–3): 157–78.

Pisano, G. (1990), "The R&D Boundaries of the Firm: An Empirical Analysis," *Administrative Science Quarterly* 35: 153–76.

* Ronstadt, R. C. (1978), "International R&D: The Establishment and Evolution of Research and Development Abroad by Seven US Multinationals," *Journal of International Business Studies* 9(1): 7–24.

Santangelo, G. (2000), "Corporate Strategic Technological Partnerships in the European Information and Communications Technology Industry," *Research Policy* 29: 1015–31.

Serapio, M., and Dalton, D. (1999), "Globalisation and Industrial R&D: An Examination of Foreign Direct Investment in R&D in the United States," *Research Policy* 28: 303–16.

Siotis, G. (1999), "Foreign Direct Investment Strategies and Firms' Capabilities," *Journal of Economics and Management Strategy* 8(2): 251–70.

Sjöholm, F. (1996), "International Transfer of Knowledge: The Role of International Trade and Geographic Proximity," *Weltwirthschaftliches Archiv* 132: 97–115.

Stopford, J. M., and Wells jr., L. T. (1972), *Managing the Multinational Enterprise: Organisation of the Firm and Ownership of Subsidiaries*, New York: Basic Books.

* Teece, D. J. (1986), "Profiting from Technological Innovation: Implications for Integration, Collaboration, Licensing and Public Policy," *Research Policy* 15: 285–305.

—— (1992), "Competition, Cooperation and Innovation. Organizational Arrangements for Regimes of Rapid Technological Progress," *Journal of Economic Behavior and Organization* 18: 1–26.

—— (1996), "Firm Organisation, Industrial Structure and Technological Innovation," *Journal of Economic Behavior and Organization* 31: 193–224.

—— Rumelt, R., Dosi, G., and Winter, S. G. (1994), "Understanding Corporate Coherence: Theory and Evidence," *Journal of Economic Behavior and Organization* 23: 1–30.

UNCTAD (2001), *World Investment Report: Promoting linkages*, New York: United Nations.

—— (2002), *World Investment Report: TNCs and Export Competitiveness*, New York: United Nations.

Vernon, R. (1966), "International Investment and International Trade in Product Cycle," *Quarterly Journal of Economics* 80: 190–207.

Veugelers, R., and Cassiman, B. (1999), "Make and Buy in Innovation Strategies: Evidence from Belgian Manufacturing Firms," *Research Policy* 28: 63–80.

Warrant, F. (1991), *Le Déploiement mondial de la R&D industrielle*. Brussels, Commission des Communautes Européennes—Fast, December.

Zander, I. (1995), *The Tortoise Evolution of the Multinational Corporation—Foreign Technological Activity in Swedish Multinational Firms 1890–1990*, Stockholm: IIB.

*—— (1999), "How do you Mean 'Global'? An Empirical Investigation of Innovation Networks in the Multinational Corporation," *Research Policy* 28: 195–213.

Zanfei, A. (1994), "Technological Alliances Between Weak and Strong Firms: Cooperative Ventures with Asymmetric Competences," *Revue d'Economie Industrielle* 7: 255–79.

*—— (2000), "Transnational Firms and the Changing Organisation of Innovative Activities," *Cambridge Journal of Economics* 24: 515–42.

PART III

HOW INNOVATION DIFFERS

Introduction to Part III

THE existence and significance of differences across industry and over time in the structure and organization of the innovation process has been a central topic in the literature on innovation. A well-known distinction in many discussions of innovation processes is between so-called "Schumpeter Mark I" industries, characterized by numerous small, entrepreneur-led firms, and "Schumpeter Mark II" industries, which are dominated by large, oligopolistic firms with extensive organized R&D. As Bruland and Mowery point out in Chapter 13, however, such differences are the results of lengthy processes of historical change and depend on both technological and institutional factors. Malerba, in Chapter 14, extends this analysis to a sample of contemporary "high-tech" industries, and shows how differences in innovation and industrial dynamics may be analyzed as the interplay within different "sectoral innovation systems" among technology, actors and institutions. Chapter 15 by von Tunzelman and Acha, in contrast, look at so-called "low-tech" and "medium-tech" industries, which are often assumed to be less innovative than their "high-tech" counterparts. The authors, however, emphasize that innovation (often involving different processes) is pervasive within these sectors as well. Similar findings are reported by Miles in Chapter 16, which surveys the rapidly expanding literature on innovation in services. Hall, in Chapter 17, discusses the diffusion of technology, which is closely related to innovation, and underpins the pervasiveness of innovation throughout the economy.

CHAPTER 13

INNOVATION THROUGH TIME

KRISTINE BRULAND
DAVID C. MOWERY

13.1 INTRODUCTION

MOST analysts of innovation emphasize the importance of a historical approach, with good reason. First, innovation is time consuming, based on conjectures about the future, and its outcomes typically are uncertain for long periods. Analysis of any innovation therefore requires an understanding of its history. Second, innovative capabilities are developed through complex, cumulative processes of learning. Finally, innovation processes are shaped by social contexts, as Lazonick has pointed out: "The social conditions affecting innovation change over time and vary across productive activities; hence theoretical analysis of the innovative enterprise must be integrated with historical study" (Lazonick 2002: 3).

Historical patterns of innovation are characterized by complexity, reflecting the heterogeneous nature of economic activity, and the diversity of processes of technology creation across sectors and countries. These characteristics make it problematic to construct overarching schemas of historical development. Nevertheless, some historians and analysts of innovation have developed taxonomies of epochs, often based on "critical technologies" that define whole periods of development. One form of this is the wave theory proposed by Schumpeter in *Business Cycles*, in which steam power drove the First Industrial Revolution, electricity the Second Industrial Revolution, and so on. Other work that does not rely on wave theories also stresses the role of a small number of technologies in driving broader processes of economic growth. Although valuable, many of these frameworks overemphasize the importance of the allegedly critical technologies while slighting other areas of innovation and economic activity that are no less important. In what follows we challenge some of the historical discussions that stress the transformative effects of "critical innovations." Instead, we emphasize the complex multisectoral character of innovation, and hence the need to take seriously the coexistence of a range of innovation modes, institutional processes, and organizational forms.

Our discussion of innovation through time highlights changes in the structure of the innovation process in successive periods, and is informed by the innovation system concept (discussed in Ch. 7 by Edquist, Ch. 11 by Asheim and Gertler, and Ch. 14 by Malerba). In adopting this framework, we focus on the changing structure of economic activity, changes in relevant institutions, and changes in patterns of knowledge generation and flows within emergent industrial economies.[1]

We begin the discussion below by reviewing recent historical interpretations of the impulse toward industrialization in the world economy. We then discuss the changing structure of the innovation process in different phases of industrialization, focusing on the First Industrial Revolution in Britain from roughly 1760 to 1850, the so-called Second Industrial Revolution during the late nineteenth and early twentieth centuries, and what might be called the Third Industrial Revolution after World War II. Our discussion thus includes the widespread appearance of shop-floor-driven technological innovation in eighteenth-century Great Britain and moves forward to consider the invention of the art of invention, to use the philosopher A. N. Whitehead's phrase (Whitehead 1925), in the late nineteenth and early twentieth centuries, with the emergence of organized industrial R&D within the firm. The Third Industrial Revolution, most clearly illustrated by the post-war United States, is one in which private and public institutions compete and collaborate in new fields of innovation, a mode of innovation that is not yet exhausted.

13.2 THE FIRST INDUSTRIAL REVOLUTION

13.2.1 Institutions, Innovation and the Impulses to Growth

Sustained innovation-based development is a recent and unevenly distributed historical phenomenon. A substantial literature on "world history" has sought to explain the rise of the West, and particularly the European breakthrough to sustained productivity growth, in the late eighteenth century. Why were some human societies able to break out of a Malthusian trap, shifting from "extensive" economic growth that relied on increased labor input and a wider division of labor to innovation-based intensive growth with sustained rises in real output per head? An important contribution to this historical debate is Pomeranz (2000), who argues that prior to the mid-eighteenth century, Europe, Japan, China, and India were at a broadly similar level of economic development—this was "a world of surprising resemblances." Why did only Northwestern Europe make the transition to innovation-based growth? Pomeranz suggests that two factors were crucial: the acquisition by the major European powers of colonies as markets for manufactures and sources of food and raw material, and the development within Europe of coal as a new energy source.[2]

An alternative explanation for the industrialization of Northwest Europe stresses institutional changes (see Braudel 1984; Wallerstein 1974; Landes 1998), focusing on the emergence of property rights as impulses to innovation. A variant of this institutional analysis is provided by Jones (2003) who argues that technology-based growth has occurred at several points in world history; the challenge is less to understand growth than to understand the forces that prevent growth. He stresses the inhibitory role of political institutions that are based on surplus extraction by political and military elites. Only when such rulers are weakened by crisis do opportunities arise for gain from innovation. Since the political power of established political elites in Northwest Europe eroded during the fourteenth to seventeenth centuries, the emergence of sustained, innovation-based economic growth first occurred in this region of the world economy.

There is disagreement within this literature over the timing of the divergence, as well as the relative importance of different factors in supporting the growth of such institutions as private property rights and the weakening of rent-seeking political and military elites. But all of the scholars adopting this approach emphasize institutional change as an indispensable precondition for sustained innovation-led growth.

13.2.2 Innovation in the First Industrial Revolution

Most economic historians regard the developments in Britain and Northwestern Europe from around 1760 as an economic and technological watershed. Innovation during this period is best conceptualized as an economy-wide process that involved technological, organizational, and institutional change, spanning many sectors and product groups. This view of British industrialization contrasts with the classic historical accounts that emphasize epochal technological breakthroughs in steam power and textile technologies (see e.g. Mantoux 1961). The debate is a significant one for the broader study of innovation, since important scholarly pieces in the field of innovation studies have followed the "key innovations" interpretation of the First Industrial Revolution (e.g. Freeman and Louçã 2002; for an economic history of industrialization in this framework, see Lloyd-Jones and Lewis 1998).

13.2.3 Sectoral Patterns of Technological Advance: The Patenting Evidence

One important source of evidence on the pace and sectoral distribution of innovative activities during the Industrial Revolution is patent statistics from the period. Although the high cost of patenting (approximately £120 for England, at a time when the annual income of a skilled worker was about £50) and limited access to patent attorneys by many inventors arguably make patent data a biased source of evidence, no other comparably comprehensive sources exist on innovative activity during the Industrial Revolution. MacLeod (1988) finds that patenting grew rapidly after 1750, especially in capital goods. The two technologies for which patenting grew most rapidly during this period are power sources and textile machinery. But patenting also expanded significantly in other capital-goods categories; including agricultural equipment, brewing, shipbuilding, canal building, and metallurgy. Although the share of all patenting accounted for by capital goods grew during the 1750–1800 period, this category nevertheless accounted for no more than 40 per cent of British patents in the half century between 1750 and 1800.

A great deal of inventive activity during this period focused on consumer goods. According to Berg (Berg 1998; see also Sullivan 1990), much of this consumer-goods patenting affected a vast number of small, novel products such as buckles and fasteners, cabinets and furniture, and spectacle frames. Indeed much of the patent activity within the textiles sector—roughly one-third—involved new products (Griffiths et al. 1992). Much of the inventive activity in this key sector within the Industrial Revolution involved new thread types and fabrics, and focused on a consumer market.

Patent evidence thus suggests that the period of the Industrial Revolution was a period of broad technological change. Nevertheless, in recognition of the limitations of patent data for tracking innovative activity, we turn now to more qualitative evidence on the sectoral structure of innovation.

13.2.4 Sectoral Patterns of Change: Technological Histories

13.2.4.1 *Steam Power and Textiles*

Four innovations—the spinning jenny, the water frame, the spinning mule, and the automatic mule—were associated with dramatic growth in the British textiles industry of the First Industrial Revolution. Between the late eighteenth and the middle of the nineteenth century, the cotton textile industry grew spectacularly in the size of its output, in labor productivity, in the scale of enterprises, in capital employed, and in its share of national income. Value added in cotton rose from less than £500,000 in 1760 to about £25,000,000 by the mid-1820s. In spinning, the number of direct labor hours required to process 100 pounds of cotton declined from 300 in 1790 to 135 in 1820 (Mokyr 2002: 50–1), and the average annual input of raw cotton per factory rose by over 1,000 per cent during 1797–1850 (Chapman, 1972: 70). Dramatic as these changes were, they should be kept in proportion: textiles made up about 25 per cent of manufacturing output at their peak. Innovation and productivity were growing elsewhere as well.

Another critical innovation of this period was the steam engine of James Watt, first introduced in 1775. Watt's innovation is commonly described as the emblematic technology of the Industrial Revolution (see Toynbee 1908; Deane 1965). Yet von Tunzelmann's study of steam power (1978) showed that the machine diffused relatively slowly, that it had only modest economic advantages over existing power technologies (and hence could not significantly affect economic growth), and had limited backward and forward linkages with the rest of the British economy, further reducing its "catalytic" effects (see Box 13.1). As we noted earlier, the innovations that

---

Box 13.1 Technological diffusion in the First Industrial Revolution

Since the economic effects of innovations depend on their widespread adoption (see Ch. 17 Hall by in this volume), it is important to recognize that many of the important innovations of the First Industrial Revolution in fact diffused relatively slowly. For example, the Watt steam engine, described above as an emblematic innovation of the First Industrial Revolution, diffused gradually through the British economy. By 1800, twenty-five years after the introduction of the Watt steam engine, Manchester (a central locus of industrialization in textiles) had about 32 engines, and Leeds (another emergent textiles center) about 20. By 1817 Glasgow had 45 engines, by 1820 Birmingham had about 60 engines, and by 1825 Bolton had 83. Growth rates of steam-generated horsepower averaged between 6 and 10 per cent per year in the late 1830s, more than 50 years after the Watt engine's introduction. Von Tunzelmann (1978) argued that this gradual pace of diffusion reflected the high costs of steam engines and their fuels through the 1850s, long after the introduction of the engine. Similar points apply to other important innovations of this period. The Roberts automatic spinning machine, said by no less an observer than Karl Marx to "open up a completely new epoch in the capitalist system," was a major innovation—the world's first truly automatic power machine. But it diffused slowly; fifty years passed before this machine accounted for a majority of the output of the UK cotton spinning industry.

contributed to British economic growth and industrialization spanned a broader group of technologies and sectors.

13.2.4.2 *Innovation in Other Sectors*

Although industrialization necessarily was associated with a fall in the share of national output flowing from agriculture, British agriculture grew in absolute terms during 1750–1850 and was highly innovative. During this period, key innovations were developed in farm tools, cultivation implements (plows, harrows, mowers), sowing implements, harvesting equipment (reapers, rakes, hoes, scythes, winnowing and threshing devices, etc.), and drainage equipment (for a detailed overview, see Bruland 2004). Agricultural innovation was associated with the emergence by the 1830s of a specialized agricultural equipment industry, which in turn supported the growth of numerous small engineering works and foundries.

Closely linked with technical change in the agricultural sector were innovations in the processing, distribution, and consumption of food, which during the Industrial Revolution (and after) dominated British manufacturing. Technological innovations in food preservation, refrigeration, baking, brewing, and grain milling supported expansion in the scale of production establishments and organizational innovation of production and firms. Baking was the first British industry to develop and use the production line, based on new techniques that supported more accurate timing of operations. Brewing and milling were the first sectors to deploy large, professionally managed enterprises with national distribution systems.

A similarly innovative sector was the glass industry, which manufactured widely used and differentiated products—windows, bottles and containers, lamps, and spectacles. Glass was one of the few large-scale production activities in early industrialization, and relied on experimentation and research to a degree not widely appreciated in many accounts of the role of science in technological innovation during this period. The most knowledge-intensive segment of glass production was optical glass, where developments of the technology deployed optical theory, pioneering the integration of science with production. Although the first Industrial Revolution overall was far from a science-based phenomenon, developments in at least some of the key innovative sectors prefigured subsequent changes in the organization of innovation.

These examples of innovation could easily be expanded to include sectors such as iron and steel, chemicals (alkalis and chlorine), pottery and ceramics, machinery and machine tools, instruments, mining, and paper and printing. The pervasiveness and extent of innovation in these and other industries once again suggests that innovation during the First Industrial Revolution was not confined to "leading sectors" of the economy, but was present in virtually all sectors during the period. We cannot ignore the sectors such as textiles and steam power that have driven so much of the historiography of industrialization; but their role needs to be kept in economic and technological perspective.

13.2.5 The Organization of Innovation and Learning in Early Industrialization

How was innovation organized during the First Industrial Revolution? Analyses of patent data suggest that virtually all inventions in the late eighteenth and early nineteenth centuries resulted from the efforts of individual inventors. These inventors may have worked alone in individual workshops or in larger enterprises, but the key point was the individuality of inventive and innovative effort, and its integration with shop-floor production. Inventors' skills and knowledge bases were rooted in existing trades, such as watchmaking, carpentry, blacksmithing, metalworking, and wood-working. Textile machinery in particular was fabricated largely within the existing textile-producing firms. A specialized capital goods sector appeared only in the 1820s.

The upsurge in inventive and innovative activity during the Industrial Revolution did not depend in any general way on "science" as we now understand it, although isolated instances of the integration of science and industry during the late eighteenth and early nineteenth centuries are apparent in such areas as glassmaking. The period saw the emergence of formal and informal scientific societies in Great Britain and wide diffusion of scientific ideas (see Uglow 2002), but this early "scientific revolution" produced few practical applications. Although the search and learning processes employed by inventors during this period are best described as "trial and error," this characterization inaccurately minimizes the extent and sophistication of the knowledge required for innovation in early industrialization. Indeed, Mokyr has proposed that a central factor in the Industrial Revolution was an "Industrial Enlightenment," associated with improvement in the quantity and accessibility of knowledge concerning industrial techniques. This "Enlightenment" included the surveying and codification of artisanal techniques in published manuals, handbooks, textbooks, and pamphlets on industrial practices (Mokyr 2002: 34–5). Patterns of learning and knowledge accumulation during the Industrial Revolution may have begun as tacit and practical, but during the late eighteenth and early nineteenth centuries, more and more of this learning was codified, accelerating the diffusion of industrially relevant knowledge across sectors.

13.2.6 Institutions and the Organization of Enterprise during the First Industrial Revolution

Institutional change that affected the organization of firms and production processes played an important role in the upsurge of innovation during early industrialization. This is a vast topic, and we focus on two crucial institutional changes—the development of new forms of company law and finance that supported the growth of corporate firms; and the rise of managerial control of production, which transformed workplace organization and scale. These institutional innovations together made possible the subsequent growth in factory production.

Most industrial enterprises operating during the eighteenth century were extremely small. Large-scale factories were uncommon before the early nineteenth century, and the small-scale workshop or production unit was the primary organizational form for most of the period of early industrialization. These small firms were individually owned or were partnerships, locally financed, in which liability for debts was the personal responsibility of owners who usually acted as managers. Two institutional forms made possible an expansion in the scale of enterprises: joint-stock (i.e. limited liability) organization and the growth of financial networks.

Joint-stock associations emerged in the medieval period in Britain, but were permitted only via the explicit authorization of the state. A series of piecemeal reforms after 1825 were followed by legislation permitting the creation of companies with separate legal identity, limited liability and tradeable shares. General legislation for the joint stock form was passed in 1844 and consolidated in the statutes of 1856 and 1862 (Mathias 1983: 325; see Harris 2000 for a comprehensive account). Although much industrial financing remained local and small in scale (see Hudson 1986 for an account of local networks' role in financing the woollen industry), these legal reforms enabled substantial growth in the financing and scale of industrial enterprises. But joint stock organization and access to finance were necessary rather than sufficient conditions for enterprise growth. Even more significant was the development of management systems and managerial control.

Managers of these early industrial enterprises confronted serious challenges in the assembly and maintenance of a suitable workforce, the control of work, and the adoption of new techniques and organizational structures for production activities by a restructured workforce. Pollard highlights "two distinct, though clearly overlapping difficulties; the aversion of workers to entering the new large enterprises with their unaccustomed rules and discipline and the shortage of skilled and reliable labour" (Pollard 1965: 160). The emergence of rule-based disciplinary methods, the laborious construction of supervisory systems, and the habituation of workers to an organized and controlled working day emerged slowly but were central developments of early industrialization. New management techniques that appeared during the Industrial Revolution permitted the development of larger, centralized production sites and of the mechanized factory. In turn, such sites permitted the application of power, the adoption of new industrial techniques, and closer managerial control over the organization and pace of work.

These organizational and managerial innovations were defining characteristics of the First Industrial Revolution. In pottery for example, the most important managerial innovator was Josiah Wedgwood, who developed a number of product innovations—new designs, new glazes and finishes, and new basic materials—and pioneered new marketing methods. But his most important innovations were organizational—the creation of an integrated workforce, the design of a plant organized around a set of production sequences, and above all the creation of a workforce subject to control and discipline (McKendrick 1961; see Box 13.2). Wedgwood's innovations strengthened managerial power over the production workforce, which formed the central context for innovation (and struggle) in the later nineteenth century.

Box 13.2 Josiah Wedgwood and "modern" management in pottery fabrication

In the second half of the eighteenth century, rising incomes and increased coffee and tea consumption accelerated growth in the market for china and other types of glazed, fired clay plates, cups, and related items. This was part of a wider growth in demand for "luxury" consumer goods (Berg and Eger 2003). The production of pottery was concentrated in Staffordshire in central England, and was dominated by small enterprises operated by craftsmen, often producing on a piecework basis. Production was controlled by individual craftsmen, and production rhythms and volumes were haphazard. Josiah Wedgwood transformed the industry by developing factory-based production techniques that supported the creation of an enterprise of unprecedented scale. Wedgwood's success rested on two achievements. First, he successfully lobbied the British government to improve regional transportation infrastructure (a publicly financed turnpike was built in 1763 and a canal, on which Wedgwood sited his factory, was completed in 1771), thereby enabling his factory to serve the British market while reducing formerly exorbitant breakage rates. Second, he introduced radical organizational innovations, developing new techniques for organizing production and managing the workforce (Bruland 1989).

Wedgwood, an acquaintance of Matthew Boulton, the entrepreneur who formed the successful steam-engine firm of Boulton and Watt, modeled his new production organization on Boulton's factory, emphasizing a physical layout that separated and sequentially organized the various operations that went into production of his china (Langton 1984). Consistent with this organization, Wedgwood assigned workers to specific tasks, relying on specialization to enhance skill and consistency in the performance of these tasks. Workmen "were not allowed to wander at will from one task to another as the workmen did in the pre-Wedgwood potteries. They were trained to one task and they had to stick to it" (McKendrick 1961: 32).

Having reorganized the structure of production and jobs within his organization, Wedgwood had to develop techniques to encourage and/or force workers to adapt to this new system. He invested heavily in the retraining of experienced workers (with mixed results) and in the training of new employees, many of whom were young women (women accounted for 25 per cent of his employees as of 1790). Even more important, however, was Wedgwood's emphasis on codification of technical guidelines for the performance of the various tasks in his factory and development of extremely detailed, written rules for worker behavior. Wedgwood also introduced sanctions and rewards for punctuality and absenteeism on the part of workers, going so far as to develop an early prototype of a timeclock for monitoring workers' attendance.

Box13.2 (cont.)

Wedgwood's new methods were significant organizational changes in the production of kitchenware and china, resting on a transformation of the nature and character
of work itself. The new methods encountered considerable resistance from experienced
workers, but he successfully created a production system without equal in the industry,
employing 200 workers. By 1790, less than twenty-five years after its foundation.
Wedgwood himself was enormously wealthy, and the firm survived as an independent
entity into the twentieth century.

An essential ingredient in the transformation of economic and innovative activity
that characterized the First Industrial Revolution thus was the development of new
techniques of economic organization and management. Another wave of institutional change and managerial innovation proved indispensable to the Second Industrial Revolution and an organizational innovation that was at its heart—the
industrial research laboratory.

13.3 THE SECOND INDUSTRIAL REVOLUTION

13.3.1 A Second Phase of Industrialization

In the late nineteenth century industrial technologies began to change, and a range
of new technologies and industries emerged. This Second Industrial Revolution
took place on the continent of Europe and in the USA. In Europe it was led by the
emergence of new industrial regions in France and Germany, such as the Ruhr. It
involved a shift away from the basic industries that had developed in Britain before
diffusing to Europe and the United States (iron, steel, coal, textiles, and mechanical
engineering), to new industrial sectors (such as chemicals, optics, and electricity),
and signaled the passing of technological leadership from Britain to the United
States and Germany.

The Second Industrial Revolution was characterized by organizational innovations that laid the groundwork for links between industry and formal science that
became stronger during the course of the twentieth century. The development of
these stronger links transformed the innovation process in several ways: (1) Formal
training for would-be inventors became far more important and the role of artisanal
ingenuity diminished;[3] (2) the role of institutions external to the firm that

conducted such formal training and research increased; and (3) bodies of empirically grounded, codified scientific and technological knowledge internal to the firm became powerful engines for expansion and diversification.

The technological shifts of the late nineteenth century were accompanied by changes in firm structure. Large-scale, vertically integrated enterprises emerged in Germany and the United States that incorporated specialized research and development departments or laboratories. Within such firms scientific work was carried out by teams of researchers and depended on networks of scientific contacts in the education (particularly university) systems. These professionally managed firms of unprecedented size became the agents of Schumpeter's "creative destruction" by the mid-twentieth century, as industrial innovation became a core component of corporate strategy.

13.3.2 Was the Second Industrial Revolution a "Science-Driven" Phenomenon?

Although important scientific breakthroughs and an expanded application of science to industry did emerge in the late nineteenth century, for most of the century these two trends were more loosely coupled to one another than is commonly thought. The construction of a bridge between recent scientific discoveries and technological innovation typically requires considerable time. For example, no significant technological applications followed Faraday's demonstration of electromagnetic induction in 1831, with the exception of the telegraph. Yet this scientific discovery laid the foundations for one of the defining industries of the Second Industrial Revolution, electrical equipment and electric power generation.

As the example of electricity suggests, technological exploitation of new scientific understanding often requires considerable time, since additional applied research is needed to translate a new but abstract formulation into economically useful knowledge. In other important cases, such as Perkin's accidental synthesis in 1856 of mauveine, the first synthetic dyestuff, exploitation of scientific advances required the development of complex process technologies for which no scientific foundation existed. Although chemical science was vitally important to industrial developments during the period, much of the actual timing of innovation, i.e. the translation of scientific breakthroughs into commercial products, depended on advances in manufacturing technologies that remained poorly understood through much of the nineteenth century.

In other industries, the linkage between science and technological innovation remained weak, simply because technological innovation did not require scientific knowledge. This was true of a broad range of metal-using industries in the second half of the nineteenth century, a period during which America took a position of technological leadership. The development of this new machine technology rested on mechanical skills of a high order, as well as considerable ingenuity in conception and design. It required little or no recourse to the scientific knowledge of the time, and US success in this and other industries, such as chemicals, meatpacking, and consumer goods, relied on access to a large domestic market.

The creation of a truly "national" market in the United States in turn was facilitated by the construction of a reliable national infrastructure for communications and transportation. The enterprises most heavily involved in the creation of this infrastructure were themselves among the largest industrial firms organized in the United States, and the organizational and financial innovations developed by firms such as Western Union and the Pennsylvania Railroad were widely emulated in other industries as new firms of unprecedented scale were created (Chandler 1977). But few if any of these economically crucial organizational innovations relied on science.

13.3.3 The Origins of Industrial Research

A defining characteristic of the "new industries" of the Second Industrial Revolution was their increased reliance on organized experimentation. The pioneers in this organizational innovation were the large German chemicals firms that grew rapidly in the last quarter of the nineteenth century, based on innovations in dyestuffs. By the first decade of the twentieth century, a number of large US firms had established similar in-house industrial research laboratories. In both nations, the growth of industrial research was linked to a broader restructuring of manufacturing firms that transformed their scale, management structures, product lines, and global reach. But the development of industrial research in the German chemicals and electrical equipment industries also relied on complementary changes in the institutional structure of the nascent "German" innovation system that occurred before and after German unification in 1870.[4]

Scientific advances in physics and chemistry during the last third of the nineteenth century created considerable potential for the profitable application of scientific and technical knowledge in industry. The first in-house industrial R&D laboratories were established by German firms seeking to commercialize innovations based on the rapidly developing field of organic chemistry. Kekule's 1865 model of the molecular structure of benzene, a key component of organic chemistry and synthetic dyestuffs, provided the first scientific foundation for developing new products. But scientifically trained personnel were needed to translate Kekule's breakthrough into new products. The rapid expansion in Germany's network of research and technical universities during the second half of the nineteenth century thus was critically important to the growth of industrial research, particularly in the chemicals industry. German universities produced a large pool of scientifically trained researchers (many of whom sought employment in France and Germany during the 1860s), university faculties advised established firms, and university laboratories provided a site for industrial researchers to conduct scientific experiments in the early stages of the creation of in-house research laboratories.

The German universities pioneered the development during the nineteenth century of the modern model of the "research university," in which faculty research was central to the training of advanced degreeholders. In addition, the German polytechnic institutes that had been founded during the 1830s by the various German principalities were by the 1870s transformed into technical universities that played a central role in training engineers and technicians for the chemicals and electrical equipment industries. By the 1870s, according to Murmann (1998), Germany had nearly thirty university and technical university departments in organic chemistry, and seven major centers of organic chemistry research and teaching. And technically trained personnel moved into senior management positions within German industry, in contrast to the situation in Great Britain, further strengthening the links between corporate strategy and industrial research.

The contrast between Germany and Great Britain in the role of universities is especially striking.[5] British universities received far less public funding, supported less technical education, and were less closely linked with domestic chemicals firms than was true in Germany by the 1880s. British university enrollment increased by 20 per cent between 1900 and 1913, far less than the 60 per cent increase in German university enrollment during the same period. Enrollment at the "redbrick" British universities (largely founded during the nineteenth century, this group excludes the ancient English universities of Oxford and Cambridge) grew from roughly 6,400 to 9,000 during 1893–1911, but only 1,000 of the students enrolled in these universities as of 1911 were engineering students, while 1,700 were pursuing degrees in the sciences (Haber 1971: 51). By contrast, the German technical universities alone enrolled 11,000 students in engineering and scientific degree programs by 1911. British government funding of higher education amounted to roughly £26,000 in 1899, while the Prussian government alone allocated £476,000 to support higher education. By 1911, these respective amounts stood at £123,000 and £700,000 (Haber 1971: 45 and 51).

The institutional transformation of Germany's national innovation system was both a cause and an effect of the growth of the chemicals and electrical equipment industries. Werner von Siemens of the Siemens electrical equipment firm was a founder of the German Association for Patent Protection in 1874, and the first national patent law in the new German state was passed in 1877. Although the law did not cover dyestuff products, stronger intellectual property protection increased the ease with which firms could appropriate the returns to their R&D, and many of the largest German chemicals firms established formal in-house R&D laboratories after its passage.

The large, profitable firms that emerged in these science-based industries actively lobbied the German government for increased support of higher education (The "Club of German Chemists," drawn largely from senior management of German chemicals firms, lobbied for additional faculty appointments in chemistry) and for other forms of support of research related to their enterprises. Werner von Siemens donated the land for the Imperial Institute of Physics and Technology, located near his firm's headquarters in Berlin and the city's technical university, and the Institute was formally established with public funds in 1887. Similar lobbying by the chemicals industry led to the announcement in 1910 by the German emperor of the foundation of the Kaiser Wilhelm Institute for Chemistry, staffed largely by academic chemists and funded by industry. Both the Institute of Physics and Technology and the Kaiser Wilhelm Institute were dedicated to "mission-oriented" fundamental research, much of which was longer-term in nature than the R&D performed in industry but nonetheless more applied than the work of university faculties (Beyerchen 1988).

The creation by German chemicals firms of in-house industrial research laboratories also was associated with change in the management and structure of these firms (see Box 13.3). Family managers were replaced by professional managers and, eventually, by professional chemists. Their in-house R&D activities produced new products in fields other than dyestuffs, e.g. Bayer's aspirin. And the importance of

---

Box 13.3 The foundation of R&D laboratories by Bayer and Du Pont

Bayer's foundation of a laboratory was triggered in part by a realization among the firm's senior management that it was unable to compete effectively with Hoechst and BASF (which had founded research laboratories respectively in 1877 and 1878), as well as the growing difficulties that Bayer faced in forming strong linkages with leading university research chemists. In 1883, Carl Duisberg, who later served as the first director of Bayer's in-house research facility and the firm's CEO, was sent by Bayer managers to work with the chemistry faculty at the University of Strasbourg (then a German university), before returning to Bayer to begin work in the firm's R&D laboratory (a small room just off the main production floor in Bayer's plant). At the same time, Bayer sought to strengthen its links with German university chemists through other tactics, including the negotiation of contracts with leading research chemists and the funding of research by new Ph.D. degreeholders in university or technical university laboratories.

Duisberg's first laboratory was at best an appendage to Bayer's main production facility, but his success in dyestuff synthesis led to an expansion in his staff. Nevertheless, Duisberg's group had important responsibilities in production engineering and problem-solving, as well as marketing, until roughly 1890. Only in 1891 was a dedicated laboratory established at Bayer and a clear distinction made within the organization between R&D and workaday technical support (see Meyer-Thurow 1982).

The US chemicals firm Du Pont established its first industrial research facility, the Eastern Laboratory, in 1902, and founded the Experimental Station in 1903. Creation of the Eastern Laboratory followed the acquisition of control of the Du Pont Company, founded in the early nineteenth century, by T. Coleman Du Pont, Pierre S. Du Pont, and Alfred I. Du Pont from other family members. The Company's transformation from a loose holding company to a multifunction, diversified industrial corporation began with this 1902 change in control.

Du Pont's Eastern Laboratory was the first laboratory to be physically and organizationally separated from the manufacturing operations of the firm. Its R&D activities were devoted almost entirely to improvements in manufacturing processes for Du Pont's existing product line of dynamite and high explosives. By contrast, the Experimental Station, founded one year later, focused on the development of new products and improved applications of Du Pont's smokeless-gunpowder products. The Experimental Station also monitored and evaluated inventions from sources outside of the Du Pont Company.

A US government antitrust suit against Du Pont forced the divestiture of a portion of its black powder and dynamite businesses in 1913, and the firm used its R&D laboratories to diversify its product lines through R&D and the acquisition of technologies from external sources during and after World War I.

close links between research personnel and the users of these new products, as well as the proliferation of new products, triggered expansion in these firms' internal distribution and marketing capabilities. Just as US firms were doing by the end of the nineteenth century, the German chemicals firms expanded their boundaries to incorporate new functions and a much broader and more diversified product line. A similar sequence occurred at roughly the same time in the German electrical equipment industry, as Siemens and AEG, among other leading firms, established in-house research laboratories during the 1870s and 1880s.

The development of industrial research within US manufacturing firms followed these developments in the German chemicals and electrical machinery industries. Many of the earliest US corporate investors in industrial R&D, such as General Electric and Alcoa, were founded on product or process innovations that drew on recent advances in physics and chemistry. The corporate R&D laboratory brought more of the process of developing and improving industrial technology into the boundaries of US manufacturing firms, reducing the importance of the independent inventor as a source of patents (Schmookler 1957).

But the in-house research facilities of large US firms were not concerned exclusively with the creation of new technology. Just as the German dyestuff firms' laboratories had, these US industrial laboratories also monitored technological developments outside of the firm and advised corporate managers on the acquisition of externally developed technologies.

As Pavitt notes in his chapter in this volume, in-house R&D in US firms developed in parallel with independent R&D laboratories that performed research on a contract basis (see also Mowery 1983). But over the course of the twentieth century, contract-research firms' share of industrial research employment declined. The complex and uncertain projects undertaken within many in-house research facilities did not lend themselves to "arm's-length" organization.

As had been the case in Germany, the development of industrial research, as well as the creation of a market for the acquisition and sale of industrial technologies, benefited from a series of reforms in US patent policy between 1890 and 1910 that strengthened and clarified patentholder rights (See Mowery 1995). Judicial tolerance for restrictive patent licensing policies further increased the value of patents in corporate research strategies. Although the search for new patents provided one incentive to pursue industrial research, the impending expiration of these patents created another important impetus. Both American Telephone and Telegraph and General Electric, for example, established or expanded their in-house laboratories in response to the intensified competitive pressure that resulted from the expiration of key patents (Reich 1985; Millard 1990: 156). Intensive efforts to improve and protect corporate technological assets were combined with increased acquisition of patents in related technologies from other firms and independent inventors.

Schumpeter argued in *Capitalism, Socialism and Democracy* that in-house industrial research had supplanted the inventor-entrepreneur (a hypothesis supported by Schmookler 1957) and would reinforce, rather than erode, the position of dominant firms. The data on research employment and firm turnover among the 200 largest US manufacturing firms suggest that during 1921–46 at least, the effects of industrial research were consistent with his predictions. Displacement of these firms from the ranks of the very largest was significantly less likely for firms with in-house R&D laboratories (Mowery 1983).

13.3.4 Innovation in the Interwar Chemicals Industry

As we noted in the previous section, one of the critical science-based industries associated with this Second Industrial Revolution was chemicals. A comparison of US and German innovative performance in this industry highlights many of the points made above concerning the new institutional and organizational underpinnings of innovation during this period. Although German and US chemicals firms had pioneered in the development of a new structure for innovation that relied on in-house R&D and the "routinization of innovation," these two groups of firms pursued somewhat different innovative strategies during the interwar period following the creation of their R&D facilities. These differences highlight the influence of cross-national contrasts in market structure and resource endowments, factors that receded somewhat in importance after 1945.

One important point of contrast was the quality of scientific research in chemistry (as opposed to technological innovation) of leading firms and universities in the two nations. Through 1939, German scientists received fifteen out of the thirty Nobel Prizes awarded in chemistry, US scientists received only three, and French and British scientists each accounted for six. Between 1940 and 1994, US scientists received thirty-six of the sixty-five chemistry Prizes awarded, German scientists received eleven, British scientists received seventeen, and French scientists received one (*Encyclopaedia Britannica* 1995: 740–7). Although the situation was beginning to change during the 1930s and would change dramatically after 1945, the United States remained a scientific backwater during this portion of the twentieth century.

Technological change in the American chemical industry was shaped by several features: (1) the large size and rapid growth of the American market; (2) the opportunities afforded by large market size for exploiting the benefits to be derived from large-scale, continuous process production; and (3) a natural resource endowment—oil and gas—that provided unique opportunities for transforming the resource base of the organic chemical industry and achieving significant cost savings, if an appropriate new technology could be developed.

The introduction and rapid adoption of the internal combustion automobile in the opening years of the twentieth century in the United States brought in its wake an almost insatiable demand for liquid fuels. This demand in turn spurred the growth of a new sector of the petroleum refining industry that was specifically calibrated to accommodate the needs of the automobile in the first two decades of the twentieth century. Petroleum refining had two important, related features. First, it was highly capital-intensive; by the 1930s it had become the most capital-intensive of all American industries. Second, productive efficiency required that small batch production, so characteristic of other chemical products, such as synthetic organic materials, be discarded in favor of high-volume production methods that required continuous-process technologies. American leadership in petroleum refining provided the critical knowledge and the engineering and design skills to support the chemicals industry's shift from coal to petroleum feedstocks in the interwar years.

The large size of the American market had introduced American firms at an early stage to the problems involved in the large-volume production of basic products, such as chlorine, caustic soda, soda ash, sulfuric acid, superphosphates, etc. The early American experience with large-scale production contributed to the US chemical industry's transition to petroleum-based feedstocks. The dominant participants in this industrial transformation were Union Carbide, Standard Oil (New Jersey), Shell and Dow. But the shift to petroleum benefited as well from the adoption by US petroleum firms, notably Humble Oil (an affiliate of Standard Oil of New Jersey), of new techniques for discovery and exploitation of petroleum deposits. The availability of low-cost petroleum and natural gas, coupled with the new, large-scale process innovations developed by US firms, enabled significant reductions in manufacturing costs.

By contrast, the German chemicals industry fashioned new technologies that compensated for the absence of such domestic feedstocks. During World War II, German tanks and airplanes were fueled by synthetic gasoline and ran on tires made from synthetic rubber derived from coal feedstocks. Only in the wake of World War II and the creation of a set of multilateral institutions governing international trade and finance did the revival of international trade and US guarantees of access to foreign petroleum sources support a shift by the German chemicals industry to petroleum feedstocks (Stokes 1994). Here as well as elsewhere, the post-1945 revival of international trade and investment flows relaxed somewhat the constraints on technological innovation imposed by reliance on domestic natural resources. As Abramovitz (1994) and other scholars (Nelson and Wright 1994) have pointed out, economic conditions in Europe and elsewhere in the global economy gradually came to resemble more closely those that had given rise to US supremacy in this sector.

13.4 A "Third Industrial Revolution"? R&D and Innovation during the Post-1945 Period

13.4.1 The Post-war Transformation

The structure of the innovation process in the industrial economies was transformed after 1945. Global scientific leadership shifted decisively from Western Europe to the United States. A new set of industries, focused on ICT and biomedical innovation, grew rapidly. As global trade and investment flows revived after the 1914–45 period of war and depression, international flows of technology also expanded, and by the 1980s and 1990s enabled economies such as Japan, South Korea, and Taiwan to advance to the front rank as sources of industrial innovation. Developments in the United States illustrate these trends most vividly, and highlight the development of a US "national innovation system" that contrasted sharply with its 1900–1940 counterpart.

The structure (if not the scale) of the pre-1940 US R&D system resembled those of other leading industrial economies of the era, such as the United Kingdom, Germany, and France—industry was a significant funder and performer of R&D and central government funding of R&D was modest. Innovation and R&D during the post-1945 period in the United States were transformed by the dramatic increase in central government spending on R&D, much of which was allocated to industry and academic research. As Lundvall and Borrás point out in Chapter 22 of this volume, this expansion in public R&D funding was motivated primarily by national security and public-health concerns and secondarily by political support for basic research. The post-war US R&D system differed from those of other industrial economies in at least 3 aspects: (1) small, new firms were important entities in the commercialization of new technologies; (2) defense-related R&D funding and procurement exercised a pervasive influence in the high-technology sectors of the US economy; and (3) US antitrust policy during the post-war period was unusually stringent.

The prominence of new firms in commercializing new technologies in the post-war United States contrasts with their more modest role during the interwar period. In industries that effectively did not exist before 1940, such as computers, semiconductors, and biotechnology, new firms were important actors in R&D and commercialization, in contrast to post-war Japan and most Western European economies. Moreover, in both semiconductors and computers, new firms grew to positions of considerable size and market share.

Several factors contributed to this prominent role of new firms in the post-war US innovation system. The large basic research establishments in universities, government, and a number of private firms served as important "incubators" for the development of innovations that "walked out the door" with individuals who established firms to commercialize them. Relatively weak formal protection for intellectual property during the 1945–80 period paradoxically also aided the early growth of new firms. Commercialization of microelectronics and computer hardware and software innovations by new firms was aided by a permissive intellectual property regime that facilitated technology diffusion and reduced the burden on young firms of litigation over inventions that originated in part within established firms. In microelectronics and computers, liberal licensing and cross-licensing policies were byproducts of antitrust litigation, illustrating the tight links between these strands of US government policy. US military procurement policies also contributed to the growth of new firms in microelectronics and contributed to high levels of technological spillovers among these firms.

13.4.2 Electronics and ICT

Advances in electronics technology created three new industries—electronic computers, computer software, and semiconductor components—in the post-war global economy. The electronics revolution can be traced to two key innovations—the transistor and the computer. Both appeared in the late 1940s, and the exploitation of both was spurred by Cold War concerns over national security.

New firms played a prominent role in the introduction of new products, reflected in their often-dominant share of markets in new semiconductor devices, significantly outstripping that of larger firms. Moreover, the role of new firms grew in importance with the development of the integrated circuit (IC). The US military's willingness to purchase semiconductor components from untried suppliers was accompanied by conditions that effectively mandated substantial technology transfer and exchange among US semiconductor firms. To reduce the risk that a system designed around a particular IC would be delayed by production problems or by the exit of a supplier, the US military required its suppliers to develop a "second source" for the product, i.e. a domestic producer that could manufacture an electronically and functionally identical product. To comply with second source requirements, firms exchanged designs and shared sufficient process knowledge to ensure that the component produced by a second source was identical to the original product. These requirements spurred interfirm "spillovers" of knowledge and knowhow within the semiconductor industry.

The development of the US computer industry also benefited from Cold War military spending, but in other respects the origins and early years of this industry differed from semiconductors. During the war years, the American military sponsored a number of projects to develop high-speed calculators to solve special military problems. The ENIAC—generally considered the first fully electronic US digital computer—was funded by Army Ordnance, which was concerned with the computation of firing tables for artillery. From the earliest days of their support for the development of computer technology, the US armed forces were anxious that technical information on this innovation reach the widest possible audience. The technical plans for the military-sponsored IAS computer were widely circulated among US government and academic research institutes, and spawned a number of "clones" (e.g. the ILLIAC, the MANIAC, AVIDAC, ORACLE, and JOHNIAC—see Flamm, 1988: 52).

Although business demand for computers gradually expanded during the early 1950s, government procurement remained crucial. The projected sale of fifty machines to the federal government (a substantial portion of the total forecast sales of 250 machines) influenced IBM's decision to initiate the development of its first business computer, the 650 (Flamm 1988). New firms played a much more modest role in the early years of the business computer industry, however, in part because established firms such as IBM had developed powerful marketing organizations for electromechanical business equipment sales that were (with some difficulty) adapted to selling computers to their business customers (Usselman 1993). The appearance of new markets for computers that resulted from the development of scientific computers and minicomputers during the 1950s and 1960s and desktop systems during the 1970s and 1980s, however, provided opportunities for entry by many new firms, such as Digital Equipment, Cray Systems, Computer Research Associates, and Apple.

The progress of computer technology since the 1950s has been driven by the interaction of several trends: (1) dramatic declines in the price–performance ratios of components, including central processing units and such essential peripherals as data storage devices; (2) resulting in part from (1), the rapid extension of computing technology into new applications; and (3) the increasing relative costs of software. Throughout the brief history of the electronic computer industry, these trends have created bottlenecks that have influenced the path of technological change. The IBM 360 mainframe computer, for example, which cemented IBM's dominance of the US computer industry during the 1960s and 1970s, created a "product family" of computers in different performance and price classes that all utilized a common operating system and other software.

The Intel Corporation's commercialization of the integrated circuit microprocessor in 1971 transformed the structure of the US computer industry during the next twenty-five years. Development of the microprocessor at Intel resulted from a search for an integrated circuit that could be used in a wide range of applications. Rather than designing a custom "chipset" for each application, the microprocessor made it possible for Intel to produce a powerful, general-purpose solution to many diverse applications, breaking a bottleneck that limited technological progress and diffusion.

The microprocessor was the foundation for a new generation of computing technology that transformed the ICT industry. Desktop computers diffused rapidly throughout most industrial economies, and the sheer size of the resulting "installed base" radically changed the economics of the computer hardware and software industries. Computer software, formerly dominated by relatively small independent vendors or subsidiaries of established computer systems producers, became a mass-market industry that supported entry by a flood of new firms, some of which proved to be enormously profitable. Established producers of computer hardware were severely affected by the rise of desktop systems, which encroached rapidly on markets for mainframe and minicomputers. By the end of the twentieth century, only a few of the dominant computer systems firms of the 1970s remained active.

The final dramatic transformation in computing technology, the Internet, depended on the appearance of the desktop computer (see Box 13.4). The development of computer networking and the Internet began during the 1960s and was sponsored by governments in the United States, France, and the United Kingdom. The large-scale and early commitment to deployment of computer networks in the United States, which relied entirely on public funds, as well as the rapid diffusion of desktop computers in the United States during the 1980s, created a huge domestic network by the late 1980s. Although the HTTP and HTML "hypertext" software protocols were invented at the European research center, CERN, in 1991, the first commercial application of these protocols as part of a "Web browser" emerged from a US university in 1993.

Box 13.4 The Internet

A collection of independent but interconnected networks built and managed by a variety of organizations, the Internet owes its success to institutional as well as technological innovations (Mowery and Simcoe 2002). Research in computer networking was supported by governments in France, the United Kingdom, and the United States during the 1960s. A central goal of Defense Department computer-networking research was to enable more effective use by researchers in universities, government, and industry of the small number of large research computers then available.

Research and early experiments in computer-networking technology in all three nations led to the development of prototype networks. But the ARPANET, deployment of which in 1969 was sponsored by the US Defense Advanced Research Projects Agency, was far larger and linked more diverse groups of researchers than the prototype networks in France and the United Kingdom. Computers attached to the ARPANET "backbone" communicated on the basis of a shared set of protocols (TCP/IP), another outcome of DARPA research. Later policy decisions by the National Science Foundation (NSF) and other federal agencies that shared responsibility for the backbone encouraged standardization of Internet infrastructure. These agencies also promoted expansion of the Internet beyond the science and engineering communities. In 1990, the US Defense Department transferred managerial control over the Internet infrastructure to NSF, and five years later NSF transferred responsibility for the core network to the private sector.

Software protocols and architectural elements critical to the Internet had been placed in the public domain from the beginning. Open standards encouraged expansion by making available the details of core innovations and lowering entry barriers for firms that supplied hardware, software, and networking services. State and federal regulation of telecommunications aided the rapid diffusion of the Internet in the United States by maintaining low, time-insensitive rates. The 1982 settlement of the federal government's antitrust suit against AT&T restructured the US telecommunications industry and encouraged entry by new service providers, spawning further innovation. But through the late 1980s, the Internet was used mainly by researchers from the academic, industrial, and government communities throughout the world.

Key inventions (HTML and HTPP, developed by Berners-Lee) from CERN, the European particle-physics installation, were used by US technology developers (a group of graduate students in computer science at the University of Illinois, among whom the best-known is Marc Andreesen, who moved to Netscape to commercialize a browser based on the MOSAIC technology developed at the University) to produce the first "browser" in 1994, which vastly expanded use of the Internet and led to the "WorldWide Web." The large "installed base" of desktop computers in the United States was a powerful impetus to the "user-led" innovation that quickly produced a vast array of new applications and eventually, a speculative bubble in the equity markets. But the "radical innovation" of the WorldWide Web in fact represented a culmination of nearly thirty years of research and innovations in networking protocols and software, as well as high-speed data transmission, routers, and computer processing and memory technologies.

Commercial exploitation of the Internet proceeded most rapidly in the United States during the remainder of the 1990s. Like previous US innovations in ICT, the Internet's development drew on a mix of public and private R&D funding, as well as entrepreneurial new firms seeking to commercialize new applications and an active (and entrepreneurial) community of university researchers. Moreover, the Internet's commercial development, as well as the development of desktop systems and software in the United States, benefited from the enormous size of the US market and the large domestic installed base of desktop computers that was rapidly connected to the Internet. Innovation in the Internet benefited from the participation of millions of users in developing or refining new applications, just as had been the case with desktop computer software during the 1980s. In other words, the development of one of the major industries of the late twentieth and early twenty-first centuries benefited from the large size of the US domestic market, a central feature of US economic development that has been prominent since the nineteenth century.

13.4.3 Innovation in Pharmaceuticals and Biotechnology

World War II initiated a transition in the United States to a pharmaceutical industry that relied on formal, in-house research and on stronger links with US universities that were also moving to the forefront of research in the biomedical sciences. The post-war era in the US pharmaceuticals industry opened with a widespread expectation in the industry that there existed a vast potential market for new pharmaceutical products, and that catering to this market, however costly, would prove to be highly profitable.

The post-war period also witnessed a remarkable expansion of federal support for biomedical research through the huge growth in the budget of the National Institutes of Health. Between 1950 and 1965, the NIH budget for biomedical research grew by no less than 18 per cent per year in real terms. By 1965, the federal government accounted for almost two-thirds of all spending on biomedical research. Although NIH funding has continued to grow rapidly, it has been outstripped by growth in privately funded R&D since the 1960s. The US Pharmaceutical Manufacturers Association estimated that foreign and US pharmaceuticals firms invested more than $26 billion in R&D in the United States in 2002, substantially above the $16 billion R&D investment by the National Institutes of Health in the same year (see Pharmaceutical Manufacturers Association 2003, for both estimates).

The identification of the double helix structure of DNA by Watson and Crick in 1953 eventually set off a new epoch of technological change in the pharmaceuticals industry. Biotechnology has created new techniques for drug discovery, as well as new techniques for production of existing drugs, such as insulin (Henderson et al.

1998). The biotechnology enterprise was supported by huge federal expenditures on R&D, including the Nixon Administration's War on Cancer of the early 1970s. As Mowery and Sampat point out in their chapter (8) on universities in this volume, biotechnology has been a key area of university–industry research collaboration, as well as university patenting and licensing, since the 1970s. And as Powell and Grodal (Ch. 3) note, much of the innovation process in the "new" pharmaceuticals industry that has been triggered by the rise of biotechnology relies on collaboration between the long-established major pharmaceuticals firms, who retain strong capabilities in marketing and management of the complex regulatory process, and new firms that specialize in biotechnology-based drug discovery.

13.4.4 A New "Resource Base" for Innovation

The creation of new industries in the post-war US economy illustrates a fundamental change in the nature of the US resource endowment and its relationship to technological innovation. The trajectory of innovation in the United States for much of the 1900–1945 period was influenced by exploitation of the nation's abundant natural resource endowment. During the post-1945 period, however, a combination of factors shifted US innovation from a natural resource-intensive path of development to one that more intensively exploited a burgeoning US "endowment" of scientists and engineers, derived from both domestic and foreign sources. The scale of US markets remained important, but even this source of national advantage became less significant outside of the ICT sector.

The post-war electronics and biotechnology industries did not rely on domestic endowments of natural resources. But the development of these industries assuredly did benefit from the abundance of scientific and engineering human capital in the post-war United States, as well as an unusual mix of public and private demand for electronics technologies. The creation of an institutional infrastructure during this century that, by the 1940s, was capable of training large numbers of electrical engineers, physicists, metallurgists, mathematicians, chemists, biologists, doctors, and other experts capable of advancing these new technologies, meant that the post-war American endowment of specialized human capital was initially more abundant than that of other industrial nations.

This shift from natural to "created" resource endowments as critical factors in international competition has obviously not been confined to the United States. Indeed, Nelson and Wright (1994) among others (Abramovitz 1994; see also Ch. 19 in this volume by Fagerberg and Godinho) argue that the early post-war economic advantages enjoyed by the United States have been eroded by the widespread investments by governments in education and the development of strong domestic R&D infrastructures. In the post-war era, the resource base for knowledge-based industries in electronics, no less than in chemicals, pharmaceuticals, or automobiles, has been transformed.

Natural resources *per se* now play a less central role, particularly by comparison with domestic stocks of human capital that can be expanded through public investments in education and training (see Ch. 7 by Edquist in this volume). In addition, many of the historic scale-based advantages enjoyed by US firms in manufacturing industries were eroded during the post-1945 period by the revival of international flows of trade and capital that made it possible for smaller nations to achieve scale economies through the export of their products. Expanded trade and capital flows have also spurred growth in cross-border flows of technology and knowhow (see Ch. 12 by Narula and Zanfei in this volume). This transformation made it possible for nations such as Taiwan and South Korea, with low pre-1945 levels of industrial development (in contrast to Japan, which had created a relatively sophisticated, albeit militarized, industrial base by the 1930s) and relatively modest endowments of natural resources, to "catch up" with the industrial economies (see Ch. 19 by Fagerberg and Godinho in this volume).

13.5 CONCLUSION

History rarely presents neat lessons for generalization, and the historical study of innovation is no exception. The primary lessons from our historical discussion concern the heterogeneity of the innovation process across time, across sectors, and across countries. Much of the surviving historical evidence that guides the historical study of innovation tends to highlight the formal and obscures the informal processes of knowledge accumulation, learning, and dissemination that underpin technological change and that contribute to its economic benefits. An important area for further research is the enrichment of our historical understanding of the informal processes for knowledge accumulation and diffusion that have been neglected in historical research on the First Industrial Revolution in particular. In addition, as we pointed out in our introductory discussion, a key historiographical mystery remains—why was Northwest Europe the locus of the first transition to sustained, innovation-led growth, rather than Asia or some other region of the global economy? Much of the discussion of this age-old question relies heavily on sweeping generalizations, and more research on the (asserted) failure of non-European economies to make the transition to sustained economic growth during this early period is needed.

One of the defining characteristics of innovation through time is change in the structure of the innovation systems that influence the development and dissemination of knowledge and innovations. The "innovation system" characteristic of the First Industrial Revolution relied on a craft-oriented, trial-and-error process, in which familiarity with basic woodworking and metalworking techniques was valuable. Inasmuch as demand factors appear to have influenced the upsurge of innovation on a broad front during this period, the institutional changes that laid the groundwork for growth in incomes and the expansion of markets for consumer goods also were important.

By contrast, the scale and organizational complexity of the innovation system that characterizes the "Second Industrial Revolution" are vastly greater. A new system of innovation emerged, pioneered by German and US firms in the electrical equipment and chemicals industries, characterized by organized innovation activities in large firms interacting with a public R&D infrastructure. The innovation process during this period was institutionalized within large enterprises of unprecedented scale, and the Second Industrial Revolution in the United States relied heavily on the creation of a national market of great size and homogeneity.

The "Third Industrial Revolution" defies summary description, but its characteristic innovation system relied on the state to an even greater extent than the Second Industrial Revolution. The role of the state as R&D funder and (in many cases) "first customer" for the high-technology industries that developed during the Cold War also contributed to another novel feature of the innovation system of the Third Industrial Revolution that is highlighted in Pavitt's chapter in this volume—increased collaboration and interaction among different institutions. State actions also contributed to the spread of innovation-led development to Asia, as the military alliances and economic institutions of the post-1945 period supported the expanded international trade and capital flows that were indispensable to "catch-up." In addition, Asian governments' strategies for technology transfer and industrial development were of great importance (see Ch. 19 by Fagerberg and Godinho in this volume).

Recent historical research stresses the wide distribution of innovation within the industrializing economies of the First Industrial Revolution, in many cases involving sectors overlooked by the previous historical research that has emphasized the "key sectors" of steam and textiles. This characterization of the process of technological innovation has not been sufficiently integrated into conceptual and theoretical work in innovation studies. The integration of recent historical evidence with the broader conceptual frameworks employed in the field of innovation studies represents an important task for future research. More research also is needed on the factors underpinning innovation in sectors other than the "leading industries" of the First Industrial Revolution. Although the patent data indicate that consumer goods were an important focus for such innovation, the factors triggering this upsurge remain poorly understood. Similarly, the development of organized innovation during the

Second Industrial Revolution outside of the electrical equipment and chemicals industries has received far too little attention.

Institutional change in areas ranging from managerial hierarchies and enterprise control to the training of scientists and engineers is of central importance to innovative performance and to change in the structure of innovation systems over time. In many cases, these institutional changes have occurred in response to pressure from innovators and entrepreneurs, and are best characterized as "coevolving" with change in industries and technologies (see Engerman and Sokoloff 2003 for a useful discussion of this point). The transformation of the German and US university systems, as well as the much more limited restructuring of British universities, is but one example of this coevolution; the development of intellectual property rights systems is another. But the conditions under which such political pressure arises, as well as the factors contributing to the success of failure of such pressure, remain poorly understood.

Although it now is widely celebrated as a hallmark of twenty-first-century "knowledge-based economies," science-based innovation is in fact a relatively recent development. Indeed, it appears well after the institutionalization of R&D within industry in the early twentieth century in the United States and Germany. Moreover, even "high-technology" sectors such as biotechnology and semiconductors continue to rely on experimental methods that at their heart are "trial and error" approaches (see Pisano 1997; Hatch and Mowery 1998).

Our approach has been limited both in time and geographical coverage. Perhaps the two most important omissions from this account are the spread of industrialization beyond Germany within nineteenth-century Europe, and a full account of the rise of the Asian economies after 1945 (see Ch. 19 by Fagerberg and Godinho in this volume for a fuller account of the process of economic "catch-up" in Asia). In the first of these cases it is important to remember that the smaller European economies, now among the wealthiest in the world, have benefited from the import and adaptation of technology during the nineteenth-century, and that Ireland since 1970 has enjoyed rapid growth from similar sources. The post-1945 growth of Japan and Korea was originally based on altogether different scales and types of innovative industrialization—adaptation of foreign technologies, largely in such mature industries as automobiles, steel, and shipbuilding (see also Ch. 19 in this volume). But both of these episodes highlight the importance of broad institutional change, rather than the "strategic importance" of any single industry or technology, in much the same way as do the three "Industrial Revolutions" described in this chapter.

NOTES

1. Chapter 2 by Lazonick in this volume also discusses the role of the institutional environment in contributing to firm-level innovative capabilities.
2. See Peer Vries (2002) for an overview and critique of the literature in this field.
3. "[S]cientific explanations proved to be reliable guides to the commercial development of new processes and products. Unlike the unrestrained inventions of myth and fable, they could not be ignored by industrial firms except at the risk of being displaced by rival firms. But to understand and apply scientific explanation required years of training in the theology of an invisible pantheon of scientific entities. That requirement professionalized industrial science and diminished the role of artisan invention" (Rosenberg and Birdzell 1986: 253).
4. This discussion draws on Beer (1959), Murmann (1998), and Murmann and Landau (1998).
5. See Murmann (1998, 2003) for a more detailed discussion.

REFERENCES

ABRAMOVITZ, M. (1994), "The Origins of the Postwar Catch-Up and Convergence Boom," in J. Fagerberg, B. Verspagen, and N. von Tunzelmann (eds.), *The Dynamics of Technology, Trade and Growth*, Aldershot: Edward Elgar, 21–52.

BEER, J. J. (1959), *The Emergence of the German Dye Industry*, Urbana, Ill.: University of Illinois Press.

BERG, M. (1998), "Product Innovation in Core Consumer Industries in Eighteenth Century Britain," in M. Berg and K. Bruland (eds.), *Technological Revolutions in Europe: Historical Perspectives*, Cheltenham: Edward Elgar, 138–58.

—— and EGER, E. (2003), *Luxury in the Eighteenth Century*, New York: Palgrave.

BEYERCHEN, A. (1988), "On the Stimulation of Excellence in Wilhelmian Science," in J. R. Dukes and J. Remak, *Another Germany: A Reconsideration of the Imperial Era*, Boulder, Colo.: Westview Press, 139–68.

BRAUDEL, F. (1984), *The Perspective of the World*, Berkeley: University of California Press.

BRULAND, K. (1989), "The Transformation of Work in European Industrialization," in P. Mathias and J. A. Davis (eds.), *The Nature of Industrialization: The First Industrial Revolutions*, Oxford: Basil Blackwell, 154–70.

*—— (2004), "Industrialisation and Technological Change," in R. Floud and P. Johnson (eds), *The Cambridge Economic History of Modern Britain*, vol. 1, *Industrialisation*, Cambridge: Cambridge University Press, 117–46.

*CHANDLER, A. D., Jr. (1977), *The Visible Hand*, Cambridge, Mass.: Harvard University Press.

CHAPMAN, S. D. (1972), *The Cotton Industry in the Industrial Revolution*, London: Macmillan.

DEANE, P. (1965), *The First Industrial Revolution*, Cambridge: Cambridge University Press.

Encyclopaedia Britannica (1995), 15th edn., vol. 8, Chicago: Encyclopaedia Britannica.

* Asterisked items are suggestions for further reading.

ENGERMAN, S. L., and SOKOLOFF, K. L. (2003), "Institutional and Non-institutional Explanations of Economic Differences," NBER Working Paper 9989.

FLAMM, K. (1987), *Targeting the Computer*, Washington, DC: Brookings Institution.

—— (1988), *Creating the Computer*, Washington, DC: Brookings Institution.

FREEMAN, C., and LOUÇÃ, F. (2002), *As Time Goes By: From the Industrial Revolution to the Information Revolution*, Oxford and New York: Oxford University Press.

GRIFFITHS, T., HUNT, P. A., and O'BRIEN, P. K. (1992), "Inventing Activity in the British Textile Industry, 1700–1800," *Journal of Economic History* 52: 881–906.

HABER, L. F. (1971), *The Chemical Industry, 1900–1930*, Oxford: Clarendon Press.

*HARRIS, R. (2000), *Industrializing English Law: Entrepreneurship and Business Organization 1720–1844*, Cambridge: Cambridge University Press.

HATCH, N., and MOWERY, D. C. (1998), "Process Innovation and Learning by Doing in Semiconductor Manufacturing," *Management Science* 44: 1471–7.

HENDERSON, R., ORSENIGO, L., and PISANO, G. (1999), "The Pharmaceutical Industry and the Revolution in Biotechnology," in D. C. Mowery and R. R. Nelson (eds.), *The Sources of Industrial Leadership*, New York: Cambridge University Press, 267–311.

HUDSON, P. (1986), *The Genesis of Industrial Capital: A Study of the West Riding Wool Textile Industry c. 1750–1850*, Cambridge: Cambridge University Press.

*JONES, E. L. (2003), *Growth Recurring. Economic Change in World History*, Ann Arbor: University of Michigan Press.

LANDES, D. (1998), *The Wealth and Poverty of Nations*, London: Little, Brown and Co.

LANGTON, J. (1984), "The Ecological Theory of Bureaucracy: The Case of Josiah Wedgwood and the British Pottery Industry," *Administrative Science Quarterly* 29: 330–54.

*LAZONICK, W. (2002), "Innovative Enterprise and Historical Transformation," *Enterprise and Society* 3: 3–47.

LLOYD-JONES, R., and LEWIS, M. J. (1998), *British Industrial Capitalism since the Industrial Revolution*, London: UCL Press.

McKENDRICK, N. (1961), "Josiah Wedgwood and Factory Discipline," *Historical Journal* 4(1): 30–55.

*MACLEOD, C. (1988), *Inventing the Industrial Revolution: The English Patent System 1660–1800*, Cambridge: Cambridge University Press.

MALERBA, F. (1985), *The Semiconductor Business*, Madison, Wis.: University of Wisconsin Press.

MANTOUX, P. (1961), *The Industrial Revolution in the Eighteenth Century: An Outline of the Beginnings of the Modern Factory System in England*, revised ed. (English translation originally published in 1928). New York: Harper & Row.

MATHIAS, P. (1983), *The First Industrial Revolution: An Economic History of Britain 1700–1914*, 2nd edn., London and New York: Methuen.

MEYER-THUROW, G. (1982), "The Industrialization of Invention: A Case Study from the German Chemical Industry," *Isis* 73: 363–91.

MILLARD, A. (1990), *Edison and the Business of Innovation*, Baltimore: Johns Hopkins University Press.

*MOKYR, J. (2002), *The Gifts of Athena: Historical Origins of the Knowledge Economy*, Princeton and Oxford: Princeton University Press.

—— (2003), "Industrial Revolution," in J. Mokyr (ed.), *The Oxford Encyclopedia of Economic History*, vol. 3, Oxford: Oxford University Press.

MOWERY, D. C. (1983), "Industrial Research, Firm Size, Growth, and Survival, 1921–46," *Journal of Economic History* 43: 953–80.

*—— (1995), "The Boundaries of the U.S. Firm in R&D," in N. R. Lamoreaux and D. M. G. Raff (eds.), *Coordination and Information: Historical Perspectives on the Organization of Enterprise*, Chicago: University of Chicago Press for NBER.

—— (1996), *The International Computer Software Industry: A Comparative Study of Industry Evolution and Structure*, Oxford: Oxford University Press.

—— (1999), "The Computer Software Industry," in D. C. Mowery and R. R. Nelson (eds.), *Sources of Industrial Leadership*, New York: Cambridge University Press, 133–68.

—— and ROSENBERG, N. (1998), *Paths of Innovation: Technological Change in 20th Century America*, New York: Cambridge University Press.

—— and SIMCOE, T. (2002), "The History and Evolution of the Internet," in B. Steil, R. Nelson, and D. Victor (eds.), *Technological Innovation and Economic Performance*, Princeton: Princeton University Press, 229–64.

MURMANN, J. P. (1998), "Knowledge and Competitive Advantage in the Synthetic Dye Industry, 1850–1914," unpublished Ph.D. dissertation, Columbia University.

—— (2003), "The Coevolution of Industries and Comparative Advantage: Theory and Evidence," unpublished MS.

—— and LANDAU, R. (1998), "On the Making of Competitive Advantage: The Development of the Chemical Industries of Britain and Germany Since 1850," in A. Arora, R. Landau, and N. Rosenberg (eds.), *Chemicals and Long-Term Economic Growth*, New York: Wiley, 27–70.

NELSON, R. R., and WRIGHT, G. (1994), "The Erosion of U.S. Technological Leadership as a Factor in Postwar Economic Convergence," in W. J. Baumol, R. R. Nelson, and E. N. Wolff (eds.), *Convergence of Productivity*, New York: Oxford University Press.

ORSENIGO, L. (1988), *The Emergence of Biotechnology*, London: Pinter.

PHARMACEUTICAL MANUFACTURERS ASSOCIATION (2003), *Industry Profile 2003* http://www.phrma.org/ publications/publications/profile02

PISANO, G. (1997), *The Development Factory*, Boston: Harvard Business School Press.

—— SHAN, W., and TEECE, D. J. (1988), "Joint Ventures and Collaboration in the Biotechnology Industry," in D. C. Mowery (ed.), *International Collaborative Ventures in U.S. Manufacturing*, Cambridge, Mass.: Ballinger.

POLLARD, S. (1965), *The Genesis of Modern Management. A Study of the Industrial Revolution in Great Britian*, London: Arnold.

*POMERANZ, K. R. (2000), *The Great Divergence: China, Europe and the Making of the Modern World Economy*, Princeton: Princeton University Press.

REICH, L. S. (1985), *The Making of American Industrial Research*, New York: Cambridge University Press.

REID, T. R. (1984), *The Chip*, New York: Simon and Schuster.

ROSENBERG, N. (1985), "The Commercial Exploitation of Science by American Industry," in K. B. Clark, R. H. Hayes, and C. Lorenz (eds.), *The Uneasy Alliance*, Boston: Harvard Business School Press, 19–51.

—— and BIRDZELL, L. E. (1986), *How the West Grew Rich*, New York: Basic Books.

SAMUEL, R. (1977), "The Work-shop of the World: Steam Power and Hand Technology in mid-Victorian Britain," *History Workshop* 3: 6–72.

SCHMOOKLER, J. (1957), "Inventors Past and Present," *Review of Economics and Statistics* 57: 321–33.

—— (1962), "Changes in Industry and in the State of Knowledge as Determinants of Industrial Invention," in R. R. Nelson (ed.), *The Rate and Direction of Inventive Activity*, Princeton: Princeton University Press for NBER.

STOKES, R. (1994), *Opting for Oil: The Political Economy of Technological Change in the West German Chemical Industry, 1945–61*, New York: Cambridge University Press.

SULLIVAN, R. (1990), "The Revolution of Ideas: Widespread Patenting and Invention During the Industrial Revolution," *Journal of Economic History* 50(2): 340–62.

TILTON, J. E. (1971), *The International Diffusion of Technology: The Case of Transistors*, Washington, DC: Brookings Institution.

TOYNBEE, A. (1969; orig. 1908), *Lectures on the Industrial Revolution in England*, New York: A. M. Kelley.

UGLOW, J. (2002), *The Lunar Men*, New York: Farrar, Straus & Giroux.

USSELMAN, S. (1993), "IBM and its Imitators: Organizational Capabilities and the Emergence of the International Computer Industry," *Business & Economic History* 22: 1–35.

VON TUNZELMANN, G. N. (1978), *Steam Power and British Industrialization to 1860*, Oxford: Clarendon Press.

*VRIES, P. (2002), "Are Coal and Colonies Really Crucial? Kenneth Pomeranz and the Great Divergence," *Journal of World History* 12(2): 407–45.

WADE, R. (1999), *Governing the Market: Economic Theory and the Role of Government in East Asian Industriaization*, Princeton: Princeton University Press.

WALLERSTEIN, I. (1974), *Capitalist Agriculture and the Origins of the Modern World Economy*, New York: Academic Press.

WHITEHEAD, A. N. (1925), *Science and the Modern World*, New York: Macmillan.

CHAPTER 14

SECTORAL SYSTEMS

HOW AND WHY INNOVATION DIFFERS ACROSS SECTORS

FRANCO MALERBA

14.1 INTRODUCTION

INNOVATION greatly differs across sectors in terms of characteristics, sources, actors involved, the boundaries of the process, and the organization of innovative activities. A focus on a "representative firm" as the main actor; on narrowly, well defined, and static boundaries as a sector delimitation; on R&D and learning-by-doing as the only two sources of innovation; on competition and formal R&D joint ventures as the only kind of interaction among firms; and on the patent system and public support for R&D as the only relevant institutions and policies that matter for innovation, would capture only part of the action that takes place in sectors and would identify only a few of the key variables that matter for innovation and performance.

A comparison of actors, sources, institutions, and policies for innovation in different sectors (e.g. in pharmaceuticals and biotechnology, chemicals, software, computers, semiconductors, telecommunications, or machine tools) shows striking differences. The role of innovation in the dynamics and transformation of these sectors is highly diverse.

How is it possible to analyze consistently these differences and their effects on sectoral growth and performance? The industrial economics approach pays a lot of attention to differences across sectors in R&D intensity, market structure, the range of viable R&D strategies and R&D alliances, the intensity of the patent race, the effectiveness of patent protection, the role of competition policy and the extent of R&D support. But, while these are very important factors, they are not the only ones nor are they the most relevant for a full understanding of the differences in innovation across sectors.

A rich and heterogeneous tradition of sectoral studies has clearly shown both that sectors differ in terms of the knowledge base, the actors involved in innovation, the links and relationships among actors, and the relevant institutions, and that these dimensions clearly matter for understanding and explaining innovation and its differences across sectors. However, these case studies are quite different in terms of methodology, variables, and countries examined.

This chapter will briefly discuss the previous literature on differences across sectors in innovation (Section 14.2) and then propose the concept of sectoral systems of innovation (14.3). In the next sections, the basic building blocks of sectoral systems will be discussed: knowledge, technological domains, and sectoral boundaries (14.4); actors, relationships and networks (14.5); and institutions (14.6). Then the dynamics and transformation of sectoral systems (14.7) is examined. Finally, some policy implications (14.8) and the challenges ahead (14.9) are discussed.

The chapter will discuss a large number of sectors that are highly innovative and technologically advanced and have strong links with science, which nevertheless organize innovation very differently: computers, semiconductors, telecommunication equipment and services, software, chemicals, pharmaceuticals and biotechnology, and machine tools. Most of the sectoral examples in this paper are drawn from Mowery and Nelson (1999) and Malerba (2004).

14.2 PREVIOUS LITERATURE ON SECTORAL DIFFERENCES IN INNOVATION

The literature has advanced some distinctions among sectors in innovation and diffusion based on different dimensions. The simplest one, widely used in international studies by the OECD, EU, and international organizations, refers to sectors that are high R&D-intensive (such as electronics or drugs) and low R&D-intensive (such as textiles or shoes).

Another distinction, coming from the Schumpeterian legacy, focuses on differences in market structure and industrial dynamics among sectors. Schumpeter Mark I sectors are characterized by "creative destruction," with technological ease of entry and a major role played by entrepreneurs and new firms in innovative activities. Schumpeter Mark II sectors are characterized by "creative accumulation" (in Keith Pavitt's words) with the prevalence of large established firms and the presence of relevant barriers to entry for new innovators. This regime is characterized by the dominance of a stable core of a few large firms, with limited entry. The distinction refers to the early Schumpeter of *Theory of Economic Development* (1911, "Schumpeter Mark I") and to the later one of *Capitalism, Socialism and Democracy* (1942, "Schumpeter Mark II"). Machinery or biotechnology are examples of Schumpeter Mark I sectors, while the semiconductor industry of the 1990s (think of microprocessors and dynamic memories) or mainframe computers in the period 1950s–1990s are examples of Schumpeter Mark II sectors.

Other differences across sectors have been related to technological regimes, a notion introduced by Nelson and Winter (1982), referring to the learning and knowledge environment in which firms operate. A specific technological regime defines the nature of the problem firms have to solve in their innovative activities, affects the model form of technological learning, shapes the incentives and constraints to particular behavior and organization, and influences the basic processes of variety generation and selection (and therefore the dynamics and evolution of firms). More generally, Malerba and Orsenigo (1996 and 1997) have proposed that a technological regime is composed by opportunity and appropriability conditions, degrees of cumulativeness of technological knowledge, and characteristics of the relevant knowledge base. More specifically, technological opportunities reflect the likelihood of innovating for any given amount of money invested in search. High opportunities provide powerful incentives to the undertaking of innovative activities and denote an economic environment that is not functionally constrained by scarcity. In this case, potential innovators may come up with frequent and important technological innovations. Appropriability of innovations summarizes the possibilities of protecting innovations from imitation and of reaping profits from innovative activities. High appropriability means the existence of ways of successfully protecting innovation from imitation. Low appropriability conditions denote an economic environment characterized by the widespread existence of externalities (Levin et al. 1987). Cumulativeness conditions capture the properties that today's innovations and innovative activities form the starting point for tomorrow innovations. More broadly, one may say that high cumulativeness means that today's innovative firms are more likely to innovate in the future in specific technologies and along specific trajectories than non-innovative firms. Cumulativeness may be due to knowledge/cognitive factors, organizational factors, or market factors of the "success breeds success" type. The properties of the knowledge base relate to the nature of knowledge underpinning firms' innovative activities. Technological knowledge involves various degrees of specificity, tacitness, complementarity, and independence and may greatly differ across sectors and technologies (Winter 1987). Differences in techno-logical regimes affect the organization of innovative activities at the sectoral level and may lead to a fundamental distinction between Schumpeter Mark I and Schumpeter Mark II models. High technological opportunities, low appropriability, and low cumulativeness (at the firm level) conditions lead to a Schumpeter Mark I pattern. By contrast, high appropriability and high cumulativeness (at the firm level) condi-tions lead to a Schumpeter Mark II pattern: think again of the semiconductor industry of the 1990s (i.e. microprocessors and dynamic memories) and mainframe computers in the period 1950s–1990s.

Technological regimes and Schumpeterian patterns of innovation change over time (Klepper 1996). According to an industry life-cycle view, a Schumpeter Mark I pattern of innovative activities may turn into a Schumpeter Mark II. Early in the history of an industry—when knowledge is changing very rapidly, uncertainty is very high, and barriers to entry very low—new firms are the major innovators and are the key elements in industrial dynamics. When the industry develops and eventually matures and technological change follows well-defined trajectories, economies of scale, learning curves, barriers to entry, and financial resources become important in the competitive process. Thus, large firms with monopolistic power come to the forefront of the innovation process (Utterback 1994; Gort and Klepper 1982; Klepper 1996). In the presence of major knowledge, technological, and market discontinuities, a Schumpeter Mark II pattern of innovative activities may be replaced by a Schumpeter Mark I. In this case, a rather stable organization charac-terized by incumbents with monopolistic power is displaced by a more turbulent one with new firms using the new technology or focusing on the new demand (Henderson and Clark 1990; Christensen and Rosenbloom 1995). Although rather archetypical, these analyses point to the direction of placing a lot of attention to differences across sectors in some key factors related to knowledge and learning regimes. As the examples discussed above suggest, change over time also reflects institutional change and the coevolution of industries and institutions.

Other distinctions refer to sectors that are net suppliers of technology and sectors that are users of technology. On the bases of the R&D done by 400 American firms and of intersectoral flows in the American economy, Scherer (1982) identifies sectors that are net sources of R&D for other sectors (such as computers and instruments), and sectors that are net users of technology (such as textiles and metallurgy). A similar analysis is done by Robson et al. (1988) who, on the basis of 4,378 innovations in the UK between 1945 and 1983, identify (a) "core sectors" (such as electronics, machinery, instruments, and chemicals) which gener-ate most of innovations in the economy and are net sources of technology, (b) secondary sectors (such as auto and metallurgy) which play a secondary role in terms of sources of innovation for the economy, and (*c*) user sectors such as services which mainly absorb technology.

A key difference among sectors refers to the sources of innovation and the appropriability mechanisms. Pavitt (1984) proposes four types of sectoral pattern for innovative activities. In supplier-dominated (e.g. textile, services) sectors, new technologies are embodied in new components and equipment, and the diffusion of new technologies and learning takes place through learning-by-doing and by-using. In scale-intensive sectors (e.g. autos, steel), process innovation is relevant and the sources of innovation are both internal (R&D and learning-by-doing) and external (equipment producers), while appropriability is obtained through secrecy and patents. In specialized suppliers (e.g. equipment producers), innovation is focused on performance improvement, reliability, and customization, with the sources of innovation being both internal (tacit knowledge and experience of skilled techni-cians) and external (user–producer interaction); appropriability comes mainly from the localized and interactive nature of knowledge. Finally, science-based sectors (e.g. pharmaceuticals, electronics) are characterized by a high rate of product and process innovations, by internal R&D, and by scientific research done at universities and public research laboratories; science is a source of innovation, and appropriability means are of various types, ranging from patents, to lead-times and learning curves, and to secrecy. The Pavitt taxonomy has been tremendously successful in empirical research and has guided the identification of firms and country advantages. Refine-ments and enrichments of the taxonomy have been proposed in the following decades. A very interesting and relevant work in this direction is the one by Marsili (2001).

Differences across sectors in appropriability conditions have been examined by Levin et al. (1987), PACE (1996) and Cohen et al. (2002) using survey questionnaires for R&D managers in the United States, Europe, and Japan, following the pioneering Yale survey. Here, major differences across sectors have been identified in terms of appropriability means—patents, secrecy, lead-times, learning curves, and comple-mentary assets. All these surveys have found major differences across sectors in the use of patents.

14.3 SECTORAL SYSTEMS OF INNOVATION

The contributions examined above focus on a specific difference among sectors. In this and the following sections, a multidimensional, integrated, and dynamic view of innovation in sectors is proposed, related to the framework of sectoral systems of innovation, which provides a methodology for the analysis and comparison of sectors.

A sector is a set of activities that are unified by some linked product groups for a given or emerging demand and which share some common knowledge. Firms in a sector have some commonalities and at the same time are heterogeneous. A sectoral system framework focuses on three main dimensions of sectors:

(*a*) Knowledge and technological domain
(*b*) Actors and networks
(*c*) Institutions

(*a*) *Knowledge and technological domain.* Any sector may be characterized by a specific knowledge base, technologies and inputs. In a dynamic way, the focus on knowledge and the technological domain places at the centre of the analysis the issue of sectoral boundaries, which usually are not fixed, but change over time.

(*b*) *Actors and networks.* A sector is composed of heterogeneous agents that are organizations or individuals (e.g. consumers, entrepreneurs, scientists). Organizations may be firms (e.g. users, producers, and input suppliers) or non-firms (e.g. universities, financial institutions, government agencies, trade-unions, or technical associations), and include subunits of larger organizations (e.g. R&D or production departments) and groups of organizations (e.g. industry associations). Agents are characterized by specific learning processes, competencies, beliefs, objectives, organizational structures, and behaviors, which interact through processes of communication, exchange, cooperation, competition, and command.

Thus, in a sectoral system framework, innovation is considered to be a process that involves systematic interactions among a wide variety of actors for the generation and exchange of knowledge relevant to innovation and its commercialization. Interactions include market and non-market relations that are broader than the market for technological licensing and knowledge, interfirm alliances, and formal networks of firms, and often their outcome is not adequately captured by our existing systems of measuring economic output.

(*c*) *Institutions.* Agents' cognition, actions, and interactions are shaped by institutions, which include norms, routines, common habits, established practices, rules, laws, standards, and so on. Institutions may range from ones that bind or impose enforcements on agents to ones that are created by the interaction among agents (such as contracts); from more binding to less binding; from formal to informal (such as patent laws or specific regulations vs. traditions and conventions). A lot of institutions are national (such as the patent system), while others are specific to sectors (such as sectoral labor markets or sector specific financial institutions).

Over time, a sectoral system undergoes processes of change and transformation through the coevolution of its various elements.

The notion of sectoral system of innovation and production complements other concepts within the innovation system literature (Edquist 1997) such as national systems of innovation delimited by national boundaries and focused on the role of non-firm organizations and institutions (Freeman 1987; Nelson 1993; Lundvall 1993), regional/local innovation systems in which the boundary is the region (Cooke et al. 1997), technological systems, in which the focus is on technologies and not on sectors[1] (Carlsson and Stankiewitz 1995; Hughes 1984; Callon 1992), and distributed innovation system (in which the focus is on specific innovations—Andersen et al. 2002).

What are the main differences between a sectoral innovation system and a national innovation system perspective? While national innovation systems take innovation systems as delimited more or less clearly by national boundaries, a sectoral system approach would claim that sectoral systems may have local, national, and/or global dimensions. Often these three different dimensions coexist in a sector. In addition, national innovation systems result from the different composition of sectors, some of which are so important that they drive the growth of the national economy. For example, Japanese growth in the 1970s and 1980s was driven by specific sectors, which were different from the sectors behind the American "resurgence" during the 1990s. Similarly, Italian economic growth is driven by specific sectors. Thus, understanding the key driving sectors of an economy with their specificities greatly helps in understanding national growth and national patterns of innovative activities.

The theoretical and analytical approach of sectoral systems is grounded in the evolutionary theory. Evolutionary theory places a key emphasis on dynamics, innovation processes, and economic transformation. Learning and knowledge are key elements in the change of the economic system. "Boundedly rational" agents act, learn, and search in uncertain and changing environments. Agents know how to do different things in different ways. Thus learning, knowledge, and behavior entail agents' heterogeneity in experience and organization. Their different competences affect their persistent differential performance. In addition, evolutionary theory places emphasis on cognitive aspects such as beliefs, objectives, and expectations, which are in turn affected by previous learning and experience and by the environment in which agents act. A central place in the evolutionary approach is occupied by the processes of variety creation (in technologies, products, firms, and organizations), replication (that generates inertia and continuity in the system), and selection (that reduces variety in the economic system and discourages the inefficient or ineffective utilization of resources). Finally, for evolutionary theory, aggregate phenomena are emergent properties of far-from-equilibrium interactions and have a metastable nature (Nelson 1995; Dosi 1997; Metcalfe 1998). Here, the environment and conditions in which agents operate may drastically differ. Evolutionary theory stresses major differences in opportunities related to science and technologies. The same holds for the knowledge base underpinning innovative activities, as well as for the institutional context. Thus the learning, behavior, and capabilities of agents are constrained and "bounded" by the technology, knowledge base, and institutional context. Heterogeneous firms facing similar technologies, searching around similar knowledge bases, undertaking similar production activities, and "embedded" in the same institutional setting, share some common behavioral and organizational traits and develop a similar range of learning patterns, behavior, and organizational forms.

One last remark regards the aggregation issue regarding products, agents or functions. For example, sectoral systems may be examined broadly or narrowly (in terms of a small set of product groups).[2] A broad definition allows us to capture all the interdependencies and linkages in the transformation of sectors, while a narrow definition identifies more clearly specific relationships. Of course, within broad sectoral systems, different innovation systems related to different product groups may exist. The choice of the level of aggregation depends on the goal of the analysis.

In the following pages we will concentrate on each block of a sectoral system of innovation and production:

- Knowledge, technological domain, and boundaries
- Agents, interaction and networks
- Institutions

14.4 KNOWLEDGE, TECHNOLOGICAL DOMAIN, AND SECTORAL BOUNDARIES

Knowledge plays a central role in innovation. Knowledge is highly idiosyncratic at the firm level, does not diffuse automatically and freely among firms, and has to be absorbed by firms through their differential abilities accumulated over time. The evolutionary literature has proposed that sectors and technologies differ greatly in terms of the knowledge base and learning processes related to innovation. Knowledge differs across sectors in terms of domains. One knowledge domain refers to the specific scientific and technological fields at the base of innovative activities in a sector (Dosi 1988; Nelson and Rosenberg 1993), while another comprises applications, users, and the demand for sectoral products. Recently a major discontinuity has taken place in the processes of knowledge accumulation and distribution with the emergence of the knowledge-based economy which has redefined existing sectoral boundaries, affected relationships among actors, reshaped the innovation process, and modified the links among sectors (Nelson 1995; Dosi 1997; Metcalfe 1998; Lundvall 1993; Lundvall and Johnson 1994).

What do we know about the main dimensions of knowledge? First, knowledge may have different degrees of accessibility (Malerba and Orsenigo 2000), i.e. opportunities of gaining knowledge external to firms, which in turn may be internal or external to the sector. In both cases, greater accessibility of knowledge may decrease industrial concentration. Greater accessibility internal to the sector implies lower appropriability: competitors may gain knowledge about new products and processes and, if competent, imitate those new products and processes. Accessibility of knowledge that is external to the sector may be related to the levels and sources of scientific and technological opportunities. Here, the external environment may affect firms through human capital with a certain level and type of knowledge or through scientific and technological knowledge developed in firms or non-firm organizations, such as universities or research laboratories (Malerba and Orsenigo 2000).

The sources of technological opportunities differ markedly among sectors. As Freeman (1982) and Rosenberg (1982), among others, have shown, in some sectors opportunity conditions are related to major scientific breakthroughs in universities; in others, opportunities to innovate may often come from advancements in R&D, equipment, and instrumentation; while in still other sectors, external sources of knowledge in terms of suppliers or users may play a crucial role. Not all external knowledge may be easily used and transformed into new artifacts. If external knowledge is easily accessible, transformable into new artifacts and exposed to a lot of actors (such as customers or suppliers), then innovative entry may take place (Winter 1984). If advanced integration capabilities are necessary (Cohen and Levinthal 1989), the industry may be concentrated and formed by large, established firms.

Second, knowledge may be more or less cumulative, i.e. the degree by which the generation of new knowledge builds upon current knowledge. One can identify three different sources of cumulativeness.

(1) Cognitive. The learning processes and past knowledge constrain current research, but also generate new questions and new knowledge.
(2) The firm and its organizational capabilities. Organizational capabilities are firm-specific and generate knowledge which is highly path-dependent. They implicitly define what a firm learns and what it can hope to achieve in the future.
(3) Feedbacks from the market, such as in the "success-breeds-success" process. Innovative success yields profits that can be reinvested in R&D, thereby increasing the probability to innovate again.

High cumulativeness implies an implicit mechanism leading to high appropriability of innovations. In the case of knowledge spillovers within an industry, however, it is also possible to observe cumulativeness at the sectoral level. Cumulativeness may also be present at the local level. In this case, high cumulativeness within specific locations is more likely to be associated with low appropriability conditions and spatially localized knowledge spillovers. Finally, cumulativeness at the technological and firm levels creates first mover advantages and generates high concentration. Firms that have a head start develop a new knowledge based on the current one and introduce continuous innovations of the incremental type.

Accessibility, opportunity, and cumulativeness are key dimensions of knowledge related to the notion of technological and learning regimes (Nelson and Winter 1982; Malerba and Orsenigo 1997), which, as seen above, may differ across sectors. Other dimensions of knowledge could be related to its tacitness, codificability, complexity, systemic features, scientific base, and so on (Winter 1987; Cowan, David, and Foray 2000).

The boundaries of sectoral systems are affected by the knowledge base and technologies. However, the type and dynamics of demand represent a major factor in the processes of transformation of sectoral systems. The same holds for links and complementarities among artifacts and activities. These links and complementarities are, first of all, of the static type, as are input–output links. Then there are dynamic complementarities, which take into account interdependencies and feedbacks, both at the demand and at the production levels. Dynamic complementarities among artifacts and activities are major sources of transformation and growth of sectoral systems, and may set in motion virtuous cycles of innovation and change. This could be related to the concept of filiere and the notion of development blocks (Dahmen 1989). Links and complementarities change over time and greatly affect a wide variety of variables of a sectoral system: firms' strategies, organization, and performance; the rate and direction of technological change; the type of competition; and the networks among agents. Thus the boundaries of sectoral systems may change more or less rapidly over time, as a consequence of dynamic processes related to the transformation of knowledge, the evolution and convergence in demand, and changes in competition and learning by firms.

In general, the features and sources of knowledge affect the rate and direction of technological change, the organization of innovative and production activities, and the factors at the base of firms' successful performance.

Great differences among sectors in the dimensions discussed above exist. Let us compare, for example, pharmaceuticals and machine tools. In the pharmaceutical industry, the knowledge base and the learning processes have greatly affected innovation and the organization of innovative activities. In the early stages (1850–1945), the industry was close to chemicals, with little formal research until the 1930s and a major use of licenses. The following period (1945–early 1980s) was characterized by the introduction of random screening of natural and chemically derived compounds. This led to an explosion of R&D and, although few blockbusters were discovered in each period, nevertheless, each period enjoyed high growth. The advent of molecular biology since the 1980s led to a new learning regime based on molecular genetics and rDNA technology, with two search regimes: one regarding specialized technologies, the other generic technologies. Nowadays, no individual firm can gain control on more than a subset of the search space. Innovation increasingly depends on strong scientific capabilities and on the ability to interact with science and scientific institutions in order to explore the search space (McKelvey, Orsenigo, and Pammolli 2004; Henderson, Orsenigo, and Pisano 1999).

In machine tools, innovation has been mainly incremental and now is increasingly systemic. Knowledge about applications is very important, and therefore user–producer relationships as well as partnerships with customers are common. The knowledge base has been embodied in skilled personnel on the shop floor level (with applied technical qualification) and in design engineers (not necessarily with a university degree but with long-term employment in the company). Internal training (particularly apprenticeships) is quite relevant. In small firms, R&D is not done extensively and R&D cooperation is not common. Recently, the knowledge base has shifted from purely mechanical to mechanic, microelectronic and information intensive, with an increasing codification and an increasing use of formal R&D. Products have increasingly being modularized and standardized. A key role is also played by information flows about components coming from producers of different technologies, such as lasers, materials, measurement, and control devices. Nowadays, many large machine tool companies operate already on an international basis making use of specific knowledge sources at their different firm sites (Wengel and Shapira 2004; Mazzoleni 1999).

14.5 ACTORS, RELATIONSHIPS, AND NETWORKS

Sectoral systems are composed of heterogeneous actors. In general, a rich, multidisciplinary, and multisource knowledge base and rapid technological change implies a great heterogeneity of actors in most sectors.

Firms are the key actors in the generation, adoption, and use of new technologies, are characterized by specific beliefs, expectations, goals, competences, and organization, and are continuously engaged in processes of learning and knowledge accumulation (Nelson and Winter 1982; Malerba 1992, Teece and Pisano 1994, Dosi, Marengo, and Fagiolo 1998, Metcalfe 1998). The extent of firm heterogeneity is the result of the opposing forces of variety creation, replication, and selection (Nelson 1995; Metcalfe 1998). Selection increases homogeneity, while entry and technological and organizational innovations are fundamental sources of heterogeneity. Firm heterogeneity is also affected by the characteristics of the knowledge base, specific experience and learning processes, and the working of dynamic complementarities.

Actors also include users and suppliers who have different types of relationships with the innovating, producing, or selling firms. Users and suppliers are characterized by specific attributes, knowledge, and competencies, with more or less close relationships with producers (VonHippel 1988, Lundvall 1993). As previously mentioned, in a dynamic and innovative setting, suppliers and users greatly affect and continuously redefine the boundaries of a sectoral system.

Other types of agents in a sectoral system are non-firm organizations such as universities, financial organizations, government agencies, local authorities, and so on. In various ways, they support innovation, technological diffusion, and production by firms, but again their role greatly differs among sectoral systems. In several high technology sectors, universities play a key role in basic research and human capital formation, and in some sectors (such as biotechnology and software) they are also a source of start-ups and even innovation. In sectoral systems such as software or biotechnology–pharmaceuticals, new actors such as venture capital companies have emerged over time. These financial organizations have played a different role according to the stage of the industry life-cycle. When industry matures or large firms are relevant, capital constraints become lighter and much investment is self-financed. By contrast, for start-ups in emerging or new high-tech sectors, capital constraints are very high and specific financial intermediaries such as venture capital firms are important (Rivaud-Danset 2001; Dubocage 2002).

Often the most appropriate units of analysis in specific sectoral systems are not necessarily firms but individuals (such as the scientist who opens up a new biotechnology firm), firms' subunits (such as the R&D or the production department), and groups of firms (such as industry consortia).

The focus on users, government agencies, and consumers puts a different emphasis on the role of demand. In a sectoral system, demand is not seen as an aggregate set of similar buyers or atomistic undifferentiated customers, but as composed of heterogeneous agents who interact in various ways with producers. Demand then becomes composed by individual consumers, firms, and public agencies, which are in turn characterized by knowledge, learning processes, and competences, and which are affected by social factors and institutions. The emergence and transformation of demand become then a very important part in the dynamics and evolution of sectoral systems. In addition, demand has often proven to be a major factor in the redefinition of the boundaries of a sectoral system, a stimulus for innovation, and a key factor shaping the organization of innovative and production activities.

Within sectoral systems, heterogeneous agents are connected in various ways through market and non-market relationships. It is possible to identify different types of relations, linked to different analytical cuts. First, traditional analyses of industrial organizations have examined agents as involved in processes of exchange, competition, and command (such as vertical integration). Second, in more recent analyses, processes of formal cooperation or informal interaction among firms or among firms and non-firm organizations have been examined in depth (as one may see from the literature on tacit or explicit collusion, or hybrid governance forms, or formal R&D cooperation). This literature has analyzed firms with certain market power, suppliers or users facing opportunistic behavior or asset specificities in transaction, and firms with similar knowledge having appropriability and indivisibility problems in R&D. Finally, the evolutionary approach and the innovation systems literature have also paid a lot of attention to the wide range of formal and informal cooperation and interaction among firms. However, according to this perspective, in uncertain and changing environments networks emerge not because agents are similar, but because they are different. Thus, networks integrate complementarities in knowledge, capabilities, and specialization (see Lundvall 1993; Edquist 1997; Nelson 1995; Teubal et al. 1991). Relationships between firms and non-firm organizations (such as universities and public research centers) have been a source of innovation and change in several sectoral systems: pharmaceuticals and biotechnology, information technology, and telecommunications have been relevant (Nelson and Rosenberg 1993).

One final observation needs to be made: the key role played by networks in a sectoral system leads to a meaning of the term "sectoral structure" different from the one used in industrial economics. In industrial economics, structure is related mainly to the concept of market structure and of vertical integration and diversification. In a sectoral system perspective, on the contrary, structure refers to links among artifacts and to relationships among agents: it is therefore far broader than the one based on exchange–competition–command. Thus we can say that a sectoral system is composed of webs of relationships among heterogeneous agents with different beliefs, goals, competencies, and behavior, and that these relationships affect agents' actions. They are rather stable over time.

In summary, the types and structures of relationships and networks differ greatly from sectoral system to sectoral system, as a consequence of the features of the knowledge base, the relevant learning processes, the basic technologies, the characteristics of demand, the key links, and the dynamic complementarities. Again, let's provide some examples.

Again, the comparison of four quite different sectoral systems, such as chemicals, computers, semiconductors, and software, illustrates this point. In chemicals, the structure of the sectoral system has been centered around large firms, which have been the major source of innovation over a long period of time. Large R&D expenditures, economies of scale and scope (Chandler 1990), cumulativeness of technical advance, and commercialization capabilities have given these firms major innovative and commercial advantages (Arora, et al. 1998). With the diffusion of the synthetic dyestuff model, firms scaled up their R&D departments and the role of universities increased. The introduction of polymer chemistry (1920s) affected the structure of the industry because knowledge about the characteristics of different market segments became important, so that firms had to develop extensive linkages with downstream markets. The other major change related to the development of chemical engineering and the concept of unit of operation led to an increasing division of labor between chemical companies and technology suppliers, with the rise of the specialized engineering firms (SEFs), which developed vertical links with chemical companies. In this period, university research continued to be important for the development of innovations, and links between universities and industry increased. In addition, advances in chemical disciplines and the separability of knowledge increased the transferability of chemical technologies. Thus, there has been a greater role of licensing also by large firms, which in turn increased knowledge diffusion.

In computers, the different stages of the evolution of the industry (related to different products) have been characterized by different actors and networks. Having been a typical Schumpeter Mark II sector for most of its history (until very recently), mainframe computers have always been dominated by large firms, with high cumulativeness of technical advance. In particular, during the 1960s and 1970s, mainframes were produced and integrated by vertically integrated firms, and IBM was the typical example. IBM was producing both components and systems and was active in the development, manufacturing, marketing, and distribution of large systems and of the key components. When minicomputers were introduced, the computers sector experienced the entry and growth of firms specialized in components or in systems (with the early years characterized by a Schumpeter Mark I pattern). The same holds for the early years of microcomputers. Later on, however, competition became characterized by groups of specialized firms related to different platforms. Each platform was characterized by divided technical leadership of several disintegrated firms. Innovation became decentralized, and the control over the direction by a single firm became very difficult. Recently, in computer networks, modularity and connectedness increased the role of networks of firms with local development and local feedbacks (Bresnahan and Greenstein 1999; Bresnahan and Malerba, 1999).

In semiconductors, the industry has been characterized by a quite different set of actors, ranging from merchant semiconductor manufacturers to vertically integrated producers. The types of actors have been quite different from period to period and from country to country during the evolution of the industry. New entrants and specialized producers were quite relevant in the United States, with entrants particularly high either early on in the history of the industry or during phases of technological discontinuities (and giving the industry a typical Schumpeter Mark I fashion in these periods of rapid and radical change). Large, vertically integrated producers were more common in Japan and Europe (Malerba 1985; Langlois and Steinmueller 1999). Thus, in these countries a Schumpeter Mark II mode characterized the industry. In semiconductors, other main actors have played a major role. The military was one of the major factors responsible for the growth of the American industry, compared to Europe and Japan, because it supported the entry of new firms and provided competent firms with a large and innovative demand. During the 1970s in Japan, MITI was a major factor in allowing the Japanese industry (composed of large producers) to close the gap with American producers in some product ranges (such as memory devices).

In software, specialization of both global players and local producers is present. In addition, the changing knowledge base has created an evolving division of labor among users, "platform" developers, and specialized software vendors (Bresnahan and Greenstein 1998). The sectoral system of innovation in software, however, is incomplete without the addition of companies that utilize these platforms to deliver enterprise-critical applications. Many of these applications continue to be produced in-house by organizations using the tools provided as part of the platform or available from the development tools markets (Steinmueller 2004).

14.6 INSTITUTIONS

In all sectoral systems, institutions play a major role in affecting the rate of technological change, the organization of innovative activity, and performance. They may emerge either as a result of deliberated planned decision by firms or other organizations, or as the unpredicted consequence of agents' interaction.

Some institutions are sectoral, i.e. specific to a sector, while others are national. The relationship between national institutions and sectoral systems is quite important in most sectors. National institutions have different effects on sectors. For example, the patent system, property rights, or antitrust regulations have different effects as a consequence of the different features of the systems, as surveys and empirical analyses have shown (see for example Levin et al. 1987). However, the same institution may take different features in different countries, and thus may affect the same sectoral system differently. For example, the well-known diversity between the first-to-invent and the first-to-file rules in the patent systems in the United States and in Japan had major consequences on the behavior of firms in these two countries. Often, the characteristics of national institutions favor specific sectors that fit better the specificities of the national institutions. Thus, in certain cases, some sectoral systems become predominant in a country because the existing institutions of that country provide an environment more suitable for certain types of sectors and not for others. For example, in France, sectors related to public demand have grown considerably (Chesnais 1993). In other cases, national institutions may constrain development or innovation in specific sectors, or mismatches between national and sectoral institutions and agents may take place. The examples of the different types of interaction between national institutions and sectoral evolution in various advanced countries in Dosi and Malerba (1996) are cases in point.

The relationship between national institutions and sectoral systems is not always one-way, as it is in the case of the effects of national institutions on sectoral variables. Sometimes, the direction is reversed, and goes from the sectoral to the national level. In fact, it may occur that the institutions of a sector, which are extremely important for a country in terms of employment, competitiveness, or strategic relevance, end up emerging as national, thus becoming relevant for other sectors. But in the process of becoming national, they may change some of their original distinctive features.

Again, major differences emerge across sectors, as in the case of pharmaceuticals, software, machine tools, and telecommunications, for example. In pharmaceuticals, national health systems and regulations have played a major role in affecting the direction of technical change, in some cases even blocking or retarding innovation. In addition, patents have played a major role in the appropriability of the returns from innovations. In software, standards and standard setting organizations are important, and IPR play a major role in strengthening appropriability. However, the emerging open source movement aims to create a new segment of the software industry which is characterized by new distribution methods and by cooperative production activities based on voluntary association. This has reduced the possibility of maintaining proprietary control over data structure, thus inducing entry and more competition (Steinmueller 2004). In machine tools, internal and regional labor markets and local institutions (e.g. local banks) have played a major role in influencing international advantages of specific areas. Trust-based, close relationships at the regional level have over a long time ensured a sufficient financing of the innovation and of the expansion plans of family businesses in Germany and Italy (Wengel and Shapira 2004). Finally, in telecommunications, the roles of regulation, liberalization/privatization, and standards have been of major importance in the organization and performance of the sector. As discussed in Dalum and Villumsen (2001), liberalization and privatization have had major effects on the behavior and performance of incumbents and have transformed the structure of the industry. An example of the role of institutions is given by GSM in Europe.

14.7 THE DYNAMICS AND TRANSFORMATION OF SECTORAL SYSTEMS

As mentioned above, at the base of the dynamics and transformation of sectoral systems lies the interplay among evolutionary processes (such as variety creation, replication, and selection) that differ from sector to sector (Nelson 1995; Metcalfe

1998). Processes of variety creation may refer to products, technologies, firms, and institutions, as well as firm strategies and behavior and could take place through entry, R&D, innovation, and so on (Cohen and Malerba 2002). Sectoral systems differ extensively in the processes of variety creation and of heterogeneity among agents. The creation of new agents—both new firms and non-firm organizations—is particularly important for the dynamics of sectoral systems. As examined by Audretsch (1996) and Geroski (1995), among others, the role of new firms differs drastically from sector to sector (in terms of entry rates, composition, and origin), and thus has quite different effects on the features of sectoral systems and their degree of change. Sectoral differences in the level and type of entry seem to be closely related to differences in the knowledge base; level, diffusion and distribution of competences; the presence of non-firm organizations (such as universities and venture capital); and the working of sectoral institutions (such as regulations or labor markets) (Audretsch 1996; Malerba and Orsenigo 1999; McKelvey 1997; Geroski 1995). Processes of selection play the key role of reducing heterogeneity among firms and may drive out inefficient or less progressive firms. They may refer to products, activities, technologies, and so on. In addition to market selection, in several sectoral systems non-market selection processes are at work, as in the cases of the involvement of the military, the health system, and so on. In general, selection affects the growth and decline of the various groups of agents and the range of viable behaviors and organizations. Selection may greatly differ across sectoral systems in terms of intensity and frequency. Theoretical work (see Metcalfe 1998) and empirical work on "competence destroying" innovation, industrial dynamics, firms' entry and exit, and mergers and acquisitions have shed light on several aspects of selection.

Changes in sectoral systems are the result of coevolutionary processes of their various elements, involving knowledge, technology, actors, and institutions. Nelson (1994) and Metcalfe (1998) have discussed these processes at the general level by focusing on the interaction between technology, industrial structure, institutions, and demand. These processes are sector-specific and often path-dependent. Here, local learning, interactions among agents, and networks may generate increasing returns and irreversibilities that may lock sectoral systems into inferior technologies.[3] In addition, the interaction between knowledge, technology firms, and institutions are also shaped by country-specific factors.

In general, one could say that changes in the knowledge base and in the relevant learning processes of firms induce deep transformations in the behavior and structure of the agents and in their relationships between one another. Overall market competition and market structure depend on the strategies and fortunes of individual companies, which are linked to different national contexts or to the international scene. Firms have diverse reactions in order to try to increase their fit and to survive in their particular environment. These environments keep changing, not least due to innovations and choices made by all the constituent competitors: some of these environments are national, others increasingly international.

Over the past decades, computers have had major coevolutionary processes, quite different from one another. In mainframes, coevolution has been characterized by large systems requiring user–producer relationships, centralization of user information systems, and extensive sales and service efforts by large vendors. Market structure was highly concentrated and suppliers were vertically integrated. A dominant design (IBM/360) emerged in the growth phase of the segment and a market leader (IBM) dominated the industry early on, with a coordinating role over the platform and the ability to steer the direction of technical change. The US government played a role in early support for technological exploration and was a major buyer of early computers. In minicomputers and microcomputers, coevolution has been characterized by technological change focused on dedicated applications in the case of minicomputers or on systems that increased ease of use and a lower price/ performance ratio (in the case of microcomputers). The relationships with customers have required much less post-sales effort and service. Market structure was characterized by high entry early on, and then by increasing concentration in platforms in both minicomputers and microcomputers. In computer networks, connectivity and compatibility led to modular, open, and multiform client/server platforms. Technical change follows a variety of directions with an upsurge in the number of potential technologies associated with the relevant platforms. Interdependencies and externalities have increased. Divided technical leadership has emerged, in that no single firm has been able to govern change and coordinate platform standards.

This example is quite different from coevolution in other sectoral systems. In pharmaceuticals, the nature of the process of drug discovery (discussed in Section 14.4) had important consequences on the patterns of competition and on market structure. Until the molecular biology revolution, dominant firms persisted as leaders. The molecular biology revolution induced deep changes in the incentive structures within firms and universities, with the advent of university spin-offs and the emergence of the specialized new biotechnology firms. In this process of adaptation and change, different dynamic processes led to different patterns of competition and performances (McKelvey, Orsenigo, and Pammolli 2004). In telecom equipment and services, the early separation of the radio spectrum for use in one-way broadcasting and two-way telephony gave rise to an oligopolistic structure that persisted for quite a long time (Dalum and Villumsen 2001). The convergence within ICT and between ICT and broadcasting-audio-visual and the emergence of the Internet originated a more fluid market structure with a lot of different actors with different specializations and capabilities, and new types of users. This in turn greatly expanded the boundaries of the sector by creating new segments and new opportunities, and also by creating national differences in the organization of innovation. Moreover, the emergence of the Internet has generated more pressure in favor of open standards and has led to the rise of new actors (such as ISP and content providers). In software, since the early 1980s, the spread of networked computing, embedded software, the Internet, the development of open-system architecture and open source, and the growth of web-based network computing has led to the decline of large computer producers as developers of integrated hardware and software systems and to the emergence of a lot of specialized software companies. Also, software distribution has greatly changed, from licensing agreements in the early days, to the rise of independent software vendors, to price discounts for package software, and, with the diffusion of the CD-ROM and the Internet, to shareware and freeware (this last one particularly relevant with Linux) (D'Adderio 2001). In machine tools, a major driving force for coevolutionary processes is the demand from advanced customer sectors, namely the automotive, aeronautics, and defense industries, and the increasing use of electronic devices.

The emergence of new clusters that span several sectors, such as internet–software–telecom, biotechnology–pharmaceuticals, and new materials, is one of the most relevant current transformation processes in sectoral systems. Here a great role is played by the integration and fusion of previously separated knowledge and technologies and by the new relations involving users, consumers, firms with different specializations and competences, and non-firm organizations and institutions grounded in previously separated sectors.

14.8 POLICY IMPLICATIONS

A sectoral system of innovation approach provides a design for innovation and technology policies. Within a system of innovation framework, identifying deficiencies in the functioning of a system is the same as identifying those systemic dimensions that are missing or inappropriate or not working and which lead to a "problem" in terms of comparative performance. When we know the causes behind a certain "problem"—for example, weak technological transfer between universities and industry—we have identified a "system failure." Not until they know the character of the system failure can policy makers know whether to influence or to change organizations or institutions or the interactions between them. Therefore, an identification of a problem should be supplemented with an analysis of its causes as part of the analytical basis for the design of an innovation policy. Benchmarking is not enough.

Thus a sectoral system approach provides the identification of "system failures" and the related variables which should be policy targets. Sectoral analyses should focus on systemic features in relation to knowledge and boundaries, heterogeneity of actors and networks, institutions and transformation (through coevolutionary processes). As a consequence, the understanding of these dimensions becomes a prerequisite for any policy addressed to a specific sector.

Given the major differences among sectoral systems examined in this chapter, the impact of general or horizontal policies may drastically differ across sectors, because the channels and ways policies have their effects differ from sector to sector. For example, cooperation and networks or non-firm organizations and institutions could have different relevance in different sectors. Therefore, policies affecting networks or non-firm organizations, such as transfer agencies, have to take these differences into account.

In addition, a sectoral system approach emphasizes that, for fostering innovation and diffusion in a sector, technology and innovation policies may not be enough. A wide range of other policies may be necessary. Innovation and technology policy could be supplemented by other types of policies, such as science policy, industrial policy, policies related to standards and IPR, and competition policy. This point highlights the importance of the interdependencies, links, and feedbacks among all of these policies, and their combined effects on the dynamics and transformation of sectors.

Relatedly, a sectoral system approach emphasizes that policy makers being within a variety of networks are an active internal (part) of sectoral systems at different levels. In fact, the policy makers intervene actively in knowledge creation, IPR, corporate governance rules, technology transfer, financial institutions, skill formation, and public procurement. As a consequence, they have to develop advanced competences and create an institutional setting in order to be effective and consistent at the various different levels.

Finally, policy has to consider the coexistence of different geographical dimensions of sectoral systems. Developments in the local, national, regional, and global levels influence the articulation of technological capabilities. Policies that focus on only one level are likely to miss constraints or opportunities that are influential in the innovative behavior of individual organizations.

The emphasis on the diversity of sectoral systems highlights also different policy measures for different sectors. In fact, policy needs are closely related to the problems faced by the various actors operating in the sectoral contexts and to the sectoral specificity of knowledge, boundaries, actors, and networks.

In sum, traditional innovation policies have been formulated as providing public resources for R&D and changing the incentives for firms to innovate. Tax breaks for R&D, innovation subsidies, and patents are typical examples of these policies. A sectoral system perspective does not deny the significance of this approach. It recognizes, however, that the effects may run rapidly into diminishing returns. To offset this, it is necessary that innovation opportunities be enhanced. Improving the organization of an innovation system within a sector is an almost certain route to improving the complementary payoffs from public and private R&D. The sectoral perspective provides a tool for policy makers to comprehend the differences in innovation systems and for identifying the specific actors that should be influenced by policy. The quid pro quo, however, is that policy makers need to invest much more effort in understanding the idiosyncrasies of the specific sectors that they use to channel the influence of policy (Edquist et al. 2004).

14.9   THE CHALLENGES AHEAD

This chapter has claimed that innovation greatly differs across sectors in terms of sources, actors, features, boundaries and organization. It has proposed an integrated and comparative way to look at sectors based on the sectoral systems framework.

Some remarks have to be advanced here in way of conclusion. The discussion of sectoral systems has shown that there could be several levels of sectoral aggregation, and that the choice of one depends on the goal of the analysis. While the discussion here has been very broad in terms of sectors in order to emphasize linkages, interdependencies, and transformation, for different research goals the level of disaggregation could be much higher, at the level of product groups. Still, we may talk about systems of innovation in this respect.

Geographical boundaries are a key dimension to be considered in analyses of sectoral systems. National boundaries are not always the most appropriate ones for an examination of structure, agents, and coevolution. Often, sectoral systems are highly localized and frequently define the specialization of local areas (as in the case of machinery, some traditional industries, and even information technology). For example, machinery is concentrated in specialized regional areas. Similarly, sectoral specialization and local agglomeration have overlapped in Route 128 (for minicomputers) and in Silicon Valley (for personal computers, software, and microelectronics) (Saxenian 1994). Moreover, in the context of transnational economic integration, the sector may matter as much or more than the national system.

Differences across countries in sectoral systems have been relevant and have affected countries' international performance. In general, one could claim that those countries that did not have effective sectoral system characteristics did not perform well in international markets. The same holds for those countries that tried to replicate the success of world leaders by mimicking some of the features of the sectoral systems of the leading countries, without having the appropriate set of actors, linkages, and institutions. By contrast, those countries that have tried to specialize in subsectors with products, knowledge, and institutional requirements that match their specific institutional framework have been successful (Coriat, Malerba, and Montobbio 2004).

Finally, this chapter has tried to show how relevant a sectoral system approach is for an understanding of the features, determinants, and effects of innovation, in terms of research and policy. The policy aspect has been discussed in the previous section and will not be repeated here, but research on sectoral systems may prove very fruitful and has to move along several lines of advancement.

(1) A sectoral system framework may allow for detailed analyses of innovation in sectors in terms of knowledge and learning processes, structure (where structure is seen here as a network of relationships), and institutions. In addition, a sectoral system approach provides a way to examine the dynamics of sectors due to innovation and technological change and the coevolutionary processes taking place among knowledge, technology, actors, and institutions. Different sectoral systems may be compared along similar dimensions (in order to try to identify similarities across sectors), and the same sectoral system may be examined across different countries (in order to focus on the interplay between sectoral and national variables).

(2) The specific mechanisms, causal relationships, and interactions among the variables composing a sectoral system have to be studied in great depth both empirically and theoretically. This requires the development of quantitative analyses, econometric studies, and formal models. Driven by empirical analyses, appreciative and formal theoretical work has to be carried out regarding the basic relationships among the elements of a sectoral system, the emergence and persistence of networks, the basic processes of variety creation and selection, and coevolution. Here, both theoretical models of industry dynamics and history-friendly models can be useful. In the best evolutionary and innovation system traditions, this work should go hand in hand with, and be continuously confronted by, empirical work.

(3) Research should focus on some key variables that are still rather unexplored. In particular:

- the extent and features of within-sector firms heterogeneity and the related processes of variety creation and selection;
- demand, in terms of emergence, structure, and role in the innovation process;
- networks, in terms of emergence, composition, structure, and evolution;
- coevolution of the various elements of a sectoral system;
- institutions, both in terms of emergence and role of sectoral institutions and in terms of the sectoral effects of national institutions.

(4) Taxonomies of sectoral systems have to be constructed. Here, comparative work is particularly relevant. These taxonomies should group sectoral systems in terms of elements, structure, and dynamics, so that regularities

may be identified among sectors. Pavitt's taxonomy (Pavitt 1984) and the Schumpeter Mark I and Schumpeter Mark II distinction could be useful starting points.

(5) Analyses of the relationship between the presence and strength of elements of sectoral systems and the international performance of countries have to be developed (see e.g. Coriat, Malerba, and Montobbio 2004).

In conclusion, as stressed above, a full understanding of the determinants, features, and effects of innovation in sectoral systems requires the integration of various types of complementary analyses: descriptive, quantitative, econometric, and theoretical.

NOTES

1. In fact sectoral systems often have more than one technology, while the same technology (as in the general purpose technology case) may be used by many different sectors.
2. Similarly, in addition to firm and non-firms organizations, also agents at lower and higher levels of aggregation such as individuals or consortia of firms may be the key actors in a sectoral system.
3. For example, sectors with competing technologies such as nuclear energy (Cowan 1990), cars (and their power sources—Foreman-Peck 1996), metallurgy (ferrous casting—Foray and Grubler 1990) and multimedia (VCR—Cusumano et al. 1992) show interesting examples of path-dependent processes.

REFERENCES

ANDERSEN, B., METCALFE, J. S., and TETHER, B. S. (2002), "Distributed Innovation Systems and Instituted Economic Processes," Working Paper ESSY, http://www.cespri.it/ ricerca/ metcalfeetal.PDF

ARORA, A., LANDAU, R., and ROSENBERG, N. (eds.) (1998), Chemicals and Long-Term Growth: Insights from the Chemical Industry, New York: John Wiley.

AUDRETSCH, D. (1996), Innovation and Industry Evolution, Cambridge, Mass.: MIT Press.

BRESCHI, S., and MALERBA, F. (1997), "Sectoral Systems of Innovation," in Edquist 1997: 130–55.

BRESNAHAN, T., and GREENSTEIN, S. (1998), "Technical Progress in Computing and the Uses of Computers," Brooking Papers on Economic activity: Microeconomics 1: 1–78.

————— (1999), "Technological Competition and the Structure of the Computer Industry," Journal of Industrial Economics 47: 1–40.

—— and MALERBA, F. (1999), "Industrial Dynamics and the Evolution of Firms' and Nations' Competitive Capabilities in the World Computer Industry," in Mowery and Nelson 1999: 79–132.

CALLON, M. (1992), "The Dynamics of Techno-Economic Networks," in R. Coombs, P. Saviotti, and V. Walsh (eds.), *Technical Change and Company Strategies*, London: Academy Press.

* CARLSSON, B., and STANKIEWITZ, R. (1995), "On the Nature, Function and Composition of Technological Systems," in B. Carlsson (ed.), *Technological Systems and Economic Performance*, Dordrecht: Kluwer.

CASPER, S., and KETTLER, H. (2002), "National Institutional Frameworks and the Hybridization of Entrepreneurial Business Models: The German and UK Biotechnology Sectors," Working Paper ESSY, http://www.cespri.it/ricerca/ESSY.htm

—— and SOSKICE, D. (2004), "Patterns of Innovation and Varieties of Capitalism: Explaining the Development of High-Technology Entrepreneurialism in Europe," in F. Malerba (ed.), *Sectoral Systems of Innovation. Concept, Issues and Analyses of Six Major Sectors in Europe*, Cambridge: Cambridge University Press.

CESARONI, F., GAMBARDELLA, A., GARCIA-FONTES, W., and MARIANI, M. (2004), "The Chemical Sectoral System: Firms, Markets, Institutions and the Processes of Knowledge Creation and Diffusion," in Malerba 2004.

CHANDLER, A. (1990), *Scale and Scope: the Dynamics of Industrial Capitalism*, Cambridge, Mass.: Bellknap Press.

CHESNAIS, F. (1993), "The French National System of Innovation," in Nelson 1993: 192–229.

CHRISTENSEN, C. M., and ROSENBLOOM, R. S. (1995), "Explaining the Attacker's Advantage: Technological Paradigms, Organizational Dynamics, and the Value Network," *Research Policy* 24: 233–57.

* CORIAT, B., MALERBA, F., and MONTOBBIO, F. (2004), "The International Performance of European Sectoral Systems," in Malerba 2004.

COHEN, W., GOTO, A., NAGATA, A., NELSON, R., and WALSH, J. (2002), "R&D Spillovers, Patents and the Incentives to Innovate in Japan and the United States," *Research Policy* 31: 1349–67.

—— and LEVINTHAL, D. (1989), "Innovation and Learning: The Two Faces of R&D," *Economic Journal* 99: 569–96.

—— and MALERBA, F. (2002), "Is the Tendency to Variation a Chief Source of Progress?" *Industrial and Corporate Change* 10: 587–608.

COOKE P., URANGE, M. G., and EXTEBARRIA, G. (1997), "Regional Innovation Systems: Institutional and Organizational Dimensions," *Research Policy*, 26: 475–91.

CORIAT, B., and WEINSTEIN, O. (2004), "The Organization of and the Dynamics of Innovation: A 'Sectoral' View," in Malerba 2004.

COWAN, R. (1990), "Nuclear Power Reactors: A Study of Technological Lock-In," *Journal of Economic History* 50: 541–66.

—— DAVID, P., and FORAY, D. (2000), "The Explicit Economics of Codification and the Diffusion of Knowledge," *Industrial and Corporate Change* 9: 211–53.

CUSUMANO, M. (1991), *Japan Software Factories*, Oxford: Oxford University Press.

—— MYLONADIS, Y., and ROSENBLOOM, R. (1992), "Strategic Maneuvering and Mass-Market Dynamics: The Triumph of VHS over Beta," *Business History Review* 66: 51–94.

D'ADDERIO, L. (2001), *Inside the Virtual Product: The Influence of Integrated Software Systems on Organisational Knowledge Dynamics*, SPRU, Brighton: University of Sussex.

—— (2002), "The Diffusion of Integrated Software Solutions: Trends and Challenges," Working Paper ESSY, http://www.cespri.it/ ricerca/ESSY.htm

* Asterisked items are suggestions for further reading.

DAHMEN, E. (1989), "Development Blocks in Industrial Economics," in B. Carlsson (ed.), *Industrial Dynamics*, Boston: Kluwer.

DALUM, B. (2002), "Data Communication—the Satellite and TV Subsystems," Working Paper ESSY, http://www.cespri.it/ ricerca/ESSY.htm

—— and VILLUMSEN, G. (2001) "Fixed Data Communications—Challenges For Europe," Working Paper ESSY, http://www.cespri.it/ ricerca/ESSY.htm

DOSI, G. (1988), "Sources, Procedures and Microeconomic Effects of Innovation," *Journal of Economic Literature* 26: 1120–71.

—— (1997), "Opportunities, Incentives and the Collective Patterns of Technological Change," *Economic Journal* 107: 1530–47.

—— and MALERBA, F. (1996), *Organization and Strategy in the Evolution of the Enterprise*, London: MacMillan.

—— MARENGO, L., and FAGIOLO, G. (1998), "Learning Evolutionary Environments," IIASA Working Paper, WP-96-124.

DUBOCAGE, E. (2002), "The Financing of Innovation by Venture Capital in Europe and in the USA: A Comparative and Sectoral Approach," Working Paper ESSY, http://www.cespri.it/ ricerca/ESSY.htm

EDQUIST, C. (ed.) (1997), *Systems of innovation*, London: Pinter.

—— (2004), "Telecommunication Equipment and Services," in Malerba 2004.

*—— MALERBA, F., METCALFE, S., MONTOBBIO, F., and STEINMUELLER, E. (2004), "Sectoral Systems: Implications for European Technology Policy", in Malerba 2004.

* ESSY (2002), "European Sectoral System and European Growth and Competitiveness, European Targeted Socio Economics Research," in ESSY, http://www.cespri.it/ ricerca/ESSY.htm

FORAY, D., and GRUBLER, A. (1990), "Morphological Analysis, Diffusion and Lock-out of Technologies: Ferrous Casting in France and the FRG," *Research Policy* 19: 535–50.

FOREMAN-PECK, J. (1996), "Technological Lock-in and the Power Source for the Motor Car," University of Oxford, Discussion Paper in Economics.

FREEMAN, C. (1968), "Chemical Process Plant: Innovation and the World Market," *National Institute Economic Review* 45: 29–51.

—— (1982), *The Economics of Industrial Innovation*, London: Pinter.

—— (1987), *Technology Policy and Economic Performance: Lessons from Japan*, London: Pinter.

GAMBARDELLA, A. (1995), *Science and Innovation: The U.S. Pharmaceutical Industry during the 1980s*, Cambridge: Cambridge University Press.

GEROSKI, P. (1995), "What do we Know about Entry?" *International Journal of Industrial Organization* 4: 421–40.

GORT, M., and KLEPPER, S. (1982), "Time Paths in the Diffusion of Product Innovations," *Economic Journal* 92: 630–53.

HENDERSON, R., and CLARK, K. (1990), "Architectural Innovation," *Administrative Science Quarterly* 35: 9–30.

—— ORSENIGO, L., and PISANO, G. (1999), "The Pharmaceutical Industry and the Revolution in Molecular Biology," in Mowery and Nelson 1999: 267–312.

HUGHES, T. P. (1984), "The Evolution of Large Technological Systems," in W. Bijker, T. Hughes, and T. Pinch (eds.), *The Social Construction of Technological Systems*, Cambridge, Mass.: MIT Press.

KLEPPER, S. (1996), "Entry, Exit, Growth and Innovation over the Product Life Cycle," *American Economic Review* 86: 562–83.

LANGLOIS, R., and STEINMUELLER, E. (1999), "The Evolution of Competitive Advantage in the Worldwide Semiconductor Industry," in Mowery and Nelson 1999: 19–78.

LEVIN, R., KLEVORICK, A., NELSON, R., and WINTER, S. (1987), "Appropriating the Returns from Industrial R&D," *Brookings Papers on Economic Activity* 3: 783–831.

LUNDVALL, B. Å. (1993), *National Systems of Innovation*, London: Pinter.

—— and JOHNSON, B. (1994), "The Learning Economy," *Journal of Industry Studies* 1: 23–42.

McKELVEY, M. (1997), "Using Evolutionary Theory to Define Systems of Innovation," in Edquist 1997: 200–22.

—— ORSENIGO, L., and PAMMOLLI, F. (2004), "Pharmaceuticals as a Sectoral Innovation System," in Malerba 2004.

—— ALM, H., and RICCABONI, M. (2002), *Does Co-location matter? Knowledge collaboration in the Swedish Biotechnology-pharmaceutical Sector, Working Paper ESSY*, http://www.ce-spri.it/ ricerca/ESSY.htm

MALERBA, F. (1985), *The Semiconductor Business*, Madison, Wis.: University of Wisconsin Press.

—— (1992), "Learning by Firms and Incremental Technical Change," *Economic Journal* 102: 845–59.

—— (2002), "Sectoral Systems of Innovation and Production," *Research Policy* 31: 247–64.

*—— (ed.) (2004), *Sectoral Systems of Innovation: Concept, Issues and Analyses of Six Major Sectors in Europe*, Cambridge: Cambridge University Press.

—— and ORSENIGO, L. (1996), "Schumpeterian Patterns of Innovation," *Cambridge Journal of Economics* 19: 47–65.

*—— —— (1997), "Technological Regimes and Sectoral Patterns of Innovative Activities," *Industrial and Corporate Change* 6: 83–117.

—— —— (1999), "Technological Entry, Exit and Survival: An Empirical Analysis of Patent Data," *Research Policy* 28: 643–60.

—— —— (2000), "Knowledge, Innovative Activities and Industry Evolution," *Industrial and Corporate Change* 9: 289–314.

—— and TORRISI, S. (1996), "The Dynamics of Market Structure and Innovation in the Western European Software Industry," in Mowery 1996: 165–96.

* MARSILI, O. (2001), *The Anatomy and Evolution of Industries: Technological Change and Industry Dynamics*, Cheltenham: Edward Elgar.

—— and VERSPAGEN, B. (2002), "Technology and the Dynamics of Industrial Structures: An Empirical Mapping of Dutch Manufacturing," *Industrial and Corporate Change* 11(4): 791–815.

MAZZOLENI, R. (1999), "Innovation in the Machine Tools Industry: A Historical Perspective of the Dynamics of Comparative Advantage," in Mowery and Nelson 1999: 169–216.

METCALFE, S. (1998), *Evolutionary Economics and Creative Destruction*, London: Routledge.

MOWERY, D. (ed.) (1996), *The International Computer Software Industry: A Comparative Study of Industry Evolution and Structure*, Oxford: Oxford University Press.

*—— and NELSON, R. (1999), *The Sources of Industrial Leadership*, Cambridge: Cambridge University Press.

NELSON, R. (1993), *National Innovation Systems: A Comparative Study*, Oxford: Oxford University Press.

*——(1994), "The Coevolution of Technology, Industrial Structure and Supporting Institutions," *Industrial and Corporate Change* 3: 47–64.

NELSON, R. (1995), "Recent Evolutionary Theorizing about Economic Change," *Journal of Economic Literature* 33: 48–90.

—— and ROSENBERG, N. (1993), "Technical Innovation and National Systems," in R. Nelson (ed.), *National Innovation Systems*, Oxford: Oxford University Press, 3–22.

—— and WINTER, S. (1982), *An Evolutionary Theory of Economic Change*, Cambridge, Mass.: Belknapp Press.

OWEN-SMITH, J., RICCABONI, M., PAMMOLLI, F., and POWELL, W. W. (2002), "A Comparison of US and European University–Industry Relations in the Life Sciences," Working Paper ESSY, http://www.cespri.it/ricerca/ESSY.htm

*PAVITT, K. (1984), "Sectoral Patterns of Technical Change: Towards a Taxonomy and a Theory," *Research Policy* 13: 343–73.

PACE (1996) (Policy, Appropriability and Competitiveness of European Enterprises), Brussels: European Commission.

RIVAUD-DANSET, D. (2001), "The Financing of Innovation and the Venture Capital, the National Financial and Sectoral Systems," Working Paper ESSY. http://www.cespri.it/ricerca/ESSY.htm

ROBSON, M., TOWNSEND, J., and PAVITT, K. (1988), "Sectoral Patterns of Production and Use of Innovation in the U.K.: 1943–1983," *Research Policy* 17: 1–14.

ROSENBERG, N. (1982), *Inside the Black Box*, Cambridge: Cambridge University Press.

——(1998), "Technological Change in the Chemicals: the Role of University–Industry Relationships," in Arora, Landau, and Rosenberg 1998: 193–230.

SAXENIAN, A. (1994), *Regional Advantages*, Cambridge, Mass.: Harvard University Press.

SCHERER, M. (1982), "Interindustry Technological Flows in the U.S.," *Research Policy* 11: 227–46.

STEINMUELLER, W. E. (2002) "Embedded Software: European Markets and Capabilities," Working Paper ESSY, http://www.cespri.it/ ricerca/ESSY.htm

——(2004), "The Software Sectoral Innovation System," in Malerba 2004.

TEECE, D., and PISANO, G. (1994), "The Dynamic Capabilities of Firms: An Introduction," *Industrial and Corporate Change*, 3: 537–56.

TEUBAL, M., YINNON, T., and ZUSCOVITCH, E. (1991), "Networks and market creation," *Research Policy* 20: 381–92.

TIROLE, J. (1988), *The Theory of Industrial Organization*, Cambridge, Mass.: MIT Press.

TORRISI, S. (1998) *Industrial Organisation and Innovation: An International Study of the Software Industry*, Cheltenham: Edward Elgar.

TUSHMAN, M. L., and ANDERSON, P. (1986), "Technological Discontinuities and Organizational Environments," *Administrative Science Quarterly* 14: 311–47.

UTTERBACK, J. (1994), *Mastering the Dynamics of Innovation*, Boston: Harvard Business School Press.

VON HIPPEL, E. (1988), *The Sources of Innovation*, Oxford: Oxford University Press.

WENGEL, J., and SHAPIRA, P., (2004), "Machine Tools: The Remaking of a Traditional Sectoral Innovation System?" in Malerba 2004.

WINTER, S. (1984), "Schumpeterian Competition in Alternative Technological Regimes," *Journal of Economic Behaviour and Organisation* 5: 287–320.

——(1987), "Knowledge and Competence as Strategic Assets," in D. J. Teece (ed.), *The Competitive Challenge: Strategies for Industrial Innovation and Renewal*, Cambridge, Mass.: Ballinger, 159–84.

CHAPTER 15

INNOVATION IN "LOW-TECH" INDUSTRIES

NICK VON TUNZELMANN

VIRGINIA ACHA

15.1 INTRODUCTION

THE title of this chapter is inherently paradoxical—low-tech industries are not supposed to be characterized by any significant amount of innovation *ex definitione*. We intend to resolve this conundrum by arguing that there are few—if any— industries in present-day advanced countries which conform to the general under- standing of what constitutes a "low-tech" industry. In our view, this is more than just a matter of semantics, and it is crucial for understanding where the comparative advantages of countries at varying levels of development may lie. We believe that a policy obsession with purported "gaps" in "high-tech" industries has distracted the attention of both policy makers and academics away from making more positive efforts to develop and sustain development in other sectoral directions which some countries might find more viable. In the OECD, high-tech industries as the OECD itself defines them account for only about 3 per cent of value-added (Hirsch- Kreinsen et al. 2003), rising to 8.5 per cent if medium-high-tech industries like motor vehicles are included (OECD 2003), so even if they could be expanded the impact on GDP would be quite small. Governments need to give more thought to the activities which generate most of the output and employment of their countries and the best targets for "dynamic comparative advantage" for growth.

In this chapter we will consider not just traditional "low-tech" industries but also those classified by the OECD as "medium-tech." We combine them hereafter as "low- and medium-tech" (LMT) industries. Our reason for banding the two together is that both are being driven by similar factors which somewhat distinguish them from the high-tech industries. These are frequently "mature" industries, where technologies and market conditions may change more slowly. They can include non-manufacturing activities, such as exploration for new resources in the oil and gas industry. We naturally avoid much discussion of services as being the subject of the following chapter, but this ignores the growing interrelationships of service and production/manufacturing activities in many relevant areas, so we include some consideration of the services sector.

Section 15.2 untangles the relationship between the technologies and markets which comprise an "industry"—effectively we reject OECD-type classifications in favor of alternative sectoral taxonomies. This permits a more constructive analysis of the key drivers of change at the industry/sector level as between demand and supply factors in Section 15.3. Section 15.4 elucidates the roles of firm strategies and structures in the LMT industries. The implications of this discussion for the evolution of industry structure, especially entry, are considered in Section 15.5, and illustrations from some "low-tech" industries are given in Section 15.6. Section 15.7 provides implications for government policy, which we believe requires radical rethinking, and Section 15.8 concludes with a call for revising the academic agenda.

15.2 THE "TECHNOLOGY PROFILE" OF INDUSTRIAL SECTORS

Sectors are generally taken to be identifiably similar aggregations of productive activities. Conventionally, sectors of all types were supposed to be recognizably different from one another not only in the goods and services they produced but also in the technologies and processes they used to produce them. However, the boundaries have blurred over historical time in both dimensions. Technologies originally developed for one set of products spill over into use in the production or "architecture" of other sets of products. Moreover, new technologies more often tend to supplement and complement old technologies rather than replace them.

One simple consequence is that even "old" products can be produced by, or partly consist of, elements drawn from what had previously been a totally different set of activities. Equally, markets have become more blurred through the bundling of goods and services (e.g. sales of music products via the Internet).

As a result, conventional classifications of sectors as high- or low-tech (etc.), as long practised by the OECD, are becoming less and less useful for academic analysis, though their sway still prevails in government policy making (see Section 15.7 below). To be fair, the OECD (2003) is coming—rightly—to place greater emphasis on the "knowledge-intensity" of industries, based on criteria such as their use of capital inputs from R&D-intensive industries. This necessitates rethinking the kinds of taxonomies that help us to comprehend structural change (see Ch. 6 by Smith, this volume).

15.2.1 Technologies vs. Products

The conventional definition of the technological profile of industrial sectors put forward by the OECD claims to measure the direct plus indirect technology content of particular industries. The majority of manufacturing industries are defined mostly according to their product range but a good number have in common their technologies rather than their products, e.g. biotechnology. Whether, say, plant biotechnology is regarded as part of the biotechnology industry (technology-defined) or of agriculture (product-defined) makes a big difference to the inferences drawn.

Allowing for these difficulties, others then arise in connection with the measures of technology content. The key issues are outlined by Smith (Ch. 6 in this volume), and need not be reiterated here. However we would emphasize that a good part of the innovation activities in LMT industries may fall outside the Frascati definition of R&D (OECD 1994), for example oil and gas exploration (see Box 15.1). Knowledge search, identification and proof, rather than basic research, are likely to be of particular importance to innovation in the non-manufacturing activities of LMT industries. Most importantly, we have to ask what part of each "industry" we are characterizing as high- or low-tech when considering their growth potential.

Meliciani (2001) found that in the 1980s the ICT industries figured prominently among the fastest-growing industries in advanced industrial countries, but in the 1970s they did not, whereas many LMT industries were present among the fastest growing group at this time. For the years after 1994, the OECD high-tech sectors show first rapid growth then rapid contraction. Governments have been attracted towards the high-tech industries for their potential for growth and structural change. The benchmarks here are however confused by the definitional problems already highlighted—for instance Denmark appears to have a comparative

---

Box 15.1 Planetary science in the North Sea

There are many who believe that all of the science and technology needed to dig oil and gas out of the ground has been already discovered and deployed. Relatively low R&D intensity figures attest to the same; over the period 1995 to 1997, the top ten oil majors worldwide averaged an R&D intensity of 0.52 per cent. However the accounting category of R&D does not capture the full scope of investment into the search for novelty and the applications thereof in the upstream petroleum industry. Many exploration activities that involve, in the words of the Frascati Manual, an "appreciable element of novelty and the resolution of scientific and/or technological uncertainty" and "contribute to the stock of knowledge" are captured under different budgetary headings, notably exploration costs (Acha 2002). Mansfield (1969: 53) was also concerned that geological and geophysical exploration were excluded from the R&D definitions for precisely these reasons.

In fact, the annually growing numbers (in their hundreds per company) of technical and academic papers produced by the leading oil operating and service companies reflect a substantially developed research program for the evolution of the science and technologies underpinning this industry. This fact is better reflected in the number of Ph.D. qualified staff from these firms actively working on a better understanding of the composition and dynamics of this planet. This scientific endeavor in planetary science does not go unnoticed even outside the industry. NASA has enlisted the help of upstream oil industry companies in developing new technologies for drilling on Mars (Babaev 2000). Likewise space scientists are collaborating with the geologists and geophysicists who have been delineating marine space impact craters on Earth, including one off the UK whose only natural analog exists on Jupiter's ice moon, Europa.

---

advantage in low-growth sectors (such as processed foods), but its production lies towards the high-tech end of these low-growth sectors (like applying biotechnology to food processing).

15.2.2 Factor intensity

The LMT industries are usually regarded as providing many points of entry for developing countries, in view of their relative labor-intensity. Not all are of this nature, however. Some branches of such "low-tech" industries as food processing are highly capital-intensive (e.g. tobacco and many beverages), as are some branches of building materials (e.g. cement). Many more take on varying shades of labor- or capital-intensity depending on the economic environment in which they find themselves—the same industry may be capital-intensive in the USA and labor-intensive in China. Areas like software production have been favored points of entry into "high-tech" production in countries like India in the 1990s and beyond precisely because they are "labor-intensive" (though intensive in cheap skilled labor rather than cheap unskilled labor).

Peneder (2001) provides a tripartite classification of manufacturing industries, at the 3-digit level of disaggregation. One of his taxonomies rests on factor intensity (mainstream i.e. average; labor-intensive; capital-intensive; marketing-driven; technology-intensive), another on labour skills (low-skill; medium-skill blue-collar; medium-skill white-collar; high-skill), and the third on external service inputs (from knowledge-based services; from retail and advertising services; from transport services; and from other industries). Only one of the ninety-nine manufacturing industries he lists (aircraft and spacecraft) has the classic high-tech profile of being technology-intensive and predominantly using high-skill and knowledge-based services; conversely there are labor-intensive industries which utilize high skills (e.g. machine tools) and others utilizing knowledge-based service inputs (some branches of metallurgy). His classification underlines the great variety of observable combinations.

15.2.3 Pavitt's taxonomy

In contrast to classifying industrial sectors according to product range, Pavitt (1984) arranged them according to technology characteristics (see Tidd et al. 2001). The suggested categories are: "supplier-dominated," "scale-intensive," "information-intensive," "science-based," and "specialized-supplier" (for further discussion of this taxonomy see the Introduction and Chapter 6 by Smith, in this volume). A modified taxonomy of this kind that is more explicitly geared to the kinds of technological paradigms (chemical, mechanical, etc.) in different sectors is given by Marsili (2001). Both Pavitt and Marsili deliberately aimed at means of classification that brought together characteristics which certain groups of technologies appeared to share, even though they might pertain to different "sectors." Generally this taxonomy does a better job of explaining technological performance than factor content, but since this is its intention that may be not too surprising. However, again the LMT industries resist easy classification, precisely because many of them are not very distinctive or singular in technological terms.

An attempt to use a modified Pavitt taxonomy to analyze changes in world export shares is given in Table 15.1. The industries given special attention in this chapter are sprinkled across the first four categories, as already implied (e.g. food products, oil and gas, paper in agricultural products and raw materials; textiles and clothing, glass in traditional manufactures; vehicles, steel in scale-intensive; machinery for many of these in specialized suppliers). The inroads made into European and US export shares during the 1970–95 period by Japan and the Asian NICs in science-based and specialized suppliers are evident, though the share of the Asian NICs also increases in

Table 15.1 Market shares, 1970–1995 (percentage ratio of national exports to world exports)

		Agricultural products and raw materials	Traditional industries	Scale-intensive	Specialized suppliers	Science-based	Total
Europe	1970	24.1	57.0	55.7	61.2	48.6	44.6
	1995	31.6	40.1	47.3	47.6	33.8	39.6
	Change	+7.5	−16.9	−8.4	−13.6	−14.8	−5.0
USA	1970	13.1	7.4	14.5	22.3	29.5	14.8
	1995	11.0	6.7	10.3	13.7	17.9	11.8
	Change	−2.1	−0.7	−4.2	−8.6	−11.6	−3.0
Japan	1970	1.2	9.3	13.8	6.4	7.7	6.7
	1995	1.4	3.2	12.8	15.7	14.3	9.0
	Change	+0.2	−6.1	−1.0	+9.3	+6.6	+2.3
Asian NICs	1970	2.0	6.1	1.0	0.8	1.0	2.1
	1995	3.4	16.2	8.7	8.8	17.8	10.8
	Change	+1.4	+10.1	+7.7	+8.0	+16.8	+8.7

Source: Fagerberg et al. 1999: 12.

traditional manufactures. The US actually loses least, in percentage points, in traditional manufactures from 1970 to 1995, though its share in such trade was very low throughout the period compared with Europe. Nevertheless the table should warn us against making oversimplified statements about technological patterns of development.

15.2.4 Sutton's taxonomy

Yet another way of classifying industries can be derived from the work of Sutton (1991, 1998), who demonstrated that what firms are prepared to spend on marketing their products on the one hand, and on developing their technologies on the other, depend on factors that were partly under a firm's control and partly beyond it. The latter means that if the firms belong to a certain industry they are committed to a certain level of "sunk costs," e.g. for their production processes, regardless of what strategies they then adopt. The nature of technologies as they relate to products sets "bounds" to market concentration in a particular line of activity. The tire industry, for example, is very capital intensive, bound by production technologies that have not essentially changed since the first large-scale use of rubber; this promotes a global oligopolistic structure (Box 15.2 discusses the tire industry more fully). But what firms, in tires or elsewhere, choose to do as a result of strategic decision making—the endogenous element—would rest on the profitability of their strategic expenditures in the face of similarly strategic investment behavior by their competitors.

Box 15.2 What's so clever about a rubber tube?

The modern tire industry has its origins in the nineteenth century and, by and large, its development has mirrored that of the automotive industry. The world tire industry has evolved to address several markets which each have different characteristics, direct customers, and potential for growth. Over a century of development, it has segmented across a wide number of markets, including automotive, aerospace, bicycle, and locomotive tires. The industry is relatively highly concentrated; in 2000, the top ten global tire manufacturers accounted for 83 per cent of the sales of the top seventy-five tire manufacturers globally (www.tirebusiness.com/statistics).

Technology is applied by the tire manufacturers to reduce costs, to differentiate the product line and to focus on greater value-adding activities (Acha and Brusoni 2003). Facing a global market where it is more and more difficult to make a profit, the leading manufacturers are continuously focusing on reducing costs through reducing throughput and labor costs (including the long-awaited introduction of robotics), innovations in processing technologies and source product (a new polyurethane tire polymer), and in the method itself.

Beyond influencing the cost and ease of production, tire manufacturers have invested in research and technology to also help them to move away from the "commodity trap," where products can only compete on price. Product differentiation has occurred as companies offer a variety of tire profiles and even colours to match cars (the latter was led by Kumho, a Korean tire manufacturer that has successfully broken into the top ten manufacturers). Such differentiation is certainly much more than cosmetic; manufacturers have successfully incorporated new sensors into tire assemblies and developed run-flat tires. Leading tire manufacturers are now looking to move up the value chain by manufacturing entire tire assembly systems rather than simply supplying the tire itself.

Of course all of this could be achieved through the support of suppliers from the high-tech sectors of electronics and chemicals, yet the tire manufacturers themselves patent in these areas and lead developments as applied to their business. Patenting in technologies related to tires has increased, and most dramatically so for the tire manufacturers (Acha and Brusoni 2003). The top ten tire manufacturers worldwide had an average R&D intensity of 4 per cent in 2000. Moreover, their patents are applicable to a large number of International Patent Classification (IPC) subclasses (an average of forty-four subclasses over the period 1990 to 2000), indicating some complexity in the nature of these firms' knowledge bases. These firms have broadened their focus to address the crucial interface between their chemical knowledge base (i.e. rubber and other chemical compounds) on the one side, and mechanical engineering and electronics (e.g. sensors) on the other.

Drawing on this reasoning, Davies and Lyons (1996) classified country strengths on the basis of dividing industries into four categories: those with high R&D (i.e. technology) expenditures; those with high advertising (i.e. sales) expenditures; those with both; and those with neither. They showed that Western Europe was relatively strong in the second and third of these categories, i.e. those where both R&D and advertising were high (like pharmaceuticals) or advertising alone was high. The latter includes particularly some industries normally regarded as "low-tech," especially food processing (the industry selected by Sutton in his 1991 book to validate his theory). The Sutton approach and associated taxonomy can be especially useful for analyzing LMT industries, because supply (technology) is combined with demand (product) aspects in a rigorous way.

15.2.5 Summary

Attempts to appraise innovation through adopting conventional sectoral classifications can be quite misleading. Innovation is rapid in particular segments of both high-tech and LMT "sectors," even if more segments of the high-tech sectors display such rapid innovation (for evidence see Ch. 6 by Smith in this volume). It is admittedly possible to detach the high-tech segments from LMT "industries" and label them as new high-tech sectors, as was done for artificial fibers in textiles when they arose to compete with natural fibers in the early twentieth century (see Section 15.6), but the final products remain very similar so this looks specious. Approaches that blend the technology dimension and the product dimension, such as those by Sutton or by Peneder, appear to be not only more analytically satisfying but better able to account for observed empirical differences between countries and regions. As suggested below, they are also better able to account for dynamic paths of industrial evolution over historical time. Furthermore, they are in our view a more advisable platform for policy than simple OECD-style definitions of high- or low-tech. They need however to be supplemented by technology-oriented distinctions among sectors such as the Pavitt taxonomy to provide a better grasp of the nature of structural change and competitiveness.

15.3 THE KEY DRIVERS

The drivers of change as they affect low, medium, and high-tech sectors can be similarly envisaged from the side of the products or from the side of the technologies, which again give rise to significant differences in interpretation and understanding. Firms hold different interpretative "frames" (see below), and in LMT industries firm-level differences in the interpretation of demand drivers are particularly important because their well-established markets necessitate a broader variety of strategic choices for differentiation. Demands change sometimes slowly but sometimes rapidly and unpredictably, negating attempts to routinize operations and generating turbulence.

15.3.1 Demand Differentiation

15.3.1.1 *Quality Innovation*

An important way in which even older industries can bounce back is by producing for new markets. Producing the same type of goods for untouched regions can work for well-known brands, like Coca-Cola, but producing different types of the same categories of good ("product differentiation") is generally necessary for such resurgence. Within given markets, product growth is heavily determined by income elasticities of demand, i.e. the responsiveness of consumers to the particular product as their incomes rise. It is usually the case that low-tech industries face somewhat inelastic demands because many produce comparative "necessities," and as consumers attain higher income levels they have satisfied most of their needs for necessities. To stave off this "satiation of wants," producers in LMT industries have to find new products to attract the custom of higher earners. The availability of advanced technologies may be an important factor for innovation strategies in LMT firms through dictating the scope for such new products, and even then may not result in products that customers find attractive, as has been the case for genetically modified (GM) foodstuffs in some countries (see below).

15.3.1.2 *New Tastes*

In addition to quality upgrading, consumers may switch their demand patterns to goods which have new characteristics. While high-tech sectors may have greater innate capacity to spawn product innovations, LMT industries may be faced with a greater necessity to do so. Sectors such as food, energy generation and automobiles have to confront intense pressure from communities and from governments to produce safer and more environmentally friendly items. The same pressures extend to the processes by which they produce these outputs. Less essential but often more lucrative for LMT industries have been shifts in their product mixes to reflect the changing composition of consumers, for example the implications of demographic change (gender relations, ageing, etc.). These create new niches in which firms in low-tech industries can experience some resurgence. For example, a leading Japanese toy manufacturer, Bandai, recently launched a new doll product range ("Primopuel") targeted not at children but at "empty nesters" (women without children at home) by embedding sophisticated electronic sensors and programming within this "real baby" doll in an effort to develop a higher value market.

Both the potential for developing higher-quality products ("quality ladders") and new products therefore offset the seemingly inevitable maturation of older industries, and give rise to new production and trade patterns (Grossman and Helpman 1991). Indeed, over time, through adding value in processing and in new products, the declines in demand for the products of these industries have been less marked than might be expected. In the case of tires, companies have developed new model profiles (e.g. "fat" tires) to shift demand options from simple requirements (four tires per vehicle) to a series of options (different tires for different occasions), and to differentiated product lines of increasing quality (e.g. "run-flat" tire systems).

The challenge for innovation strategies becomes how and how well firms in LMT industries can alter their products and services and leverage the outcomes through introducing better products or new products, as explored in Sections 15.4 to 15.6 below.

15.3.2 New Technological Paradigms

15.3.2.1 *General Purpose Technologies and Learning in LMT Firms*

Certain new technologies can spill out of their industry of origin and be recruited by older industries. Key technologies often have the property of being able to become "pervasive," through their take-up in one industry after another (Freeman and Perez 1988; Freeman and Louçã 2001). Industrial revolutions are generally constituted of several of these "general purpose technologies" (Helpman 1998); for example machinery, steam power, and iron in the First Industrial Revolution of the late eighteenth century; chemicals, internal combustion, electricity and steel in the Second Industrial Revolution of the late nineteenth century; and information and communication technologies, biotechnology and smart materials in the Third Industrial Revolution of the late twentieth century. In our view, the general-purpose technologies associated with the Third Industrial Revolution create new opportunities for LMT industries to enhance their innovative and economic performance through the effective adoption and application of ICT, biotechnology, and smart materials.

The properties of these technologies are not of concern here, being the high-tech sectors of their day, but their spread to adopting sectors is of great significance. They arise most commonly (though not invariably) in "upstream" activities—in the cases mentioned, in equipment and capital goods, in motive power and in basic materials—from which they trickle down to user industries.

In LMT industries there is usually relatively little formal learning by science and technology, at least at the firm level. Instead, innovation- and adoption-related learning activities operate in practical and pragmatic ways by doing and using. The bulk of the new technologies on which they draw, including these general-purpose technologies, are developed by separate companies (rarely subsidiaries), specialized in the relevant technological fields. However the downstream LMT industries need to have "absorptive capacities" to make productive use of these upstream developments. As shown by Cohen and Levinthal (1989, 1990), absorptive capacities are best inculcated by doing some of the innovative activities oneself, even if this means replicating what has already been done elsewhere. Thus food-processing companies involved in advanced ("third-generation") biotechnology do not carry out much of the associated research themselves but are often prominent in patenting in less advanced ("second-generation") biotechnology—this seems to provide them with the necessary absorptive capabilities. In addition, the national systems of innovation literature has drawn attention to how formal science may congregate in national or regional laboratories in such industries, instead of being internalized within firms (Nelson 1993). The applicable science produced by these organizations (e.g. public laboratories) may be generic, but the applications themselves will require firm-specific absorptive capacities, generated not just from formal R&D but from broader-based innovative activities that include engineering, continuous improvement processes, and organizational innovations such as integrated service and supply.

15.3.2.2 Carrier Industries

Describing one particular industry supplying technologies, namely machine tools, Rosenberg (1963) showed how the number of different types of tools was quite limited and as a result their principles could readily be stretched to being applied in industries other than where they were first deployed. The counterpart of this process was the equally significant role of "carrier industries," which incorporated these proliferating machine tools into making the machines they produced or used. Based on the notion of general purpose technologies, any industry can act as a carrier if its demand for the new capital good is large enough or growing fast enough. Thus even low-tech sectors can act as receptors for new process-oriented technologies, and many examples are given below.

15.4 Firms and Corporate Change in LMT Industries

This section raises issues concerning strategies and structures at the firm level, issues of scale and scope in both Carge firms and SMEs, and discusses different kinds of integration.

15.4.1 Strategy and Structure

15.4.1.1 *Strategy in LMT Industries*

According to Porter (1985), there are three main types of corporate strategy, namely cost leadership, differentiation, and focus. If viewed from the aspect of the orthodox product lifecycle, cost leadership is the likely choice of firms in a "mature" LMT industry that rely on process innovation to pare down costs even though the innovative spark has largely ceased. From the alternative points of view outlined above, however, the other two strategies also are plausible candidates for companies wishing at least to survive if not to thrive in competitive lower-tech environments. Branding is often crucial to the choice of a differentiation strategy, as consumers choosing between a Saab and a Škoda can be influenced by product reputations (Škoda is nowadays using advertising of its new technologies to change deep-seated customer prejudices).

15.4.1.2 *Functions and Structures of the Firm*

LMT firms and industries do not differ from any others in the tasks that need to be carried out, but clearly they may place less emphasis on technology functions and more on product/marketing functions than, say, a science-based industry. Building implicitly on this presumption, Chandler (1990) argued that the multidivisional ("M-form") company was an appropriate organizational structure for diversified firms in industries characterized by "branded, packaged goods" (as well as many that were science-based), such as many branches of the food industry. By organizing divisions entrusted with specific product lines in multiproduct firms, and giving middle management the job of running these divisions, there could be greater focus on the targeting of markets, as all the functions necessary for the product line were incorporated within the division (and the responsibility of lower management tiers). However this raised several problems, for example if there were inherent spillovers between the same functions located in different divisions—this could make it awkward for M-form companies to introduce information systems and other radical technological changes, which were bound to affect all their carefully separated divisions (Mowery 1995).

15.4.1.3 *"Frames" for Each Function*

In previous work, we suggested that large firms interpret competitive challenges and problems through mental "frames," organizational cognitive maps through which they view their horizons and functions, including technology (Acha 2002). In contrast to the Porter concept of corporate strategy, the frame is the filter through which strategies are conceived and chosen. This filter, held by individuals and mediated at the organizational level by senior management, comprises the main variables, functions, and contingencies of a particular phenomenon. In practice, managers consider particular activities rather than the whole at once. A technology frame, therefore, is the interpretative system of managers to understand the firm's technological position and opportunities as well as the expectations of the dynamics of their relevant innovation system(s) (Orlikowski and Gash 1994; Orlikowski 2000).[1]

For the oil and gas industry we found that frames significantly influenced performance differences among companies with similar technological competencies. That is, there is little direct correlation between technological performance (say patenting and scientific publications) and business performance (say expansion or profitability), unless this intervening variable of the frame is accounted for.

The frame concept is relevant to the study of innovation in LMT sectors. Firms in LMT industries are basically using rather than selling technology, and therefore tend to adopt technology frames that are quite different from those in high-tech industries (where technology itself is a key selling point), and often quite different from other LMT firms even in the same industry. In general, the market characteristics of LMT industries (where, over time, markets become segmented and competitive advantage depends upon product differentiation, cost efficiency and control of complementary assets) lead firms to different interpretations about the role for technology for commercial success.

In "high-tech" firms, by contrast, the role for technology is more explicitly central to commercial success, and there may be greater tendencies for consensus (general or by groups) on aspects of technology frames across these competing firms because: (1) the market is relatively "thin," with relatively fewer product and technology options as yet; (2) appropriation of technology through IPR is more aggressively pursued and this structures the role of technology for the industry; and (3) regulatory environments more frequently play a structuring role (e.g. biotech). To generalize, variation in technology frames across "high-tech" firms derives more from a focus on how the technology (broadly stated) should develop, whereas variation in technology frames amongst LMT firms pertains more to what the role for technology (broadly stated) should be.

15.4.2 Scale and Scope: Large Firms and SMEs

Although they are now mature, such "LMT" industries as meatpacking, automobile production, and consumer durables were important sources of production innovation in their early years, essentially developing the technologies of mass production. The driving force of mass production was to reap economies of scale in the production processes, and the M-form company was a very suitable organizational form for doing so (Chandler 1977). Beyond being labour-saving and capital-intensive, the main direction of technical change was to be "time-saving." This was achieved through raising throughput, reducing downtime, and improving the machinery so as to ensure the speedy throughflow in all stages of its operation. These benefits can be thought of as arising from "dynamic economies of scale," as contrasted with the usual static scale economies arising, for instance, out of having large plant and equipment. In some medium-tech industries like steel, the adoption of new production technologies (the electric furnace) has led to a decline in average establishment size (e.g. huge "integrated plants" have been replaced by "minimills"), in which the losses in terms of static scale economies have been more than offset by increases in dynamic scale economies through specialization and throughput.

Because of the high capital costs of large-scale assembly lines in automobiles, it was necessary to produce very long runs of the product in order to cover them. Competition in this industry increasingly took the form of maintaining the same major components (body, engine, etc.) for as long as possible, while giving the semblance of regular updating of the model by adding on inexpensive "frills," e.g. associated with annual product launches. In essence the assembly line, to pay off, required producing products that were as standardized as possible.

This approach was challenged by the Toyota system of "lean production" (Womack et al. 1990), which was a response to the needs of customers for variety and specialization while sacrificing as little as possible of the benefits of high throughput—in effect obtaining "dynamic economies of scope." Best known of the constituents of lean production was just-in-time production, accelerating the speed of response to changes in customer needs, but other changes worked in the same direction. The lean production system was intended to supplement the American system of interchangeable parts in the product with interchangeable parts in the process. The economies of scope obtained in the efficient production of a variety of products (the Toyota Crown car family allegedly came in over 100,000 variants) more than compensated for abandoning standardization.

At the other end of the spectrum, small and medium-sized enterprises (SMEs) have reappeared on government technology policy agendas. One of the reasons for renewed emphasis on them as sources of innovation is the perceived advantage of SMEs in responding quickly to technological change, because of the absence of complex management structures within smaller enterprises. Against that, SMEs may lack the financial clout to undertake the kinds of investments in new technologies necessitated by radical change (Dodgson and Rothwell 1995). Questions of access to new technologies and on what terms then become paramount. This becomes significant in those LMT industries which are dominated by SMEs, which include many of the more traditional areas. Whilst SMEs are frequently seen as a cause of hope in high-tech sectors, they are often (perhaps unfairly) seen as a matter for despair in LMT sectors.

15.4.3 Vertical and horizontal integration

The LMT industries have been characterized by a variety of patterns of vertical integration and disintegration through their development. The low-tech industries (plus some medium-tech ones) for the most part long pre-date the onset of industrial revolutions and some date back to prehistoric times. When industries first emerge, there is usually a high degree of vertical stratification, i.e. with different firms responsible for each stage of production (the "chain"). Some go on to acquire additional technological capabilities which may eventually spin off into newly created companies (as Rosenberg (1963) showed for machine tools). However the segments interact in systemic fashion with one another through the vertical "chains." In textiles, there were "imbalances" between each stage of production— the early mechanization of weaving speeded up the production of cloth, so pressure was placed on spinners to speed up their own production of yarns to go into the cloth, thereby giving rise to the mechanization of spinning; this in turn became so successful that renewed pressure was put on the weaving segment to develop power-based weaving, and so on. The other stages of textile production—fiber production (e.g. cotton growing), preparation and finishing—were similarly affected along the "value chain." Similar "imbalances" arose in the stages of production of iron.

In the second phase of industrialization there were pressures to link the segmented processes more directly through moves towards vertical integration. The pressures of throughput that gave rise to "mass production" in the USA in the late nineteenth century also promoted vertical integration, because this could ensure smooth production flows throughout the value chain.

More recently, the rise of the steel minimill, partly from technical change, at the expense of the large integrated mill has been one of the more dramatic demonstrations of a retreat from vertical integration. Tasks that corporations previously undertook themselves have been "outsourced" where possible, thereby returning to the traditional low-tech model of vertical disintegration even in some high-tech industries. To retain and intensify the economies of scope required in this unfolding set of circumstances, companies also chose to limit the range of their horizontal diversification. Firms, including some of the largest corporations, practised "downsizing" and in many cases stripped out large numbers of middle management in the belief that this furthered "lean production."

While financial considerations had often encouraged diversification into unrelated activities, studies demonstrated that "conglomerate" firms based on unrelated diversification were not very profitable (Rumelt 1974). Taking on board the technological and production aspects, and thereby taking into account the issues of synergies and economies of scope, many firms reoriented their structure to limit themselves to "related" diversification. Yet many of the larger companies in low-tech industries like food manufacturing continue to pursue apparently unrelated diversification. The most likely reason for this is that low technological opportunity in traditional segments of low-tech industries may also go with relatively high appropriability, especially through branding. This generates a pool of resources (Penrose 1959), which can be used to invest in other areas in which branding provides a secure financial base. The scale and scope of economies thus arise via marketing rather than through technologies.

15.5 CHANGE AT THE INDUSTRY LEVEL

15.5.1 Vertical Alignment and Networks

These changes at the firm level in terms of size, integration and diversification carry strong implications for the structure of the industries in which they are embedded. Firms have been driven to develop closer relationships with upstream suppliers and downstream customers. Among the best known of these developments are those in the motor vehicle industry. One implication of the Toyota system of flexible and just-in-time manufacture was the need for high reliance on suppliers to deliver equipment and components on time and of high quality. Rather than controlling this process hierarchically, Toyota undertook joint development with the suppliers, spending long periods of time negotiating the exact specifications and costings required. Some of the most critical suppliers became "first-tier" suppliers and worked in close association with Toyota wherever the production was located. Others were seen as less critical or supplied the suppliers: these became "second-tier" and "third-tier" suppliers and were involved less and less directly in negotiations with Toyota itself.

Many of these firms saw themselves as developing into "system integrators," acting as the hub for operations immediately connected to their central activity. Depending on how far they outsourced these related activities, they might end up as directly producing only a minority of the value added by their "system." Moreover system integrators could emerge at varying points on the value chain, so that vertical links might increasingly be occurring between various system integrators. That is, each integrator would be surrounded by a network of suppliers and related activities.

Within these complex structures of interrelationships, new power balances could emerge. In the lower-tech industries especially, like textiles and some branches of food, the manufacturing stages of the "chain" were squeezed as power tended to shift downstream to the final stages and even to the retailers. The growing competition in diversified products in conjunction with the rising incomes and influence of consumers, often dissatisfied with what they are getting, is driving this trend. It remains to be seen whether this pattern will proceed further in the direction of buyer-driven rather than producer-driven chains (Gereffi 1999).

15.5.2 Industry Differences

These observations raise the issue of how industries differ in their behavior; in particular, can one observe different patterns in LMT industries than high-tech industries at the industry and sector levels? Many studies have sought to classify industry behavior by resorting to the two broad ideal types usually identified with the name of Schumpeter (see Ch. 14 by Malerba in this volume).

One group of such studies identifies the key conditions as opportunity, appropriability, cumulativeness, and knowledge base (Malerba and Orsenigo 1996, 1997). Using these criteria, the great majority of industries in Europe are classified as Mark I or Mark II using objective indicators. Thus among the LMT industries, clothing falls obviously into the category of Mark I, characterized by low technological opportunity, weak appropriability of any innovations, small firms and rapid entry and exit, and a practical rather than scientific knowledge base. Motor vehicles as a medium-tech industry have greater technological opportunities, greater appropriability and the persistence of large firms. High-tech industries naturally have high technological opportunities and a scientific knowledge base, though they may be variously typified by large or small firms.

From the viewpoint of LMT industries, where demand plays such a large role (see Section 15.3.1 and case studies below), market opportunity can be as important as technological opportunity, and may be very different in extent as well as in nature. Fast-growing areas of consumption are not always the same as fast-growing areas of technology, as noted above. Moreover, for our immediate purposes, some of the low-tech industries are Mark I but others are nearer to Mark II, while food-processing resists any easy classification since its sub-branches operate in a whole variety of ways.

We have indicated that technological opportunities may be enlarging again for "low-tech" industries (see Section 15.3.2 above), although firms in such industries will for the most part outsource the development of these new technologies. Outsourcing may limit opportunities for "user" firms to appropriate the returns from innovation that relies on this approach (recall that appropriability was a key factor in the Sutton analysis discussed above). In the supplier-dominated low-tech industries (as defined by Pavitt) the appropriability of technologies rests upon the division of power between the technology developers—the upstream suppliers—and the users, like food or clothing companies. These activities are rarely vertically integrated, because the suppliers usually wish to supply a variety of users both within the same activity and outside. The appropriability of the products depends on both the marketing endeavours of the companies concerned, and the power balance vis-à-vis downstream distributors and retailers. Links with technology suppliers tend to be much more distant than in high-tech sectors.

15.5.3 Industrial Dynamics

15.5.3.1 *Entry, Exit, and Technological Accumulation*

Models of industrial dynamics in the Schumpeterian tradition often contrast Mark I and Mark II industries in respect of the technological advantage of incumbent firms over possible entrants (Marsili 2001). The deciding factor can be the extent to which individual firms in an industry—incumbents or entrants—can earn economies of scope in combining or differentiating technological trajectories (Sutton 1998).

Many LMT industries are however characterized by high levels of "turbulence," with a churning of entry and exit. These pose the issue of learning in turbulent environments—if the individuals cannot be retained in the industry when exit occurs, new entrants may simply replicate their predecessors' mistakes. However, in North American environments at least, the individuals concerned do tend to go on to form another firm (Baldwin 1995). Alternatively the continuity can be maintained by technological dependence on a large supplier or an industrial district, as is often found in clothing. In complex products systems, where alliances are reconstituted for each new project, learning is achieved by the flux of interactions of the constituent firms, although a high level of "forgetting" also seems to be common (Hobday 1998). The dichotomous Mark I and II categorizations may be too restrictive to portray the main patterns of evolution of different industries, and more complex schema such as that of Pavitt (1984) or Marsili (2001) may be preferable for understanding the impact of differing technologies.

15.5.3.2 *Dynamic Competition in Time*

Across the full range of industries, the modern era is supposed to be characterized by competition that has intensified because of globalization and because of more rapid change in market demand. In LMT industries the pace of change and competition may be less intense, as market leaders seek to retain their predominance by brand loyalty. In the tobacco industry, the trajectory of a new product launch is normally to begin with a new brand marketed at low prices and with high tobacco content. Once loyalty is obtained the producer relaxes the prices and/or the material content until eventually consumers lose faith, whereupon a new brand is launched on the same terms to eclipse it. Equally, brand loyalty can be extended through globalized marketing.

Thus marketing-based appropriability offsets some of the pressures to develop new products. However firms in LMT industries are arguably almost as susceptible as those in higher-tech industries to innovations that accelerate the development times ("cycle time") and rate of application and diffusion for new technologies. In the clothing industry, Benetton combined its production network of small suppliers with a sophisticated ICT system for feedback from customers, achieving rapid global growth in a relatively mature industry (Belussi 1987). For the more basic production technologies, the LMT industries are distinguished less by the pressure to innovate than by the difficulties encountered in applying those basic technologies to purposes that may be very different from their original intentions. Having to confront such difficulties amplifies rather than diminishes the importance of speed to market for competitiveness even in LMT industries. The key issue here is surely the real-time achievability of economies of scope—do integrated firms or outsourced networks effect this better?

15.6 INNOVATION IN SOME LOW-TECH INDUSTRIES

15.6.1 Textiles and Clothing

As the original "leading sector" of the First Industrial Revolution in Britain, textiles have often been regarded as the quintessential low-tech industry in the modern era. First, however, most low-tech industries today are not structured like textiles and clothing. Second, textiles themselves have shown a repeated ability to update by bringing in contemporary developments in technology. In the early twentieth century the chemicals technologies at the forefront of the Second Industrial Revolution were recruited to launch "artificial fibers" (rayon etc.); in the middle of the century "synthetic fibers" (nylon etc.) drew upon synchronous advances in plastics. Both artificial and synthetic fibers remained branches of the chemicals industry in terms of corporate structures as well as technologies and are thus often thought of not as textiles proper, but this hardly seems reasonable. In more recent times computerized technology is making inroads into the still more fragmented clothing sector, while genetics is assisting advances in textiles. "Microfibers" of exceptional fineness have evolved for particular market niches and have begun penetrating standard clothing segments.

It is indeed the case that the recent developments have been slow to diffuse, but this probably has less to do with technological limits than with organizational aspects. The textile–clothing industry is still largely based on a pre-industrial vertical structure that is highly segmented and allows countries with low wage rates to enter through the use of relatively simple technologies, such as sewing machines. To leverage new textile technologies in the way we have described, change in production technologies may have to be linked to product changes, as for instance with microfibers.

The demand side normally offers greater prospects for leverage than the supply (technologies) side in the textiles and clothing sector. As incomes rise, consumers are willing to pay extra for fashionable brand-names. Many of the observed changes in what is fashionable rest on styling rather than technological innovation, though naturally that does not imply any lesser need for creativity. Even here, however, new technology may help bridge the gap between designers, producers, and their markets. An important innovation by Benetton was the firm's use of ICT to learn quickly about changes in market tastes in order to instruct their suppliers (Belussi 1987).

15.6.2 Food-processing

A second industry frequently, and often inappropriately, classified as low-tech is food manufacturing. The "industry" is characterized by a huge variety of organizational forms, related principally to the extent to which marketing can be leveraged (Sutton 1991). It has been customary to see the industry as "supplier-dominated" according to Pavitt's taxonomy, resting on a dependence on suppliers of the machinery for production, but this is changing nowadays in some critical ways. As such, the industry is becoming more market-driven, but one cannot reasonably leave technology out of the picture.

Like all industries, the technical efficiency of this industry comes from the knowledge bases it draws upon, but here there exists an unusual amount of variety to go with the variety of organizational structures. The range of expertise from science (e.g. food microbiology) through to engineering has to span not just production conditions but sanitation, quality assurance, environmental acceptability, and so on. The range of new technologies currently being drawn upon in this industry covers almost the whole spectrum. The traditional reliance on suppliers of machinery is being overtaken by needs for technologies from advanced instrumentation (e.g. lasers), electronics and computing (problematic because of the irregular shape of some of the leading products), biotechnology (for both materials and production processes), pharmaceuticals, and smart materials (especially in packaging), supplied by high-tech firms or public laboratories. The seemingly simple packaging of readymade and microwavable foods for sale in supermarkets in fact required very sophisticated analyses of smart materials to combine heat responsiveness, gas release (controlled oxygenation), ease of production, ease in filling during processing, as well as ease of consumer use.

Many of these new techniques remain controversial, particularly in the field of biotechnology (genetic modification, GM). At the heart of these controversies lies an issue already noted—is a GM food the same product or a different one? Around such points trade wars rage. As already implied, the nature of the product as a staple of human existence means that safety and quality procedures play a substantial role. The difficulty with prevailing methods for testing food safety (wet chemistry) is that they involve cutting off a piece of the product—which may or may not be representative—taking it to the laboratory and waiting about three weeks for the results. By this time the rest of the product could long since have been sold and consumed. Biotechnological methods and other procedures are therefore being sought to effect testing in "real time," i.e. synchronously with the processing. Again the main motivation for technological innovation is time-saving—reducing downtime and waste and increasing throughput in the interests of systemic efficiency.

Changes on the demand side (from socioeconomic factors like rising wealth, growing female employment, ageing of the population) and the above technological changes on the supply side are channeled in part through a changing vertical structure of the industry, in which growing power is accruing to giant retailers vis-à-vis large processors. At the time of writing, the US food retailer Wal-Mart has become the world's largest company according to the *Fortune* list, superseding the high-tech and medium-tech companies of the recent past. Similarly in Europe, supermarket chains have been proactively utilizing information technologies for robust expansion of their power base. These developments make the predominance of demand-side influences on innovation here abundantly clear.

15.7 ROLE OF GOVERNMENT POLICY

Government technology policies at national and supranational levels have on the whole tended to give the highest priority to high-tech sectors and activities. One of the most commonly used benchmarks for the success of a government's technology strategy is the proportion of total output emanating from high-tech sectors, in which dimension East Asian countries tend to score very highly (see Ch. 19 by Fagerberg and Godinho in this volume). Such a benchmark falls into the trap, emphasized throughout this chapter, of confusing the sector with the technology level. Success in low-end activities in high-tech industries, such as "screwdriver" assembly activities, may or may not lead on to success in high-end activities in those sectors—often it may simply generate competition from new rivals in the low-end activities, based on providing still cheaper labour.

As noted in Chapter 22 by Lundvall and Borrás in this volume, government policy towards innovation was for long dominated by the so-called "linear model" of technology-push. Although this approach does not preclude applications of high technology to more traditional fields, the bias in practice usually veers towards new fields as well as new technologies. For many decades the bias towards such technologies was also coupled to one towards consolidation of firms and industries, seen as "national champions" for technological and commercial success. This view has been weakened in recent times as small and medium-sized enterprises (SMEs) have been increasingly thought of as progenitors of high technology, in fields such as biotechnology and genomics, software, advanced instrumentation, and so on. Yet in low-tech industries, SMEs are most widely seen as dragging down overall performance.

The antithesis of the "supply–push" contribution of the linear model is the "demand–pull" approach to innovation, in which the causal sequence is roughly reversed. Governments, though, tend to be loath to "leave it all to market forces," in an arena in which "market failures" are so evidently present as they are in innovative activities. An intermediate position is for governments to foster diffusion rather than upstream invention and innovation (Stoneman and David 1986). While this may not overcome all the market failure shortcomings, there is much to be said for rebalancing policy in such a way as to place greater stress on the diffusion aspects. Of particular relevance to this chapter would be the diffusion of high-tech activities into supposedly low-tech sectors. As we have observed, there is a much wider scope for such migration of technologies across sectoral boundaries than is often supposed.

Many countries have in fact based their development on starting in supposedly low-tech activities like the food-processing or textile industries surveyed above—Denmark, Switzerland, and Australia are evident examples (von Tunzelmann 1995). At the same time, a perhaps even larger number of countries have become locked into the low-tech activities and never adequately escaped, as is often alleged for Latin America. The difficulty remains that countries' comparative advantages often remain indissolubly linked to the low-tech sectors. Our view that there are no true low-tech sectors in the modern world however overrides this. It is perfectly possible to diffuse high technologies into the "low-tech sectors," as our illustrations of chosen low-tech sectors aimed to underline. In that way intermediate and developing countries do not need to face a dilemma when choosing between static comparative advantage in traditional fields and dynamic comparative advantage from techno-logical opportunity—they can have their cake and eat it.

This would involve moving away from usual defeatist views of traditional sectors as "sunset industries." As Ernest Hemingway observed, "the sun also rises." The sun can rise both from downstream diffusion of high-tech activities into "low-tech sectors," and from the lateral diffusion of old technological fields into new ones. Countries like Switzerland developed, consciously or unconsciously, a pattern of evolution that went—in this case—from textiles to dyeing and chemicals and then into strength in pharmaceuticals on the one hand, and into machinery and thence to advanced engineering on the other.

A clear implication of these policy considerations is that strength in low-tech sectors does not have to act as a "block to development," although national and supranational governments frequently embody this error in their technology pol-icies. The EU, for instance, counterposes the Framework Programs and the Struc-tural Funds as means for overcoming development-oriented market failures, but it is conceivable to think of reconciling the two, and using the Structural Funds as means for implementing innovation in catching-up regions (Fagerberg et al. 1999). The "block to development" view can be replaced by a "development block" view, in which LMT sectors act as "carrier industries" for diffusing the gains from new technologies across the industrial spectrum, as they have so often managed to do in the past (Rosenberg 1976).

We do not pretend that this has yet happened on any major scale, though it is significant in some services and a number of older branches of manufacturing. On the contrary, on analogy with past "long waves" of technological innovation and industrial growth, it is the key challenge for the next few decades.

15.8 Conclusions

The underlying argument of this chapter is that low-tech sectors do not lack for technological opportunities, nor indeed for appropriability and other factors asso-ciated with benefiting from technological innovation. Indeed, our principal analyt-ical conclusion is that in the modern world there are no true "low-tech sectors." Instead what we observe is a varying degree of permeation of high technologies into low-tech and medium-tech as well as into high-tech sectors.

The primary issue for further research is thus the need to explore much more critically how industries and sectors ought to be classified. The OECD itself appears to be moving towards replacing its conventional accounting of direct and indirect R&D with a more subtle assessment of the knowledge intensity of industries (OECD 2003; see Ch. 6 by Smith, for further discussion). This will have substantial implications for government policy making but also for analytical understanding of structural change and the sources of long-term development. We would suggest going further and aiming to think not just of improving our conception of how technologies can be clustered (as for instance in the Pavitt taxonomy) but also of how technologies map into products. Some pioneering work has been published by individual scholars in this regard (e.g. Piscitello 2003) but this work needs support from national and supranational agencies to instil the notion that demand as well as supply factors drive technological change and long-term prosperity. The LMT industries bulk far too large in total output to be relegated to the scrap-heap of stagnation.

NOTES

1. Frames should not be confused with organizational "culture"; the latter comprises the values established at the group level and interpreted by individuals, whereas frames comprise the interpreted relationships between phenomena (and their associated values) and are composed of individuals, mediated at a group level.

REFERENCES

ACHA, V. L. (2002), "Framing the Past and Future: The Development and Deployment of Technological Capabilities in the Upstream Petroleum Industry," Unpublished D.Phil. Thesis, SPRU, University of Sussex.

—— and BRUSONI, S. (2003), "Complexity and Industry Evolution: New Insights from an Old Industry," in Best Papers Proceedings of 2002, European Association for Evolutionary Political Economy Conference, 2003.

BABAEV, H. (2000), "Scientists, Engineers Teaming up to Develop Mars Drilling Technology," reprinted in Oil and Gas Journal 98 (17).

BALDWIN, J. R. (1995), The Dynamics of Industrial Competition: A North American Perspective, Cambridge: Cambridge University Press.

—— and SABOURIN, D. (2002), "Advanced Technology Use and Firm Performance in Canadian Manufacturing in the 1990s," Industrial and Corporate Change 11: 761–90.

BELUSSI, F. (1987), "Benetton, Information Technology in Production and Distribution: A Case Study of the Innovative Potential of Traditional Sectors," SPRU Occasional Papers 25, SPRU, University of Sussex.

CHANDLER, A. D. jr. (1977), *The Visible Hand: The Managerial Revolution in American Business*, Cambridge Mass,: Belknap Press.

—— (1990), *Scale and Scope: The Dynamics of Industrial Capitalism*, Cambridge, Mass.: Belknap Press.

COHEN, W. M., and LEVINTHAL, D. A. (1989), "Innovation and Learning: The Two Faces of R&D," *Economic Journal* 99: 569–96.

—— —— (1990), "Absorptive Capacity: A New Perspective on Learning and Innovation," *Administrative Science Quarterly* 35: 128–52.

*DAVIES, S., and LYONS, B. (1996), *Industrial Organization in the European Union: Structure, Strategy and the Competitive Mechanism*, Oxford: Oxford University Press.

DODGSON, M., and ROTHWELL, R. (eds.) (1995), *The Handbook of Industrial Innovation*, Cheltenham: Edward Elgar.

*FAGERBERG, J., GUERRIERI, P., and VERSPAGEN, B. (eds.) (1999), *The Economic Challenge of Europe: Adapting to Innovation-Based Growth*, Cheltenham: Edward Elgar.

*FREEMAN, C., and LOUÇÃ, F. (2001), *As Time Goes By: From the Industrial Revolutions to the Information Revolution*, Oxford: Oxford University Press.

—— and PEREZ, C. (1988), "Structural Crises of Adjustment: Business Cycles and Investment Behaviour," in G. Dosi, C. Freeman, R. Nelson, G. Silverberg, and L. Soete (eds.), *Technical Change and Economic Theory*, London: Pinter, 38–66.

GEREFFI, G. (1999), "International Trade and Industrial Upgrading in the Apparel Commodity Chain," *Journal of International Economics* 48: 37–70.

GROSSMAN, G. M., and HELPMAN, E. (1991), *Innovation and Growth in the Global Economy*, Cambridge, Mass.: MIT Press.

HELPMAN, E. (ed.) (1998), *General Purpose Technologies and Economic Growth*, Cambridge, Mass.: MIT Press.

* HIRSCH-KREINSEN, H., JACOBSON, D., LAESTADIUS, S., and SMITH, K. (2003), "Low-Tech Industries and the Knowledge Economy: State of the Art and Research Challenges," mimeo, EU 5th Framework project, "Pilot: Policy and Innovation in Low-tech."

HOBDAY, M. (1998), "Product Complexity, Innovation and Industrial Organisation," *Research Policy* 26: 689–710.

MALERBA, F., and ORSENIGO, L. (1996), "Schumpeterian Patterns of Innovation," *Cambridge Journal of Economics* 19: 47–65.

*—— —— (1997), "Technological Regimes and Sectoral Patterns of Innovative Activities," *Industrial and Corporate Change* 6: 83–117.

MANSFIELD, E. (1969), *Industrial Research and Technological Innovation: An Econometric Analysis*, New Haven: Yale University Press.

*MARSILI, O. (2001), *The Anatomy and Evolution of Industries: Technological Change and Industrial Dynamics*, Cheltenham: Edward Elgar.

MELICIANI, V. (2001), *Technology, Trade and Growth in OECD Countries: Does Specialisation Matter?* London: Routledge.

MOWERY, D. C. (1995), "The Boundaries of the US Firm in R&D," in N. R. Lamoreaux and D. M. G. Raff (eds.), *Coordination and Information: Historical Perspectives on the Organization of Enterprise*, Chicago: University of Chicago Press, pp. 147–82.

* Asterisked items are suggestions for further reading.

Nelson, R. R. (ed.) (1993), *National Innovation Systems: A Comparative Analysis*, New York: Oxford University Press.

OECD (1994). *The Measurement of Scientific and Technical Activities: Proposed Standard Practice for Surveys of Research and Experimental Development* (Frascati Manual: 1993), Paris: OECD.

—— (2003), *Science, Technology and Industry Scoreboard 2003 – Towards a Knowledge-Based Economy*, Paris: OECD.

Orlikowski, W. J. (2000), "Using Technology and Constituting Structures: A Practice Lens for Studying Technology in Organizations," *Organization Science* 11: 404–28.

—— and Gash, D. C. (1994), "Technological Frames: Making Sense of Information Technology in Organizations," *ACM Transactions on Information Systems* 2: 174–207.

*Pavitt, K. (1984), "Sectoral Patterns of Technical Change: Towards a Taxonomy and a Theory," *Research Policy* 13: 343–73.

*Peneder, M. (2001), *Entrepreneurial Competition and Industrial Location*, Cheltenham: Edward Elgar.

Penrose, E. T. (1959), *The Theory of the Growth of the Firm*, 3rd edn., Oxford: Oxford University Press, 1995.

Piscitello, L. (2003), "Generation and Applicability of Technological Competencies: Another way of Measuring Coherence in Corporate Diversification," paper for Conference in Honour of Keith Pavitt, Politecnico di Milano.

Porter, M. E. (1985), *Competitive Advantage: Creating and Sustaining Superior Performance*, New York: Free Press.

Rosenberg, N. (1963), "Technological Change in the Machine Tool Industry, 1840–1910," *Journal of Economic History* 23: 414–46.

—— (1976), *Perspectives on Technology*, Cambridge: Cambridge University Press.

Rumelt, R. (1974), *Strategy, Structure and Economic Performance*, Boston: Harvard Business School Press, rev. edn., 1986.

Stoneman, P. L., and David, P. A. (1986), "Adoption Subsidies vs. Information Provision as Instruments of Technology Policy", *Economic Journal*, Conference papers, 96: 142–50.

Sutton, J. (1991), *Sunk Costs and Market Structure*, Cambridge, Mass.: MIT Press.

*—— (1998), *Technology and Market Structure*, Cambridge, Mass.: MIT Press.

Tidd, J., Bessant, J., and Pavitt, K. (2001), *Managing Innovation: Integrating Technological, Market and Organizational Change*, Chichester: Wiley.

*von Tunzelmann, G. N. (1995), *Technology and Industrial Progress: The Foundations of Economic Growth*, Cheltenham: Edward Elgar.

Womack, J. P., Jones, D. T. and Roos, D. (1990), *The Machine that Changed the World*, New York: Rawson.

CHAPTER 16

INNOVATION IN SERVICES

IAN MILES

16.1 INTRODUCTION

SERVICES innovation is a topic of growing interest for innovation researchers and policy makers alike. This has particularly been so in the last decade, as services have grown to constitute the larger part of employment and output in most industrial countries. The services sectors of these economies are important for their productivity, economic competitiveness, and quality of life. But innovation in services is important for other reasons beyond the economic importance of the sector. Innovation in services extends beyond the services sectors to affect service activities in all sectors of the economy. Second, some services play central roles in innovation processes throughout the economy, as agents of transfer, innovation support, and sources of innovations for other sectors.

16.2 SERVICES—GROWTH, CHARACTERISTICS, AND INNOVATION

Output and employment in the service sectors have grown significantly throughout the industrial world since the 1950s. Table 16.1 shows that as early as the 1970s, services

Table 16.1 Share (%) of gross value added in services in total GDP in EU countries

	EC-9 (1973)	EC-10 (1981)	EC-12 (1986)	EU-15 (1995)	EU-15 (2001)
EC/EU	50.7	56.3	59	63.1	65.3
Ireland	44.5	49.4	50.4	48.6	49
Finland	n/a	49.8	53	56.4	57.1
Portugal	n/a	48.8	54.1	59.1	61
Austria	n/a	55.5	58.6	63.3	63.6
Denmark	58.8	60.9	60.3	64	64
Spain	47.8	57.3	58	63.7	64.2
Germany	48.1	54.2	55.9	62.5	64.9
Italy	51.1	56.4	60.6	62.8	64.9
Greece	40.3	47.7	48.9	62.6	65.4
Sweden	n/a	60.1	59.7	62.2	65.5
Netherlands	52.2	58.8	60.4	63.9	65.5
France	50.8	57.4	61.1	65.3	66.8
UK	54.7	54.5	57.1	62.1	67.2
Belgium	54.6	60.7	62.8	66.1	67.3
Luxembourg	46.9	65.2	76.4	83.7	83.8

Source: Eurostat (2003), *50 years of figures on Europe*, Luxembourg: European Communities.

constituted more than half of the value added in European Union countries, and by the new century they contributed over two-thirds.

16.2.1 Services Diversity—and Distinctiveness

The rates of growth in services output and employment vary among the nations in Table 16.1, and in the case of Ireland, the share of output attributable to services declined during 1986–2001. These differences reflect contrasts in the composition of the services sector in different countries, highlighting the fact that the category "service sector" comprises a huge range of different activities with very different characteristics. At one extreme are personal services like hairdressing that involve basic technologies and often are organized on a small-scale basis. In contrast, the FIRE (finance, insurance, and real estate services) sector is dominated by very large firms using advanced information technologies intensively. Other technologies are used in distributive services, which include transport in all its varieties and retail and wholesale trade and, in some classifications, telecommunications and broadcasting.

HORECA (hotels, restaurants and catering) is dominated by food preparation and delivery, and includes other elements of hospitality, entertainment, and comfort. Social and collective services such as public administration and health and educational services are delivered largely or entirely through the state, though patterns of organization vary a great deal over time and across countries: the back-office operations of such bodies can be highly IT-intensive. Business services include practical support with logistics and office and building services, as well as support for administrative matters (such as law and accountancy) and technology support, such as computer and engineering services. This summary by no means exhausts the variety of activities and technologies included in the services sector.

Services markets also are diverse, spanning consumers, businesses, and the public sector and its clients. The transformations effected by these services industries operate on such diverse "raw materials" as human clients (as well as some other biological organisms, e.g. veterinary services), physical artifacts (they may be repaired, maintained, stored, transported, tested, integrated into larger systems . . .), and data, symbols and information (that may be processed, stored, telecommunicated, etc. by services like financial industries as well as by computer and communications services, etc.). The technical skills demanded of the workforce range from the minimal ones used in fast food outlets and office cleaning to the professional qualifications of market researchers and architects, and the scientific and engineering credentials of staff in specialized R&D firms.

This diversity means that any generalizations about the nature of services and innovation in services must be qualified by numerous exceptions. Some services are more like manufacturing in terms of some parameters—some are technology-intensive (e.g. media, telecommunications), some work with material artifacts (e.g. rapid prototyping, repairs). And the operations of many manufacturing firms include a great many "services" activities (e.g. transport and logistics, office work, marketing and aftersales). Nevertheless, a set of common features characterizes many services and differentiates them from manufacturing.

For example: many services products are intangible ones, which makes it harder to store, transport, and export them than is true of manufacturing products. Historically, many service innovations have been difficult to protect via patent mechanisms, although this situation has begun to change in some sectors such as computer software and services, or FIRE (See Hall 2003; Graham and Mowery 2003, FhG. 151 2003). Services are typically interactive, involving high levels of contact between service supplier and client in the design, production, delivery, consumption, and other phases of service activity. Service products are often produced and consumed in the course of supplier–client interaction at a particular time and place ("coterminality"). Innovations may focus on this interaction as much as on conventional product and process characteristics, and may rely less on technical knowledge and more on social and cultural nous. Many services are highly information-intensive, with a preponderance of office-based work or communicative and transactional operations, such as telemarketing. Some service products are deliverable electronically, such as text-based reports, TV programs, music recordings, computer software, and websites. The informational components of many other services are subject to IT-based innovation. But in other respects, the contrasts within the service sector are as significant as those differentiating the sector from manufacturing. After all, the sector includes the most concentrated, knowledge-intensive, and IT-intensive sectors in modern industrial economies (banking, professional services, etc.), as well as the least (retail, cleaning, etc.).

16.2.2 Research on Innovation in Services

Despite the economic importance of the sector, innovation in services received little systematic attention until the 1980s, when a trickle of studies appeared that focused primarily on services and new IT. In the 1990s, a number of major research projects on services innovation were launched, and some services were included in R&D and innovation surveys by the end of the twentieth century.

Although many recent surveys reveal growth in R&D investment within services, R&D surveys may underrepresent the innovative activities of service firms. The increases in service-sector R&D revealed by many of these surveys may reflect (*a*) increased coverage of services firms by surveys (services were excluded from many national R&D surveys until the 1990s), and (*b*) reclassification of some large firms' activities into the services sector. The US National Science Foundation's summary of recent trends in service-sector R&D investment, trends that incorporate both of the effects noted above, nevertheless suggests some systematic change in service-sector R&D activities, as well as important differences among the Triad economies in service-sector R&D:

R&D performance by the U.S. service-sector industries underwent explosive growth between 1987 and 1991, driven primarily by computer software firms and firms performing R&D on a contract basis. In 1987, service-sector industries performed less than 9 per cent of all R&D performed by industry in the United States. During the next several years, R&D performed in the service sector raced ahead of that performed by U.S. manufacturing industries, and by 1989, the service sector performed nearly 19 per cent of total U.S. industrial R&D, more than double the share held just two years earlier. By 1991, service-sector R&D had grown to represent nearly one-fourth of all U.S. industrial R&D. Since then, R&D performance in U.S. manufacturing industries increased and began growing faster than in the burgeoning service sector... Unlike the United States, Japan has yet to see a dramatic growth in service-sector R&D... R&D in Japan's service-sector industries reached 4.2 per cent of the total R&D performed by Japanese industry in 1996 and 4.5 per cent in 1997...

R&D within the EU's service sector has doubled since the mid-1980s, accounting for about 11 per cent of total industrial R&D by 1997. Large increases in service-sector R&D are apparent in many EU countries, but especially in the United Kingdom (19.6 per cent of its industrial

R&D in 1997), Italy (15.3 percent), and France (10.0 percent).   (National Science Foundation 2002: ch. 6)

This picture of service-sector R&D investment undermines the view that innovation in services results solely from the sector's adoption of manufacturing innovations.[1] Abundant case-study evidence also highlights innovation by services firms (see e.g. Andersen et al. 2000; Tidd and Hull 2003). Indeed, some service industries invest heavily in R&D and pursue R&D programs that are at least as sustained as are those of manufacturing. Some useful information on corporate R&D can be gleaned from company annual reports and accounts, and Company Reporting Ltd. produces an annual analysis of these data for the UK's Department of Trade and Industry (DTI). The 2003 analysis (DTI 2003) reports that IT services rank fifth among all industries in worldwide R&D spending in 2003, and sixth among UK industries in 2003 R&D spending. The analysis in this report of R&D spending by IT firms indicates that Microsoft ranks eleventh in worldwide tabulations of corporate R&D spending, British Telecom is the fifth largest R&D spender in the UK, and Reuters the eleventh. Fully 6 per cent of the total R&D spending by the world's largest R&D investors—the top 700 companies, whose R&D spending exceeded £35 million in 2002—is accounted for by firms classified as mainly involved in software and IT services. (The equivalent figure for the UK alone is 5 per cent.)

Data from the Community Innovation Survey (CIS) also indicate that services firms are major innovators, and that it is not just "high-tech" services (like software and telecommunications) which play significant roles in technology development. Although there are innovative activities in all branches of services, the innovation surveys depict lower levels of innovation and R&D investment within services, on average, than within manufacturing. The adoption by services firms of technologies produced in other sectors is indeed a major form of innovation in the sector, much of which thus displays "supplier-driven" characteristics.

16.2.3 Services Attributes and Innovation Trajectories: Industrialization and Modularization

Such patterns of innovation may be related to the ways in which (most) services differ from (most) manufacturing firms. How might the distinctive characteristics of service-sector firms produce distinctive patterns of innovation? Services' interactivity means that their products are often customized to particular client needs. Historically this characteristic has entailed the provision of services on a small scale and local basis. Over thirty years ago Theodore Levitt (1972) argued that service firms needed to adopt a "production line approach," emulating industrial practices and moving toward mass production of standardized products, a more refined division of labor and higher levels of technology. In fact many services have been highly standardized and technology-intensive for a long time—consider railways and conventional telecommunication and broadcast services.[2] But it is apparent that the growth of large-scale firms in other service sectors is associated with a form of increased standardization. McDonald's and other fast food restaurants are one familiar example of such firms in the service sector. Fast food chains also display a measure of customization, in that their products are composed of various components, or modules, which can be combined in numerous ways according to customer demand; new modules may be added to increase variety and support other forms of innovation.

The industrialization of services has been criticized as generating low-quality, low-skill jobs (McDonaldization, or McJobs, to some commentators). Other types of innovation and reorganization in services may produce new forms of social exclusion. For example, the use of call centers and other elements of banking automation often occurs in tandem with the closure of traditional outlets, such as high street branches. Out-of-town hypermarkets have had an adverse effect on high street shopping areas and their environs. Fears have been expressed that consumer e-commerce will have similar effects, since these new channels of customer contact may not be available to all consumers. Such concerns may trigger consumer, regulatory or legislative responses that will affect service markets and innovation strategies.

But exactly what is connoted by the "industrialization" of services cited by Levitt? After all, many manufacturing firms now emphasize flexible specialization, mass customization, reintegration of highly atomized division of labor, and the like. Some of these trends are making manufacturing more like services and some are being emulated by large services companies, even as other service organizations continue to follow more classical industrialization trajectories.

One way in which services have emulated manufacturing is in the adoption and development of an organizational innovation—quality control procedures. In service firms, as has been true of many manufacturing firms, attention to quality has often served as a trigger to innovation, by requiring firms to view their services as consisting of a number of component parts to which quality control principles can be applied. This is typically used to identify areas where there is weakness in service performance and where change thus needs to be engineered. Information Technology is often introduced in the context of improving customer service quality, especially in speeding responsiveness via means such as call centers. Understanding the components of a service also often generates insights into ways in which these components can be transformed or reconfigured into new service bundles (Sundbo 1998). Modularization underpins much services innovation, since decomposition of service processes and/or products may spur process innovation and the identification of new products and product combinations. This type of innovative activity does not necessarily rely on R&D investment,[3] though in the sectors where there has been much talk of "unbundling"—software and telecommunications—there is much conventional R&D.

Another major impetus to "services industrialization" has been the application of IT. This has made it possible to automate elements of the back-office work of many service firms and large firms in other sectors—for instance through the use of document processing, email, Enterprise Resource Planning software and systems, etc. IT is widely applied in back-office work, and in the management and execution of customer-facing services such as mail and telephony. Many of these IT applications make possible the provision of customized services through recombinations of standard service modules. Back-office automation through IT also has changed the spatial location of service activities. Telephone call centers—dedicated offices where the work revolves around the computer-assisted answering of telephone calls, normally for the provision of routine customer service information—are a case in point.[4] Over the last decade we have seen the relocation of call centers in the UK to lower-wage areas of the country, and more recently to the Indian subcontinent—a specific manifestation of the process of "offshoring." There is considerable debate as to the extent to which higher-level office service work will also be subject to offshoring—some quite sophisticated software activities have followed more basic programming work in the move to overseas locations. The new international division of labor clearly involves a redistribution of service as well as manufacturing and extractive activities.[5]

16.2.4 Services Diversity and IT-related Innovation Trajectories

IT has been very widely applied across service sectors—indeed there is disproportionately more investment in IT from services than from manufacturing (which has given rise to some discussion of the "productivity paradox" as a services phenomenon).[6] In some cases, these applications involve little more than adopting mobile telephones, personal computers, and similar devices and their supporting software and services. In many cases, however, IT applications require significant innovative effort, as substantially new applications of the IT are developed. Large service users in sectors like finance and retail have invested huge sums in developing sophisticated networks and new systems for capturing, archiving and analyzing data. Large numbers of graduates in IT-related subjects have been absorbed by these sectors, to the extent that sometimes there have been complaints from manufacturers that a skill shortage has been created.[7]

In many ways IT has provided a technology that can be applied to the generic information-processing activities of services, much as earlier revolutionary innovations in energy technology (e.g. the steam engine, or electric power) could be applied to generic materials-processing activities in manufacturing. IT is by no means the only technology employed in services, nor the only technological field in which services firms are active innovators. Medical services and specialized biotechnology service firms are major users of genomics and post-genomics knowledge and techniques, for example. But IT is pervasive across services, and the uptake of new IT stimulated a recognition that services were often users of innovation and indeed innovators in their own right. Richard Barras (1986, 1990) noted that in many ways the IT revolution was an industrial revolution in the service sectors. He argued that IT-based services innovation has followed a pattern that differs from that usually depicted for manufacturing. In contrast to the classic product cycle in manufacturing, he suggested that services innovation—or more precisely, their IT-based innovation—typically follows a "Reverse Product Cycle" (RPC).

Barras argued that the RPC involved three phases—Improved Efficiency, Improved Quality, and New Services. IT was first introduced to improve existing processes, and only later became the basis for service product innovation, reversing the "product cycle" model of manufacturing innovation popularized by Abernathy and Utterback (1978).[8] For "vanguard services" (the technologically sophisticated sectors such as financial services, for instance), Barras suggested that these three phases roughly characterized the 1970s, 1980s, and 1990s respectively. Thus, insurance services moved from computerization of policy records, to providing online policy quotations, and then to supplying complete online services during these three decades. A similar account applies as well to service functions outside of the service sector. Many new telematics service firms—such as EDS, which emerged from General Motors—were organized around manufacturing firms' in-house innovations in communications and data management processes that enabled these firms to streamline their internal communication activities. Eventually, these service activities were spun off as IT service companies. Some online information services originated from in-house data management services, e.g. for publishing firms.

The analysis of Barras has influenced a large body of scholarship that highlights the contrasts between services innovation trajectories and those of manufacturing (Barras 1986, 1990). Many services may remain non-innovative, but increasing numbers of IT adopters are experimenting with applications and related innovations, thereby providing new arenas for testing the implications of the Barras framework.[9]

Critics of the RPC model have highlighted a number of issues:

- counterexamples (e.g. in IT-based services, which often begin with product innovations);
- conceptual difficulties (the blurring of production and consumption makes it difficult to establish a point in time at which innovation shifts from efficiency enhancement to product innovation);

- the historically specific nature of the story (after assimilating new IT and going through an RPC process, will services then follow the classic product cycle?).

Uchupalanan (1998, 2000) mounted a systematic critique of the RPC approach, tracing the history of five IT innovations[10] through all firms in the banking services sector in Thailand. He uncovered a diverse range of innovation strategies and trajectories that were far richer than the RPC account. The banks were influenced by the strategies of competitors with respect to each given innovation, by their experiences with earlier innovations (and their plans for others), and by pressures from regulators and the market. The interrelation of market competition, firm circumstances and innovation dynamics meant that the RPC "story" of innovation processes was rarely applicable in this context. At best it was one of a number of possible patterns of development.

The RPC approach, which has proven influential in many recent analyses of services innovation, may neglect non-IT innovations and innovations undertaken in earlier periods of service innovation. But this framework's attention to the increasing innovative activities of service firms, as well as the historic omission of their innovative activities from most scholarship, means that the RPC is an important contribution, providing a starting point for further research on services innovation.

Other studies of services innovation place more emphasis on the relation between service firms and their clients. One line of work uses the term "servuction" (services production, intended to highlight analogies with manufacturing production activities) to describe the activities and procedures involved in producing and sustaining supplier–client relations and delivering "service"—roughly what we call "interactivity" above. Belleflame et al. (1986) classified a sample of innovations from service companies in terms of whether these centered on servuction, production, or a combination of both (without assuming that these would necessarily be IT-centered innovations),[11] and found numerous examples of all types of innovation. Tordoir (1996) studied professional services, and noted very different patterns of interaction between suppliers and clients, introducing a distinction between "jobbing" relations (where the service supplier provides a relatively standardized service), and "sparring" ones (where the supplier and client negotiate the details of what service is to be provided, and how). More recent studies have addressed the question of how service innovation differs according to the degree to which services are standardized or specialized to specific clients (cf. Hipp et al. 2000, 2003). The more interactive services would seem to require greater exchange of knowledge between service firm and client, and the learning processes involved can be a fertile basis for innovation: however, many of the innovations are liable to be one-off solutions for particular problems, or elaborate customizations of more generic solutions.

The RPC approach argues that IT-based innovation begins with back-office processes and then moves forward into functions involving more customer contact.

These latter functions may involve new service products (such as new types of bank accounts more closely tailored to the circumstances of individual clients), or delivery innovations (e.g. online banking and cash machines). There is also scope for innovation in other customer-facing functions, for example in targeted marketing and in computer-assisted helplines, etc. Interaction around service relations certainly does involve information flows—especially in the case of informational services like counseling, consultancy, and education. But interaction may involve many physical elements, not all of which are primarily carriers of informational content (consider services performing physical and biological functions like catering, transport, cleaning, surgery, and hairdress-ing).

IT-based innovation may apply to the informational elements of such service provision—marketing, ordering, transactions, etc. IT may be an important adjunct to other types of services innovation—without telephones, pizza delivery services would be much less attractive! But it is not the only vector. Other types of interaction are also involved in services, including physical and other elements, and these may benefit from non-IT-based innovation—motorcycles and insulated boxes in the case of pizza delivery. The RPC model, which emphasizes the catalytic effect on services innovation of the introduction of IT equipment and software from manufacturing sectors, could be extended to apply to these technologies in the service sector as well. Examples include supermarkets investing in the development of refrigeration tech-nology (as well as general refrigeration design inputs, supermarkets played import-ant roles in determining what sort of shift was made to alternatives to CFCs in equipment), railway companies supporting the development of better trains (before privatization, British Rail had a substantial R&D facility conducting work on topics such as faster engines, more environmentally sound carriage painting practices, and safety issues), and so on.

Another approach to examining the way in which services differ considers the key transformations that they effect, and distinguishes among three broad groups of services (Miles 1993). This classification highlights the adoption and innovation dynamics of services with respect to a number of major technologies in addition to IT.

A first set of services, *Physical Services* (transport, domestic services, catering etc.) involve physical transformations. These have been particularly suitable for the application of automotive and electric power technologies—e.g. in freight haulage, laundries, cooking equipment, and so on. The adoption of new technologies in this sector has frequently created competition among different modes of provision: between road, rail, and air services; laundries and launderettes; traditional restaur-ants and fast food outlets; etc. But the competition between modes of provision can go much deeper. In particular, consumer services such as laundry or food prepar-ation have been subject to competition from what Gershuny (1978) described as "self-services," production by consumers in the household. Such technologies as the automobile, home refrigeration and other household appliances have enabled consumers to use manufactured goods to provide services in the household in competition with service firms. The adoption by consumers of these manufactured goods also has given rise to some new services, such as automotive repair establishments.

New IT is now widely used in physical services, such as transport, logistics, retailing, and warehousing. Computers have long been applied to their back-office accounting and transactional functions, and electronic cash registers and scanners[12] introduced into supermarkets and smaller shops. These IT-based devices are linked to the office systems being used for stocktaking, and in some cases data are fed to automated warehouses and much wider systems of supermarket automation. New "transport informatics" systems do more than just document timetables and the locations of vehicles, providing advanced routing and tariffing procedures, backed up by mobile communications, "smart cards" and other innovations.

The large-scale *Human Services* subsector has often been organized under the auspices of the welfare state, and many of the front-office tasks in this subsector have had little scope for application of generic IT. Many of the *Social Welfare* activities of this subsector have utilized office and communications systems, while the *medical* field has exploited successive generations of medical technology—surgical equipment, pharmaceuticals, radiology, etc. Typically both sets of services have had to combine large-scale administrative data processing applications (payroll, pensions, passports, driving licenses, and the like) and planning (for example, managing housing systems and waste disposal services, monitoring epidemiological and environmental statistics) functions, with tasks that are much more customized in terms of the characteristics of specific citizens or clients. The large-scale tasks were early pioneers of computer use. Now, PCs and data networks are being widely adopted, allowing decision support in the context of the details of individual clients—for example, expert systems to aid medical diagnosis and prescription, or to speed up the task of assessing individual entitlement to welfare benefits. There are some self-service-type applications, too, where the clients use public access terminals or home-based equipment to gather information on service provision, or access the services directly (e.g. interactive teaching aids, databases on available jobs or benefits, etc.) Some integration within and across public services is possible as information on the same client held in different databases can be combined—but privacy rules have often impeded such integration, and there are many other instances where perceived threats to privacy and civil liberties have restricted innovations involving capture and distribution of information.

Finally, *Information Services* are relatively less dependent on motor power technologies, although electricity is important to them. They have made considerable use of information technologies, including those that predate new IT and its foundations in microelectronics. Some of these services, such as broadcasting and cinema, are founded on these technologies, while others, like consultancies and technical services, have used them as important tools for producing and delivering their outputs. In some of the consumer services here, new and even older generations of IT display self-service trends similar to those that motors induced in physical services. Thus, in the field of entertainment, traditional consumer services like theater and cinema have been subject to competition from new ones such as TV and other audiovisual equipment and, more recently, videogames, PCs, and online entertainment. Professional and business services are among the most IT-intensive sectors of industry, as are financial services, which are also largely concerned with information processing (in this case processing and manipulating data about property relationships).

New IT is thus enormously important—and often very visible—in information services. Such innovations as automated teller machines and smart cards, new telephone and telematics services, and shifts from analog broadcasting to interactive digital media and "narrowcasting", are all the focus of considerable investment and activity. Most national broadcasting services in the West now offer opportunities to access archived radio and sometimes TV services through the Internet, and often provide a good deal of additional content, discussion fora, and the like.[13] Some of the service firms involved play a prominent role in guiding the innovations, as in the case of banks who are heavily involved in defining the characteristics of new teller machines. The pace of innovation in these industries is quickened by the shift in regulatory policy in many countries, which has led to new entrants and increased international competition confronting many firms and sectors (Sauvé and Mattoo 2003).

This sort of analysis provides a useful mapping of different types of services and innovation trajectories, but reveals little about the dynamics and processes of innovation in the various services that have been distinguished. The RPC model is not the last word on this subject. One potentially fruitful avenue for further research applies evolutionary and characteristics-based approaches to services innovation (see Gallouj 2002).

16.3 Innovation Systems and the Organization of Innovation

IT provides a generic technology that can be widely applied across service sectors—a substantial break with past experience, where even technologies like automobiles and typewriters, almost universally adopted in certain services branches, were rarely applied in others.[14] Though services vary in the extent and speed with which they are adopting these innovations, and some of them are more relevant to certain branches of services than to others, many technological opportunities are opening up for all services. Even personal services such as hairdressing are subject to innovations—there are shops where video technology and PCs are used to give clients an impression of how they would look with different hairstyles, for instance. And IT is a very configurational technology, which challenges users to develop new software and interfaces, and to bring diverse elements of technologies together in new ways.

Their increased technology-intensity has not yet transformed all of the features of innovation in services. Many services are arguably laboring under a heritage derived from past periods where few generic technologies found ready application in their activities. The low technology-intensity that previously characterized services meant that many services firms paid little attention to strategies for the management or adoption of innovation. Limited use of advanced technology would also mean that most service firms would have little incentive to be linked into innovation systems that would connect them with those responsible for generating the new technologies. And in the wider innovation systems themselves, few of the innovation-related facilities offered by institutions such as university departments, research institutes, and government laboratories are tailored to the requirements of services. It is thus not surprising that few service firms make much use of these resources currently (Institute of Innovation Research 2003).

There are, as always when discussing services, exceptions. For example rail, broadcasting, and telecommunications services, many of which were state-owned and very large organizations for much of their existence, were closely related to manufacturers. These large services firms frequently ran their own laboratories, testing sites, technical training programs, etc. Other exceptions include very large-scale service firms in sectors like financial and retailing services. Especially in the latter case, supermarket chains have become adept managers of their own supply chains, and are often active in dictating production processes and products—including innovations—in their agricultural and manufacturing suppliers. For example, their suppliers may be requested to adopt specific environmental or animal husbandry practices, to use ecommerce techniques, etc. As well as influencing their suppliers, such firms have been pioneers in some sorts of technology for their own use—for example, data warehousing and data-mining methods.[15] Another, and particularly important, class of exceptions, are the business services that are, as we shall see, important actors in innovation systems, contributing to innovation across the economy. Consultancies, training organizations, and many firms helping to service new technologies are just a few of the agents involved here—not to mention specialized R&D and design services themselves!

But relatively few service firms and sectors have strong links with national or regional innovation systems. The organization of innovation is a new development in many service firms—for instance, in a set of recent interviews the author found that many large service firms in the UK were only just beginning to examine the application of IPR arrangements to their innovations, as opposed to their trademarks and copyright content.[16] In part, this new orientation reflects a general emphasis on intangible assets, and in part it is in response to the emergence of "business methods" patenting in the USA, which are seen as impacting upon many service operations.

One result of this heritage is that many service firms are unable to apply relevant knowledge, and, more particularly, are unable to learn effectively. They can rapidly adopt "off the shelf" technologies like PCs, but find it harder to develop more customized or innovative solutions. In part this is because services innovation is rarely organized in terms of the "standard" models of R&D management structures, and is typically conducted on a more ad-hoc, project management basis. Service firms tend to stress human resources and technology acquisition rather than formal R&D (Tether et al. 2001; Sundbo 1998), resulting in limited coordination of learning experiences—the innovations that are made are often not reproduced in subsequent projects, and flows of knowledge within and between services firms as to technological opportunities, good practice in innovation, and the like, may be limited.[17] CIS data discussed below show that many services sectors display somewhat lower levels of innovation than their parallels in manufacturing. This may result from weak integration into innovation systems, the orientation of existing technology support institutions toward manufacturing activities, and weak internal organization for innovation in many services—even if the firms themselves do not report exceptional problems in innovative efforts.

Such features of services innovation partly reflect the historical legacy discussed above. But they also partly reflect the fact that the features of different types of economic activity can shape innovation processes and trajectories in distinctive ways. Arguably, the nature of service innovations demands different approaches to management and organization of the innovation process—for example, requiring more emphasis on the service workers (especially if professionals) and clients, and on the interaction process between them. The traditional R&D lab may not be well suited to such innovations, though we find large technology-intensive services firms have been running such laboratories (in some cases, such as telecommunications and railways, for many decades).

The differences between services and manufacturing innovation may reflect historical legacies stemming from the low technology-intensity of many services activities (which may be overcome in time), alongside of the significance of characteristics such as interactivity to many services (which may demand specific processes and practices). But these generalizations need to be qualified to take into account the enormous heterogeneity within services—they do not apply to large segments of the services sector. Developing a more nuanced framework that is sensitive to the enormous heterogeneity among the various components of the services sector is an important task for future research.

16.3.1 Innovation Surveys and Services

Despite substantial efforts to improve the situation in recent years, data on many aspects of services and services innovation are less detailed and less comprehensive than those for manufacturing. Even the Community Innovation Survey excludes public and personal services. Many features of services innovation are quantified in such innovation surveys and similar data, but qualitative features are harder to detect. Few available innovation indicators were designed with services in mind, and thus may fail to adequately capture the dynamics of services innovation.

With these caveats in mind, as well as the discussion above on the differences between innovation in services and manufacturing, it is noteworthy that the evidence from these surveys indicates that services innovation does not seem to follow dramatically different paths from those displayed in manufacturing. The differences appear to be more of degree than of kind.

Tether et al. (2001) present the most extensive analysis to date of services innovation, based on the CIS2 (second Community Innovation Survey) data.[18] Just under half of the service enterprises reported undertaking innovative activities between 1994 and 1996. This share is slightly smaller than that reported for manufacturers, and this difference between manufacturing and service firms' innovation propensities remains even when size is controlled for. Most service branches include a higher share of smaller businesses than do manufacturing branches (though financial services are dominated by very large firms). And larger enterprises are more likely to engage in innovative activities in (most) service branches, as elsewhere. But this is not enough to account for the difference between services and manufacturing firms in reported levels of innovation. (Looking at the size distributions, we find 36 per cent of small services and 48 per cent of small manufacturers; 48 per cent and 55 per cent of medium-sized services and manufacturers, respectively; and 73 per cent and 79 per cent of large services and manufacturers, respectively, are classified as innovative enterprises.)

The proportion of innovators is high amongst the technology-oriented services, many of whom are comparable in terms of reported innovation levels to high-tech manufacturing—68 per cent of computer services, 64 per cent of telecommunication, and 55 per cent of technical services are classified as "innovative enterprises" in CIS2. More traditional services appear to be particularly low innovators (only 24 per cent of transport services, for example)[19].

Such results should be interpreted cautiously, since as this chapter has pointed out, the measures of innovation employed in the CIS are probably less than ideal for studying services. Services firms are less likely to see what they are doing as technological innovation, rather than customization or one-off service production. Organizational innovation may be important in services, but is not examined in the survey. The CIS2 also fails to specify the sorts of technology involved in innovations, meaning that the role of IT in services innovation cannot be traced.

But the role of IT in services innovation is difficult to overstate—Licht and Moch (1999) found that *all* innovating services firms in Germany undertook IT innovations, even if they also applied other technologies. Unfortunately, the survey they worked with only covered services, so we cannot say whether this conclusion would also apply to other sectors.

Contrary to some expectations, just under half the innovating service enterprises covered by the survey analyzed in Licht and Moch reported that they had engaged in R&D between 1994 and 1996. Indeed, a quarter of these firms reported having engaged in R&D on a continuous basis. However, R&D is less common in innovating services firms than amongst manufacturers of similar characteristics, again controlling for size.[20] R&D is more common in large service enterprises, and in technology-oriented services.

Of course, R&D is not strictly speaking the most common or most important source of innovation for many services. Acquisition of machinery and equipment, acquisition of other external technologies (including software) and training directly linked to innovation were the most widely undertaken innovation-related activities in services, according to CIS2 data (Tether et al. 2001). These sources illustrate the importance of technology adoption in services innovation. The importance of the human element for services is also underlined by the training expenditures of these firms. On average, acquired technologies accounted for the largest share of expenditures on innovation; while in-house R&D accounted for another quarter of total expenditures on innovation, with this share higher amongst technology-oriented services. Technology-oriented services tend to spend most on innovation, but all sectors surveyed contain some very high spending (and some very low spending) enterprises.

Another set of CIS questions concerned sources of information for innovation. Tether and Swann (2003) contrast services and manufacturing firms' use of information sources, using CIS3 data for the UK. The broad pattern of results is fairly similar across the sectors, with rather more manufacturing firms typically reporting use of each information source. Sources within the enterprises were the most commonly used, being cited by 85 per cent of manufacturers and 81 per cent of services. A substantial proportion of non-R&D performing innovators reported using such sources. Suppliers are also widely used by both sectors (83 per cent and 77 per cent of manufacturers and services firms respectively cited these sources), followed by customers (80 per cent and 73 per cent). Rather less important sources of information are the technical press (65 per cent for both sectors), competitors (66 per cent and 62 per cent), and trade fairs (72 per cent and 58 per cent). Two information sources where services firms report making more use than manufacturers are meetings and conferences (manufacturing 52 per cent, services 62 per cent), and consultants (48 and 56 per cent). Other sources, such as standards, government offices, other parts of the enterprise, and universities (36 and 24 per cent) are more often cited by manufacturers.

The sectoral differences are less acute than might have been expected (and closer inspection of variations across different services indicates considerable diversity here). But the greater use of consultants and lower use of sources such as Universities tends to confirm the notion that many services are poorly linked into wider innovation systems, and the formal institutions that support them.[21]

Many specialized services firms are carrying out functions that are also undertaken within firms in other sectors of the economy. Office work, transport, commercial transactions, security, and catering, and similar activities—services—are undertaken in all sectors, though to differing degrees. Just as innovation surveys were not designed with services in mind at the outset, so there may be some overlooking of service functions, which may well slip between the "product innovation/process innovation" categories of CIS-type instruments, and may even be matters that the respondents to the surveys are unaware of. A number of recent studies consider services supplied in support of the core products of manufacturing firms (e.g. Kuusisto 2000; Lay 2002; Mathé and Shapiro 1993); commentators such as Davies (2003) and Howells (2001) have argued that the service component of such products is growing dramatically in many sectors. The issues that confront adoption and innovation of service-related technologies in services firms may well be liable to affect innovation in service functions in manufacturing firms. Service functions can in principle be significant loci of organizational learning and innovation in all sectors. It remains to be established whether the organization of innovation in such service functions more resembles conventional manufacturing innovation, or more informal systems used for services innovation.

16.3.2 Innovation-Supporting Services in Innovation Systems

Business services have risen dramatically in economic significance—for instance, from little over 3 per cent of US value-added and employment in 1970 to 9.9 and 13.8 per cent respectively in 2000, with equally striking increases in other countries (ECORYS-NEI 2003). They have also become highly evident contributors to innovation across the economy.[22] They provide intermediate inputs to industry and other organizations—they are business services as much because they are servicing business processes as because their clients are often private firms themselves. Knowledge-intensive business services (KIBS), in particular, play important roles in innovation systems (Leiponen 2001; Miles 1999b; and several chapters of Gadrey and Gallouj 2002). Some are transnational firms, dealing with knowledge from the frontier of practice. But many are smaller; locally based KIBS may be important agents of transfer of locally-specific knowledge, embedded in local networks, between actors in regional innovation systems (cf. Kautonen 2001). Several studies suggest that the presence and use of KIBS enhances the performance of economic sectors and regions.[23]

KIBS include all kinds of business services that are founded upon highly specialized knowledge—social and institutional knowledge in many of the traditional professional services, or more technological and technical knowledge. These firms typically have very high levels of qualified staff, such as university graduates.[24] Some KIBS are based on administrative, legal, marketing, or similar knowledge. Others are directly based on scientific and technological knowledge—testing, prototyping, environmental services, engineering consultancy, etc. The knowledge requirements for technology users are bound to be more challenging where new technology is involved. Firms are less likely to have acquired the knowledge necessary to understand, master, and utilize new product and process opportunities.[25] Thus, many technical KIBS focus on new technological opportunities—examples include Web and Internet, software and computer services; others on the production and transfer of knowledge about new technology: information and training services, for example.

Technology-related KIBS sectors are among the most active innovators in the economy, as indicated by CIS and other data (e.g. DTI 2003, which includes data indicating the high levels of R&D expenditure, patenting, etc. of software and IT services).[26] We saw above that Tether and Swann (2003), using CIS3 data for the UK, found that most services are poorly linked with the public elements of innovation systems (e.g. in terms of sourcing information from and collaborating with universities). But they also report that some technical KIBS were outstandingly well linked to these components of innovation systems. Some technical services (e.g. contract R&D) had uniquely high levels of interaction with the public science base, higher than any manufacturing sector; however, IT services and more professional services tended to have low levels of contact, relying more on professional associations and the like to refresh their knowledge.

Many KIBS play important roles in innovation processes in their client firms and sectors. These roles may not be altogether new—some have existed for many decades. But there has been considerable growth of employment and output in KIBS sectors, which can only mean that their use has also expanded, and suggests that their significance for innovation across the economy will also have mushroomed.

What are these roles? KIBS may provide the firm with general information about its internal operations and external environments. Technology and innovation-related information often forms part of this. KIBS may simply play a role in identifying the nature of a particular problem or class of problems confronting the firm (for instance, that competitors are launching products with new functionalities, that regulations may mean that processes will have to generate less of a particular pollutant, etc.). KIBS may propose ways of solving a technological problem (for example recommending that particular strategies are undertaken for product or process innovation). They may provide advice (for example, recommending a specific technological solution), or actually implement such a solution on a "turnkey" or long-term basis (as in the case of systems integrators and facilities managers). The KIBS relate technological knowledge to the specific problems encountered by the client. German survey data indicate that technology-related KIBS are more likely to produce specialized products that are tailored to client needs than were the other services studied (Hipp et al. 2003).

In addition to expertise in specific industrial, technological, or functional domains, KIBS professionals require skills in interpersonal communication, presentation of materials, "impression management," and the like. These are fairly rare capabilities, and their combination is rarer still: for this reason labor costs and wages are high in most KIBS. Toivonen (2001) finds that in Finland effectively all of the KIBS studied require combinations of generic and sector-specific skills. Common requirements were marketing and sales skills; social skills such as sensitivity to others, willingness to share knowledge and motivate others, capacity for self-renewal, IT-related skills, and sector-specific knowledge related to one's particular expertise, to the processes and business mechanisms characteristic of the KIBS, and knowledge about the industries and organizations of clients.

16.4 CONCLUSIONS

The growth of services to their dominant position in industrial economies means that we can no longer ignore services innovation, or simply assume that it follows the patterns and processes depicted in manufacturing production processes. Understanding "service innovation" may widen our approaches to explaining, measuring, and managing innovation. Innovation studies will have to take on board the issues of organizational and market innovation, interorganizational and client-facing innovation, and even aesthetic and cultural innovation. Many of the mostly widely remarked features of the evolution of technological innovation in the late twentieth century depended centrally on combined material and non-material innovation. This suggests that our models of innovation should put less emphasis on artifacts and technological innovation, and more on seeing innovation as involving changes in market relationships that can be effected at least partly through artifact and service innovations, with organizational and technological dimensions.

Service innovation now occupies a more prominent position in innovation studies over the last few years, but this has yet to be reflected in an accumulation of knowledge about services in innovation policy. There are few obvious efforts to give services a prominent role in such policies, and their specific requirements may be overlooked in current programs. (There is a dearth of research concerning how innovation policies affect services, let alone analysis of service-related policies.) One exception is Finland, where KIBS are being treated as important actors in the innovation process. We have seen that KIBS are significant elements of innovation systems, and as policy adopts a more systemic viewpoint, we can expect their role to become more of a focus in policy.

The greater policy salience of services innovation is, in turn, likely to stimulate more detailed research into the area. It will be important to move beyond documenting the growth of KIBS, to examine precisely how they function as knowledge sources and intermediaries: what sorts of exchange of knowledge take place in what ways and at what steps of the innovation process; how these are managed on service and client sides of the equation; what skills and capabilities are required for effective and innovative solutions to be implemented. There are still very few studies that address these fundamental questions.[27]

Notes

1. This view has been widely expressed, for instance, with services being classified as supplier-driven in Pavitt's original taxonomy of types of innovation (1984)—though in a subsequent paper (1994) he put software services into the "specialized supplier" group, and added a category of "information intensive" firms which included finance, retailing, travel and publishing.
2. Hipp et al. (2000) show that a surprisingly large proportion of German service firms consider their outputs to be largely standardized. Least standardization was reported by business services such as technical and computer services.
3. For just one example, consider Jones (1995) on in-flight catering services.
4. By 2000, about 1 per cent of the UK workforce was employed at telephone call centers.
5. For a series of studies on service internationalization and innovation—still a largely unexplored topic—see Miozzo and Miles (2003).
6. Cf. Miles and Matthews (1992), Roach (1988) for early analyses. The data, derived from input–output tables, allow us to examine investment patterns across different industries. IT investment from services sectors constitutes a greater share of this investment than services' output constitutes of all output. And for the UK in the 1980s, Miles and Matthews found that the proportion of sectoral investment devoted to computers and telecommunications equipment was 5.7 and 4.6 per cent respectively for services, 4.1 and 4.0 per cent for manufacturing. (The expenditure on telecommunications equipment among services at this time was heavily dominated by telecommunication services firms; and in general the high levels of IT investment are driven by specific sectors, such as finance.)
7. Not that the problem is not reported by manufacturers—I found UK public sector organizations in the 1980s to be vociferously complaining that their projects were plagued by departure of IT staff to financial services. Beyond the affluent West similar concerns are

also raised, for instance *Computerworld Singapore* (7(3): 20) documented in October 2000 that financial services computer professionals there were consistently paid more than comparable staff in other sectors.

8. This model, while proving extremely influential and a useful starting point for analysis, is widely viewed as of limited applicability even to manufacturing.

9. US researchers have also suggested such developments: see e.g. Faulhaber, Noam, and Tasley (1986). But see Bronwyn Hall's Ch. 17 in this volume, which indicates that "followers" may spend a long time in that state before becoming innovative in their own right.

10. Interbranch on-line services; automated teller machines; credit cards and associated services; remote banking, and electronic funds transfer at point of sale.

11. Gallouj and Weinstein (1997) provide a useful review, comparing "servuction" to a number of other formulations. Though there have been few "servuction" studies focusing on innovation, the conceptual approach continues to be developed, e.g. by Gadrey and de Bandt (1994).

12. Note that bar-code scanners require the cooperation of manufacturers in bar-coding their products. Similarly, financial service innovations like credit and debit cards require the cooperation of retailers in accepting these cards and using validation systems.

13. The BBC website at http://www.bbc.co.uk is a case in point. New BBC radio and TV channels are only available digitally, and remarkable volumes of archived content and text are also online.

14. Probably the only universal innovations were ones involving the construction and maintenance of buildings and technologies of heating and lighting—and telephones. Such innovations typically involved relatively little user learning. Even where new skills were required (e.g. automobile driving skills—which in any case were typically acquired for everyday purposes) these technologies required relatively little configuration to meet the needs of specific users. Indeed, limited scope for configuration was presented, and the relevant engineering services (e.g. garages) were a matter of high-street crafts rather than industrial laboratories.

15. NIST in the USA (TASC 1998) examined how technology-intensive services in several sectors deal with technology barriers. Some barriers were technology-specific—thus the high risks of complex technology led to needs for technical expertise not available to most individual firms. Collaborative R&D was undertaken with other firms in their industry (and often with manufacturing sector partners, too), typically to gain access to complementary research or technical skills. IT development barriers were overcome by codevelopment projects with manufacturing suppliers. Such collaborations were not in general oriented to knowledge resulting from basic research. IT implementation barriers were associated with high costs of configuring and employing systems (often IT implementation cost four to five times more than acquiring the hardware and software). Other barriers were "market-related"—thus high transaction costs associated with the systemic nature of new IT, which led to emphasis on standards-related activities to reduce barriers to IT development and implementation. Standards and protocols were often central to innovation strategies.

16. See the report by FhG-ISI (2003), which deals especially with services' patenting activities. The general argument is that the patent mechanism is not very appropriate to many service innovations, and thus the firms are not particularly oriented to these types of

intellectual property. Nevertheless, the study found numerous service firms active in patenting activity and the development of new IP strategies.

17. Again, there are striking exceptions to this rule—some service firms in areas like IT and consultancy are pioneers in innovation and knowledge management techniques.

18. The sectors studied excluded public and consumer services such as retail and HORECA; and microbusinesses and very small firms were also excluded.

19. Tether and Swann (2003) present detailed information on UK firms in CIS3 data, where different types of manufacturing and service are contrasted in some detail.

20. Nearly 70 per cent of the innovating manufacturers conducted R&D.

21. Other evidence supporting this view is presented in Miles 1999*a*; Tether and Swann 2003.

22. One—but only one—of the sources of growth in business services, in particular, is the outsourcing of activities previously undertaken in-house by firms and organizations in other sectors.

23. Some authors even describe them as forming a "second knowledge infrastructure." The traditional primary knowledge infrastructure is mainly a matter of Higher Education Institutions (HEIs) and government laboratories/public research and technology organizations (RTOs).

24. It is difficult to attribute causality to the correlational data involved here, but some very different types of study yield broadly similar results. Researchers like Antonelli (2000) and Tomlinson (2000)—using slightly different input–output datasets, and statistical methods—show an association between the use of KIBS as intermediate inputs, and the performance of the user sectors. Doubt is cast on the methodology by ECORYS-NEI (2003), however. Peneder (in European Commission 2000: ch. 4) found that clusters of industries characterized by high KIBS use performed particularly well. Hansen (1994) reported that the growth performance of the economies of US cities was related to the size of the KIBS sectors in these economies. An interesting study by Muller (2001) examines relations between KIBS and SMEs at regional level, suggesting benefits in terms of innovation on both sides.

25. Tether and Swann (2003) show that in the UK, at least, these are the sectors with the highest graduate-intensity. There are clear differences between those KIBS with high levels of science and engineering graduates (e.g. Technical and IT services) and those with other classes of University graduate (e.g. consultancy and marketing services).

26. CIS2 data show that acquisition of external technology through use of consultancy services is the second most frequent mechanism used by manufacturing firms. The most important mechanism is direct equipment purchases.

27. Tomlinson (1999) analyzed UK survey data, finding that KIBS staff are more likely than are others to learn new things, to receive training, to work with computers, and to move between different types of work. Labor mobility is often emphasized as a means for diffusing knowledge around the economy. Tomlinson argued that KIBS provide an alternative—perhaps a superior—means. (The survey suggests that people moving between jobs fare poorly on these indicators of "life long learning." Whether this result is peculiar to the UK and/or to the recession underway at the time of the study and the downward mobility it induced, requires further study, since the argument is very provocative.)

28. Cf. ECORYS-NEI (2003) for a review of literature on business services and their clients. Several relevant studies are presented in the recent collections by Dankbaar (2003) and Tidd and Hull (2003).

REFERENCES

ABERNATHY, W., and UTTERBACK, J. (1978), "Patterns of Innovation in Technology," *Technology Review* 80: 40–7.

* ANDERSEN, B., HOWELLS, J., HULL, R., MILES, I., and ROBERTS, J. (eds.) (2000), *Knowledge and Innovation in the New Service Economy*, Cheltenham: Edward Elgar.

ANTONELLI, C. (2000), "New Information Technology and Localized Technological Change in the Knowledge-Based Economy," in Boden and Miles 2000: 170–91.

BARRAS, R. (1986), "Towards a Theory of Innovation in Services," *Research Policy* 15(4): 161–73.

—— (1990), "Interactive Innovation in Financial and Business Services: The Vanguard of the Service Revolution," *Research Policy* 19: 215–37.

BELL, D. (1973), *The Coming of Post-Industrial Society*, London: Heinemann.

BELLEFLAMME, C., HOUARD, J., and MICHAUX, B. (1986), *Innovation and Research and Development Process Analysis in Service Activities*, Brussels, EC, FAST. Occasional papers no 116.

BESSANT, J., and RUSH, H. (2000), "Innovation Agents and Technology Transfer," in Boden and Miles 2000: 156–69.

BODEN, M., and MILES, I. (eds.) (2000), *Services and the Knowledge Based Economy*, London: Continuum.

BOLISANI, E., SCARSO, E., MILES, I., and BODEN, M. (1999), "Electronic Commerce Implementation: A Knowledge-Based Analysis," *International Journal of Electronic Commerce* 3(3): 53–69.

*BRYSON, J. R., and DANIELS, P. W. (eds.) (1998), *Service Industries in the Global Economy*, 2 vols., Cheltenham: Edward Elgar.

COX, D., GUMMETT, P., and BARKER, K. (2001), *Government Laboratories—Transition and Transformation*, Amsterdam: IOS Press.

*DANKBAAR, B. (ed.) (2003), *Innovation Management in the Knowledge Economy*, London: Imperial College Press.

DAVIES, A. (2003), "Are Firms Moving 'Downstream' into High-Value Services," in Tidd and Hull 2003: 321–41.

DEN HERTOG, P. (2000), "Knowledge Intensive Business Services as Co-Producers of Innovation," *International Journal of Innovation Management* 4(4): 491–528.

—— and BILDERBEEK, R. (2000), "The New Knowledge Infrastructure: The Role of Technology-Based Knowledge-Intensive Business Services in National Innovation Systems," in Boden and Miles 2000: 222–46.

DEPARTMENT OF TRADE AND INDUSTRY (2003), *The 2003 R&D Scoreboard* (data provided by Company Reporting Ltd), London: DTI available at: http://www.innovation.gov.uk/projects/rd_scoreboard/introfr.html

ECORYS-NEI and CRIC (2003), *Business Services: Contribution to Growth and Productivity in the European Union*, Report to European Commission DG Enterprise, Rotterdam: ECORYS.

EUROPEAN COMMISSION (2000), *European competitiveness report 2000*, Brussels: Commission Staff Working Paper ENTR DT 2000/045/A1 Competitiveness, available at: http://europa.eu.int/comm/enterprise/enterprise_policy/competitiveness/doc/compet_rep_ 2000/cr-2000_en.pdf

* Asterisked items are suggestions for further reading.

EVANGELISTA, R., and SAVONA, M. (1998), "Patterns of Innovation in Services: The Results of the Italian Innovation Survey," paper presented to the 7th Annual RESER Conference, Berlin, 8–10 October; revised version forthcoming in *Research Policy*.

FAULHABER G., NOAM, E., and TASLEY, R. (eds.) (1986), *Services in Transition: The Impact of Information Technology on the Service Sector*, Cambridge Mass.: Ballinger.

*FHG-ISI (Fraunhofer Institute for Systems and Innovation Research) (2003), *Patents in the Service Industries*, Karlsruhe: FhG-ISI; report to the EC; available at: ftp://ftp.cordis.lu/pub/indicators/docs/ ind_report_fraunhofer1.pdf

FUCHS, V. (1968), *The Service Economy*, New York: NBER.

—— (1969), *Production and Productivity in the Service Industries*, New York: NBER.

GADREY, J., and DE BANDT, J. (1994), *Relations de service, Marchés de service*, Paris: CNRS.

*—— and GALLOUJ, F. (eds) (2002), *Productivity, Innovation and Knowledge in Services*, Cheltenham: Edward Elgar.

GALLOUJ, C., and GALLOUJ, F. (2000) "Neo-Schumpeterian Perspectives on Innovation in Services," in Boden and Miles 2000: 21–37.

GALLOUJ, F. (2000), "Beyond Technological Innovation: Trajectories and Varieties of Services Innovation," in Boden and Miles 2000: 129–45.

*—— (2002), *Innovation in the Service Economy: The New Wealth of Nations*, Cheltenham: Edward Elgar.

—— and WEINSTEIN, O. (1997), "Innovation in Services," *Research Policy* 26: 537–56.

GERSHUNY, J. I. (1978), *After Industrial Society? The Emerging Self Service Economy*, London: Macmillan.

—— and MILES, I. D. (1983), *The New Service Economy: The Transformation of Employment in Industrial Societies*, London: Pinter.

GRAHAM, S. J. H., and MOWERY, D. C. (2003), "Intellectual Property Protection in the U.S. Software Industry," in W. Cohen and S. Merrill (eds.), *Patents in the Knowledge-Based Economy*, Washington, DC: National Academy Press, 217–58.

GREENFIELD, H. C. (1966), *Manpower and the Growth of Producer Services*, New York: Columbia University Press.

HALL, B. H. (2003), "Business Method Patents and Innovation," presented at the Atlanta Federal Research Bank conference on Business Method Patents, Sea Island, Georgia, 3–5 April.

HIPP, C., TETHER, B., and MILES, I. (2000), "The Incidence and Effects of Innovation in Services: Evidence from Germany," *International Journal of Innovation Management* 4(4): 417–54.

—— —— —— (2003), "The Effects of Innovation in Standardized, Customized and Bespoke Services: Evidence from Germany," in Tidd and Hull 2003: 175–210.

HOWELLS, J. (1999), "Research and Technology Outsourcing and Innovation Systems: An Exploratory Analysis," *Industry and Innovation* 6: 111–29.

—— (2001), "The Nature of Innovation in Services," in D. Pilat (ed.), *Innovation and Productivity in Services*, Paris: OECD, 55–79.

INSTITUTE OF INNOVATION RESEARCH (2003), "*Knowing How, Knowing Whom: A Study of the Links between the Knowledge Intensive Services Sector and the Science Base*," Mimeo, IoIR (Manchester); Report to the Council for Science and Technology. available at: http://www.cst.gov.uk/cst/reports/files/knowledge-intensive-services-study.pdf

JONES, P. (1995) "Developing New Products and Services in Flight Catering," *International Journal of Contemporary Hospitality Management* 7(2/3): 24–8.

KAUTONEN, M. (2001), "Knowledge-Intensive Business Services as Constituents of Regional Systems: Case Tampere Central Region," in M. Toivonen (ed.), *Growth and Significance of Knowledge Based Services*, Helsinki: Uusimaa TE Centre Publications 3.

KUUSISTO, J. (2000), "The Determinants of Service Capability in Small Manufacturing Firms," Ph.D. thesis, Kingston University Small Business Research Centre, Kingston.

—— and MEYER, M. (2002), *Insights into Services and Innovation in the Knowledge-intensive Economy*, Helsinki: Finnish Institute for Enterprise Management, National Technology Agency, Technology Review 134/2003.

LAY, G. (2002), *Serviceprovider Industrie: Industrial Migration from Manufacturing to Selling Products and Services*, Karlsruhe: Fraunhofer Institute for Systems and Innovation Research (FhG-ISI) 8 S (ISI-A-13-02).

LEIPONEN, A. (2001), *Knowledge Services in the Innovation System*, Helsinki, Etla: Working Paper B185; SITRA 244 (publisher: Taloustieto Oy).

LEVITT, T. (1972), "Production Line Approach to Service," *Harvard Business Review* 50(5): 41–52.

LICHT, G., and MOCH, D. (1999), "Innovation and Information Technology in Services," *The Canadian Journal of Economics* 32(2): 363–83.

*MATHÉ, H., and SHAPIRO, R. D. (1993), *Integrating Service Strategy into the Manufacturing Company*, London: Chapman & Hall.

*MIOZZO M., and MILES, I. (eds.) (2003), *Internationalization, Technology and Services*, Aldershot: Edward Elgar.

MILES, I. (1993), "Services in the New Industrial Economy," *Futures* 25(6): 653–72.

—— (1999a), "Services and Foresight," *Service Industries Journal* 19(2): 1–27.

—— (1999b), "Services in National Innovation Systems: from Traditional Services to Knowledge Intensive Business Services," in G. Schienstock and O. Kuusi (eds.), *Transformation towards a Learning Economy: the Challenge to the Finnish Innovation System*, Helsinki: SITRA (Finnish National Fund for R&D).

—— KASTRINOS, N. (with K. FLANAGAN), BILDERBEEK, R., and DEN HERTOG, P. (with W. HUITINK and M. BOUMAN) (1995), *Knowledge-Intensive Business Services: Users, Carriers and Sources of Innovation*, Luxembourg: European Innovation Monitoring Service, EIMS Publication no. 15 (ed./d-00801 mas).

—— and MATTHEWS, M. (1992), "Information Technology and the Information Economy," in K. Robins (ed.), *Understanding Information*, London: Pinter.

*MULLER, E., (2000), *Innovation Interactions between Knowledge-Intensive Business Services and Small and Medium-Sized Enterprises*, Heidelberg and New York: Physica Verlag.

NATIONAL SCIENCE FOUNDATION (2002), *Science and Engineering Indicators 2002*, Washington, DC: National Science Foundation (online version).

PAVITT, K. (1984), "Sectoral Patterns of Technical Change: Towards a Taxonomy and a Theory," *Research Policy* 13(6): 343–73.

—— (1994), "Key Characteristics of Large Innovation Firms," in M. Dodgson and R. Rothwell (eds.), *The Handbook of Industrial Innovation*, Aldershot: Edward Elgar.

PREST (2002), *A Comparative Analysis of Public, Semi-Public and Recently Privatised Research Centers*, Manchester: PREST available at: http:// les.man.ac.uk/PREST/Research/ Final% 20 Summary%20Report.pdf

ROACH, S. S. (1988), "Technology and the Services Sector: America's Hidden Competitive Challenge," in B. R. Guile and J. B. Quinn (eds.), *Technology in Services*, Washington, DC: National Academy of Engineering.

SAUVÉ, P., and MATTOO, A. (eds.) (2003), *Domestic Regulation and Service Trade Liberaliza-tion*, Oxford: Oxford University Press/World Bank.

SOETE, L., and MIOZZO, M. (1989), *Trade and Development in Services: A Technological Perspective*, Working Paper No. 89-031. Maastricht: MERIT.

SUNDBO, J. (1998), *The Organization of Innovation in Services*, Roskilde: Roskilde University Press.

—— (2000), "Organization and Innovation Strategy in Services," in Boden and Miles 2000: 109–28.

—— and GALLOUJ, F. (2000), "Innovation as a Loosely Coupled System in Services," in S. Metcalfe and I. Miles (eds.), *Innovation Systems in the Service Economy*, Dordrecht: Kluwer.

TASC (1998), The Economics of a Technology-Based Service Sector, National Institute of Standards & Technology Program Office, Strategic Planning and Economic Analysis Group, January 1998 NIST Planning Report: 98–102.

TETHER, B. S. et al. (2001), "Analysis of CIS Data on Innovation in the Service Sector: Final Report," report to European Commission DG12, CRIC, University of Manchester (112 pp.) available at: http:// www.kiet.re.kr/files/econo/20021230-inno.pdf

—— HIPP, C., and MILES, I. (2001), "Standardization and Particularization in Services: Evidence from Germany," *Research Policy* 30: 115–38.

—— and SWANN, G. M. P. (2003), "Services, Innovation and the Science Base: An Investigation into the UK's 'System of Innovation' Using Evidence from the UK's Third Community Innovation Survey," presented at the International Workshop: *Innovation in Europe: Empirical Studies on Innovation Surveys and Economic Performance*, Institute of Socio-Economic Studies on Innovation and Research Policy, National Research Council, and University of Urbino, Faculty of Economics; Rome, 28 January 2003.

*TIDD, J., and HULL, F. M. (eds.) (2003), *Service Innovation: Organizational Responses to Technological Imperatives and Market Opportunities*, London: Imperial College Press.

TOIVONEN, M. (2001), "Megatrends and Qualification Requirements in the Finnish Knowledge Intensive Business Service Sector," in M Toivonen (ed.), *Growth and Significance of Knowledge Based Services*, Helsinki: Uusimaa TE Centre Publications 3.

TOMLINSON, M. (2000), "Information and Technology Flows from the Service Sector: a UK–Japan Comparison," in Boden and Miles 2000: 209–221.

TORDOIR, P. P. (1986), "The Significance of Services and Classifications of Services," in P. Coppetiers, J.-C. Delaunay, J. Dyckman, J. Gadrey, F. Moulaert, and P. Tordoir, *The Functions of Services and the Theoretical Approach to National and International Classifications*, Lille: Johns Hopkins University Center.

—— (1996), *The Professional Knowledge Economy: The Management and Integration of Professional Services in Business Organizations*, Dordrecht: Kluwer.

UCHUPALANAN, K. (1998), *Dynamics of Competitive Strategy and IT-based Product–Process Innovation in Financial Services: The Development of Electronic Banking Services in Thailand*, D.Phil. thesis, University of Sussex, Falmer, Brighton.

—— (2000), "Competition and IT-based Innovation in Banking Services," *International Journal of Innovation Management* 4(4): 455–90.

CHAPTER 17

INNOVATION AND DIFFUSION

BRONWYN H. HALL

17.1 INTRODUCTION[1]

In 1953, a young female Macaque monkey in the south of Japan washed a muddy sweet potato in a stream before eating it. This obvious improvement in food preparation was imitated quickly by other monkeys and in less than 10 years it became the norm in her immediate group; by 1983, the method had diffused completely. In 1956, the same monkey innovated again, inventing a technique in which handfuls of mixed sand and wheat grains were cast upon the sea, so that the floating cereal could be skimmed from the surface. Again, by 1983, this method of gleaning wheat had diffused almost completely throughout the local populations of Macaques.[2] Besides the obvious fact that humankind does not have a monopoly on innovation, these examples illustrate a couple of things about the diffusion of innovations: first, when they are clearly better than what went before, new ideas of how to do things will usually spread via a "learning by observing" process, and second, the process can take some time; in these cases it took thirty years, and the life cycle of the Macaque monkey is somewhat shorter than ours (Kawai, Watanabe, and Mori 1992).

Turning to the world of humans, it is safe to say that without diffusion, innovation would have little social or economic impact. In the study of innovation, the word diffusion is commonly used to describe the process by which individuals and firms in a society/economy adopt a new technology, or replace an older technology with a newer. But diffusion is not only the means by which innovations become useful by being spread throughout a population, it is also an intrinsic part of the innovation process, as learning, imitation, and feedback effects which arise during the spread of a new technology enhance the original innovation.[3] Understanding the diffusion process is the key to understanding how conscious innovative activities conducted by firms and governmental institutions (activities such as funding research and development, transferring technology, launching new products or creating new processes) produce the improvements in economic and social welfare that are usually the end goal of these activities. For entities which are "catching up," such as developing economies, backward regions, or technologically laggard firms, diffusion can be the most important part of the innovative process.[4]

Thirty years ago, an economic historian (Rosenberg 1972) made the following observation about the diffusion of innovations:

in the history of diffusion of many innovations, one cannot help being struck by two characteristics of the diffusion process: its apparent overall slowness on the one hand, and the wide variations in the rates of acceptance of different inventions, on the other. (Rosenberg 1972: 191)

Empirical measurement and study since then has confirmed this view. This chapter and the references included in it review the diffusion of a number of inventions and innovative processes, from the boiling of water to prevent diarrheal diseases to mobile telephony in Europe. Both these studies and the figures showing diffusion rates in various countries demonstrate the truth of Rosenberg's statement. The studies go further than simply noting the speed and variation of diffusion, in that they correlate the rates of adoption with characteristics of the technologies and their potential adopters in an attempt to explain the speed of diffusion and the ultimate acceptance of the new product. Besides the wide variation in acceptance of innovations, a second important characteristic of the diffusion process is the way in which it interacts with the innovative process. This has perhaps been a somewhat less studied aspect of diffusion, owing to the difficulty of collecting systematic data, but case studies abound. Rosenberg (1982), among others, has emphasized the fact that the diffusion of innovations is often accompanied by learning about their use in different environments, and that this in turn feeds back to improvements in the original innovation.

Why is diffusion sometimes slow? Why is it faster in some countries or regions than others, and for some innovations than for others? What factors explain the wide variation in the rate at which it occurs? This chapter provides a historical and comparative perspective on diffusion that looks at the broad determinants, economic, social, and institutional. The ways in which the different social scientific disciplines think about diffusion is discussed and a framework is presented for studying its determinants. Some of the empirical evidence on these determinants is reviewed, and a range of examples given. The chapter concludes with a discussion of gaps in our understanding and future research questions.

17.2 CONCEPTUAL FRAMEWORKS

The diffusion of innovations has been studied from a number of different perspectives: historical, sociological, economic (including business strategy and marketing), and network theoretical. The choice of approach is often dictated by the use to which the results will be put, but there is no doubt that insights from one perspective can inform the research in another discipline. Perhaps a key example of this is the way in which historical study of the development and spread of certain major inventions has affected how economists understand the role of the diffusion process in determining the dynamics of productivity change, a topic I return to later in this chapter. First, I lay out some of the frameworks that have been used by different disciplines for the analysis of diffusion.

The sociological and organizational literature is exemplified by Rogers' well-known book, *Diffusion of Innovations*, now in its fourth edition. In this book, he reviews the subject primarily from a sociological perspective, but one that is informed by research on organizations, the role of economic factors, and the strategies of firms and development agencies. Rogers provides a useful set of five analytic categories that classify the attributes that influence the potential adopters of an innovation:

(1) The relative advantage of the innovation.
(2) Its compatibility, with the potential adopter's current way of doing things and with social norms.
(3) The complexity of the innovation.
(4) Trialability, the ease with which the innovation can be tested by a potential adopter.
(5) Observability, the ease with which the innovation can be evaluated after trial.

Most of these attributes are recognizable in one form or another in the many analyses of specific innovations that have been undertaken by researchers in the past, albeit under different names. For example, both trialability and observability are characteristics that speak directly to the level of uncertainty faced by a potential adopter. The latter characteristic is a key feature of the real options model of technology choice which is discussed later in this chapter and which underlies some of the work on technology adoption by business firms. Complexity as a determinant is clearly related to the economist's notions of cost and complementary investment, as is relative advantage, which an economist might consider to be determined primarily by the benefit/cost ratio of adopting the new technology.

But understanding the way in which the diffusion process unfolds, in addition to simply identifying features that determine its ultimate success or failure, requires a larger framework, one also provided by Rogers later in the same volume. In addition to the attributes listed above, which influence the adoption decision at the individual level, he points to a variety of external or social conditions that may accelerate or slow the process:

(1) Whether the decision is made collectively, by individuals, or by a central authority.
(2) The communication channels used to acquire information about an innovation, whether mass media or interpersonal.
(3) The nature of the social system in which the potential adopters are embedded, its norms, and the degree of interconnectedness.
(4) The extent of change agents' (advertisers, development agencies, etc.) promotion efforts.

Like so many students of the diffusion process, Rogers implicitly assumes that neither the new innovation nor the technology it replaces changes during the diffusion process and that the new is better than the old. These assumptions have been challenged strongly by Rosenberg (1972, 1982), who argued that not only was the new technology improved as user experience and feedback accumulated, but also that frequently the replaced technology experienced a "last gasp" improvement due to competitive pressure and that this fact could slow the diffusion of the new. A frequently given example is the rapid productivity increase in sailing ships during the nineteenth century documented by Gilfillan (1935a, 1935b).

In contrast to the focus on the external environment favored by sociologists and students of organizational behavior, many economists have tended to view the process as the cumulative or aggregate result of a series of (rational) individual calculations that weigh the incremental benefits of adopting a new technology against the costs of change, often in an environment characterized by uncertainty (as to the future path of the technology and its benefits) and by limited information (about both the benefits and costs and even about the very existence of the technology). Although the ultimate decision is made by demanders of the technology, the benefits and costs are often influenced by decisions made by suppliers of the new technology. The resulting diffusion rate is then determined by summing over these individual decisions.

The virtue of this approach to thinking about the adoption of innovations is that it is grounded in the decision making of the microeconomic unit, but this virtue comes with a cost, in that it ignores the social feedback effects (or externalities, to use the economists' term) that might result from one individual adopting and therefore encouraging another. Naturally, in the recent past, economists have risen to this challenge and included such concepts as network effects in their models (see the discussion in Box 17.1). Nevertheless, the factors and mechanisms considered in most of their studies typically fall short of many that other disciplines might consider important, such as social connectedness. An interesting early debate on this topic that reflected different views of the determinants of hybrid-corn adoption in the United States was conducted by a pioneering economist in the study of diffusion and a number of sociologists including Rogers in the pages of *Rural Sociology* (Babcock 1962; Griliches 1960*a* and *b*, 1962; Havens and Rogers 1961; Rogers and Havens 1962).[5] Looking back at this debate from today, a reasonable conclusion is that both economic and non-economic factors probably mattered for the diffusion of hybrid corn, although economic factors by themselves did a pretty good job explaining variation across states.

As an example of microeconomic analysis of the adoption decision in a modern technological setting, consider the decision to replace a wired physical connection to the Internet with a wireless one, either at home or in an office. Benefits might include the ability to work on the network throughout one's house or workplace rather than

---

Box 17.1 The QWERTY controversy—diffusion with network externalities

In an influential article published in 1985, Paul David proposed an answer to the question of why most keyboards have the QWERTYUIOP layout today, even though studies done in the first half of the twentieth century show that those trained on a keyboard with the Dvorak layout are able to type more quickly. He attributed this outcome to the importance of lock-in where there are network externalities. The argument is that the invention of touch typing in the late 1880s made typewriters a network good because of the interrelatedness between the keyboard layout and the typist's skills, the economies of scale in the user costs of typewriting due to training, and the quasi-irreversibility of investment in learning how to type. By the 1890s, these factors led to a significant lock-in to QWERTY layout, because it was easier to reconfigure the keyboard than to retrain the typist. The conclusion from this story of the diffusion of a new technology with network characteristics is that it is possible that the version of technology adopted (the "standard") was not the necessarily the "best" available, because of path dependence in the diffusion process induced by network externalities. That is, small accidents early in the choice of technologies can lead to the adoption of an inferior standard because the existence of an installed base makes that technology more attractive to new adopters. This point was also made by Brian Arthur (1989) using probability models of stochastic diffusion processes developed by Arthur, Ermoliev, and Kaniovski 1983).

David's view has been challenged forcefully by Liebowitz and Margolis (1990), on at least two grounds: First, they show that the historical evidence that the Dvorak keyboard was preferable may be weak. Second, they argue that if society faces large enough costs from adopting the wrong standard, it will pay individuals to change the standard via some form of collective action. One version of David's response to this critique was published by the *Economist* magazine in 1999

at a fixed location such as a desk, and the absence of wires. They might also include the fact that several members of the household can be online at the same time using a single telephone connection. The costs include the purchase of a base station and the services of a technician to install it, but they may also include the time of the user (adopter) spent reconfiguring his or her computer and ensuring that all the communication tools needed are working. Costs might also include the acquisition of new software, or the time spent training other members of the household or office in its use. Were we to enrich this story to include the adoption environment, we might focus on such factors as whether neighbors or colleagues already had undertaken such an installation, the extent to which it has been advertised by the supplier of the technology (or the extent to which it has been "sponsored" by a government agency or leading firm), and even the state of development of the new technology and the operating system necessary to use it (a complementary input). Note also that most of these factors have been changing rapidly over time.

As alluded to earlier, the first empirical study of the diffusion of technology by an economist was Griliches' (1957) study of the diffusion of hybrid corn seed in the Midwestern United States. This study emphasized the role of economic factors such as expected profits and scale in determining the varying rates of diffusion across the Midwestern states. At the same time, it found that the variation in initial start dates for the process depended on the speed with which the seed was customized for use in particular geographic areas. That is, diffusion depended to a certain extent on the activities of the suppliers of the technology in adapting it to local conditions, again highlighting the tendency for the fundamental characteristics of the technology to change somewhat during the adoption process. This theme is repeated throughout the history of innovation. Bruland (1998, 2002) finds that the nineteenth-century development of the Norwegian textile industry was greatly facilitated by the technology transfer activities undertaken by the mostly British machinery suppliers in the form of training, increasing the supply of skilled workers in Norway.

The marketing literature on diffusion is primarily focused on two questions: how to encourage consumers and customers to purchase new products or technologies, and how to detect or forecast success in the marketplace. That is, it often looks for factors that can be influenced in order to increase the number of agents that will choose a particular product. For this reason, the literature tends to emphasize factors such as media information or the role of social networks and change agents, as well as the characteristics of the product itself, rather than individual adopter factors such as education and income levels that are less subject to manipulation by the marketing organization. The workhorse model in marketing for many years has been the Bass (1969) model, which assumes that mass media are important early on in the diffusion process but that as time passes, interpersonal communication becomes far more important. Estimation of this model on a number of consumer durables has revealed that interpersonal communication plays a much bigger role than the media in diffusion (Rogers 1995). For an interesting discussion of the contrast between the economic and marketing views and a comparison of models from the two literatures, see Zettelmeyer and Stoneman (1993). Recent work on identifying and forecasting success in the marketing literature is illustrated by Golder and Tellis (1997). I defer discussion of their model to later in this chapter when I discuss some of the findings obtained by Tellis, Stremersch, and Yin (2002) using this methodology.

The activist view of diffusion taken by the marketing literature is also that pursued by specialists in technology policy, who are generally interested in encouraging the adoption of particular new technologies for welfare-enhancing reasons, either because it serves particular public policy goals (such as encouraging the boiling of water to reduce disease in less-developed countries) or because certain technologies are viewed as conferring externalities on society as a whole (such as the adoption of Internet use or vaccination against a communicable disease). In understanding the variation across countries in diffusion, variables describing their institutions and culture have proved essential in some cases (but not all, see the discussion of Tellis, Stremersch, and Yin 2002 in Section 17.5).

17.3 MODELING DIFFUSION

The most important thing to observe about the decision to adopt a new invention is that at any point in time the choice being made is not a choice between adopting and not adopting but a choice between adopting now or deferring the decision until later. It is important to look at the decision in this way because of the nature of the benefits and costs. By and large, the benefits from adopting a new technology, as in the wireless communications example given above, are flow benefits that are received throughout the life of the acquired innovation. However, the costs, especially those of the non-pecuniary "learning" type, are typically incurred at the time of adoption and cannot be recovered. There may be an ongoing fee for using some types of new technology, but it is usually much less than the initial cost. Economists call costs of this type "sunk." That is, *ex ante*, a potential adopter weighs the fixed costs of adoption against the benefits he expects, but *ex post*, these fixed costs are irrelevant because a great part of them have been sunk and cannot be recovered.

The argument that adoption is characterized by sunk costs implies two stylized facts about the adoption of new technologies: first, adoption is usually an absorbing state, in the sense that we rarely observe a new technology being abandoned in favor of an old one.[6] This is because the decision to adopt faces a large benefit minus cost hurdle; once this hurdle is passed, the costs are sunk and the decision to abandon requires giving up the benefit without regaining the cost, so even if the gross benefit is reduced relative to what was expected, the net benefit is still likely to be positive. Second, under uncertainty about the benefits of the new technology, there is an option value to waiting before sinking the costs of adoption, which may tend to delay adoption.[7]

An important exception to the rule that adoption is normally an absorbing state is the possibility of fads or fashions, which might be defined as things such as the "hula hoop" craze or various types of weight-loss diets, which diffuse rapidly and then disappear after a time. The experience of a wave of adoption followed by a wave of disuse seems to be somewhat more likely in the case of innovations in "practice," such as medical practice or business practice, than in the case of physical products, possibly because in the latter case the costs that are sunk are out of pocket costs paid to others, whereas in the former much of the cost (although by no means all) comes in the form of the adopter's time and effort. That is, the possibility of sunk costs may loom larger to the adopter when denominated in dollar or euro symbols. Nelson et al. (2002) discuss this phenomenon more fully and give some examples (such as the quality circle movement). These authors place considerable emphasis on the difficulty in these cases of getting feedback that the innovation truly is an improvement. Relatively low sunk costs combined with uncertain benefits will mean that the decision to adopt is more easily reversible in the case of practices. Strang and Soule (1998) also discuss the cyclicality of fashions in business practices.

It is a well-known fact that when the number of users of a new product or invention is plotted versus time, the resulting curve is typically an S-shaped or ogive distribution. The not very surprising implication is that adoption proceeds slowly at first, accelerates as it spreads throughout the potential adopting population, and then slows down as the relevant population becomes saturated. In fact, the S-shape is a natural implication of the observation that adoption is usually an absorbing state. Figure 17.1, which represents the diffusion of electric motors in US manufacturing between 1898 and 1955, shows such a curve. In 1898, the share of manufacturing horsepower produced by electric motors was about 4 per cent. It increased steadily and smoothly between 1900 and about 1940, at which point nearly all horsepower is produced by electricity. Saturation appears to be reached at around 90 per cent, presumably because for some specialized uses, other types of motors are preferred.

Looked at in terms of the benefits and costs of technology adoption, a range of simple assumptions will generate this curve. The two leading models explain the dispersion in adoption times using two different mechanisms: consumer heterogeneity, or consumer learning. The heterogeneity model assumes that different consumers expect to receive different benefits from the innovation. If the distribution of benefits over consumers is normal (or approximately normal, that is unimodal with a central tendency), the cost of the new product is constant or declines

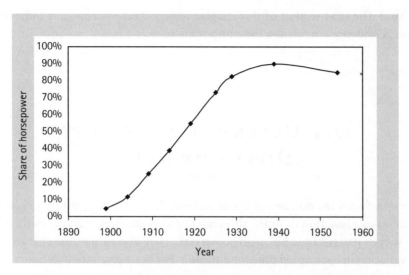

Figure 17.1 Diffusion of electric motors in US manufacturing

monotonically over time, and it is assumed that consumers adopt when the benefit they receive for the product is greater than its cost, the diffusion curve for the product will have the familiar S-shape.

An important alternative model is a learning or epidemic model, which is more popular in the sociological and marketing literatures (the Bass model is an exemplar), but has also been used by economists. In this model, consumers can have identical tastes and the cost of the new technology can be constant over time, but not all consumers are informed about the new technology at the same time. Because each consumer learns about the technology from his or her neighbor, as time passes, more and more people adopt the technology during any period, leading to an increasing rate of adoption. However, eventually the market becomes saturated, and the rate decreases again. This too will generate an S-shaped curve for the diffusion rate.[8] In general, combining this model with the previous model simply reinforces the S-shape of the curve. Golder and Tellis (1997) define a concept they call "take-off," which is their attempt to identify the point at which the empirical diffusion curve appears to have its greatest inflection relative to the initial growth rate.[9] For the data in Figure 17.1, this point would be in about 1910. Because for many consumer products the existence of such a point is a good predictor of eventual success, the focus of their work is to identify predictors of this point.

Regardless of the details of the mechanism generating the probability distribution of adoption times, the question which concerns both social scientists and those interested in encouraging the spread of new technologies is the question of what factors affect the rates at which these events occur. A second and no less interesting question is what are the determinants of the ceiling at which the S-curve asymptotes. That is, when would we expect this ceiling to be less than 100 per cent of the potential user base? The next section of this chapter reviews these factors and some of the empirical evidence concerning their importance.

17.4 DETERMINANTS OF THE DIFFUSION RATE

Figure 17.2 shows the number of US households that have adopted particular new inventions as a function of time. Although not smooth, these curves clearly follow the S-shaped pattern noted by many observers. They also exhibit the characteristic wide variation in the elapsed time for diffusion. For example, it took over forty years for the clothes washer to go from one-quarter of all households to three-quarters, whereas it took less than ten years for the video cassette recorder or color television (not shown) to make the same leap. Table 17.1 shows the diffusion of common household electronic appliances in Japan between 1989 and 1995. It is noteworthy that there is considerable variation in the diffusion rates for different products even during the same six-year period, and this variation is not explained by the level of diffusion that was already achieved in 1989 (compare the refrigerator to the air conditioner, or the CD/cassette/radio player to the video camera).

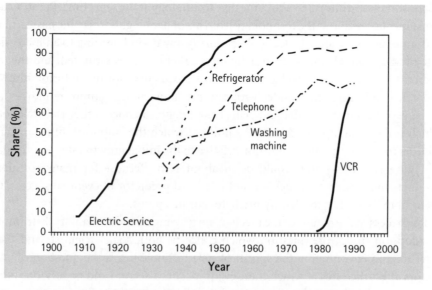

Figure 17.2 Diffusion of major innovations in the United States

Source: Dallas Federal Reserve Bank.

Table 17.1 Diffusion in Japanese Households (%)

New Product	1989	1995	Change
Cordless phones	NA	43.7	NA
CD/radio/cassette player	31.5	68.2	36.7
Convection heater/cooler	34.7	57.3	22.6
Washing machine	34.7	55.4	20.7
Word processor	25.1	43.7	18.6
Microwave oven	72.9	89.5	16.6
Video camera	17.5	34.0	16.5
Air conditioner	64.8	79.3	14.5
Automobile	76.6	82.1	5.5
Personal computer	12.4	16.6	4.2
Television	98.4	99.3	0.9
Refrigerator	62.9	63.6	0.7

Source: Japan Echo, Inc. Information Bulletin No. 18.

From the considerations reviewed earlier in the chapter, one can derive a list of factors that might be expected to influence the diffusion of innovations. These can be classified into four main groups, those that affect the benefits received, those that affect the costs of adoption, those related to the industry or social environment, and those due to uncertainty and information problems. Alternatively, using the classification system of Rogers, one can identify the first and second as combining to yield relative advantage and complexity, the third as compatibility, and the fourth as being determined by trialability and observability.

17.4.1 Benefit Received from the New Technology

Clearly the most important determinant of the benefit derived from adopting a new technology is the amount of improvement which the new technology offers over any previous technology. This is to a great extent determined by the extent to which there exist substitute older technologies that are fairly close. For example, in Figure 17.2, we see that radio and the automatic clothes washer were both introduced in the United States in the early 1920s, but that diffusion of the former was much more rapid than the latter. This may be partly because a fairly good substitute for the automatic clothes washer in the form of manual clothes washing machines existed whereas there was no very good substitute for radio. It is also consistent with the Tellis et al. (2002) finding that across European countries during the latter half of the twentieth century, the single most important factor that explains speed of diffusion is whether the good in question is "white" (household appliance) or "brown" (entertainment or information consumer durable). These authors hypothesize that the general explanation for this finding is that "brown" goods are more status-enhancing, in that they are more readily observable to non-members of the household. Unfortunately they did not control for the prices of the goods because of lack of consistent data across countries, so it is difficult to know whether this finding might also be related to differences in the full costs of adoption across goods and countries.

An important factor in explaining the slowness of technology adoption is the fact that the relative advantage of new technologies is frequently rather small when they are first introduced. As many authors have emphasized, as diffusion proceeds learning about the technology takes place, the innovation is improved and adapted to different environments, thus making it more attractive to a wider set of adopters (Rosenberg 1972; Nelson et al. 2002). The implication is that the benefits to adoption generally increase over time; if they increase faster than costs, diffusion will appear to be delayed (because the number of potential adopters will increase over time, expanding the size of the adopting population). In the Rosenberg (1982) study, the leading example was the airframe, specifically the stretching of the Boeing 747, but in fact one could argue that any technology in which learning by doing or using is an important aspect of its development will display feedback between diffusion and innovation. A good example might be applications software, most of whose development after initial launch is dictated by the experience and demands of users, or the worldwide web, where enhancements after the first web browser was created were dramatic.

17.4.2 Network Effects

Increasingly, the value of some new technology to the consumer depends partly on the extent to which it is adopted by other consumers, either because the technology is used to communicate with others (such as the Internet, or instant messaging) or because the provision of software and services for the technology depends on the existence of a large customer base. Goods of this type are usually termed network goods by economists: their chief characteristic is that they rely on standards to ensure that they can communicate either directly or indirectly. For these goods, an important determinant of the benefit of adoption is therefore the current or expected network size.

For example, Saloner and Shepard (1995) examine the adoption of ATM machines by banks, under the assumptions that consumers prefer a larger network of ATM machines to a smaller and that banks respond to consumer preference. These authors do indeed find that banks with more branches adopt an ATM network sooner, even after controlling for overall bank size, and argue that this confirms that a higher network value leads to earlier adoption of a new technology, other things equal.[10] This example illustrates both the importance of networks and also the role of large firms as intermediates between technology and consumers in sponsoring particular standards for networks.

A famous example of the role of "network externalities" in consumer adoption of new technologies is the VHS/Beta competition, which resulted eventually in a single standard for video recorder/players in a large part of the world. Most observers attribute this outcome to the consumer desire for a large range of software in the form of pre-recorded tapes to go with this hardware, and to the fact that VHS had an initial early advantage in the length of program that could be recorded. See Park (2002) for details on the diffusion of this technology to consumers.

Although network effects (particularly those from networks that diffuse knowledge about or experience with an innovation) have always been viewed as important for the diffusion of innovations, especially in the sociological literature, recent work in economics has focused on the role played by standards in accelerating or slowing the diffusion process, as in the VHS/Beta example (David 1985; Katz and Shapiro 1985; Arthur 1989; Economides and Himmelberg 1995). The central message of the modern economic literature on standards and network externalities is that consumers and firms receive benefits from the fact that other consumers and/or firms have chosen the same technology that they have. These benefits are viewed as being of two kinds, direct and indirect. Direct network benefits are those that arise because they allow the adopter to communicate with others using the same technology. Examples are the choice of fax communication technology or the choice of word processor document format. Indirect benefits arise from the fact that adoption of a product that uses a particular technological standard by a greater number of people increases the probability that the standard will survive and that goods compatible with that standard will continue to be produced. The VHS/Beta example alluded to earlier can be viewed as an example where indirect network benefits were very important, although direct benefits presumably also play a part (the benefits from being able to loan a video made on one's own machine to a friend or neighbor).

The close connection between technological standards and network externalities comes from the fact that standards create a number of effects all of which go in the direction of making it more likely that a good will exhibit network externalities. First, a technological standard increases the probability that communication between two products such as telephones, instant messaging services, or a CD player and a CD, will be successful. Second, standards ease consumer learning and encourage adoption when the same or similar standards are used in a range of products. The use of a particular standard, such as a Windows operating system, by others in a consumer's network, also helps learning and will encourage adoption, because of the relative ease with which a new adopter can obtain advice from those nearby. Third, a successful standard increases the size of the potential market for a good, which can be important in lowering the cost of its production and in increasing the variety and availability of complementary goods. Besides the VHS/ Beta example referred to earlier, an example of this latter effect might be the wider availability of software for the Windows operating system, in comparison to Macintosh OS or Linux.

Although standards have always mattered for diffusion, the increasing importance of digital and information technologies have increased their salience and led to a variety of "standards battles" and to strategic behavior on the part of firms that hope to influence their adoption. Earlier examples of standards battles are the competition between AC and DC methods of distributing electricity (David 1990a), and the failure of gas-powered refrigerators to succeed in the market despite their apparent efficiency, because of the sponsorship of electric power by GE and Westinghouse (Rogers 1995). Nevertheless, it is clear that the importance of this phenomenon has increased recently, with increase in information and communication technologies. Consider for example, the battle between Netscape and Microsoft Internet Explorer for dominance in the web browser market.

The increase in the importance of standards that has accompanied the growth in importance of the information and telecommunications industries has led to a wave of economic modeling. These models incorporate the increasing returns phenomenon that results from the positive feedback from installed base to adoption by other consumers. An early effort is that by Arthur, Ermoliev, and Kaniovski (1983), which emphasizes David's insight that where there are multiple possible standards, small events early in the process that favor one of the standards can lead to an adoption process that settles on an inferior standard. By adding heterogeneity in consumers' tastes or localization in information spillovers, later researchers have produced more complex models of diffusion in the presence of network externalities that results in more than one standard surviving in the market even in the presence of increasing returns in adoption (Bassanini and Dosi 1998; Wendt and van Westarp 2000).

Industrial organization and strategy theorists have centered their modeling efforts on the implications of increasing returns in adoption for competitive strategy and market structure. Examples of this literature include Katz and Shapiro (1985, 1986, 1994), Farrell and Saloner (1992), and Shapiro and Varian (1999). In a series of papers, Katz and Shapiro have explored the implications of consumer adoption behavior in the presence of network externalities for the strategic interactions among firms offering competing products. In general, the theoretical literature of which these papers are an example identifies multiple possible equilibria among firms competing in such environments, so that it is difficult to draw firm conclusions.

Farrell and Saloner study the speed of diffusion (relative to the socially optimal rate) when the good in question is subject to network externalities, so that early adopters ignore the consequence of their adoption on future adopters and on the users of the previous technology. They show that in this setting, diffusion can be either too fast (excess momentum) or too slow (excess inertia). Finally, the book by Shapiro and Varian draws out the implications of these various theoretical models for the production and marketing of information goods (broadly defined), many of which exhibit the properties that give rise to network externalities. They describe strategies for competing in markets where network externalities are important and where it is important to win standards battles because losing them means business failure.

17.4.3 Costs of Adopting the New Technology

The second main class of factors affecting the decision to adopt new technology are those related to its cost. This includes not only the price of acquisition, but more importantly the cost of the complementary investment and learning required to make use of the technology. Such investment may include training of workers and the purchase of necessary capital equipment (whose diffusion is therefore affected by the same factors). It is difficult to overemphasize the importance of this point about the need for complementary investment, especially for complex modern technology that requires the reorganization of the process that will use it (see Ch. 5 by Lam in this volume, for more on this topic).

For example, in a series of recent papers Eric Brynjolfsson has argued that the full cost of adopting new computer information systems based on networked personal computers is about ten times the cost of the hardware.[11] Greenan and Guellec (1998) use data on French firms and workers to make a similar point, that the effective adoption of ICT requires organizational change as well, and that this raises the cost of adoption, which slows diffusion. Caselli and Coleman (2001) compare the rates of computer investment across OECD countries between 1970 and 1990 and highlight the importance both of worker skill level and of complementary capital investments in determining the rate of purchase of new computing systems. The implication of this work is that the use of new computing technology requires both the training of workers and the installation of related equipment (for example, remodeling expenses for space to install servers, along with the necessary cooling equipment). The need for complementary investment therefore has two effects: it slows diffusion because it raises the cost, and because this type of investment usually takes time, it slows down the rate at which the benefits of the new technology are seen by the firm and the economy in the form of increased productivity.

David (1990b) has argued that a similar adjustment took place in manufacturing industry use of electric power, which took 40 years to diffuse completely in the

United States (also see Figure 17.1 and Mowery and Rosenberg 1998). The installation of electric power in a factory required a complete redesign of its layout and a change in task allocation, which meant that adopting this new technology was a rather costly process, and tended to occur slowly, or when greenfield investment was being undertaken. David argues that a similar reorganization of workflow takes place when computer technology is introduced into the workplace or when Internet-based processing replaces telephone or mail order processing. Recent productivity growth evidence in the United States appears to confirm the view that major technological-organizational change takes time for its effects to be felt (Gordon 2003; *Economist* 2003).

Shaw (2002) has documented this kind of phenomenon in the replacement of manual monitoring of production lines in continuous hot steel production lines by computerized pulpit operation. Not only does this involve a substantial investment in high technology equipment, but it also requires fewer workers with substantially higher cognitive skills. Where they used to be on the production line working physically with the machinery, they are now in small rooms ("pulpits") above the line, monitoring and adjusting the process using computer technology.

Technology producers often try to subsidize the adoption of new technologies by providing free training and other help to (potential) users and by charging reduced introductory rates for a certain period. Another symptom of the desire of innovating firms to reimburse new customers for their sunk costs in previous technologies is the widespread practice among software firms of offering competitive upgrades to owners of rival products as well as to the owners of their own products. For a more complete discussion of strategies used by technology producers to encourage diffusion and increase the installed base of their product, see Shapiro and Varian (1999).

Because most of the costs of adoption are fixed, firms' choices to change or introduce technologies may be influenced by their own scale and by the market structure of the industry within which they operate. An interesting example of this phenomenon is given by Paul David in a series of papers on the introduction of the mechanical reaper in US and British agriculture in the nineteenth century (David 1975a and b). He argues persuasively that adoption was delayed in Britain relative to the United States for two reasons: first, because the reaper was a fixed cost investment, profitability required a farm and fields of a certain size; second, because it was incompatible with the typically British pattern of small fields divided by hedgerows. In addition to the difference between countries, he also finds that diffusion was delayed in the US itself until the price of labor rose to a level that made the investment in the reaper (a labor-saving device) profitable.

In the present-day context, a similar empirical finding can be found in many studies of diffusion. Majumdar and Venkataraman (1998) looked at the replacement of mechanical switching by electronic switching in the US telecommunications industry and found that larger firms adopted first, presumably because the costs per customer were somewhat lower. Note that even when technology adoption involves an investment in equipment that is proportional to the existing size of the firm, the requirement that the firm have sufficient absorptive capacity, and the need for worker training or other complementary changes may create a fixed cost that is not proportional to firm size.

As in the case of investment in innovation, firm investment in new technologies is also sensitive to financial factors. As was suggested earlier, the decision to adopt new technology is fundamentally an investment decision made in an uncertain environment, and therefore we should not be surprised to find that all the arguments for a relationship between sources of finance and choice of investment strategy that have been advanced in the investment literature have a role to play here. Chapter 9 by O'Sullivan in this volume reviews these financial factors in some detail. For example, Mansfield (1968) reports that the adoption of diesel locomotives by railways depends somewhat on their liquidity, implying these firms faced a higher cost of external than internal finance.

17.4.4 Information and Uncertainty

The choice to adopt a new technology requires knowledge that it exists and some information about its suitability to the potential adopter's situation. Therefore an important determinant of diffusion is information about the new technology, which may be influenced by the actions of the supplier of the new technology. Obviously in many cases this takes the form of advertising, which influences the cost of the new technology directly. The choice to adopt may also depend on the information available about experience with the technology in the decision maker's immediate environment, either from those in geographic proximity or from those with whom he or she interacts.

Because benefits for adoption are spread over time while costs are usually incurred at the beginning, expectations about the length of life of either the technology or the adopter will matter. Uncertainty about benefits, costs, or length of life will slow the rate of adoption, and may often turn the decision problem into an options-like computation. As discussed earlier, the latter is a consequence of the fact that in most cases, once a new technology has been chosen, the costs are sunk and cannot be recovered. That is, the potential adopter has an option on new technology; if he sees the uncertain payoff reach a certain value (the strike price), he will exercise the option by adopting the technology (see Stoneman 2001b for a theoretical development).

Empirical work on diffusion that incorporates real options is rather scarce, although descriptive work that confirms the role of trialability and observability is widespread (for some recent examples, see Nelson et al. 2002). One notable example of an investigation of technology adoption as the exercise of an option is that of Luque

(2002). She looks at the decision by US plants to adopt three advanced manufacturing technologies, and finds that plants operating in industries with lower degrees of demand and technological uncertainty and a thicker resale market (higher resale prices for used machinery) are more likely to adopt these technologies. She argues that this confirms the importance of uncertainty in the decision; if adopting a new technology corresponds to the exercise of an option, we expect adoption to happen more often in industries with lower uncertainty and lower sunk costs.

17.4.5 Market Size, Industry Environment, and Market Structure

The relationship between firm size or industry concentration and the adoption of new technology by a firm is subject to many of the same considerations as the relationship of these factors to innovation. As discussed above, large dominant firms can spread the costs of adoption over more units, but also may not feel the pressure to reduce costs that leads to investment in new technologies. Empirically, in the case of technology adoption, most studies have found that large firms adopt any given technology sooner, but there are some exceptions. Oster (1982) found that small firms in the steel industry replaced the open hearth furnace with the basic oxygen furnace during the post-World War II period sooner than large firms. In a study of twelve major innovations in the coal, rail, iron and steel, and brewing industries, Mansfield (1961) found weak evidence that firms in competitive, less concentrated industries adopted new technologies sooner, as did Romeo (1977) in a study of the diffusion of numerically controlled machine tools.

In some cases the adoption of new technology is determined by firms, acting for the benefit of consumers and for their own benefit. As an example, consider airline adoption of computerized reservation systems. Consumers have little say in this decision although they ultimately benefit in the form of lower prices for air travel or better service, such as seat reservations. In other cases, the decision fundamentally rests with the consumer, for example the choice of video recording technology such as VHS, Beta, and now DVD. Although the same considerations of cost versus benefit apply broadly in both cases, the role of market structure may be more important in the former case than in the latter, because the adopting firms are likely to be few in number and therefore able to interact strategically with respect to the adoption decision itself. In the latter case, the strategic interaction occurs in choosing the technologies that are offered; in principle, firms can produce the same set of strategic outcomes as in the former case (via penetration pricing, etc.), but lack of perfect information about consumers' tastes and limits on their ability to segment the market sufficiently may prevent the firms from fully internalizing consumers' preferences.

Market structure can affect the decision to adopt in two distinct ways: via seller behavior and via buyer behavior. Highly concentrated providers of new technology will tend to have higher prices, slowing adoption, but they also have the ability to determine a standard more easily, increasing the benefit of adoption. If two or more oligopolistic firms are competing to offer different standards, we may in fact get too rapid adoption of a new technology, because of the incentives they face to price below cost in order to build market share (Farrell and Saloner 1992). In the case of potential adopting firms, market concentration affects both their ability to pass through any costs to consumers and also the incentives they face in incurring the costs of adoption. Many of the issues raised by the tension between the fear of displacement and the exercise of market power here are familiar from the literature on monopolists' incentives to innovate (for example, see Gilbert and Newberry 1982).

Along with market size and structure, the general regulatory environment will have an influence, tending to slow the rate of adoption in some areas due to the relative sluggishness of regulatory change and increasing it in others due to the role of the regulator in mandating a particular technological standard. As an example of the former situation consider the use of plastic pipe for plumbing, which lowers construction cost, but has been slow to diffuse in many localities due to existing building codes. As an example of the latter, Mowery and Rosenberg (1982) have written about the extent to which airline regulation by the Civil Aeronautics Board in the United States was responsible for promoting the adoption of new innovation in airframes and jet engines, in its role as standard setter and coordinator for the industry.

An important example of the unintended consequences of regulation for diffusion is the difference between the United States and Europe (and Japan to some extent) in the diffusion of household Internet use. Historically, pricing in the US telecommunications industry has permitted unlimited local calling at a single monthly rate, whereas pricing for local calls in other countries has usually been proportional to usage. These policies are largely determined by regulatory bodies, but once in place, are difficult to change because consumers and firms adapt to them. In the absence of direct connection to the network such as is available in large institutions, household Internet use requires the ability to connect over local phone lines for extended periods of time. The marginal cost of the Internet for households is therefore to a great extent determined by the cost of local calling, so diffusion of the Internet along with email and instant messaging use has been far more rapid in the United States than in other countries that are just as developed. Only with the recent advent of ISDN service charged by the month in some European countries has household Internet use begun to spread there. In contrast, the diffusion of various "text-messaging" services on wireless phones, which are a form of communication popular with teenagers and similar to the Internet instant messaging widely used in the US, has been more rapid in Japan and Europe. Relative costs of the two forms of instant communication, which in turn are due to regulatory reasons, are probably the main explanation for the differences.

17.5 CULTURAL AND SOCIAL DETERMINANTS

Economic factors like these can go a long way toward explaining differences in rates of diffusion (Griliches 1957 and subsequent authors) but other factors may also be important. For example, many have stressed differences in cultural attitudes towards risk and simple "newness."[12] These characteristics can vary within cultures as well as between them, leading to dispersion in adoption rates that are not accounted for by the economic variables. Among others, Strang and Soule (1998) provide a useful discussion of the cultural basis of diffusion.

Rogers (1995) cites a number of situations where compatibility with existing social norms has strongly influenced the adoption of health-related innovations such as the boiling of water for consumption or various types of contraceptives in under-developed countries, whose relative popularity depends greatly on local religious and cultural mores. He cites as example an instance where a strong traditional distinction made between the qualities of cold and hot water discouraged the use of the very simple preventive measure of boiling water destined for human con-sumption in order to prevent diarrheal diseases.

On the other hand, for consumer household durables, Tellis et al. (2002) find that variables such as gender, cultural attitudes, religion, etc. have little predictive power for "take-off" on average (across European countries) in the presence of lagged market penetration. When these variables are considered separately as predictors, "industriousness" (which is measured by a climate variable) and "need for achieve-ment" (which is measured by the ratio of Protestants to Catholics in the country) speeds diffusion, and a measure of "uncertainty avoidance" slows diffusion. This study is noteworthy in that it includes economic, cultural, and communication variables jointly in the same predictive equation.

17.6 CONCLUSIONS

Traditionally, diffusion is one of the three pillars on which the successful introduc-tion of new products, processes, and practices into society rests, along with inven-tion (a new idea) and commercialization/innovation (reducing the invention to practice). In some ways it is the easiest part of the process to study, because it is more predictable from observable factors than the other two. Certainly countless studies of the diffusion of individual innovations exist, and even exhibit some commonalities (see the references in this chapter and in Rogers 1995), such as the familiar s-shaped curve, and the importance of both economic factors and social networks.

Although many have criticized the linear model that lies behind the division of innovative activity into three parts as oversimplified, it remains true that without invention it would be difficult to have anything to diffuse, so that the model still serves us as an organizing principle, even if we need to be aware of its limitations. Nevertheless, an important insight from the many historical case studies of individual inventions has been the extent to which the diffusion process enhances an innovation via the feedback of information about its operation or utility under varying conditions and across different users, information that can be used to improve it. A second major finding from this literature has been the possible feedback from differences in the rate or scale of adoption across geographic areas to the rates of improvement in the innovation.

In the introduction to this chapter, Rosenberg's observations on the slowness and variability of the diffusion of different innovations were cited. The studies reviewed in this chapter have identified some explanations for these observations, such as the size of sunk costs (trialability), the adaptations and improvement necessary to make the invention useful after its initial conception, and the inherent slowness of interpersonal communication networks in spreading information. In the case of major innovations such as electricity or the computer, some studies have emphasized that the necessity of reorganizing the workplace to take advantage of the new innovation means that diffusion will be greatly delayed, and also that the expected gains from innovation may take time to be realized.

Several areas stand out as potentially fruitful for future research. First, most of the studies conducted to date have been methodologically rather simple; the most ambitious have used a hazard model to correlate the time until adoption with various characteristics of the innovation and the adopter (depending on the particular dataset). There is room for an approach that is more structural and grounded in the choice problem actually faced by the adopter. One promising avenue for modeling is the real options approach suggested by Stoneman (2001b); such a model would yield a hazard or waiting time model rather naturally, while explicitly incorporating the effects of uncertainty on the decision.[13] The cumulative distribution for adoption derived from a hazard model has the familiar S-shape.

Second, although many studies have described the process of innovation enhancement during its diffusion qualitatively, there has been relatively little systematic collection of data or modeling of the process. Investigations of this type would be very helpful in quantifying the importance of this effect, which is similar to but not the same as the well-known learning curve. One technological area where this process has been very important and might be worth study is the area of user-driven software development.

Finally, an area of research that is receiving increasing attention in a globalizing economy is that concerned with international technology transfer.[14] This literature is generally positive (as opposed to normative) in approach and empirically based, focused on identifying the mechanisms through which technology diffuses from more developed to less developed countries rather than on the adoption choice itself. That is, this analysis is conducted at the aggregate level rather than at the level of an individual decision maker. It is probably safe to say that there is room for further research in this area, as the diffusion of technology is an important source of economic and social development. Indeed, from a welfare perspective, one of the most important areas for further study is the comparative diffusion of various health and medical practices across developing countries, especially because it is apparent that there are wide variations even among similar low income countries in rates of adoption.

Notes

1. University of California at Berkeley, Scuola Sant'anna Superiore Pisa, NBER, and the Institute of Fiscal Studies, London. I am grateful to Beethika Khan for contributing some of the literature review that lies behind the issues discussed in this paper, and other contributors to this volume, especially my discussants, Kristine Bruland, John Cantwell, and Ove Granstrand, for their very helpful comments. Finally I owe an immense debt to the editors for their careful reading of multiple drafts of this chapter.

2. I am grateful to Chris E. Hall for calling this example to my attention. It is described in McGrew (1998), where a more complete set of references to the anthropological literature is given. A third feature of this example, perhaps not directly relevant to this chapter, may be noted: the fact that once having innovated, innovators tend to innovate again.

3. As discussed in the introduction to this volume, the view that every adopter develops and adapts an invention to his own use has led some of the literature to refer to adoption itself as "innovation." I will follow the more conventional practice of reserving the term innovation for the first "public" use of a new product, process, or practice.

4. See Godinho and Fagerberg (Ch. 19 in this volume) on the role of adoption of new technology in the catch-up process and in long run economic growth.

5. I am grateful to Paul David for calling some of these references to my attention.

6. Although see Rogers (1995) for some examples of innovations that failed to diffuse because they were rejected after trial.

7. An option is a choice between doing nothing and paying a fixed amount to purchase an uncertain return. It is real (as opposed to financial) if it involves investment in real assets. In this setting, the investment is the adoption of a new technology, which has uncertain benefits and costs that may change over time. The option value arises from the fact that waiting may reduce the chance that the wrong decision is made.

8. For a good presentation of this class of models and their extensions, see Geroski (2000). David (2003) provides an evolutionary interpretation of this mechanism.

9. For any particular parametric distribution function, this point might be defined at the point where the curvature of the cumulative distribution (the second derivative) is

maximized. Such a point is well defined if it exists. It occurs when about 20 per cent of the population has adopted in the case of a logit and when about 15 per cent have adopted in the case of a normal. Golder and Tellis (1997) define a non-parametric discrete version of this measure by looking at the current rate of adoption as a share of adoption to date.

10. On the adoption of ATM systems, see also Hannan and McDowell (1984*a* and *b*), who emphasize the role of bank size and industry concentration, which are chiefly cost side and market structure considerations.

11. See Brynjolfsson (2000) for a summary of this work and further references.

12. For a discussion of various cultural explanations, see Mokyr (1990).

13. In the labor economics literature, Lancaster and Nickell (1980) developed a similar model for the probability of obtaining a job when unemployed (see also Lancaster 1990).

14. See Keller 2001 for a review of this literature.

References

ARTHUR, W. B. (1989), "Competing Technologies, Increasing Returns, and Lock-in by Historical Events," *The Economic Journal* 99 (March): 116–31.

—— ERMOLIEV, Y., and KANIOVSKI, Y. (1983), "Generalized Urn Problem and Its Applications," *Cybernetics* 19: 61–71.

BABCOCK, J. M. (1962), "Adoption of Hybrid Corn—A Comment," *Rural Sociology* 27: 332–8.

BASS, F. M. (1969), "A New Product Growth Model for Consumer Durables," *Management Science* 13(5): 215–27.

BASSANINI, A., and DOSI, G. (1998), "Heterogeneous Agents, Complementarities, and Diffusion. Do Increasing Returns Imply Convergence to International Technological Monopolies?" in D. D. Gatti, M. Gallegati, and A. Kirman (eds.), *Market Structure, Aggregation, and Heterogeneity*, Berlin: Springer, 163–85.

BRULAND, K. (1998), "Skills, Learning and the International Diffusion of Technology," in M. Berg and K. Bruland (eds.), *Technological Revolutions in Europe*, Cheltenham: Edward Elgar, 45–69.

—— (2002), *British Technology and European Industrialization: The Norwegian Textile Industry in the Mid-nineteenth Century*, Oxford: Oxford University Press.

BRYNJOLFSSON, E. (2000), "Beyond Computation: Information Technology, Organizational Transformation and Business Performance," *Journal of Economic Perspectives* 14: 23–48.

CASELLI, F., and COLEMAN, W. II (2001), "Cross-country Technology Diffusion: The Case of Computers," *American Economic Review* 91(2): 328–35.

DAVID, P. A. (1975*a*), "The Mechanization of Reaping in the Ante-bellum Midwest," in P. A. David, *Technical Choice, Innovation, and Economic Growth*, Cambridge: Cambridge University Press, 195–232.

—— (1975*b*), "The Landscape and the Machine: Technical Interrelatedness, Land Tenure, and the Mechanization of the Corn Harvest in Victorian Britain," in P. A. David, *Technical Choice, Innovation, and Economic Growth*. Cambridge: Cambridge University Press, 233–90.

DAVID, P. A. (1985), "Clio and the Economics of QWERTY," *American Economic Review* 75: 332–7.

—— (1990), "At last, a Remedy for Chronic QWERTY-Skepticism!" Oxford: All Souls College, Oxford University. Manuscript.

—— (1990a), "Heroes, Herds, and Hysteresis in Technological History: Thomas Edison and the Battle of the Systems Reconsidered," *Industrial and Corporate Change* 1(1): 129–80.

*(1990b), "The Dynamo and the Computer: An Historical Perspective on the Modern Productivity Paradox," *American Economic Review* 80: 355–61.

—— (1999), " 'Myth' – informing the public about the public goods and QWERTY," (a reply to "Lock and Key"), Economic Focus, The *Economist*, 18 September.

—— (2003), "Zvi on Diffusion, Lags and Productivity Growth... Connecting the Dots," Paper presented at the Conference on R&D, Education and Productivity held in memory of Zvi Griliches in Paris, August 2003.

DAVIES, S. (1979), *The Diffusion of Process Innovation*, Cambridge: Cambridge University Press.

DIXIT, A., and PINDYCK, R. (1994), *Investment under Uncertainty*, Princeton: Princeton University Press.

ECONOMIDES, N., and HIMMELBERG, C. (1995), "Critical Mass and Network Size with Application to the U.S. Fax Market," New York University, Salomon Brothers Working Paper S/95/26 (August).

Economist (2003), "The New 'New Economy,'" 11 September.

FARRELL, J., and SALONER, G. (1992), "Installed Base and Compatibility: Innovation, Product Preannouncements, and Predation," *American Economic Review* 76: 940–55.

*GEROSKI, P. A. (2000), "Models of Technology Diffusion," *Research Policy* 29(4/5): 603–25.

GILBERT, R. J., and NEWBERRY, D. M. G. (1982), "Preemptive Patenting and the Persistence of Monopoly," *American Economic Review* 72(3): 514–26.

GILFILLAN, S. C. (1935a), *Inventing the Ship: A Study of the Inventions made in her History between Floating Log and Rotorship*, Chicago: Follett.

—— (1935b), *The Sociology of Invention: An Essay in the Social Causes of Technic Invention and Some of its Social Results; Especially as Demonstrated in the History of the Ship*, Chicago: Follett.

GOLDER, P. N., and TELLIS, G. J. (1997), "Will It Ever Fly? Modeling the Takeoff of Really New Consumer Durables," *Marketing Science* 16(3): 256–70.

GORDON, R. J. (2003), "Five Puzzles in the Behavior of Productivity, Investment, and Innovation," in *Global Competitiveness Report, 2003–2004*, World Economic Forum.

GREENAN, N., and GUELLEC, D. (1998), "Firm Organization, Technology, and Performance: An Empirical Study," *Economics of Innovation and New Technology* 6: 313–47.

*GRILICHES, Z. (1957), "Hybrid Corn: An Exploration in the Economics of Technological Change," *Econometrica* 25: 501–22.

—— (1960a), "Hybrid Corn and the Economics of Innovation," *Science* 132: 275–80.

—— (1960b), "Congruence Versus Profitability: A False Dichotomy," *Rural Sociology* 25: 354–6.

—— (1962), "Profitability Versus Interaction: Another False Dichotomy," *Rural Sociology* 27: 325–30.

HANNAN, T., and McDOWELL, J. (1984a), "The Determinants of Technology Adoption: The Case of the Banking Firm," *Rand Journal of Economics* 15(3): 328–35.

* Asterisked items are suggestions for further reading.

—— —— (1984*b*), "Market Concentration and the Diffusion of New Technology in the Banking Industry," *The Review of Economics and Statistics* 66(4): 686–91.

HAVENS, E. A., and ROGERS, E. M. (1961), "Profitability and the Interaction Effect," *Rural Sociology* 26: 409–14.

JOVANOVIC, B., and STOLYAROV, D. (2000), "Optimal Adoption of Complementary Technologies," *American Economic Review* 90(1): 15–29.

KATZ, M. L., and SHAPIRO, C. (1985), "Network Externalities, Competition, and Compatibility," *American Economic Review* 75(3): 424–40.

—— —— (1986), "Technology Adoption in the Presence of Network Externalities," *Journal of Political Economy* 94: 822–41.

*—— —— (1994), "Systems Competition and Network Effects," *Journal of Economic Perspectives* 77: 93–115.

KAWAI, M., WATANABE, K., and MORI, A. (1992), "Pre-Cultural Behaviors Observed in Free-Ranging Japanese Monkeys on Koshima Islet over the Past 25 Years," *Prim. Rep.* 32: 143–53.

KELLER, W. (2001), "International Technology Transfer," NBER Working Paper Number w8573.

LANCASTER, T. (1990), *The Economic Analysis of Transition Data*, Cambridge: Cambridge University Press.

—— and NICKELL, S. (1980), "The Analysis of Reemployment Probabilities for the Unemployed," *Journal of the Royal Statistical Society* A 143(2): 141–65.

LIEBOWITZ, S. J., and MARGOLIS, S. E. (1990), "The Fable of the Keys," *Journal of Law and Economics* 33: 1–26.

LUQUE, A. (2002), "An Option-Value Approach to Technology Adoption in U.S. Manufacturing: Evidence from Microdata," *Economics of Innovation and New Technology* 11(6): 543–68.

McGREW, W. C. (1998), "Culture in Nonhuman Primates?" *Annual Review of Anthropology* 27: 301–28.

MAJUMDAR, S., and VENKATARAMAN, S. (1998), "Network Effects and the Adoption of New Technology: Evidence from the U.S. Telecommunications Industry," *Strategic Management Journal* 19: 1045–62.

MANSFIELD, E. (1961), "Technical Change and the Rate of Imitation," *Econometrica* 29(4): 741–66.

*—— (1968), *Industrial Research and Technological Innovation*, New York: Norton.

*MOKYR, J. (1990), *The Lever of Riches*. Oxford: Oxford University Press.

MOWERY, D., and ROSENBERG, N. (1998), *Paths of Innovation, Technological Change in 20th-Century America*, Cambridge: Cambridge University Press.

—— —— (1982), "Government Policy and Innovation in the Commercial Aircraft Industry, 1925–75," in R. R. Nelson (ed.), *Government and Technical Progress: A Cross-Industry Analysis*. Oxford: Pergamon Press.

NELSON, R. R., PETERHANSL, A., and SAMPAT, B. N. (2002), "Why and How Innovations Get Adopted: A Tale of Four Models," New York: Columbia University (Photocopied).

OSTER, S. M. (1982), "The Diffusion of Innovation among Steel Firms: The Basic Oxygen Furnace," *Bell Journal of Economics* 13(1): 45–56.

PARK, S. (2002), "Quantitative Analysis of Network Externalities in Competing Technologies," New York: SUNY at Stony Brook. Photocopied.

*ROGERS, E. M. (1995), *Diffusion of Innovations*, 4th edn., New York: The Free Press.

—— and HAVENS, A. E. (1962), "Rejoinder to Griliches' 'Another False Dichotomy,'" *Rural Sociology* 27: 332–4.

ROMEO, A. A. (1977), "The Rate of Imitation of a Capital-embodied Process Innovation," *Economica* 44: 63–9.

*ROSENBERG, N. (1972), "Factors Affecting the Diffusion of Technology," *Explorations in Economic History* 10(1): 3–33.

—— (1982), "Learning by Using," in N. Rosenberg, *Inside the Black Box*, Cambridge: Cambridge University Press, 120–40.

SALONER, G., and SHEPARD, A. (1995), "Adoption of Technologies with Network Effects: An Empirical Examination of the Adoption of Automated Teller Machines," *Rand Journal of Economics* 26(3): 479–501.

SHAPIRO, C., and VARIAN, H. (1999), *Information Rules*, Boston: Harvard Business School Press.

SHAW, K. (2002), "By What Means Does Information Technology Affect Employment and Wages," in Greenan, N., Y. D'Horty, and J. Mairesse, (eds.), *Productivity, Inequality, and the Digital Economy*. Cambridge, Mass.: The MIT Press, 229–68.

*STONEMAN, P. (2001*a*), *The Economics of Technological Diffusion*, Oxford: Blackwells.

—— (2001*b*), "Financial Factors and the Inter Firm Diffusion of New Technology: A Real Options Model," University of Warwick EIFC Working Paper No. 2001-08 (December).

*STRANG, D., and SOULE, S. A. (1998), "Diffusion in Organizations and Social Movements," *Annual Review of Sociology* 24: 265–90.

TELLIS, G. J., STREMERSCH, S., and YIN, E. (2002), "The International Takeoff of New Products: The Role of Economics, Culture, and Country Innovativeness," *Marketing Science* 22(2): 188–208.

WENDT, O., and VAN WESTARP, F. (2000), "Determinants of Diffusion in Network Effect Markets," Paper presented at the 2000 IRMA International Conference, Anchorage, Alaska.

ZETTELMEYER, F., and STONEMAN, P. L. (1993), "Testing Alternative Models of New Product Diffusion," *Economics of Innovation and New Technology* 2: 283–308.

PART IV

INNOVATION AND PERFORMANCE

INTRODUCTION TO PART IV

....

THE literature on the relationship between innovation and economic performance has been dominated by economists, and the first chapter in this section focuses on an issue that has received more attention than any other in this area, the relationship between innovation and economic growth (Chapter 18 by Verspagen). As the author shows, economists have adopted different frameworks for the analysis of this relationship; the two most relevant approaches being "evolutionary" and "new growth" theory. These approaches differ less in their views of the importance of innovation for growth, which is acknowledged in both, than on the precise mechanisms through which innovation affects growth. One important way through which innovation spurs growth is through the diffusion of technology from the developed to the less-developed world (so called "late-comers"). Fagerberg and Godinho (Chapter 19) provide a historical and interpretative survey of the literature on catching-up by "late-comers", with special focus on the role played by innovation for the outcome of such processes. A related issue is that of changing patterns of competitiveness, and the role of innovation in this context. Cantwell (Chapter 20) examines the literature on innovation and competitiveness. Although the competitiveness issue has spawned numerous controversies since the 1980s, a perhaps even more hotly debated issue, particularly within Europe, concerns the employment effects of innovation. Pianta (Chapter 21) provides an extensive survey of the large empirical literature on this subject. The section—and the entire volume—concludes with Chapter 22 by Lundvall and Borrás on science, technology, and innovation policies.

CHAPTER 18

INNOVATION AND ECONOMIC GROWTH

BART VERSPAGEN

18.1 INTRODUCTION

ACCORDING to Maddison (2001), the world economy began to grow in roughly the year AD 1000 following a long period of stagnation (Figure 18.1).[1] A marked increase in growth rates occurred in the period 1600–1800, and, as the figure suggests, growth has been increasing ever since.

Economic history addresses the issue of the way in which the record of economic growth in Figure 18.1 is related to historical developments such as the Industrial Revolution (see also Bruland and Mowery, Ch. 13 in this volume). Although the exact impact that this has had on growth rates, particularly at the sectoral level, remains a subject of debate (e.g. Crafts 1985), it seems beyond dispute that a change of technology in the pure sense, coupled with organizational changes at various levels of aggregation, are the main driving factors behind the continuous increase of living standards entailed by this process.

The historical perspective also nicely illustrates the fact that there is much more to economic growth than just the data on per capita income that are so widely used by economists. Economic growth is an historical process of structural change in the

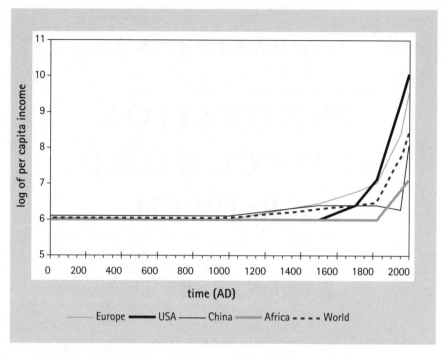

Fig. 18.1 Long-run growth in the world economy, according to the data in Maddison (2001)

broadest sense, and only the most elementary aspects of this process can be measured by the data on production and income. The form of structural change most visible in the statistics is the changing sectoral mix of the economy. Chenery, Syrquin, and Robinson (1986) have illustrated the regularity between the changing sectoral composition of the economy and the increasing level of productivity, while a "deeper" manifestation of structural change in the long-run process of economic growth remains largely the domain of historical research.

Although the argument that technological and organizational innovation are responsible for this lengthy period of gradually accelerating growth is appealing, in fact economic theories explaining any such relationship are far from straightforward. Growth theory, especially when focused on the issue of technology, is a field characterized by spirited scholarly debate. An important current debate is that between the evolutionary approach and the more neoclassically inspired "endogenous growth theory".

This chapter argues that the gap between these two approaches is rooted in fundamental differences in their basic worldviews. While the neoclassical tradition adheres to a worldview in which cause and effect are clearly separable, and growth is an ordered, steady state phenomenon, the evolutionary worldview is one of historical circumstances, complex causal mechanisms that change over time, and, above all, turbulent growth patterns that appear to be far from a steady state.

Before these two approaches are compared (in Section 18.3), I discuss some of the perspectives found in the earlier literature (Section 18.2). This discussion includes both highly applied methods from the mainstream toolbox of economics (growth accounting and the literature on R&D and productivity) and applied work from a post-Keynesian or Schumpeterian perspective. Section 18.4 outlines a few lines for future research.

18.2 GROWTH AND TECHNOLOGY: TRADITIONAL ECONOMICS APPROACHES

While technological change and economic growth were at the core of the work of the classical economists (think of Adam Smith or Karl Marx), these topics largely vanished from the scene with the neoclassical revolution in economic thinking in the late nineteenth and early twentieth centuries. The neoclassical growth models that appeared half a century ago (Solow 1956) treated technological change as an exogenous phenomenon. Technology was an explanatory factor "of last resort," in the sense that growth not explained by the variables included in the model was assumed to be the result of exogenous technological change. However, when empirical work—so called "growth-accounting" (Abramovitz 1956; Solow 1957)[2]—indicated that the unexplained share of long run economic growth tended to be very high, the interest in technological change and other possible explanatory factors not taken into account by the modelers increased.

Following Solow (1957), growth accounting commonly starts from the assumption of so-called "neutral technological change," implying that technological change improves the productivity of both labor and capital equally. Moreover, all markets are assumed to be "perfectly" competitive and in equilibrium. Economies of scale are assumed to be insignificant.

These assumptions support the following approach to calculating the contribution of "technological progress" to economic growth: subtract from the growth rate of GDP the weighted growth rates of the capital stock and employment, using the share of wages in GDP as a weight for employment, and subtract from one to get the weight for the capital stock. What remains, the "residual," is labeled "total factor productivity" (TFP) growth. This should, following Solow, be seen as the result of technological progress. Although convenient, the strong assumptions underlying these calculations are likely to be violated in practice, and the residual almost certainly includes many more factors than just the contribution of technology. This is why Abramovitz (1956) called the residual "a measure of our ignorance."

Over the years, the growth accounting method has been greatly refined. First, the collection of more refined statistical data allowed more production factors to be distinguished, e.g. human capital, various types of labor (different by educational level), different types of capital, etc. In this way, the residual shrinks, attributing a larger part of it to the factors that are now better measured (Denison 1962).[3] The second line of extension has been to refine the concept in a theoretical way, for example by assuming that some factors (capital) are quasi-fixed, i.e. cannot be reduced or increased as a result of short-run fluctuations in output growth (e.g. Morrison 1986).

The TFP concept remains important in studies of growth by economists, as it provides a "proximate" indicator of the impact of technological change on growth. Nonetheless, the problems that remain in conceptualization and measurement have made many scholars in the field critical of its use. Perhaps the most fundamental critique is that many of the factors going into the growth accounting calculations are interrelated by causal links not accounted for by the underlying theory.[4]

Growth theory in the 1950s and 1960s was based on a simplistic view of technology as a "public good." Technological knowledge obviously has some characteristics of a public good, i.e. more than one firm can use the same piece of knowledge at the same time (non-rivalry), and once knowledge is in the open, it is hard to exclude specific firms from using it (non-excludability). In its extreme form, this view leads to the conclusion that all knowledge can be acquired externally as "general knowledge," and firms need not develop knowledge themselves.

Other important aspects of technology, however, make it a private rather than a public good (see also Fagerberg, Ch. 1 in this volume). Pure public goods do not require any special effort or special skills on the side of the consumer or receiver of the services of the good. This is obviously not the case for technological knowledge. Using technological knowledge, even if it stems from the public domain, requires considerable skills and efforts on the side of the receiver of this knowledge. The reason for this is that knowledge has a strongly cumulative and often tacit character. Every piece of new knowledge builds to a large extent on previous knowledge, and to apply knowledge requires that one have command over the older knowledge on which the new knowledge builds.

A number of models developed during the 1950s and 1960s made technology endogenous. In Kaldor (1957) this took the form of a so-called "technical progress function," which assumed a linear relation between growth of labor productivity and the growth of capital per worker. Kaldor's work gave rise to a specific tradition, often labeled "Post-Keynesianism." Work in this tradition takes the role of demand into account explicitly.[5]

The Post-Keynesian tradition also emphasizes the role of "cumulative causation" or "positive feedback." Contrary to the neoclassical idea of knowledge as a public good, these models assume that knowledge is specific to the agents that develop it and does not spill over easily to other agents or nations. This idea was applied to regional growth in Kaldor (1970), and goes back to Verdoorn (1949), Fabricant (1942) and Young (1928). In this view, generating knowledge is mainly a learning process deeply rooted in gaining experience with specific production processes and products: learning-by-doing and learning-by-using are key concepts. Only those engaged in the actual learning experiences will gain from it, and others, who do not profit from experience, will be left behind.

The consequence of this is a tendency for "success to breed success": those nations (or regions, or agents) that are growing rapidly accumulate experience and hence learn faster than others. This leads to a better competitive position for those already ahead and enables them to move further ahead. Hence, the crucial tendency here is one of divergence, in which some nations (regions) are able to grow rapidly while others are left behind. A model of regional growth along these lines was presented in Dixon and Thirlwall (1975).[6]

An important contribution in this post-Keynesian tradition is Cornwall (1977), who argues that manufacturing is the leading sector in economic growth because of the externalities it generates for other sectors. The motivation for this hypothesis is consistent with the Schumpeterian idea that large innovations have a broad impact across many sectors (see also Section 18.4.2 below). This is coupled with a view that, for many countries, the inflow of knowledge from abroad is paramount (see also Fagerberg and Godinho, Ch. 19 in this volume).

Attempts to generate models of endogenous technological change were also formulated in the neoclassical tradition in the 1960s. Arrow (1962) introduced a model of learning-by-doing as the source of technological progress, and Uzawa (1965) and Shell (1967) formulated full-fledged growth models with endogenous technological change, which in many respects can be considered as the front-runners of the wave of "endogenous growth models" that emerged in the late 1980s and early 1990s (see Section 18.3.3).

The work on growth accounting also contributed to the emergence during the 1970s of a purely empirical approach to the issue of growth and technology that formulated and estimated econometric models of the relationship between GDP and R&D investment (e.g. Griliches 1979, 1984). These studies employ a production function that adds a "knowledge stock" measure (typically, cumulative, depreciated R&D investment) to the traditional factors of labor and capital. Estimates of the elasticities of output with regard to the various production factors suggest that knowledge (R&D) has a significant impact on productivity growth. This approach has been used at various levels of aggregation: firms (e.g. Griliches and Mairesse 1984), sectors (e.g. Verspagen 1995) or countries (e.g. Griliches 1986).

An important issue in this literature is the empirical identification of so-called R&D spillovers. This goes back to the notion that knowledge is at least partly a public good and can be used by others than the firm that developed it. In the context of a production function, spillovers are incorporated by introducing two R&D

knowledge stocks: one formed by R&D undertaken by the firm (or nation, or sector) itself, and another one formed by R&D undertaken by other firms (nations, sectors; see Los and Verspagen 2000, for a micro-level application). These studies generally conclude that the social rate of return to R&D is larger than the private rate of return, at any level of aggregation. Firms thus tend to benefit from other firms' R&D, and the same holds at the international level: one nation's productivity growth is to an important extent determined by that of others. Despite their econometric sophistication, however, these studies reveal little about the exact channels through which spillovers operate. These channels may include traded goods, employee mobility, technology alliances, or even knowledge that is "simply in the air."[7]

18.3 COMPETING PARADIGMS FOR EXPLAINING THE RELATION BETWEEN GROWTH AND TECHNOLOGY

Two major approaches emerged during the 1980s and 1990s as the dominant approaches to the analysis of the relationship between technology and growth. These are the neoclassical approach, which is also dominant in other fields of economics, and the neo-Schumpeterian or evolutionary approach. While the neoclassical approach consists of a relatively homogenous set of interrelated sub-approaches (models), the field of neo-Schumpeterian or evolutionary economics consists of a more loosely connected set of contributions. The evolutionary approach includes formal models as well as more "appreciative" or historical approaches, as will be explained in more detail below. Even the label used to describe this approach is not yet common understanding. Here, we will use, mainly for convenience, the short description of "evolutionary economics," but we include under this heading a broad category of work, including what some have called neo-Schumpeterian economics.

Both of these approaches agree on basic issues such as the importance of innovation and technology for economic growth, as well as the positive role that can be played by government policy for science and technology. Yet they disagree on the behavioral foundations underlying these respective theories. These differences can be characterized by saying that the neoclassical theory sacrifices a significant amount of realism in terms of describing the actual innovation process in return for a quantitative modeling approach that favors strong analytical consistency, while the evolutionary approach embraces the micro complications of the innovative process and applies a more eclectic approach. Given these differences, it is useful to start with an overview of their analytics in terms of the microeconomic aspects of endogenous technological change and innovation.

18.3.1 Microeconomic Aspects of Technology and Innovation of Importance for the Analysis of Economic Growth

We focus in this section on two important aspects of the micro-foundations of innovation and technological change: uncertainty and differences in the significance of innovations. Economists typically deal with uncertainty by postulating a probability distribution for a certain range of events. Using these probability distributions, the economic consequences of decisions can be weighted by their probability. Rational actors can make calculations that are more complex than those in an environment of certainty, but the results do not differ appreciably. We refer to such a situation as a case of weak uncertainty.

The situation changes, however, when the possible outcomes of an uncertain process are not known in advance, i.e. the events for which a probability distribution is needed cannot be identified. Arguably, this is a better description of the innovation process, at least where more radical innovations are concerned (Box 18.1 discusses some examples of this in the history of computing). We will refer to the situation in which the possible outcomes of an uncertain process are not known in advance as strong uncertainty. Under strong uncertainty, the elegant calculations using probability-weighted outcomes to calculate the expected value of a stochastic process no longer apply. As will be seen below, the treatment of uncertainty as either weak or strong is an important distinction between neoclassical and evolutionary approaches to economic growth.

The second issue to be discussed in this section is the technological or economic significance of innovations. The history of technology is filled with innovations that have transformed the world—a non-exhaustive list includes the steam engine, electricity, the automobile, the computer, and genetic engineering. Each of these innovations had an almost immeasurable impact on the economy. But there are many examples of less significant innovations that have had far less economic significance.

One may argue that the above comparison is not a fair one, since "the computer" or "the steam engine" never existed. All of the above examples of radical innovations took decades to develop, and were the result of a combination of radical technological breakthroughs as well as many cumulative incremental innovations. Although one can therefore not speak of "the computer" or "the steam engine," it still remains true that some innovations, no matter at which level they are defined, are much more valuable than others. In fact, a large share of innovations eventually

Box 18.1 Technological change and uncertainty

Uncertainty with regard to technologies may be understood in different degrees. Consider, for example, the difference between the first conception of a computer in the days of Turing, Mauchly, and Eckert (i.e. the 1940s and early 1950s), and the introduction of the Pentium chip by Intel in 1992. In the first case, according to a history of the early days of computers in the United States (Katz and Philips 1982), the leading business men of the day saw no commercial possibilities for the computer. They quote Thomas J. Watson Sr., CEO of IBM, as having expressed the feeling that "the one SSEC machine which was on exhibition in IBM's New York offices could solve all the scientific problems in the world involving scientific calculations." The same T. J. Watson, by the way, quickly led IBM into leadership in the global computer industry in the late 1950s.

These pessimistic views of the commercial potential of computers reflect the fact that businessmen such as Watson had no familiarity with computer technologies in their modern form. Under such circumstances, it was impossible to appreciate the many new uses that were to be found, or the possibility that the functions and capabilities of the room-sized computers of the early 1950s could be made to fit in the space of a desktop. One of several problems in recognizing the commercial opportunities of a major technological breakthrough, and a factor contributing to uncertainty, is the lack of any frame of reference for judging these impacts.

The situation was very different in respect of the introduction of the Pentium chip by Intel in 1992. By that time, Intel and other firms, as well as users, had accumulated knowledge about the applications for computers and the devices attached to them. Intel also knew that its products were a major input in the small computers that were being purchased by a large population of consumers and firms. But Intel still faced some degree of technological uncertainty, because of the complex nature of the new design. Indeed, Intel's engineers had made a small mistake that could produce errors in the Pentium's calculations. The publicity over the so-called "Pentium Bug" eventually forced Intel to take back all faulty chips and offer free replacements. The example shows that even for a relatively mature technology, some degree of uncertainty remains.

turns out to be useless, simply because a promising technological idea never makes it into a successful commercial application.

This has given rise to a distinction in the literature between incremental and radical innovations. But this distinction obscures the fact that the size distribution of innovations is not a dichotomy, but instead covers a continuous range of innovation sizes. Moreover, there is an important interaction and interdependence between radical and incremental innovations. For example, the first workable steam engine (the so-called Newcomen engine) was very large and had a limited applicability as well as efficiency. It took more than fifty years for the next step to be taken, i.e. James Watt's engine with a separate condenser. If we can characterize the impact of some innovations as "major," "basic," or "radical," it is only because of a continuous stream of incremental innovations following the introduction of a basic new design.

Box 18.2 Evolution and the blind watchmaker

Let us use Richard Dawkins's metaphor of the blind watchmaker to illustrate the general idea behind economic growth as an evolutionary process. Dawkins's story starts from the idea of William Paley, an eighteenth-century theologian. Paley argued that certain objects, like a watch, are by their nature obviously created by conscious design, whereas for others, like a rock, it is easy to believe that they "have always been around." His argument then went on to stress that nature contains many such objects that are obviously created by conscious design. The most famous of such objects discussed by Paley is the human eye. He then used this argument to offer the proposition that the world must have been created by a conscious being (God).

Dawkins uses Paley's examples to argue that the watch may look as if it was carefully designed (and in the case of a watch it really was), but it might just as well have been created by an evolutionary process that can be thought of as a *blind watchmaker*. This blind watchmaker is unable to design the watch by carefully planning it on a drawing board and then implementing it using precision instruments. Instead, he operates through the processes of random mutation and natural selection. His approach is to start with a simple device and add small and simple changes in a random way. These changes are subjected to a real-world test, i.e., whether or not they lead to an improvement in keeping the time. Only if they do so are they kept; otherwise they are discarded. From a new design that incorporates such a successful small change, the process may start again, and step-by-step a more complicated design emerges. In the end, after a long and gradual process, a complicated artifact such as a watch may result. Although this artifact looks as if it were carefully designed, it was instead the blind watchmaker and his tools of random mutation and natural selection that created it.

Carrying the metaphor over to economic growth and technology, our watchmaker is blind because of the strong uncertainty facing the individual economic decision maker. No businessman can perfectly foresee the huge potential of a new innovation when it first emerges. But it is through a process of incremental innovations, each one of which is implemented by an entrepreneur who sees some market for the newly resulting artifact, that the full potential of the technology unfolds. The incremental innovations are the economic counterpart of biological mutation. Natural selection has its counterpart in economic selection, i.e. markets that decide whether or not certain innovations become successful. Just as in biology, many of the "mutations" (incremental innovations) are not successful, and the selection process erases them from history.

The metaphor is thus concluded by arguing that, although the individual entrepreneur has to cope with strong uncertainty and therefore cannot design a process that we may call a technological revolution, the capitalist system, working by means of a combination of the creation of novelty (innovation) and economic selection (markets), can create "objects" that seem as if they have been carefully designed. With hindsight, technological revolutions, such as the diffusion of steam power or Information and Communication Technologies (ICTs) may look as if they were planned from the very beginning to create a "new economy," but in reality, so it is argued by evolutionary theory, these technological systems were created by the trial-and-error method of the blind watchmaker.

18.3.2 The Evolutionary Approach to Technology and Growth

18.3.2.1 *The Evolutionary Philosophy*

The evolutionary approach to the analysis of economic growth is based in part on the axiom that individual humans are unable to cope in a fully maximizing way with the complexities of technology that were discussed in Section 18.3.1. A single economic decision maker, be it an entrepreneur from the early days of the Industrial Revolution or a large multinational corporation from the twenty-first century, simply cannot see all business opportunities that result from technological possibilities and/or manage them in a way that maximizes profits. These decision makers thus operate under a scheme of bounded rationality, in which relatively simple and occasionally adaptive behavioral rules ("rules of thumb" or "routines") are used to make decisions. These are not fixed, but can be changed over time, especially so under the influence of feedback from economic performance.

Although these simple behavioral rules help economic decision makers in a turbulent and complex world cope with strong uncertainty, their role sheds little light on the mechanisms through which complex modern economies remain on a path of constant technological improvement that we call economic growth. The explanation of aggregate economic performance in evolutionary economics relies on two forces: selection and the generation of novelty. Over time, the variety present in the system is reduced by selection—i.e. the growth of those entities that are better adapted to circumstances, and the decline of those that are not. Novelty is constantly added to the system, however, and thus evolution is the outcome of a constant interaction between variety and selection. Innovation is an important novelty-generating process, and the market and other economic institutions are among the most important selection mechanisms in modern economies.

In biology, the generation of novelty (mutation) is purely random, and there is no way in which the mechanism of mutation itself can learn to generate "smarter" mutations. Each mutation is truly "blind" in the sense that there is no *ex ante* way of telling whether or not it will improve the performance of the organism. In economic evolution, however, decision makers at the micro level are not "completely blind"— they plan their actions in order to generate potentially successful innovations in a process that more closely resembles the Lamarckian view of evolution. Thus, innovations introduced by profit-seeking, "satisficing" entrepreneurs will have at least some commercial potential; in other words, they are most likely biased in a "positive" direction. Nevertheless, uncertainty remains important, since it is difficult to foresee the cumulative effects of numerous small, incremental improvements, and because of the systemic nature of knowledge that results from knowledge spillovers among fields. An actor operating in one field may invent something for which he does not see the full potential in other fields.

The evolutionary approach is particularly suited for analyzing historical processes. Evolution and history are both a complex mixture of random factors, or contingencies, and more systematic tendencies. It is a well-known error to think that the biological evolutionary process is goal-oriented, i.e. that is strives to achieve a predefined aim. Our discussion of the blind watchmaker metaphor may have misled the reader into thinking that such a goal exists, i.e. that it would be the aim of evolution to create a complex artifact such as a watch or a human eye. Instead, it is only the individual mutation that has a sense. The accumulation of incremental innovations may seem to have a purpose, but in fact there is no force in the system that has formulated or even tried to achieve such a goal. The same applies to economic evolution.

Such a view of the world as a mixture between chance and necessity is shared between the historical view of the world, the evolutionary view of the world, and the dialectic (Hegelian) view of the world. It is opposed to the Newtonian or Laplacean view that portrays the world as a clockwork in which future states of the system can be predicted with full accuracy if only enough information about the present state is known. We will argue below that the neoclassical economic growth theory is much more similar to the latter view.

18.3.2.2 *Non-Formal Evolutionary Theorizing about Economic Growth and Technology*

The evolutionary approach to economic growth also draws heavily on economic history and the history of science and technology in its analysis of economic development. Historical analysis often is used by evolutionary scholars to develop heuristic patterns that can be used to describe and categorize these developments in a more general way. "In the appreciative and applied evolutionary literature much has been made of the concepts of technological paradigm (Dosi 1982) and natural trajectories (Nelson and Winter 1982). This is indeed an attempt to impose additional structure on technology and differentiate discrete interrelationships in technological space from one another, if only ex post ... This should be contrasted with the smooth, substitutable, unbounded production possibility sets of neoclassical theory" (Silverberg 2001: 1277).[8]

Dosi (1982) defines a technological paradigm as a "model and pattern of solution of selected technological problems, based on selected principles from the natural science and on selected material technologies." The term is borrowed from Kuhn's philosophy of science (Kuhn 1962), which posits that the normal development path of scientific knowledge relies heavily on a dominant framework jointly adhered to by the leading scientists in the field. The paradigm thus limits the possible directions technological development may take.

In the interpretation of Freeman and Louçã (2001), a small number of basic innovations set out a technological paradigm that may dominate techno-economic developments for a long time. Within the paradigm, the basic design of the innovation is constantly altered by incremental innovations, but the direction of technological development is limited by the paradigm. Still, there is some room for choice within the paradigm, and these choices are governed by the specific circumstances (e.g., scarcity of a particular resource) in which the technology develops. This development is termed a "technological trajectory."

Thus, in the paradigm/trajectory heuristic, a basic innovation can be thought of as setting out developments in the techno-economic domain for a number of years to come, but the success of the paradigm, and hence of the basic innovation, depends crucially on how well incremental innovation is able to adapt the paradigm to local (e.g. industry, geographical and temporal) circumstances. These circumstances include the skills and capabilities of the workforce that has to work with new machinery, as well as factors such as cultural aspects of the society in which the paradigm develops.

Another set of heuristics developed in the historical part of evolutionary economics relates to the temporal clustering of innovations. This part of the literature starts from Schumpeter's observation that innovations "are not evenly distributed in time, but that on the contrary they tend to cluster, to come about in bunches, simply

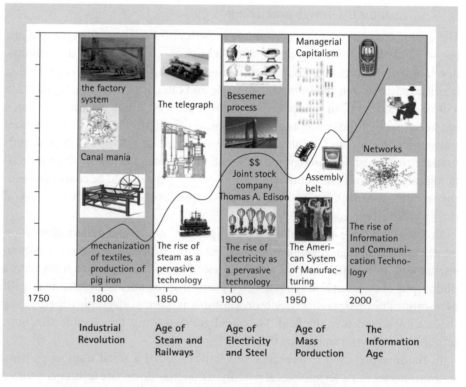

Figure 18.2 Approximate chronology of technological revolutions, based on Freeman and Soete (1997) (dates are approximate)

because first some, and then most firms follow in the wake of successful innova-tion"(Schumpeter 1939: 75). Although Schumpeter was in fact referring to a ten-dency for incremental innovations to cluster following a large innovation (this is an idea not incompatible with the paradigm view summarized above), his idea has been interpreted in the literature as implying that large (or "basic") innovations cluster in time (e.g. Mensch 1979: Kleinknecht 1987). In this view, some historical periods are characterized by an above average rate of (basic) innovations, while other periods show a relatively low rate of such activity.

Together, these two sets of heuristics have interesting implications for growth. They suggest that technological innovation can introduce an uneven temporal pattern into economic growth. In the early, exploratory stages of a paradigm, the technology progresses rapidly, but the pace of change slows when the paradigm goes into its phase of "normal" development, and slows still further when technological opportunities become less numerous (and the paradigm may start to break down as a result of this). The clustering-heuristic suggests variations over time in the rhythm of growth simply because the rate at which large, influential innovations occur differs over time.

One extreme interpretation of this temporal pattern of innovation is the idea of a "long wave" in economic growth, in which periodicity is bounded in a short range of 50–60 years (e.g. Kleinknecht 1987; Freeman and Louçã 2001). Another view claims that growth patterns are inherently turbulent, but with little regularity in terms of strict cycles. In any case, the evolutionary view argues that the uneven temporal rates of technological change mean that the economy is almost always away from anything that could be characterized as a steady state.

Theories and historical analyses of this type propose a view of the interactions among technology, the economy, and the institutional context. The institutional environment is important because it is both a facilitator of and an impediment to technological change. Moreover, the institutional context is itself an endogenous factor that changes under the influence of technological and economic develop-ments. Although it is sometimes claimed that theories of this type suffer from "technological determinism" (i.e. a tendency for one-way causality from technology to growth: see e.g. Bijker et al. 1987), work such as that of Perez (1983) proposes an interactive relationship among institutions, the economy, and technology that emphasizes mutual causality.

18.3.2.3 *Formal Evolutionary Growth Models*

Evolutionary ideas have also been used to formulate models of economic growth and technology. The starting point of this tradition is the model in Nelson and Winter (1982), in which heterogeneity is defined in terms of firms, using production techniques that employ a fixed ratio of labor and capital (so-called Leontief-technology). The generation of novelty (new fixed proportion techniques) occurs as a result of search activities by firms, but search is initiated only when the firm's rate of return falls below a certain (arbitrarily set) value. Search may take two different forms: local search or imitation. In the first case, firms search for new, yet undiscovered techniques, each of which has a probability of being discovered which linearly declines with technological distance from their current technology (hence the term *local* search). In the second search process, imitation, a firm searches for techniques currently employed by other firms but not yet used in its own production process.

Like most models in this tradition, the Nelson and Winter model has to be simulated on a computer to obtain an impression of its implications. The model, which is calibrated with the Solow (1957) data on total factor productivity for the United States in the first half of the century, yields an aggregate time path for capital, labor input, output (GDP), and wages (or labor share in output) that corresponds in a qualitative sense to those observed by Solow. Based on these results, Nelson and Winter argue that "it is not reasonable to dismiss an evolutionary theory on the grounds that it fails to provide a coherent explanation of . . . macro phenomena" (p. 226). More specifically, they argue that although both the neoclassical explanation of economic growth offered by Solow and the Nelson and Winter model seem to explain the same empirical trends, the causal mechanisms underlying the two perspectives differ greatly:

the neoclassical interpretation of long-run productivity change . . . is based upon a clean distinction between "moving along" an existing production function and shifting to a new one. In the evolutionary theory . . . there was no production function. . . . We argue . . . that the sharp "growth accounting" split made within the neoclassical paradigm is bothersome empirically and conceptually.   (Nelson and Winter 1982: 227)

Evolutionary models following Nelson and Winter (1982), such as Chiaromonte and Dosi (1993) and Silverberg and Verspagen (1994), extend these conclusions. A more complete overview is in Silverberg and Verspagen (1998). The model by Chiaromonte and Dosi shows how growth rates in a cross-section of nations may differ. The models by Silverberg and Verspagen show how "routines" of R&D investment may arise endogenously in a population of firms, and how growth patterns vary along the history of an economy that learns in such a "collective" way.

One of the rare models in this tradition that is solved analytically rather than by numerical simulation is that of Conlisk (1989). Under the assumption that technology advances are random, Conlisk constructs a model in which the growth rate of the aggregate economy is a function of three variables: the standard error of the productivity distribution of new plants (which can be interpreted as the average innovation size), the savings rate (which is defined somewhat unconventionally), and the speed of diffusion of new knowledge. Moreover, by changing some of the assumptions about the specification of technical change, the model emulates three standard specifications of technical change found in growth models in the neoclassical tradition. In this case, the first and third factors no longer have an impact on growth (they are specific to the "evolutionary" technical change specification of the model). However, the impact of the savings rate can be compared between the various model setups. Conlisk finds that using purely exogenous technical change (as in the Solow model), or learning by doing specifications (as in Arrow 1962), the savings rate does not have an impact upon (long-run) economic growth. This result, which is in fact also well known from standard neoclassical growth theory, marks an important difference between these models and his more evolutionarily-inspired specification.

The recent so-called "history-friendly models" (Malerba et al. 1999) aim to bring evolutionary models closer to empirical reality by reproducing the historical evolution of a particular industry, e.g. the computer industry. To this end, they start with a descriptive analysis of industry variables such as growth, concentration, and employment, and incorporate the insights from this analysis into a model the behavioral foundations of which are consistent with the evolutionary view. This model is calibrated and simulated to reproduce real-world trends as closely as possible. While this approach generates empirically relevant models, the simulations employ a relatively narrow set of parameter values. The work devotes little attention to a more open-ended investigation of which *minimal* set of assumptions is necessary to generate certain aspects of the structural evolution of specific industries.

These more open-ended uses of evolutionary micro models could lead to a new class of models that employ relatively simple, evolutionary microeconomic foundations to generate a broader range of phenomena in the evolutionary interpretation of technology and growth, rather than increasing the sophistication of the microfoundations. A much clearer focus on the salient macro features and what really drives them at the micro level may result from this approach, which is necessary to close the gap between the historical, evolutionary view and model building.

18.3.3 Neoclassical Views of Economic Growth and Technology

18.3.3.1 *Endogenous Growth Models*

How has mainstream economic theory coped with the complexity of technological change? The literature on neoclassical models of endogenous technology grew rapidly in the 1980s and 1990s following the publication of Romer (1986). Romer's model and others in this tradition were motivated by the apparent flaws associated with the assumption in the Solow model of decreasing marginal returns to capital: holding all other production factors (labor, land, infrastructure, buildings) fixed, the productivity of an extra (marginal) unit of investment would fall with growth in the existing capital stock. Decreasing marginal returns to investment could cause growth to slow down or even cease in the long run. As growth proceeds, capital accumulates, i.e. the capital stock increases, and hence an extra unit of investment generates less and less growth. Exogenous growth or productivity (knowledge) had been the traditional answer, but Romer (1990) and Grossman and Helpman (1991) proposed to make technology endogenous by modeling the R&D process. Abstracting from technicalities (a survey is provided by Verspagen 1992), this can be summarized as follows.

All the models assume that R&D is essentially a lottery in which the prize is a successful innovation. In the model by Aghion and Howitt (1992), this innovation-prize buys the firm a temporary monopoly of supplying the best-practice capital good used for production of consumption goods. The temporary monopoly vanishes when the next firm makes an innovation. Hence, the innovation process is modeled as a "quality ladder" of innovations, in which each new innovation supersedes the old one. In the industrial organization literature, this is called "vertical differentiation" of products.

In the model by Romer (1990), the innovation prize buys the successful firm a new variety of capital that will be demanded by producers of consumption goods forever, but has to compete with all other varieties (invented in the past, with the range continuing to expand in the future as a result of R&D). In this model, varieties of goods (innovations) do not go out of the market. Substitution between variations of goods is governed by a utility function or production function (depending on whether innovation takes place in consumer goods or intermediate goods) with a "constant elasticity of substitution." This is called "horizontal differentiation."

More tickets for the R&D lottery can be bought by doing more R&D, which is of course a costly process. Relative to the evolutionary models considered above, the crucial assumption is that the outcomes of the R&D process can be characterized realistically by weak uncertainty, i.e. the firm is able to estimate the probability that it will get the innovation-prize given its level of R&D spending. With expected benefits and costs of R&D known, the firm may make a cost–benefit analysis and derive an optimal level of R&D spending. This will, on average, correspond to a given amount of innovation, and produce a given growth rate. Although additional assumptions are necessary (e.g. with regard to the working of capital markets in which R&D expenditures have to be financed), this mechanism is the key to generating endogenous growth.

Before endogenous growth is possible in these models, there is one essential assumption about the nature of technology that needs to be made. This is related to the (partly) public good nature of technology. In the new growth models, this is represented by the assumption that there are technology spillovers between firms in the R&D process. The assumption takes two forms, depending on which flavor of model is used. In the horizontal differentiation type models (also called "love-of-variety" models), each innovation increases the level of general knowledge available in the economy, and this increases the productivity of the R&D process itself (Romer 1990). This assumption is necessary because of the ever more severe competition between the varieties of capital goods, and the falling profit rates that this causes. A tendency for R&D to be more productive (i.e. the costs of R&D to fall) offsets this falling profit rate, and keeps R&D feasible in the long run (see Grossman and Helpman 1991).

In the quality ladder models (vertical differentiation), each new innovation destroys the monopoly of the old innovator. However, the new innovator also builds on the previous innovation, because the quality of the new capital good is a fixed increase over the previous one. In other words, each new innovator is "standing on the shoulders of giants," and knowledge spills over intertemporally from one innovator to the next one. Without this spillover, endogenous growth would not be possible.

The technological spillovers in endogenous growth models lead to increasing returns to scale at the aggregate level. Even though the production functions of firms at the micro level are characterized by constant returns to scale, the R&D spillovers that flow from one firm to the rest of the economy imply increasing returns at the aggregate level. In terms of the expression for the aggregate growth rate of the economy, this feature of the endogenous growth models implies that growth at the country level depends (*ceteris paribus*) on the size of the country. Taken literally, this means that (*ceteris paribus*) larger countries will grow more rapidly. Related to this issue is the fact that the basic endogenous growth models are quite sensitive to small changes in the model specification with regard to technology spillovers. A slightly different specification of the impact of "general knowledge" on R&D productivity will lead to either zero growth in the long run, or to increasing growth rates in time (Grossman and Helpman 1991).

Technological spillovers make endogenous growth possible, but pose a challenge for policy makers. When technology generates positive externalities, the social benefits of R&D are larger than the private benefits (a rational firm investing in R&D does not consider the benefits of its R&D for its competitors). Hence the amount of R&D investment "generated by the market" will be too low from a social point of view. Technology policy in the form of R&D subsidies may bring the economy to a higher, socially optimal growth path. A similar conclusion is reached in a model of human capital and growth in Lucas (1988). In Aghion and Howitt (1992), there is also a negative externality: each new innovator destroys the rents of the existing monopolist (this is called "business stealing," or, in line with Schumpeter (1939), "creative destruction"). In this model, private R&D investment also can be too high from a social welfare perspective, depending on which of the two forms of externalities (creative destruction or standing on the shoulders of giants) is stronger.

The development of this new class of models raises promise and problems. On the positive side, it can be argued that this new growth theory takes seriously a number of arguments about technological change previously championed by evolutionary theorists but ignored by mainstream economists. These include the notion that R&D and technology are essentially stochastic phenomena (although evolutionary theory would argue that the type of uncertainty, i.e. weak uncertainty in which the probability distribution is known, is still not very adequate), and the importance of technology flows between agents (spillovers) for growth in the long run. The implication in many of these models that technology policy matters for growth also is relatively consistent with evolutionary theory, but may be less easily accommodated by mainstream economic theories that emphasize the efficiency of market forces.

On the negative side, these new growth models still propose a view of the interaction between economic growth and technology that differs substantially from that of evolutionary theory. The evolutionary view is one in which contingence and more systematic factors blend together in the dialectical process of historical time, but the new growth theory is still much closer to a Newtonian clockwork world in which there is a certain degree of "weak" uncertainty. In other words, the new growth theory still portrays the relationship between technology and growth as one of a steady-state growth pattern, which can be "tweaked" relatively easily by turning the knobs of the R&D process.

The evolutionary inclination, on the other hand, is that the nature of the growth process is more complex and variable over time. While the importance attached to the technology factor is shared with the new growth models, the belief that the relation between technology and growth is easily tweaked is not. In the evolutionary view, it is hard to predict exactly the impact of a policy measure, because it impacts on a complex range of interrelated factors. Moreover, while relations between a number of factors may have been revealed by careful research for a specific instance in time, it is to be expected that the nature of this relationship will change over time, exactly because of the (co)evolutionary nature of the process.

A more recent branch of new growth theory is the group of models that comes under the heading of "general purpose technologies" (GPT, Helpman 1998). A GPT is defined in essentially the same way as a basic innovation or paradigm in the evolutionary tradition. It consists of a basic technology (radical breakthrough), but this needs to be developed in the form of a range of intermediate (capital) goods. Within each GPT, the determinants of productivity are essentially the same as in one of the variants of the new growth models discussed above: technological change takes the form of an ever-expanding range of capital goods, but this is time-specific to the GPT. Thus, we see that at least two ideas from the evolutionary tradition are captured: the idea of differences in innovation size, and the idea that incremental innovations are responsible for the diffusion of a basic technology.

The GPT model generates cyclical growth. In its simplest form, the cycle consists of two phases. In the "low growth phase," the new GPT has been discovered, but is not yet in operation. New capital goods are being developed for it, and this activity has been halted for the old GPT. Thus, economic growth is low, because the main technology in use is no longer being developed. Once enough capital goods are available for the new GPT, its productivity outperforms that of the old GPT, the old GPT vanishes, and the economy shifts into a "high growth phase."

The GPT model resembles the evolutionary, Schumpeterian idea of long waves in economic growth. But scholars in the latter tradition have moved away from the fixed and deterministic cycle that characterizes the GPT model. Its clockwork view of economic growth has been dominant in the neo-classical tradition since the Solow model. One illustration of the limitations resulting from this view is the fact that, in the GPT view of the world, there is only room for substitution between subsequent paradigms. But economic and technological histories are filled with examples of the adaptation and survival, often in modified form, of old paradigms. For example, although the automobile is typical of the mass-production paradigm, it still plays a crucial role in the modern "Information Economy," although ICT has indeed been applied in the production of cars.

In conclusion, the evolutionary tradition and the neo-classical tradition have converged somewhat in the phenomena deemed central within each analytic approach. But they disagree on the essential nature of the growth process. The neoclassical theory conceptualizes growth as a deterministic process in which causality is clear-cut, and policies can be built on an understanding of time-invariant determinants of growth patterns. In the evolutionary view, on the other hand, contingencies and specific historical circumstances play a larger role, and causal mechanisms that prevail in one period may be subject to endogenous change in the next. In such a world, designing policy is harder, but not impossible.

18.3.3.2 *Empirical Work on Growth and Technology Following the Endogenous Growth Models*

The new growth models led to a tidal wave of empirical work on growth. Temple (1999) provides a detailed overview of this literature. The source of data for nearly all of this work was either the data by Maddison (1995) or the so-called Penn World Tables (PWT, Summers and Heston 1991). The PWT provides a broad cross-section of data for over a hundred countries. A crucial topic in the empirical debate following the endogenous growth models is the respective roles of steady state growth rates and convergence toward them. While the Solow model predicts that countries will converge to identical steady states (dependent on the exogenous rate of technological progress available to everyone), endogenous growth models predict that steady states will generally differ between countries. Empirical work on this issue has used a wide range of variables in regressions of growth rate differentials between countries, in order to examine cross-national differences in steady state growth rates.

Unfortunately, this approach is data-driven rather than theory-driven: an overall framework that governs and justifies the selection of factors is lacking. Also, many of the estimation results are sensitive to a small number of observations in the large sample (Levine and Renelt 1992). Nonetheless, this work leads to the conclusion that steady state growth rates differ between nations. Growth rates may converge toward a country-specific steady state growth path at best (so-called conditional convergence), leading to the divergence of growth paths among countries. Growth seems to be heterogeneous among countries starting from low levels of GDP per capita, with some countries falling behind, and some countries being able to catch up. This phenomenon is discussed in more detail in Fagerberg and Godinho (Ch. 19 in this volume).

Jones (1995*a* and *b*) has argued that the observed empirical record on R&D and growth is inconsistent with the theoretical predictions of endogenous growth models (see Box 18.3 on the "Jones critique" and semi-endogenous growth models). He observes that the postwar empirical evidence does not confirm the relationship proposed by R&D-based endogenous growth models that an increase in the number of R&D workers leads to higher rates of economic growth. Jones notes that the number of R&D workers has increased since the 1960s, but growth rates (of total factor productivity) have either been constant or declining during the same period. The so-called "Jones critique" has led to still more work in the endogenous growth tradition since its publication. Jones (1995*a*) suggests a so-called semi-endogenous growth model, which appears to be more consistent with the empirical facts, but in which endogenous growth only takes place when the population grows.

Box 18.3 The "Jones critique" and semi-endogenous growth models

Figure 18.3 illustrates the Jones critique for the United States and the European Union during the 1980s and 1990s. The R&D-based endogenous growth models predict that the growth rate of an economy, which is here approximated by total productivity factor growth, depends on the number of researchers in R&D. We see a steady increase in the latter, both in the US and the EU, but total factor productivity growth does not display a clear trend—instead it fluctuates widely around a roughly constant level. Does this constitute evidence against the relationship between innovation and economic growth?

Jones suggests an alternative model, which differs from the R&D-based endogenous growth models by Romer, Grossman and Helpman, and Aghion and Howitt by a different specification of the invention process. Whereas these original R&D-based growth models assumed that the growth rate of knowledge depends on the number of R&D workers in a linear way, Jones assumes that there are decreasing returns to R&D labor. This assumption is based on the idea that "the most obvious ideas are discovered first, so that the probability that a person engaged in R&D discovers a new idea is decreasing in the level of knowledge ... [and] the possibility that at a point in time the duplication and overlap of research reduce the total number of innovations" (Jones 1995*a*: 765). In this so-called semi-endogenous growth model, endogenous growth is only possible when the population grows.

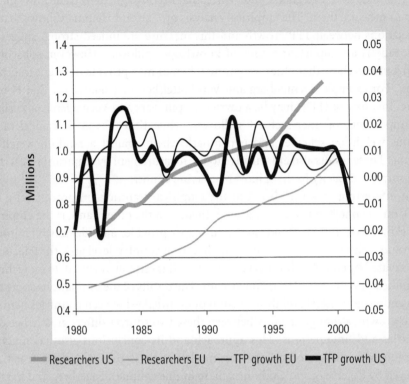

Fig. 18.3 The Jones critique. Total factor productivity growth trends are flat while the number of researchers in R&D increases (R&D researchers on left scale, tfp on right scale).

Source: Source for tfp data: Groningen Growth and Development Centre Total Economy Growth Accounting Database. Source for R&D researchers data: OECD Main Science and Technology Indicators Database.

From the point of view of evolutionary growth theory, the Jones critique appears to be the result of the misguided emphasis on steady-state growth states in the R&D-based endogenous growth models. The assumed relationship between R&D labor, the number of innovations, and resulting economic growth is based on assumptions of equilibrium behavior and weak uncertainty. In the less mechanistic evolutionary world, innovation, R&D, and growth are linked in a less rigid relationship that may change over time as a result of new and radical technological developments. In this view, the specific relationship between R&D labor and TFP growth observed by Jones may well be specific to the historical circumstances of the period, and may be subject to change in the future.

International endogenous growth models have provided other new inputs for the empirical tradition of research on R&D and productivity initiated by Griliches. This recent research focuses on the channels for international transmission of R&D spillovers. The assumption by Coe and Helpman (1995) is that these R&D spillovers are embodied in traded goods, and, hence, that R&D weighted by trade flows may be used to measure them. The empirical analysis by Coe and Helpman shows that the correlations between TFP growth and this measure are indeed strong, suggesting that trade is an important source of knowledge spillovers. However, subsequent contributions show that other weighting schemes may provide different interpretations. For example, Lichtenberg and Van Pottelsberghe (1996) show that Foreign Direct Investment (FDI) may be a carrier of spillovers and Verspagen (1997) shows the importance of inter-sectoral spillovers, while Keller (1998) is critical of the various weighting schemes and benchmarks them against a random weighting scheme. These results also are sensitive to the measurement by empirical researchers of absorptive capacity in the spillover-receiving countries.

An interesting "merger" between the empirical tradition on productivity and R&D, on the one hand, and new growth theory on the other hand, is the empirical model by Eaton and Kortum (1999). This paper provides a model in which innovation and technology diffusion are both drivers of country-level growth. The model is motivated by empirically observed trends, and is estimated with data on technology indicators (patents, R&D) and growth. The results of the estimations show that both endogenous R&D and the diffusion of knowledge between countries contribute to growth, although the mix between these two sources differs greatly between countries and time periods. This approach and its conclusions also has much in common with earlier technology gap models such as that of Nelson (1968), as surveyed by Fagerberg and Godinho (ch. 19 in this volume). Fagerberg and Verspagen (2002) recently reassessed the post-war evidence for these types of models, and concluded that, over time, innovation has become a more important source of growth as compared to the "pure" imitation of foreign technology. Models such as that of Eaton and Kortum thus have great promise to guide new growth theory in a direction that has much in common with the historically-inspired evolutionary approach.

18.4 OUTLOOK FOR THEORETICAL RESEARCH ON INNOVATION AND GROWTH

Neoclassical work in "new growth" or "endogenous growth" recently has shifted toward more "realistic" models that can accommodate a range of phenomena previously of interest only in the evolutionary tradition. Heertje described this convergence as follows:

neo-Schumpeterians [i.e., the evolutionary tradition] have been productive in their criticism of the neoclassical scheme on the basic of an evolutionary approach, but the questions they have raised have been addressed more or less successfully by many scholars, who have close links with the neoclassical tradition . . . I would not be surprised to see the present Schumpeterian mood to be part of mainstream economics before the end of this century.   (Heertje 1993: 273–5)

Is further convergence of the two traditions likely, as Heertje predicted for the end of the (previous) century? One avenue for convergence is in the further analysis of the intertemporal variability of growth patterns. At least some new growth models (e.g. Aghion and Howitt 1992) argue that time series of economic growth show variability, and this is a main topic in evolutionary models. The application of Pareto-type probability distributions, in which very large innovations have non-negligible probability, may bring the two approaches closer together, since they provide an intuitive way of modeling "strong uncertainty" (see e.g. Sornette and Zajdenweber 1999).

Each of the two approaches also contains a range of important and interesting lines of research to be pursued. In the endogenous growth tradition, the returns to purely theoretical work seem to have slowed down, but important empirical challenges remain open. The most fruitful avenue of research here seems to be further theoretical refinement induced by empirical work on technology and growth, with the explicit aim of developing empirically relevant models instead of new explorations motivated by technical problems with the existing models. For a long time, empirical research has led the way in the mainstream analysis of technology and growth, and this approach still seems to be the way forward.

Two main challenges confront the evolutionary tradition. The first is to develop a research program that goes beyond just emulating, although with a more plausible micro-foundation, the results of neoclassical analysis. Such an extension of the evolutionary research agenda could benefit from closer interaction with the non-formal work in the evolutionary tradition and greater reliance on historical research. Evolutionary modelers could seek to explain observed historical regularities in the relation between growth and technology.

A second challenge for evolutionary theorists is the development of more practically relevant models, for example, with regard to specific policy advice. Evolutionary theory rarely generates precise policy advice (see also Lundvall and Borrás, Ch. 22 in this volume), mostly as a result of the nature of the theory that points to complex interactions and rather unpredictable dynamics as important ingredients of the economic environment. To a certain extent, evolutionary theory will argue for a change in the way policy is viewed, but more precise work on how this can be implemented to achieve higher or more sustainable economic growth remains crucial.

Notes

1. The (older) data are necessarily rather imprecise, but the general trends are plausible on the basis of historical evidence. Note that since the vertical axis displays the logarithm of per capita income, a straight line would correspond to growth at a fixed rate, the slope of the line indicating the growth rate.
2. Solow (1957) is often quoted as the standard reference on growth accounting, but the ancestry of the method lies earlier (e.g. Tinbergen 1943 and Abramovitz 1956; for an overview see Abramovitz 1989: 13–15).
3. Well-known studies in this tradition are Denison (1962, 1966), Jorgensen (1967) and Maddison (1987, 1991): see Nadiri (1970) for an early overview of the methodology.
4. Critical surveys of the method can be found in Nelson (1973, 1981) and Fagerberg (1988b).
5. Pasinetti 1993 analyzes growth and technology from a demand perspective.
6. An elaborate overview of (empirical as well as theoretical) work on growth in the post-Keynesian tradition is in McCombie and Thirlwall (1994). A specific application to the issue of technology dynamics and growth is in Fagerberg (1988a).
7. Griliches (1992) provides a broad overview of empirical studies estimating R&D spillovers; Cincera and Van Pottelsberghe (2001) provide a survey on international spillovers, Van Pottelsberge (1997) on intersectoral spillovers.
8. An early attempt to develop a heuristic similar to the ones cited by Silverberg is in Sahal (1981).

References

Abramovitz, M. A. (1956), "Resources and Output Trends in the United States since 1870," *American Economic Review* 46: 5–23.

—— (1989), *Thinking About Growth*, Cambridge: Cambridge University Press.

Aghion, P., and Howitt, P. (1992), "A Model of Growth Through Creative Destruction," *Econometrica* 60 (1992): 323–51.

Arrow, K. J. (1962), "The Economic Implications of Learning by Doing," *Review of Economic Studies* 29: 155–73.

Barro, R. J. (1991), "Economic Growth in a Cross-Section of Countries," *Quarterly Journal of Economics* 106: 407–43.

Bijker, W. E., Hughes, T. P., and Pinch, T. (eds.) (1987), *The Social Construction of Technological Systems*, Cambridge, Mass.: MIT Press.

Chenery, H. B., Syrquin, M., and Robinson, S. (1986), *Industrialization and Growth: A Comparative Study*, Oxford: Oxford University Press.

Chiaromonte, F., and Dosi, G. (1993), "Heterogeneity, Competition, and Macroeconomic Dynamics," *Structural Change and Economic Dynamics* 4: 39–63.

Cincera, M., and Van Pottelsberghe, B. (2001), "International R&D Spillovers: A Survey," in M. Cincera, *Cahiers Economiques de Bruxelles* 169: 3–32.

Coe, D. T., and Helpman, E. (1995), "International R&D Spillovers," *European Economic Review* 39: 859–87.

*Conlisk, J. (1989), "An Aggregate Model of Technical Change," *Quarterly Journal of Economics* 104: 787–821.

CORNWALL, J. (1977), *Modern Capitalism: Its Growth and Transformation*, London: Martin Robertson.

CRAFTS, N. F. R. (1985), *British Economic Growth During the Industrial Revolution*. Oxford: Oxford University Press.

DENISON, E. (1962), *The Sources of Economic Growth in the United States and the Alternatives Before Us*, Washington: Committee for Economic Development.

—— (1966), *Why growth rates differ*, Washington: Brookings Institution.

DIXON, R. J., and THIRLWALL, A. P. (1975), "A Model of Regional Growth-Rate Differences on Kaldorian Lines," *Oxford Economic Papers* 11: 201–14.

DOSI, G. (1982), "Technological Paradigms and Technological Trajectories," *Research Policy* 11: 147–62.

EATON, J., and KORTUM, S. (1999), "International Technology Diffusion: Theory and Measurement," *International Economic Review* 40: 537–70.

FABRICANT, S. (1942), *Employment in Manufacturing 1899–1939*, New York: NBER.

FAGERBERG, J. (1988a), "International Competitiveness," *Economic Journal* 98: 355–74.

—— (1988b), "Why Growth Rates Differ," in G. Dosi, C. Freeman, R. R. Nelson, G. Silverberg, and L. Soete (eds)., *Technical Change and Economic Theory*, London: Pinter, 87–99.

—— and VERSPAGEN, B. (2002), "Technology-gaps, Innovation-diffusion and Transformation: an Evolutionary Interpretation," *Research Policy* 31: 1291–304.

FREEMAN, C., and LOUÇÃ, F. (2001), *As Time Goes By: From the Industrial Revolutions to the Information Revolution*, Oxford: Oxford University Press.

—— and SOETE, L. (1997), *The Economics of Industrial Innovation*, 3rd edn., London and Washington: Pinter.

*—— —— (1990), "Fast Structural Change and Slow Productivity Change: Some Paradoxes in the Economics of Information Technology," *Structural Change and Economic Dynamics* 1: 225–42.

GRILICHES, Z. (1979), "Issues in Assessing the Contribution of Research and Development to Productivity Growth," *The Bell Journal of Economics* 10: 92–116.

—— (1980), "R&D and the Productivity Slowdown," *American Economic Review* 70: 343–8.

—— (1984), *R&D, Patents and Productivity*, Chicago: Chicago University Press.

—— (1986), "Productivity, R&D and Basic Research at the Firm Level in the 1970s," *American Economic Review* 76: 141–54.

*—— (1992), "The Search for R&D Spillovers," *Scandinavian Journal of Economics* 94: S29–S47.

—— (1996), "The Discovery of the Residual: A Historical Note," *Journal of Economic Literature* 34: 1324–30.

—— and MAIRESSE, J. (1984), "Productivity and R&D at the Firm Level," in Z. Griliches (ed.), *R&D, Patents and Productivity*, Chicago: Chicago University Press, 339–74.

*GROSSMAN, G. M., and HELPMAN, E. (1991), *Innovation and Growth in the Global Economy*, Cambridge, Mass.: MIT Press.

GRUEBLER, A. (1990), *The Rise and Fall of Infrastructures: Dynamics of Evolution and Technological Change in Transport*, Heidelberg: Physica-verlag.

HEERTJE, A. (1994), "Neo-Schumpeterians and Economic Theory," in L. Magnusson (ed.), *Evolutionary Approaches to Economic Theory*, Dordrecht: Kluwer, 265–76.

* Asterisked items are suggestions for further reading.

HELPMAN, E. (ed.) (1998), *General Purpose Technologies and Economic Growth*, Cambridge, Mass.: MIT Press.

*JONES, C. (1995a), "R&D Based Models of Economic Growth," *Journal of Political Economy* 103: 759–84.

—— (1995b), "Time Series Tests of Endogenous Growth Models," *Quarterly Journal of Economics* 110: 495–525.

JORGENSON, D. W., and GRILICHES, Z. (1967), "Explanation of Productivity Change," *Review of Economic Studies* 34: 249–83.

*KALDOR, N. (1957), "A Model of Economic Growth," *Economic Journal* 67, December: 591–624.

—— (1970), "The Case for Regional Policies," *Scottish Journal of Political Economy* 67: 591–624.

KATZ, B. G., and PHILLIPS, A. (1982), "Government, Economies of Scale and Comparative Advantage: The Case of the Computer Industry," in H. Giersch (ed.), *Proceedings of Conference on Emerging Technology*, Kiel Institute of World Economics, Tuebingen: J. C. B. Mohr.

KELLER, W. (1998), "Are International R&D Spillovers Trade-Related? Analyzing Spillovers Among Randomly Matched Trade Partners," *European Economic Review* 42: 1393–612.

KLEINKNECHT, A. (1987), *Innovation Patterns in Crisis and Prosperity. Schumpeter's Long Cycle Reconsidered*, London: Macmillan.

KUHN, T. (1962), *The Structure of Scientific Revolutions*, Chicago and London: The University of Chicago Press.

LEVINE, R., and RENELT, D. (1992), "A Sensitivity Analysis of Cross-Country Growth Regressions," *American Economic Review* 82: 942–63.

LICHTENBERG, F., and VAN POTTELSBERGHE, B. (1996), "International R&D Spillovers: A Re-Examination," *NBER Working Paper 5668.*

LOS, B., and VERSPAGEN, B. (2000), "R&D Spillovers and Productivity: Evidence from U.S. Manufacturing Microdata," *Empirical Economics* 25: 127–48.

LUCAS, R. E. B. (1988), "On the Mechanics of Economic Development," *Journal of Monetary Economics* 22: 3–42.

*MADDISON, A. (1987), "Growth and Slowdown in Advanced Capitalist Economies: Techniques of Quantitative Assessment," *Journal of Economic Literature* 25: 649–98.

—— (1991), "Economic Stagnation Since 1973, its Nature and Causes: A Six-Country Survey," *De Economist* 131: 585–608.

—— (1995), *Monitoring the World Economy 1820–1992*. Paris: OECD Development Centre.

—— (2001), *The World Economy: A Millennial Perspective*, Paris: OECD Development Centre.

MALERBA, F., NELSON, R., ORSENIGO, L., and WINTER, S. (1999), " 'History-Friendly' Models of Industry Evolution: The Computer Industry," *Industrial and Corporate Change* 8: 3–40.

McCOMBIE, J. S. L., and THIRLWALL, A. P. (1994), *Economic Growth and the Balance-of-Payments Constraint*. London: St. Martin's Press.

MENSCH, G. (1979), *Stalemate in Technology: Innovations Overcome Depression*, Cambridge: Ballinger.

MORRISON, C. J. (1986), "Productivity Measurement with Non-Static Expectations and Varying Capacity Utilization," *Journal of Econometrics* 33: 51–74.

NADIRI, M. I. (1970), "Some Approaches to the Theory and Measurement of Total Factor Productivity: A Survey," *Journal of Economic Literature*: 1137–77.

NELSON, R. R. (1968), "A Diffusion Model of International Productivity Differences in Manufacturing Industry," *American Economic Review* 58: 1219–48.

—— (1973), "Recent Exercises in Growth Accounting: New Understanding or Dead End?" *American Economic Review* 63: 462–8.

—— (1981), "Research on Productivity Growth and Productivity Differences: Dead Ends and New Departures," *Journal of Economic Literature* 19: 1029–64.

*—— and WINTER, S. G. (1982), *An Evolutionary Theory of Economic Change*, Cambridge, Mass.: Harvard University Press.

*PASINETTI, L. L. (1993), *Structural Economic Dynamics*. Cambridge: Cambridge University Press.

PEREZ, C. (1983), "Structural Change and the Assimilation of New Technologies in the Economic and Social Systems," *Futures* 15: 357–75.

ROMER, P. (1986), "Increasing Returns and Long Run Growth," *Journal of Political Economy* 94: 1002–37.

—— (1990), "Endogenous Technological Change," *Journal of Political Economy* 98: S71–S102.

SAHAL, D. (1981), *Patterns of Technological Innovation*, New York: Addison Wesley.

SCHUMPETER, J. A. (1939), *Business Cycles: A Theoretical, Historical and Statistical Analysis of the Capitalist Process*, New York: McGraw-Hill.

SHELL, K. (1967) "A Model of Inventive Activity and Capital Accumulation," in K. Shell (ed.), in *Essays on the Theory of Optimal Growth*, Cambridge, Mass.: MIT Press, 67–85.

SILVERBERG, G. (2001), "The Discrete Charm of the Bourgeoisie: Quantum and Continuous Perspectives on Innovation and Growth," *Research Policy* 31: 1275–89.

*—— and VERSPAGEN, B. (1994), "Learning, Innovation and Economic Growth: A Long-Run Model of Industrial Dynamics," *Industrial and Corporate Change* 3: 199–223.

—— —— (1994), "Economic Growth and Economic Evolution: A Modeling Perspective," in F. Schweitzer and G. Silverberg (eds.), *Selbsorganisation. Jahrbuch für Komplexität in den Natur-, Sozial- und Geisteswissenschaften*, Berlin: Duncker & Humblot, 265–96.

—— (1957), "Technical Progress and the Aggregate Production Function," *Review of Economics and Statistics* 39: 312–20.

Solow, R.M. (1956), 'A contribution to the Theory of Economic Growth', *Quarterly Journal of Economics*, vol. 70, 65–94

SORNETTE, D., and ZAJDENWEBER, D. (1999), "The Economic Return of Research: The Pareto Law and its Implications," *European Physical Journal* B 8(4): 653–64.

SUMMERS, R., and HESTON, A. (1991), "The Penn World Table. Mark 5: An Expanded Set of International Comparisons 1950–1988," *Quarterly Journal of Economics* 6: 361–75.

Temple, J. (1999), "The New Growth Evidence," *Journal of Economic Literature*, March, 37(1): 112–56.

TINBERGEN, J. (1943), "Zur Theorie der Langfristigen Wirtschaftsentwicklung," *Weltwirtschaftliches Archiv* 55: 511–49.

UZAWA, H. (1965), "Optimum Technical Change in an Aggregative Model of Economic Growth," *International Economic Review* 6: 18–31.

VAN POTTELSBERGHE DE LA POTTERIE, B. (1997), "Issues in Assessing the Effect of Inter-industry Spillovers," *Economic Systems Research* 9: 331–56.

VERDOORN, P. J. (1949), "Fattori che Regolano lo Sviluppo della Produttivitá del Lavoro," *L'Industria* 1: 45–53.

VERSPAGEN, B. (1992), "Endogenous Innovation in Neo-Classical Growth Models: A Survey," *Journal of Macroeconomics* 14: 631–62.

—— (1995), "R&D and Productivity: A Broad Cross-Section Cross-Country Look," *Journal of Productivity Analysis* 6: 117–35.

—— (1997), "Estimating International Technology Spillovers Using Technology Flow Matrices," *Weltwirtschaftliches Archiv* 133: 226–48.

YOUNG, A. (1928), "Increasing Returns and Economic Progress," *Economic Journal* 38: 527–42.

CHAPTER 19

....

INNOVATION AND CATCHING-UP

....

JAN FAGERBERG

MANUEL M. GODINHO

19.1 INTRODUCTION

....

THE history of capitalism from the Industrial Revolution onwards is one of increasing differences in productivity and living conditions across different parts of the globe. According to one source, 250 years ago the difference in income or productivity per head between the richest and poorest country in the world was approximately 5:1, while today this difference has increased to 400:1 (Landes 1998). However, in spite of this long-run trend towards divergence in productivity and income, there are many examples of (initially) backward countries that—at different times—have managed to narrow the gap in productivity and income between themselves and the frontier countries, in other words, to "catch up." How did they do it? What was the role of innovation and diffusion in the process? These are among the questions that we are going to discuss in this chapter.[1]

The "catch-up" question should be seen as distinct from the discussion of "convergence," although the two issues partially overlap. "Catch-up" relates to the ability of a single country to narrow the gap in productivity and income vis-à-vis a leader country, while "convergence" refers to a trend towards a reduction of the overall differences in productivity and income in the world as a whole. The issue of convergence has been central to the economists' research agenda, in part because some prominent theoreticians formulated models of long-run growth implying such convergence (Solow 1956).[2] Of course, if all countries below the frontier catch up, convergence will necessarily follow. But if only some countries catch up (and perhaps forge ahead), while others fall behind, the outcome with respect to convergence is far from clear (Abramovitz 1986). What the empirics show is that, at best, such convergence is confined to groups of countries—or "convergence clubs" (Baumol et al. 1989)—in specific time periods. Arguably, to explain such differences in the conditions for catch-up through time, it is not enough to rely on general mechanisms. Some historical perspective is required.

During most of the nineteenth century, the economic and technological leader of the capitalist world was the United Kingdom, with a GDP per capita that was 50 per cent above the average of other leading capitalist countries. However, during the second half of the century, the United States and Germany both started to catch up and substantially reduced the UK lead. They did not achieve this growth by merely imitating the more advanced technologies already in use in the leading country, but rather did so by developing new ways of organizing production and distribution, e.g. by innovating (Freeman and Soete 1997; Freeman and Louçã 2001). In the case of the US, this led to the development of a historically new and dynamic system, based on mass production and the distribution and exploitation of economies of scale. Germany introduced new ways of organizing production, particularly with respect to R&D in the chemical and engineering industries, that in the long run would come to have a very important impact. More recently, the very rapid catch-up of Japan to Western productivity levels during the first half of the twentieth century was associated with a number of very important organizational innovations (such as the "just-in-time system," see Box 19.1) that, among other things, totally transformed the global car industry. These innovations did not only benefit Japan, but diffused (with a lag) to the established leader (the USA) and contributed to increased productivity there.

As these brief examples show, successful catch-up has historically been associated not merely with the adoption of existing techniques in established industries, but also with innovation, particularly of the organizational kind, and with inroads into nascent industries. However, as is equally clear, this has been done in different ways and with different consequences. If we extend the perspective to the most recent decades, as we will do in this chapter, this diversity in strategies and performance becomes even more striking. In the next section, we discuss some of the perspectives that have emerged in the catching up literature. Section 19.3 extends the perspective to the most recent decades, compares cases of successful catch-up to less successful ones, and considers the lessons that may be drawn. Finally, Section 19.4 raises, by way of conclusion, the question of what present day developing countries can learn, particularly with respect to policy, from the literature on innovation and catching up.

19.2 LESSONS FROM THE LITERATURE

The potentially relevant literature on why growth differs (and why some countries catch up while others do not) is arguably very large. Providing an overview of all the literature would be beyond the scope of a short essay. However, the work in this area focusing specifically on catch-up—as distinguished from economic growth more generally—and, in particular, on concepts and theories that may be helpful for understanding catch up (or lack of such), is appreciably smaller. We will in the following limit the discussion to three central cases within the latter. First, there are the contributions of Thorstein Veblen, Alexander Gerschenkron, and others on European catch-up prior to World War I.[3] The main point of interest here is the interpretation of the German catch-up with the UK, and the role of policy and institutions in this context. Second, there is a large literature on the Asian catch-up, particularly Japan, but increasingly also South Korea, Taiwan, and other countries that, to varying degrees, have attempted to follow the Japanese route. The argument that an activist, "developmental state" is an efficient mean to successful catch-up has been a central focal point in much of this literature. Third, there is a strand of macrohistorical and macroeconomic analysis focusing on interpretations of long-run data on economic growth, and on the role of technology and innovation in this context. A central contributor here has been Moses Abramovitz. In what follows next, these three perspectives will be briefly reviewed.

19.2.1 Lessons from European Catch-Up

The discussion of continental Europe catch-up illustrates nicely some of the central issues in the catch-up literature. Veblen (1915), who initiated the discussion, put forward the argument that recent technological changes altered the conditions for industrialization in latecomer economies. In earlier times, he argued, the diffusion of technology had been hampered by the fact that technology was mostly embodied in persons, so that migration of skilled workers was a necessary prerequisite for its spread across different locations. However, with the advent of "machine technology," this logic changed (ibid. 191). In contrast to the conditions that had prevailed previously, Veblen argued, this new type of knowledge "can be held and transmitted in definite and unequivocal shape, and the acquisition of it by such transfer is no laborious or uncertain matter" (ibid.). Although Veblen did not use the terminology that is now commonly applied to the process he described, it is pretty clear what he had in mind. Effectively, what he was arguing is that, while technology was previously "tacit" and embodied in persons, it later became more "codified" and easily transmittable. Hence, catch-up should be expected to be relatively easy, and was, under "otherwise suitable circumstances,"[4] largely "a question of the pecuniary inducement and... opportunities offered by this new industry" (ibid. 192). Since the latecomers could takeover the new technology "ready-made," without having to share the costs of its development, this might be expected to be a very profitable affair (ibid. 249). This being the case, Veblen predicted that other European countries, e.g. France, Italy, and Russia, would soon follow suit (he also mentioned the case of Japan).

While in Veblen's interpretation, the German catch-up was a relatively easy affair, the economic historian Alexander Gerschenkron (1962) took a different view, emphasizing the difficulty of the matter. While, he argued, technology at the time Britain industrialized was small scale, and hence institutionally not very demanding, these conditions were radically altered in the nineteenth century when Germany started to catch up. What Gerschenkron particularly had in mind was the seemingly inbuilt tendency of modern technology to require ever larger and more complex plants (static and dynamic economies of scale), with similarly changing requirements with respect to the physical, financial, and institutional infrastructure. He argued that, because of the high potential rewards from successful entry, and the heavy transformation (modernization) pressure on the rest of the economy it helped to generate, it was of paramount importance for the latecomer to target such progressive, dynamic industries, and to compete globally through investing in the most modern equipment/plants.[5] However, to succeed, catching-up countries had, in Gerschenkron's view, to build up new "institutional instruments for which there was little or no counterpart in the established industrial country" (1962: 7). The purpose of these institutional instruments would be to mobilize resources to undertake the necessary changes at the new and radically enlarged scale that modern technology required. His favorite example[6] was the German investment banks (and similar examples elsewhere in Europe), but he also admitted that, depending on the circumstances, other types of institutional instruments, such as, for instance, the government (in the Russian case),[7] might conceivably perform the same function.

Gerschenkron's work is often identified with his focus on the role of banks in industrialization, although as pointed out by Shin (1996), it is possible to see it as an attempt to arrive at a more general theory about catch-up, focusing on certain requirements that must be met for successful catch-up to take place, as well as different, though "functionally equivalent," institutional responses (or catch-up strategies). An important chain in Gerschenkron's argument is the emphasis on the advantages of targeting rapidly growing, technologically advanced industries. It should be pointed out, however, that for him this was a generalization based on historical evidence. Thus it is not obvious that his recommendations would be equally relevant for later time periods/technologies. Neither did he rule out that there might be other paths to successful industrialization than the one he recommended, although he held that to be rather exceptional. For instance, he pointed to Denmark as an example of a country that managed to catch up without targeting the progressive industries of the time, and explained this with its close links to the rapidly growing British market for agricultural products.

We may use Veblen's and Gerschenkron's accounts of the German catch-up to make a preliminary classification of catch-up strategies. The type described by Veblen assumes that technology is easily available/transferable, not very demanding in terms of skills or infrastructure, and that market forces are able to take care of the necessary coordination without the large-scale involvement of external "change agents." In contrast, there is the Gerschenkronian case in which technology transfer is so demanding in terms of skills/infrastructure that market forces, if left alone, are considered unlikely to lead to success, and so some degree of active intervention in markets by outsiders, whether private organizations or parts of government, is deemed necessary.

19.2.2 The Asian Experience

Similar perspectives to those of Veblen and Gerschenkron have also played a role in the discussions of the Asian catch-up in the post-World War II period. The primary examples are, in addition to Japan, the cases of South Korea, Singapore, and Taiwan. Although some observers have attempted to classify these as Veblen-type catch-up stories (World Bank 1993), there is by now an abundant literature showing that the catch-up strategies applied are much closer to the Gerschenkronian scheme (Johnson 1982; Amsden 1989; Wade 1990; Shin 1996).

There are many accounts of the Japanese catch-up. The so-called Meiji restoration in 1868 provides a natural starting point. What happened in 1868 was that a fraction of the ruling elite established a new regime, with the explicit purpose of strengthening the economy and the military strength of the state, which at the time was strongly challenged by Western imperialism (Beasley 1990). "A rich society and a strong army" was the slogan of the day. Since Japan lacked other "moderniza-tion" agents, the government (bureaucracy) took on the challenge. It modernized the legal system, the physical infrastructure, and the educational system, initi-ated new businesses (that later on were privatized) in industries that were deemed strategically important, etc.[8] Universities, colleges, and research centers were also founded, often with a bent towards engineering and applied science. While, particu-larly in the initial phase, the public sector played a vital role, private initiatives and cooperation between public and private actors became gradually more important. Much of the initiative came to rest with a number of emerging family-owned business groups, the Zaibatsus, in interaction with the bureaucracy and the military. Initially, the dominant industries were food processing and textiles, but, during World War I and the period that followed, the Japanese economy underwent a rapid transformation, with machinery and other "heavy industries" taking over as leading sectors. R&D activity also flourished, partly for military needs, and was, according to one source (Odagiri and Goto 1996), well above 1 per cent of GDP in the early 1940s.

The defeat of Japan in World War II changed the power structure in Japanese society by eliminating two of the three contending power centers, the military and the (owners of the) Zaibatsus, hence giving a boost to the bureaucracy that once more took on the challenge of gearing the economy and the society at large towards economic catch-up with the West. The sequence of events from the late nineteenth century somehow repeated itself, with a very important role for the state (and—in particular—the Ministry for Trade and Industry, MITI) in the early phase, and a growing role for private initiatives (and business groups) as the economy grew stronger (with no role left for the military). The new business groups that emerged, the Keiretsus, were in some cases based on the pre-war groupings (which had been dissolved by the American occupation forces), while in other cases they were totally new constructions. This reorientation differed from the pre-war groupings in respect of a stronger role for banks, and a smaller role for private investors/family ownership. In the early phases, the banks were very dependent on credit from the state, reinforcing the power of the bureaucracy in getting business to cooperate with the government in its preferred catch-up path. After a few decades of rapid growth, the banks (and business more generally) grew more independent, and the role of the state diminished and took on more "normal," Western proportions.

The exact role of the government versus private actors in the various phases of Japanese economic growth is a matter of considerable controversy, and we shall not attempt to resolve it here. Suffice it to say that government/bureaucracy intervention, through activist economic, industrial, and trade policy (protectionism), was very important, especially in the early phases. Although not everything it touched turned into "gold," and sometimes its interventions were strongly resisted by private business (and for perfectly good reasons), there is no doubt that it contributed significantly to focusing the attention of private business to catch-up with the West. An important element in this catch-up process (and the policies that were pursued) was a very rapid but orderly process of structural change, through which industries "of the past" were gradually phased out in favor of technologically more progressive industries, emphasizing in particular the combination of economies of scale, product differentiation, and rapidly growing demand, on the one hand, and continuous improvements of products and processes through learning, on the other. In this way, Japanese industry soon rose to the productivity frontier in its chosen fields, first in the steel industry and in ship-building, and later in cars and (consumer) electronics.[9] Although Japanese innovation in the catch-up phase also included a large number of product innovations, especially of the minor type (adaptations to demand), the main emphasis was on process innovations, particularly of the organizational type, that allowed for simultaneous exploitation of scale economices and flexibility, leading to high through-put, efficient inventory management, high quality/reliability, and a proven ability to adjust to the needs of the end-user (see Box 19.1).

Not surprisingly, the Japanese experience generated a lot of interest in other developing countries, particularly in Asia, that considered the policies and practices pursued as a possible model for their own catch-up towards Western levels. The prime examples are, as noted, South Korea, Singapore, and Taiwan, although the Japanese influence is also recognizable in other countries. Focusing in particular on the former, what these countries have in common is that they have caught up very

---

Box 19.1 Organizational innovation in Japan

Henry Ford allegedly once said that a customer could get a car in any color he wanted as long as it was black. This was the quintessential logic of the American system of manufacturing, based on standardized products, produced in long series for mass consumption, by low-skilled (often immigrant) labor, controlled by a hierarchy of foremen, engineers, and managers.

The attempt to adapt this system to Japanese conditions after World War II led to important modifications. First, the Japanese market was much smaller, so critical mass could only be reached through exploiting demand diversity. Second, the Japanese labor force was well educated, trained, and culturally homogeneous, and the differences in status and pay between blue and white collar workers small. As a result of these differences, the production system that evolved in Japan came to look very different from that of the USA (Freeman 1987).

The kanban or "just-in-time" system, developed by the Japanese auto industry, combines the advantages of mass production with flexibility in adjusting to changes in the composition and level of demand (Aoki 1988). What is going to be produced (and when) is decided by the part of the firm close to the end users (market). Orders are placed on a daily basis at the firm's production units, which have to deliver the requested products "just-in-time." This also holds for suppliers of parts, and the system is referred to as the "zero inventory" method. However, "zero inventory" implies that defect parts cannot be tolerated (because otherwise production would be halted). To increase quality and eliminate defects, organizational practices such as "total quality control," originally borrowed from US industry, and "quality circles" were introduced. Eventually, a new organization of work emerged, with workers rotating through different tasks and a much greater role for the individual worker (and work-team) in surveying production and quality, than what was common in the US auto industry. This new organization of work also meant more competent, committed, and motivated workers.

Important efficiency improvements stemmed from these organizational innovations. By the late 1980s, Japanese manufacturing, particularly in the car industry, was unrivalled in its efficiency (Womack et al. 1990). The time needed to produce a car in Japan in 1989 was 16.8 hours, while the equivalent figures for the US and Europe were, respectively, 25.1 and 36.2 hours (*The Economist*, 17 October 1992).

rapidly, undergone extensive structural change, and—finally—established themselves as among the major producers (and exporters) internationally in the most
technologically progressive industry of the day, electronics (broadly defined). The
government appears to have played a very important role in these processes.[10]
Everywhere, emphasis has been placed on the expansion of education, particularly
that of engineers (Lall 2000). In the early phases, governments in South Korea and
Taiwan intervened heavily with tariff protection, quantitative restrictions, financial
support, etc. to benefit the growth of indigenous industries in targeted sectors.
Singapore is a special case, since its government has relied heavily on inward Foreign
Direct Investment (FDI) in its industrialization efforts, and thus targeting has had to
be achieved through selective FDI policies (Lall 2000). In all countries, targeting
production for exports and rewarding successful export performance was very
important. More recently, all countries have placed emphasis on policies supporting
R&D and innovation. However, when it comes to the industrial structure, there is a
considerable element of diversity. In South Korea, large, diversified business groups
(chaebols), similar to the family-owned groups in pre-war Japan, have been and
continue to be very important, while in Singapore, foreign multinationals dominate
the scene. Taiwan, by contrast, is characterized by an industrial structure dominated
by small and medium-sized private firms.

With respect to the Gerschenkronian scheme, the experiences of these Asian
economies fit well with the emphasis on targeting the technologically most progressive industries. On a general level, in all four countries, the state (bureaucracy) has
played a very important role in the early stages. However, as noted above, this has
been achieved in different ways in different countries. For instance, in both Japan
and South Korea, credit rationing by the state (so-called "directed credit") was
extensively used to persuade private business to go along with the government's
objectives, while this mechanism played virtually no role in Taiwan (which underwent a financial liberalization early on). In the Taiwanese case, the government had
to rely on other instruments such as state-owned firms (which came to play an
important role) and, in particular, heavily supported "intermediate institutions"
(R&D infrastructure etc.) with mixed public/private sector participation. Moreover,
while industrialization in Japan, and in the USA and Germany before it, was
mainly geared towards the home market, exports played a similar role in the
catch-up strategies of the three "tigers." This may, arguably, have to do with
the fact that the domestic markets in the latter were in many cases too small to
support large-scale industrialization efforts, but the gradual reduction in barriers
to trade during the post-World War II period also played an important role
(Abramovitz 1994).

The freeing of international capital flows, and the deregulation of financial
markets towards the end of that period, placed the "tigers" and other late-latecomers
in a somewhat different situation from that of Japan 50–100 years earlier, with a
greater potential role for external finance, whether in the form of FDI or lending.

For instance, while Japanese catch-up was largely self-financed, the South Korean catch-up came to depend heavily on foreign lending. However, such increased debt exposure, while providing opportunities for catch-up, may also make countries vulnerable, as shown by the financial crisis in Korea (and to some extent in other Asian countries) towards the end of the 1990s (see Box 19.2). Although controversial (Shin and Chang 2003), the crisis may also be seen as an illustration of the important point that policies and institutions that worked well during the catch-up phase may not be equally well suited when this phase is completed and the former catch-up country has to compete with other developed countries on an equal footing. Another example of this, also from the financial sector, comes from Japan. The Japanese financial system was designed to generate large savings among the general public and to funnel these to the large industrial conglomerates, which were the vanguards of the catch-up process, on preferential terms, and as such it was very effective. However, when the catch-up was completed, this financial machine continued to generate large savings, even though the profit opportunities created by the potential for catch-up were largely gone. This led to excesses, crises, and depression. Hence, from being a very valuable asset, the country's financial system actually turned into a considerable burden for the Japanese economy.

Box 19.2 The financial crisis in Korea

South Korea went through a deep financial crisis in 1997–8. A factor contributing to this was the, by international standards, very high debt-exposure of Korean chaebols, which had to do with the way catch-up in Korea traditionally had been financed (as in Japan, through loans, often by state-controlled banks on concessionary terms, rather than equity). As long as international capital flows were subject to strict government control, as occurred in most of the post-World War II period, the system may have been said to fulfill its purpose (rapid catch-up). But when these restrictions eased, several Korean chaebols and financial institutions exploited their new freedom to increase their financial exposure and, as a result, substantially increase the country's national debt. This paved the way for the crisis. To resolve it, the Korean government went into a settlement with the IMF, involving, among other things, expectations of extensive "structural reforms," intending to bring the Korean system closer to the Anglo-American model (see Shin and Chang (2003) for an extended discussion). The crisis was relatively short-lived, as the Korean economy underwent a rapid recovery in 1999. It had some repercussions elsewhere in Asia, although Taiwan and Singapore were much less exposed, simply because debt financing and foreign lending did not play the same prominent role as in Korea.

19.2.3 A Macro-View

The third strand of catch-up research mentioned above operates on the macro-level and asks questions of the extent to which the catch-up or convergence actually occurred, for whom and how this may be explained. As mentioned in the introduction, an important finding in this literature is that the long-run trend since the British Industrial Revolution points to divergence, not convergence, among capitalist economies. It has also been shown that these trends differ considerably between time periods. For example, one such period in which the conditions for catch-up appear to have been especially favorable (and during which many countries managed to narrow the gap in productivity and income vis-à-vis the leader) is comprised of the decades following the end of World War II, what Abramovitz (1986, 1994) has called "the post-war catch-up and convergence boom." He suggested that such differences in performance over time and across countries might, to some extent, be explained with the help of two concepts, technological congruence and social capability. The first concept refers to the degree to which leader and follower country characteristics are congruent in areas such as market size, factor supply, etc. For instance, the technological system that emerged in the USA around the turn of the century was highly dependent on access to a large, homogeneous market, something that hardly existed in Europe at the time, which may help explain its slow diffusion there. The second concept points to the various efforts and capabilities that developing countries have to develop in order to catch-up, such as improving education, infrastructure and, more generally, technological capabilities (R&D facilities etc.). Abramovitz explained the successful catch up of Western Europe in relation to the US in the first half of the post-war period as the result of both increasing technological congruence and improved social capabilities. As an example of the former, he mentioned the manner in which European economic integration led to the creation of larger and more homogeneous markets in Europe, facilitating the transfer of scale-intensive technologies initially developed for US conditions. Regarding the latter, he pointed to, among other things, such factors as the general increases in educational levels, the rise in the share of resources devoted to public and private sector R&D, and the ability of the financial system in mobilizing resources for change.

There have also been attempts to develop testable models of cross-country differences in growth performance (or productivity) that includes the potential for catch-up as one of the explanatory factors. Classical papers on the subject are Nelson (1968) and Gomulka (1971). In a highly innovative contribution, Cornwall (1977) analyzed economic growth in the first half of the post-World War II period as driven by catching-up processes, the ability to mobilize resources for change (investment), demand, and endogenous technological change (through the working of the so-called "Verdoorn's law"). Baumol et al. (1989) presented and tested a model of cross-country growth, including the potential for catch-up and social capability (proxied by education), for a large number of countries and different time-spans, and since then there has been a plethora of such exercises confirming (or questioning) the importance of such factors (see Fagerberg 1994 and Temple 1999 for overviews). However, most of these studies have ignored Abramovitz's emphasis on the importance of technological congruence as well as the role of innovation. More to the latter, Fagerberg (1987, 1988) has suggested an empirical model based on Schumpeterian logic that includes innovation, imitation, and other efforts related to the commercial exploitation of technology as driving forces of growth. Following this approach, catch-up or convergence is by no means guaranteed, as it depends on the balance of innovation and imitation, how challenging these activities are, and the extent to which countries are equipped with the necessary capabilities. According to Verspagen (1991), who implemented similar ideas into a non-linear setting that allows for both catch-up and a "low-growth trap," poor countries with a low "social capability" are the ones at risk of being "trapped."

Abramovitz's work has been criticized by Shin (1996) for not being sufficiently historically specific. In particular, he argues that the "social capability" concept is very difficult to operationalize, a fact admitted by Abramovitz himself. However, Abramovitz's emphasis on technological congruence clearly points to an awareness of the importance of changes in technological dynamics over time, although it is clear that he himself did little to substantiate it. In this he clearly sided with Gerschenkron, who also focused almost exclusively on catch-up in scale-based technologies. There is, however, no scarcity of contributions that argue that the dynamics of the scale-based system is waning (Nelson and Wright 1992; Fagerberg et al. 1999). If so, this may have strong implications for the conditions for catch-up. We return to this issue in the final section of this chapter.

19.3 CATCHING UP: A REVIEW OF RECENT EVIDENCE

We will now take a closer look at the global catching-up process (or lack of such) during the four last decades. While other studies have provided highly aggregated analyses of differences in growth across large samples of countries (for overviews, see Fagerberg 1994 and Temple 1999), we will in this section limit the analysis to a selection of countries that we find particularly relevant for the study of such processes, and for which good data on relevant factors, such as R&D and innovation, are available. This includes the countries discussed so far, such as the previous—and present—world leaders (US and UK) and Gerschenkron's favorite object of study, Germany (we also include data for two other countries he studied, France and Italy, for comparison). Moreover, we include a group of Asian countries, which, in addition to those mentioned so far, also contains China, Hong Kong, India, Malaysia, and the Philippines. Finally, we introduce two other country groupings, with which the experience of the catch-up countries of Asia may be compared, a group of European catch-up countries (Finland, Greece, Ireland, Portugal, and Spain) and a group of potential catch-up countries from Latin America (Argentina, Brazil, Chile, and Mexico).

Table 19.1 ranks the countries in our sample by initial GDP per capita level (1960). The countries that industrialized a century or more ago are, not surprisingly, at the top, headed by the USA, while the seven countries at the bottom of the list are all Asian. In the middle, we find the remaining European and Latin American countries, joined by two Asian economies, Hong Kong and Japan. Figure 19.1 illustrates the way in which these changes in the distribution during the last four decades came along. Evidently, there is a group of Asian countries that have caught up very rapidly. In fact, the top seven performers in terms of per capita growth are all from Asia. Annual per capita growth for these countries ranges from 6.5 per cent (South Korea) to 4.2 per cent (Japan). The European catch-up countries follow, headed by Ireland (4.1 per cent) and Portugal (3.9 per cent). The more established countries, which were in the lead in the early 1960s, cluster towards the lower half of the distribution, with growth-rates in the 2–3 per cent area. At the very bottom, we find three potential "catch-up" countries that have experienced very dismal performance, the Philippines, Argentina, and Mexico, each of which evidently has "fallen behind," to use the terminology suggested by Abramovitz. The remaining potential "catching-up" countries of Latin America, plus India, although performing slightly better, also failed to reduce the gap vis-à-vis the leader, the USA. The result of this dynamic is that, while the seven Asian countries at the top of the list all improved their relative positions during this period, by moving to a higher quartile of the distribution, all the Latin American countries moved down one or more quartiles (Table 19.1).

We will now explore the manner in which these differences in performance relate to differences in relevant "social capabilities." Although, potentially, there may be many variables of interest, we have in the present context chosen to limit the discussion to three that we believe are of particular relevance, these being skills (education), R&D, and innovation (as reflected in patents). Traditionally, many analyses of differences in cross-country growth, based on large cross-country samples, have focused on differences in the extent of primary and secondary education as a possible factor behind the observed differences in performance (see e.g. Baumol et al. 1989). However, while relevant for understanding the failures of some developing countries, in Sub-Saharan Africa for instance, to enter the catch-up phase, these variables discriminate less well between the countries in our sample, which, with very few exceptions, all have relatively extensive primary and secondary

Table 19.1 Income groups, 1960–1999 (GDP per capita, 10^3 $US, 1990 constant PPPs)

	1960	GDPpc	1999	GDPpc
1st Quartile	US	11.3	US	28.1
	(West) Germany	10.1	Japan	21.0
	UK	8.6	Singapore	20.7
	France	7.5	France	20.1
	Finland	6.2	Hong Kong	19.9
	Italy	5.9	Ireland	19.7
2nd Quartile	Argentina	5.6	UK	19.2
	Chile	4.3	Finland	19.1
	Ireland	4.2	(Unified) Germany	19.0
	Japan	3.9	Italy	18.2
	Spain	3.4	Taiwan	16.6
	Mexico	2.2	Spain	14.6
3rd Quartile	Greece	3.1	Portugal	13.5
	Hong Kong	3.1	South Korea	13.2
	Portugal	3.0	Greece	11.5
	Brazil	2.3	Chile	10.0
	Singapore	2.1	Argentina	8.7
	Malaysia	1.5	Malaysia	7.7
4th Quartile	Taiwan	1.5	Mexico	6.9
	Philippines	1.5	Brazil	5.4
	South Korea	1.1	China	3.3
	India	0.8	Philippines	2.3
	China	0.7	India	1.8

Source: Calculations based on Angus Maddison/Groningen Growth and Development Centre and The Conference Board, Total Economy Database, July 2003, http://www.ggdc.net.

education systems.[11] We have, therefore, chosen to focus on third-level education (universities, colleges, etc.).

Figure 19.2 confirms that the established industrialized leaders, with the USA in a comfortable lead, place strong emphasis on higher (third level) education.[12] But some catching-up economies also figure relatively high. Finland, for instance, is

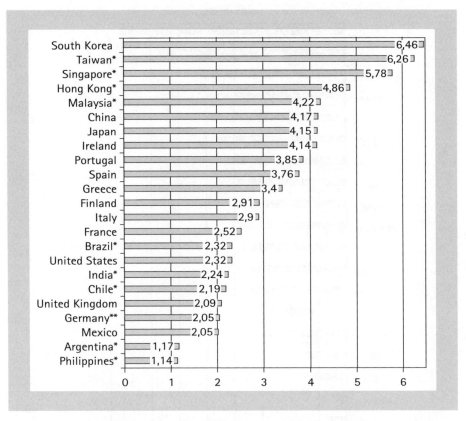

Figure 19.1 GDP per capita growth 1960–2001

Notes: All calculations based on 1990 constant prices. For countries with (*) the calculation period is 1960–1999. The German growth rate (**) refer to West Germany, 1960–1997.

Source: Calculations based on Angus Maddison/Groningen Growth and Development Centre and The Conference Board, Total Economy Database, July 2003, http://www.ggdc.net.

second, followed by South Korea. However, one should not overemphasize such differences, because today, in sharp contrast to the situation thirty to forty years ago, the great majority of countries under study share a strong emphasis on higher education. The increase in higher education is especially impressive in some of the catching-up economies in Asia and Europe, such as, for instance, Finland, South Korea, and Spain. The deviants from this strong emphasis on third level education consist of a group of (low-income) Latin American and Asian countries (Mexico, Brazil, Malaysia, India, and China) that continue to have very low levels of higher education.

Higher education is, however, a mixed bag, and not every element is necessarily equally essential for innovation or catch-up in technology. In Figure 19.3, we focus

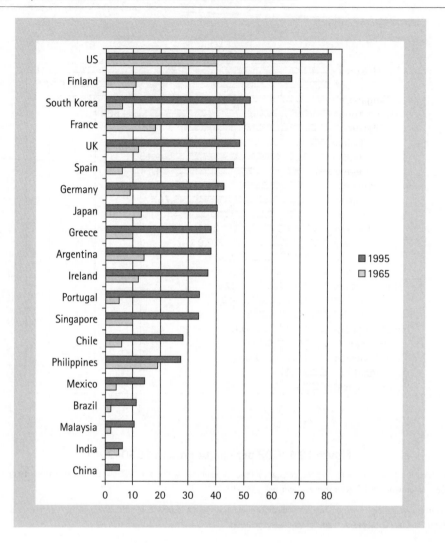

**Figure 19.2 Third level enrollment in relation to age group,
20–24 years old (1965–1995)**

Source: UNESCO, Education Statistics, various years.

more narrowly on the production of (undergraduate) university degrees in natural
sciences and engineering (as a percentage of 24 year-olds in the population). In this
case, we see a much clearer divide between the countries in the upper half of the
distribution, in which between 6 and 9 per cent of the cohort take such education,
and the countries in the lower half, in which—in all but one case—less than 3 per
cent of the cohort get such degrees. As is evident from the figure, the countries that
place most emphasis on education in natural sciences and engineering are the

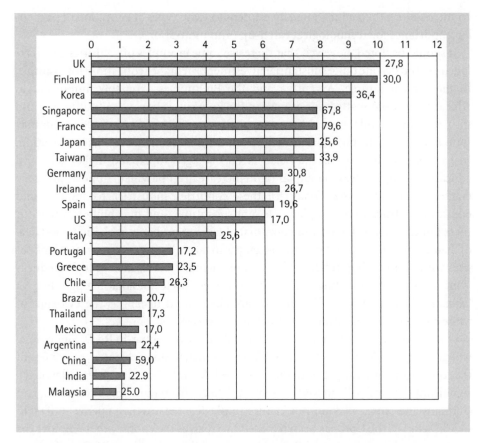

Figure 19.3 Ratio of first university degrees in natural sciences and engineering to 24–year–olds in the population, 1999 (all values in %)

Notes: All figures for 1999 or most recent year. The numbers to the right of the horizontal bars refer to the percentage share of "1st University Degrees in Natural Sciences and Engineering" in "Total 1st University Degrees" (French and Greek numbers refer only to "long" degree courses, and are therefore not directly comparable to the numbers for other countries).

Source: NSF, Science and Engineering Indicators 2002, http://www.nsf.gov/sbe/srs/seind02/start.htm

developed countries (the early industrializers) and the four Asian countries discussed in the previous section, joined by some of the catch-up economies in Europe (Finland, Ireland, and Spain, in particular). The lower half of the distribution, those with low investments in this area, includes all the Latin American countries, the less-developed countries of Asia and—closer to the mean of the sample—some of the catch-up countries in Europe (Portugal and Greece). It is also noteworthy that, while in the USA, one out of six students graduate in natural sciences or engineering, in South Korea the equivalent number is one-third and in Singapore two-thirds. Hence, countries such as Korea, Taiwan, and Singapore not only place strong emphasis on higher education in general, but to a larger extent than most other countries direct their educational investment towards types of education of particular importance for technological catch-up (and innovation).

It should be noted, however, that there are some examples of countries that have fallen behind despite quite substantial investments in higher education, for example, in the present sample, Argentina and the Philippines. Arguably, important as education is, what matters for growth in the long run is how it is put into use, and the failure to expand the employment opportunities for highly educated labour may seriously impede the potential growth effects from investments in higher education. In fact, one of the reasons the Asian NICs managed to expand higher technical education so rapidly was the similar rapid increase in employment opportunities for engineers (and scientists). Thus, for these countries, industrial, technology, and education policies were complements, not substitutes, and the ability to carry out these policies in a sustained and coordinated fashion probably explains a good deal of their economic success. Similarly, attempts to target high-growth, strategic industries without investing sufficiently in complementary assets, such as higher education, or without providing sufficient incentives for technological upgrading (a "dynamic" competitive environment), are also bound to fail, as the evidence of some countries in, for instance, Latin America shows.

One important use of highly competent labour is, of course, in R&D. Figure 19.4, which focuses on R&D as a share of GDP, shows that, in the early 1960s, only a few of the countries in our sample, with the USA, the UK, and France in the lead, devoted a significant share of GDP to R&D activities.[13] Apart from these three countries, and Germany and Japan, all countries in our sample devoted less than 1 per cent of GDP to R&D. Today, the USA has been replaced by Japan as the country that employs the largest share of its income on R&D activities, and the club of high R&D performers has been enlarged by a number of new members, with South Korea, Finland, and Taiwan deserving of particular mention. However, Singapore, Ireland, and Italy have also increased expenditures on R&D beyond the 1 per cent of GDP level. The remaining countries, including those from Latin America, many Asian and most of the catching-up economies in Europe, remain low R&D performers, although R&D investments have in several cases increased significantly compared to the situation a few decades ago. Data on patents reveal a very similar pattern (Figure 19.5).

Another indicator that is often invoked in analyses of catching-up and technology transfer is inward foreign direct investments (FDI), on the grounds that those who do such investments are assumed to control, and are willing to share, superior technology. The available evidence, however, indicates that the distribution of FDI is highly skewed, with a disproportionately high amount invested in two small economies, Hong Kong and Singapore, and, more recently, and to a lesser extent, in a number of other lower income countries, such as Ireland, Chile, Malaysia, and China. However, some of the most successful catching-up economies, such as Japan, Taiwan, and South Korea, have received very little inward FDI. This does not imply,

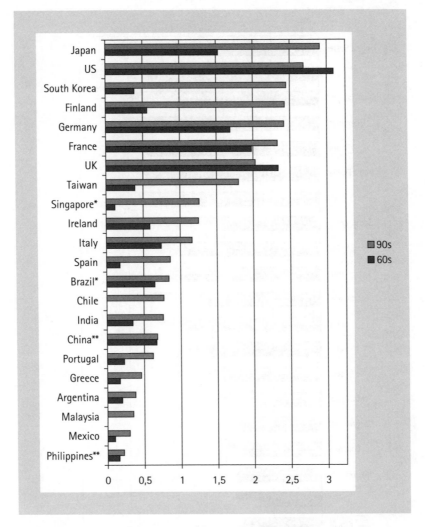

Figure 19.4 R&D as percentage of GDP, 1960s and 1990s

Note: Countries with (*)–1970s; Countries with (**)–1980s.

Source: Calculations based on OECD, UNESCO and national statistics.

of course, that these countries did not benefit from international technology flows. It is just that they found other—and perhaps equally or more efficient—ways of absorbing foreign technology (see Box 19.3).

The evidence presented here confirms the relevance of the Gerschenkronian scheme referred to earlier, in the sense that the countries that have been most successful in catching up, namely South Korea, Taiwan, and Singapore (and Japan before them), have all—after initially having acquired some capabilities through

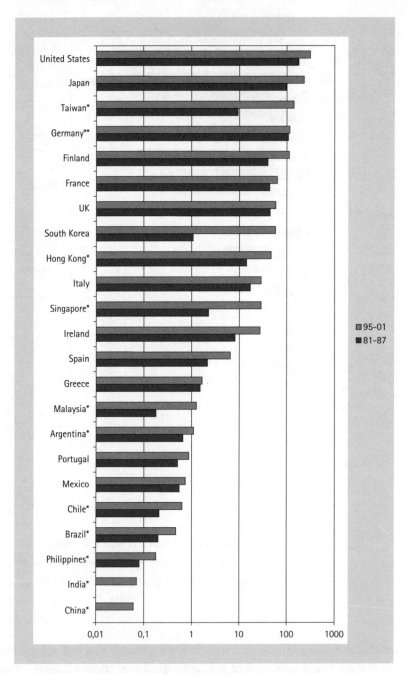

Figure 19.5 US patenting per million inhabitants (log scale)

Notes: (*) For these countries the second period is 1995–1999; (**) Up to 1989 the data refers to the Federal Republic of Germany.

Source: USPTO, Patent counts by country/state and year—All patents, all types, http://www.uspto.gov/web/offices/ac/ido/oeip/taf/reports.htm

Box 19.3 How to access foreign technology? The OEM system

Asian catch-up has benefited greatly from technology developed elsewhere. However, the mechanisms used to tap foreign technology sources differ. One central mechanism, used extensively by Singapore, is inward Foreign Direct Investment (FDI). By contrast, Taiwan and especially South Korea relied mostly on a form of subcontracting, "Original Equipment Manufacturing" (OEM). As suggested by Hobday (2000), OEM might be seen as an organizational innovation, facilitating learning and technological upgrading in latecomer firms.

Under an OEM contract, a product is produced according to a customer's specifications, normally a transnational corporation (TNC), that markets and sells the product under its own brand-name (such as, for instance, "NIKE" or "IBM"). From the 1970s onwards, many US and Japanese firms, particularly in the ICT sector, used this mechanism to contract out their production to Korean and Taiwanese firms. This allowed the latter to acquire basic producing capabilities in electronics, since the TNCs normally "helped with the selection of equipment; the training of managers, engineers and technicians; and advice on production, financing and management. ... Local learning was encouraged because the TNC depended on quality, delivery, and price of the final output" (Hobday 2000: 134).

As successful OEM arrangements evolved into closer long-term relationships, the Korean and Taiwanese firms gradually acquired more advanced capabilities, first in process engineering and later in product design. This led OEM to evolve into a more advanced stage, ODM (Own Design and Manufacturing), with a greater emphasis on R&D. A next step, OBM (Own Brand Manufacturing), occurs if a firm uses the acquired capabilities to produce and to market products under its own brand-name. This requires new capabilities in marketing, and very substantial investments in distribution. Hence, it is a difficult step, but potentially very rewarding, since a lot of the value added is generated at this stage. Several Korean and Taiwanese firms have tried, with mixed success, although a few (e.g. Samsung) have managed quite well.

more traditional activities—aggressively targeted the most technologically progressive industries of the day, in which they today play an important role. This transformation of the economy has been accompanied by extensive investments in higher education, particularly in engineering and natural sciences, and big increases in the resources devoted to R&D and innovation. As discussed in the previous section, proactive governments—and policies—have played an instrumental role in these processes, though in different ways, reflecting different historical backgrounds and conditions. However, not far behind these success stories, measured in terms of economic performance, we have a more diverse group of countries that also have managed to reduce substantially the gap vis-à-vis the frontier. Some of these, such as Finland, Ireland, and Malaysia, share the focus on targeting the technologically most progressive industries of the day (ICT), although with considerable differences between them with respect to the instruments pursued in achieving that goal.

Others, such as Portugal, Spain, and Greece, have preferred to pursue catch-up without similarly ambitious goals for changing the industrial structure, and, arguably, with more modest results, both in terms of economic performance, accumulation of skills, and technological capabilities. Still others, such as the Latin American countries considered here, have failed to invest sufficiently in skills and technological capabilities, and have as a consequence fallen further behind.

19.4 CATCHING-UP AND POLICY

The literature on catch-up processes in Europe, particularly in Germany, led to a strong focus on the relationship between catch-up, "institutional instruments," and policy. Similarly, the more recent idea of a "developmental" state, modeled on the experiences of Japan, has brought increased attention to the important role played by policy in catch-up processes. This led, among other things, the World Bank to publish a study on East Asian catch-up in which it sought to emphasize the advantages of its so-called "market friendly" approach, and to downplay the role that interventionist politics had played in the catch-up of these countries (World Bank 1993, for rebuttals see Rodrik 1994 and Cappelen and Fagerberg 1995). However, the discussion of catch-up and policy arguably has older roots (Chang 2002). Two hundred years ago, when the Americans began to consider how to reduce the gap vis-à-vis the UK, this issue was lively debated. Some, basing themselves on the great authority of Adam Smith, argued that the best thing would be to practice free trade, refrain from governmental intervention in economic affairs, and stick to America's acquired advantage in agriculture. Others, such as the first Secretary of the US Treasury, Alexander Hamilton, doubted the wisdom in this approach, and advocated an industrialization policy based on so-called "infant industry protection." The German economist Friedrich List, who came to be known as the chief protagonist of this approach, pointed out that Britain—the main advocate of free trade at the time—had itself used infant industry protection intensively during its rise to economic and technological leadership (List 1841). Its more recent advocacy of free trade, List argued, was simply an attempt to "kick away the ladder" the country itself had used to industrialize.

Since that time, the issue of catch-up and policy has been highly contentious. Recently Chang (2002) has taken a fresh look at the evidence on this issue. Based on an extensive overview of policies carried out in the industrialization phase of various (developed) countries, he demonstrated that the interventionist policies applied by Japan and other Asian countries during their catch-up were not historically unique. On the contrary, most (though not all) present-day developed countries applied such policies when in the same situation. Many different policy instruments were used to support the growth of new industries, trade protection (tariffs, etc.) being only one, and not always the most important. However, during the last few decades, Chang points out, there has been a concerted effort, led by the USA, with international organizations such as the World Bank, the IMF, and the WTO as central players, to reduce the room of maneuver for such interventionist politics by catching-up countries.[14] He argued, as List before him, that these efforts may be seen as an attempt by the present economic and technological leaders to "kick away the ladder" their own countries used to arrive at their present levels of development.

Do the changes in international rules and regulations during the last few decades, making certain types of policies (or practices) previously applied by catching-up countries more difficult (or even impossible) to pursue, imply that catch-up is becoming progressively more difficult? This is an important question that deserves a place on the research agenda in this area. But it needs to be emphasized that what is a suitable policy nowadays depends not only on the characteristics of the policies that seemed to work well in the past, but also on the economic, technological, institutional, and social context today (which may be quite different from those of previous times). However, a cautionary look at the empirical evidence from the last few decades suggests that many developing countries have found it increasingly difficult to exploit the potential for catch-up. This reading of the evidence is also confirmed by a recent empirical study by Fagerberg and Verspagen (2002), which found that the conditions for catch-up have become more stringent over time, with ever-greater demands on the technological capabilities and innovative efforts of countries striving to narrow the gap vis-à-vis the frontier. While in the 1960s and 1970s, the main factors supporting catch-up were found to be capital accumulation and a sufficient manufacturing base, in the 1980s and 1990s, the accumulation of technological capabilities and specialization in services were shown to be more relevant. These findings indicate that what has happened cannot be explained solely by changes in institutions and policies, but also has to do with a shift in the underlying technological conditions, an area clearly in need of further research. Fagerberg and Verspagen suggested that the observed shift in the conditions for catch-up "may be a reflection of the radical technological change in the last decades, with ICT-based solutions substituting earlier mechanical and electromechanical ones, and the derived change in the demand for skills and infrastructure" (2002: 1303). This is, of course, the kind of change that Abramovitz hinted at with his concept of "technological congruence." Following this, one might hypothesize that, compared to the situation three or four decades ago, the progressive technologies have become less "congruent" with the economic conditions (particularly skill-base and R&D infrastructure) that prevail in many developing countries. In fact, as shown in the previous section, today only countries that have invested massively in the formation of skills and R&D infrastructure seem to be able to catch up (while those that have not fall further behind).

A weakness of much of the existing discussion on catch-up and policy has been an excessive focus on the policy level (government) at the expense of the recipients of these policy initiatives, e.g. the firms of the potential catching-up country. As pointed out by Teece (2000: 124): "If firms are indeed the instruments of development, the study of economic development cannot take place separate from the study of the theory of the growth of the firm." We suggest this as an important area for further research, theoretical as well as applied. Although an extensive treatment of the role of firms in catching-up processes is beyond the scope of this chapter, we will nevertheless try to emphasize a few points that we believe may be useful for further work.[15] Research on the role of firms in innovation and long-run economic change commonly stress that, in most cases, firms only have imperfect knowledge of the relevant options in front of them, and that they tend to be myopic, searching in the neighborhood of their existing competence for relevant information, suggestions, and solutions (Nelson and Winter 1982; Dosi 1988; Fagerberg Ch. 1 and Lam Ch. 5, both in this volume). These characteristics are, of course, common for developed and developing country firms, but, being far from both the technology frontier and the potential market, greatly accentuate these problems. Moreover, the developing country firm may be, to a much larger extent than developed country firms, constrained by its environment: it may have a wish (and perhaps even the capability) to introduce a new product or process, but the possibility to do so may depend on capabilities in other firms or skills that are simply not there (or require substantial investments to occur). Arguably, to avoid being stuck along an inferior path and never catch up, "institutional instruments" may be needed to compensate for some of these "latecomer disadvantages," to use a Gerschenkronian term. In particular, what the developing country firm may need are "institutional instruments" that improve:

- links with the technology frontier,
- links with markets (and sophisticated users),
- supply of needed skills, services and other inputs,
- the local innovation system/network.

Arguably, much of what firms and governments in catching-up countries have done can be understood from this perspective. For instance, the diversified business-groups that developed in Japan and South Korea might be seen as "institutional instruments" fulfilling some of those needs (Shin 1996). The OEM system (original equipment manufacture, see Box 19.3) that has developed in the electronics industries of East Asia may also be seen as an "institutional instrument"—or "organizational innovation" (Hobday 2000)—geared towards simultaneously improving links with the technology frontier and the market. Similarly, attracting inward FDI may be seen as a "functional equivalent" to OEM, which, however, judged by the empirical evidence, seems to be less favourable for indigenous innovation. Other, more demanding, but perhaps also more rewarding ways—since it allows the latecomer firm to reap a larger share of the profit generated—include technology licensing, investments in own brands (OBM), etc. Improving the supply of needed skills has, of course, been a central preoccupation of many latecomer governments, as illustrated in Section 19.3 above. Moreover, we have witnessed sustained efforts by several latecomer governments in accommodating the needs of firms for a high quality R&D infrastructure (innovation system).

What can the extraordinary success that some catching-up countries have had, and the failure of others, teach present-day developing countries? One important lesson is that there is no one unique way to successful catch-up that every country has to emulate. Every country has to find its own way based on an understanding of (a) the contemporary global technological, institutional, and economic dynamics, (b) the behavior (and needs) of the relevant agents (of which the firm arguably is the most important), and (c) the specific context in which the catch-up takes place and the broader factors that influence it, being economic, technological, institutional, political, or cultural (Freeman and Louçã 2001). There are, one may suggest, big potential rewards in following the Gerschenkronian strategy of targeting technologically progressive sectors, as part of a broader attempt to transform the economy, stimulate learning, and enhance the creation of new skills (or assets). However, not every country is equipped with the necessary capabilities to pursue such a strategy. For instance, when, in the mid-nineteenth century, Japan initiated its efforts to catch up with the West, the technology gap vis-à-vis the more advanced countries was much smaller (compared to what developing countries face today), and its population already had a level of education that compared favourably with most other countries at the time (Odagiro and Goto 1996). Given that educational standards have been rising ever since, investing in education may be a good place to start for countries that have not succeeded in catering to those needs already. For those that have, there is a wider set of options, and for them some of the experiences outlined in this chapter may be highly relevant (see Box 19.4).

Box 19.4 A tale of two countries

Ireland and Portugal are two small countries on the western periphery of Europe. Both failed, for various reasons, to exploit fully the possibilities for catch-up and industrialization during the first half of the twentieth century, and were, therefore, at the beginning of our period of study, among the poorest in Europe, their economies characterized by an industrial structure dominated by traditional, low-skill activities.

However, during the second half of the century, both countries took steps to integrate their economies with the more dynamic economies of Western Europe, first through membership of EFTA (1961), and later by joining the EU (1981 and 1986, respectively). As members of the EU, Ireland and Portugal received substantial economic support through the so-called "Structural Funds," designed to induce technological and economic catch-up, and to facilitate the necessary structural changes, in poorer areas in the Union (Cappelen et al. 2003). During the 1980s and 1990s, both countries grew rapidly, and the gaps in GDP per capita between them and the more advanced Western economies decreased substantially. In tandem with this, both economies underwent extensive structural changes, with traditional ("low-tech") activities contracting and more advanced activities expanding. In Portugal, as a result of inward FDI, a rapid expansion of the automobile sector occurred. In addition, one of the traditional activities, footwear, underwent substantial technological upgrading, and managed to expand in spite of raising relative labor costs (Godinho 2000). Ireland, by contrast, leapfrogged into one of today's most progressive technologies, ICT, almost exclusively due to inward FDI (O'Sullivan 2002). As a consequence, Ireland has become one of the most "high-tech" economies in the world today, comparable to the Asian "tigers," with an export structure strongly geared towards ICT and so-called "high-tech" manufacturing.

Why did the two countries develop differently? A common language, and traditional strong ties, may have helped to focus the attention of US multinationals on Ireland. Another commonality is education. Although Irish educational efforts were not particularly high by developed country standards, Portugal lagged much further behind. This was the legacy of the authoritarian regime of Salazar, which did not wish to equip its people with more skills than strictly necessary. As a consequence, two-thirds of the Portuguese today have primary school as their highest level of education, and only one out of ten has higher education of some sort. Among the OECD countries, only Turkey is at a similarly low level.

Both countries face challenges today, but these differ, and so will, arguably, the adequate policy response. In the Irish case, worries have emerged about the weak links between the foreign-owned ICT firms and the economy at large (in spite of, it might be argued, a blooming software sector), and the allegedly negligent attitudes of many of these firms when it comes to developing technological capabilities and undertaking R&D in Ireland (O'Sullivan 2002). The Portuguese, naturally, are concerned about the increased competition for its products stemming from the enlargement of the European Union to many Eastern European countries, many with a better educated labor force.

NOTES

1. We thank Fulvio Castellacci, Sandro Mendonca and the authors and editors of this volume for helpful comments and suggestions, retaining sole responsibility for remaining errors and omissions.
2. More recently, economists have formulated theories (so called "new growth theory") that do not necessarily predict convergence. See Fagerberg (1994, 2000) and Verspagen, Ch. 18 in this volume, for an extended account.
3. For instance, although there is a sizeable literature on US economic growth, most of it is not written from a catch-up perspective, e.g. focusing on what the US experience may tell us about what other countries should do (or not do) to succeed in catching up. See,

however, the recent book by Chang (2002) and the discussion in Section 19.4 of this chapter.

4. Veblen mentions factors such as the "funds available for investment"(Veblen 1915: 186), a sufficient supply of "educated men" (ibid. 194), as well as a "sufficiently well-instructed force of operative workmen" (192). The latter, he noted, did not have to be particularly well educated or trained (188).

5. "To the extent that industrialization took place, it was largely by the application of the most modern and efficient techniques that backward countries could hope to achieve success, particularly if their industrialization proceeded in the face of competition from the advanced country" (Gerschenkron 1962: 9).

6. Surprisingly, perhaps, he did not (nor did Veblen) put much emphasis on the achievements made by Germany in other areas, such as the educational sector, and in pioneering the development of an R&D infrastructure.

7. See Gerschenkron (1962: 16–20).

8. For instance, as a result of these efforts, illiteracy—which was relatively low by international standards even before the Meiji-restoration—was almost eliminated among Japanese youth by the turn of the century, and by 1920, more than half of the children graduating from elementary school proceeded to secondary schools (Odagiri and Goto 1996).

9. Productivity (and productivity growth) remained low, however, in sheltered industries (agriculture and services), and this imposed growing burdens on the economy during the 1990s and the early twenty-first century.

10. See Johnson (1982) on Japan, Amsden (1989) on Korea and Wade (1990) on Taiwan.

11. Most countries in our sample have between 80 per cent and 100 per cent of their youth attending secondary schooling. However, some (but not all, Malaysia, for instance, has 93 per cent) of the poorer economies have less. For instance, Argentina and Chile were reported to be in the 70–75 per cent range, and China and Mexico lower still, between 50 per cent and 60 per cent. Brazil is reportedly very low, only 15 per cent. Note, however, that the data for Argentina and Brazil quoted here were from the 1985–7 period, while for the other countries it was 1998 (source UNDP, Human Development Report 2002).

12. UNESCO defines the "gross enrolment ratio in tertiary education," which is the indicator used in Figure 19.2, as "total enrolment in tertiary education, regardless of age, expressed as a percentage of the population of the five year age group following on from the secondary-school leaving age." If many tertiary students are older than this, as may be the case in several advanced countries (with a high emphasis on "long" degree courses), this indicator may give a too optimistic view on the actual share of an age group enrolled in tertiary education. Comparisons with other available indicators of tertiary education, based on educational attainment, indicates that such a bias may be present, and more so for the USA than other countries. Hence, the figure probably exaggerates the difference between the USA and other countries in the emphasis on tertiary education.

13. Note that the data reported here refer to total R&D, including the public part. If we had focused on only the part undertaken (and/or financed) by the business sector, the ranking of the countries would have been approximately the same, but the differences between the top and the bottom would have increased (in general, the more a country invests in R&D, the higher the share financed by business).

14. A recent example, mentioned by Chang, is the so-called TRIPS agreement, forcing developing countries, which used to have a very lax attitude to protection of intellectual property rights, to accept developed countries' standards and institutions in this area. See Granstrand (Ch. 10 in this volume) for an extended discussion.

15. See Granstrand (1999: ch. 6), and references therein, for an attempt to link discussion of catch-up and firm strategies.

References

* ABRAMOVITZ, M. (1986), "Catching Up, Forging Ahead, and Falling Behind," *Journal of Economic History* 46: 386–406.

—— (1994), "The Origins of the Postwar Catch-Up and Convergence Boom," in J. Fagerberg, B. Verspagen, and N. von Tunzelmann (eds.), *The Dynamics of Technology, Trade and Growth*, Aldershot: Edward Elgar, 21–52.

AOKI, M. (1988), *Information, Incentives and Bargaining in the Japanese Economy*, Cambridge: Cambridge University Press.

AMSDEN, A. H. (1989), *Asia's Next Giant: South Korea and Late Industrialization*, New York: Oxford University Press.

BAUMOL, W. J., BLACKMAN, S. A. B., and WOLFF, E. N. (1989), *Productivity and American Leadership*, Cambridge, Mass.: MIT Press.

BEASLEY, W. G. (1990), *The Rise of Modern Japan*, New York: St. Martin's Press.

CAPPELEN, A., CASTELLACCI, F., FAGERBERG, J., and VERSPAGEN, B. (2003), "The Impact of EU Regional Support on Growth and Convergence in the European Union," *Journal of Common Market Studies* 42: 621–44.

—— and FAGERBERG, J. (1995), "East Asian Growth: A Critical Assessment," *Forum for Development Studies* 2: 175–95 (repr. as ch. 3 in J. Fagerberg (2002), *Technology, Growth and Competitiveness: Selected Essays*, Cheltenham: Edward Elgar)

* CHANG, HA-JOON (2002), *Kicking Away the Ladder: Development Strategy in Historical Perspective*, London: Anthem Press.

CORNWALL, J. (1977), *Modern Capitalism: its Growth and Transformation*, London: St. Martin's Press.

DOSI, G. (1988), "Sources, Procedures and Microeconomic Effects of Innovation," *Journal of Economic Literature* 26: 1120–71.

*FAGERBERG, J. (1987), "A Technology Gap Approach to Why Growth Rates Differ," *Research Policy* 16: 87–99 (repr. as ch. 1 in J. Fagerberg (2002), *Technology, Growth and Competitiveness: Selected Essays*, Cheltenham: Edward Elgar).

—— (1988), "Why Growth Rates Differ," in G. Dosi, et al. (eds.), *Technical Change and Economic Theory*, London: Pinter, 432–57.

*—— (1994), "Technology and International Differences in Growth Rates," *Journal of Economic Literature* 32: 1147–75.

—— (2000), "Vision and Fact: A Critical Essay on the Growth Literature," in J. Madrick (ed.), *Uncovential Wisdom: Alternative Perspectives on the New Economy*, New York: The Century

* Asterisked items are suggestions for further reading.

Foundation Press, 299–330, 350–4 (repr. as ch. 6 in J. Fagerberg (2002), *Technology, Growth and Competitiveness: Selected Essays*, Cheltenham: Edward Elgar).

——— GUERRIERI, P., and VERSPAGEN, B. (eds.) (1999), *The Economic Challenge for Europe: Adapting to Innovation Based Growth*, Cheltenham: Edward Elgar.

——— and VERSPAGEN, B. (2002), "Technology-Gaps, Innovation-Diffusion and Transformation: An Evolutionary Interpretation," *Research Policy* 31: 1291–304.

* FREEMAN, C. (1987), *Technology Policy and Economic Performance: Lessons from Japan*, London: Pinter.

——— and LOUÇÃ, F. (2001), *As Times Goes By. From the Industrial Revolutions to the Information Revolution*, Oxford: Oxford University Press.

——— and SOETE, L. (1997), *The Economics of Industrial Innovation*, 3rd edn., London: Pinter.

*GERSCHENKRON, A. (1962), *Economic Backwardness in Historical Perspective*, Cambridge, Mass.: Belknap Press.

GODINHO, M. (2000), "Desenvolvimento competitivo do sector de calçado em Portugal: Lições de um caso de *clustering*," in I. Salavisa Lança (ed.), *Trajectórias competitivas na indústria portuguesa*, Oeiras: Celta.

GOMULKA, S. (1971), "Inventive Activity, Diffusion and the Stages of Economic Growth," Report No. 24, Institute of Economics, Aarhus University, Aarhus.

GRANSTRAND, O. (1999), *The Economics of Management and of Intellectual Property*, Cheltenham: Edward Elgar.

HOBDAY, M. (2000), "East versus Southeast Asian Innovation Systems: Comparing OEM- and TNC-led Growth in Electronics," in L. Kim and R. Nelson (eds.), *Technology, Learning & Innovation: Experiences of Newly Industrializing Economies*, Cambridge: Cambridge University Press, 129–69.

JOHNSON, C. A. (1982), *MITI and the Japanese Miracle: The Growth of Industrial Policy, 1925–1975*, Stanford: Stanford University Press.

*LALL, S. (2000), "Technological Change and Industrialization in the Asian Newly Industrializing Economies: Achievements and Challenges," in L. Kim and R. Nelson (eds.), *Technology, Learning & Innovation: Experiences of Newly Industrializing Economies*, Cambridge: Cambridge University Press, 13–68.

LANDES, D. (1998), *The Wealth and Poverty of Nations*, London: Abacus.

LIST, F. (1885), *The National System of Political Economy*, trans. from the original German edn. (1841), London: Longmans, Green and Company.

NELSON, R. R. (1968), "A 'Diffusion' Model of International Productivity Differences in the Manufacturing Industry," *American Economic Review* 58: 1219–48.

——— and WINTER, S. G. (1982), *An Evolutionary Theory of Economic Change*, Cambridge, Mass.: Harvard University Press.

*——— and WRIGHT, G. (1992), "The Rise and Fall of American Technological Leadership: The Postwar Era in Historical Perspective," *Journal of Economic Literature* 30: 1931–64.

ODAGIRO, H., and GOTO, A. (1996), *Technology and Industrial Development in Japan*, Oxford: Clarendon Press.

O'SULLIVAN, M. (2002), "Industrial Development: A New Beginning?" in John O'Hagan (ed.), *The Economy of Ireland: Policy and Performance of a European Region*, Gill and Macmillan, 260–85.

RODRIK, D. (1994), "King Kong Meets Godzilla: The World Bank and the East Asian Miracle," in A. Fishlow et al. (ed.), *Miracle or Design? Lessons from the East Asian Experience*, Policy Essay No. 11, Washington, DC: Overseas Development Council, 13–53.

*SHIN, JANG-SUP (1996), *The Economics of the Latecomers: Catching-up, Technology Transfer and Institutions in Germany, Japan and South Korea*, London: Routledge.

——— and HA-JOON CHANG (2003), *Restructuring Korea Inc.*, London: Routledge.

SOLOW, R. M. (1956), "A Contribution to the Theory of Economic Growth," *Quarterly Journal of Economics* 70: 65–94.

TEMPLE, J. (1999), "The New Growth Evidence," *Journal of Economic Literature* 37: 112–56.

TEECE, D. J. (2000), "Firm Capabilities and Economic Development: Implications for the Newly Industrializing Economies," in L. Kim and R. Nelson (eds.), *Technology, Learning & Innovation: Experiences of Newly Industrializing Economies*, Cambridge: Cambridge University Press, 105–28.

VEBLEN, T. (1915), *Imperial Germany and the Industrial Revolution*, New York: Macmillan.

VERSPAGEN, B. (1991), "A New Empirical Approach to Catching Up or Falling Behind," *Structural Change and Economic Dynamics* 2: 359–80.

*WADE, R. (1990), *Governing the Market: Economic Theory and the Role of Government in East Asian Industrialization*, Princeton: Princeton University Press.

WOMACK, J., JONES, D. T., and ROOS, D. (1991), *The Machine that Changed the World*, Cambridge, Mass.: MIT Press.

WORLD BANK (1993), *The East Asian Miracle: Economic Growth and Public Policy*, New York: Oxford University Press.

CHAPTER 20

INNOVATION AND COMPETITIVENESS

JOHN CANTWELL

20.1 INTRODUCTION

TRADITIONALLY, economists and economic historians since Adam Smith have discussed economic growth principally in the context of the national level—why some countries grow faster (in modern terms, acquire the capabilities for sustained growth that make them more competitive) and so become wealthier than others. While in neoclassical economics questions of national competitiveness came to assume a lesser degree of importance, as attention was shifted away from issues of growth towards those of static resource allocation and efficiency, there was even less concern with the notion of competitiveness at the firm level. The theory of the (comparative) growth of the firm was a minority interest of those such as Downie (1958), Penrose (1959), and Marris (1964), typically treated as a rather esoteric sub-branch of industrial economics, that was to be accorded a lesser status in the discipline than the conventional theory of the firm (which was really a theory of the relationship between the firm and markets). In recent years two related changes in economics and allied areas of research have been under way. One is a revival of a more widespread interest in the classical issues of competitiveness at a national level, and the other is the growing attention now paid to competitiveness at the level of industries, regions, and firms, in which fields of research a substantial new literature has emerged. In Section 20.2 the contribution is assessed of the new literature on competitiveness across countries. Section 20.3 examines innovation and competitiveness at the industry level that connects together firms and their environment, and Section 20.4 looks at the regional and firm level. Section 20.5 draws some conclusions with respect to the interaction between innovative actors, between the different levels of analysis of competitiveness, and opportunities for future research.

Competitiveness is here taken to mean the possession of the capabilities needed for sustained economic growth in an internationally competitive selection environment, in which environment there are others (countries, clusters, or individual firms, depending upon the level of analysis) that have an equivalent but differentiated set of capabilities of their own. The term competitiveness is also sometimes taken to necessarily imply as a result a continuing rise in the living standards of the individuals that are members of a social group with the required capabilities (notably in this context, to imply a sustained increase in the living standards of the citizens of the country that is suitably competitive in world markets—see Tyson 1992). While it is indeed necessarily true that productivity growth increases incomes on average (i.e. per capita income), it may well be that the process of capability generation and growth also has a disruptive effect on the distribution of incomes. This issue is not addressed directly here, since the way in which innovation affects the employment opportunities of individuals, which is a major influence upon their respective earning capacities, is the subject of Pianta (Ch. 21 in this volume).

The winners from innovation are those that construct appropriate capabilities, but capabilities are localized and nationally differentiated (as explained by Edquist, Ch. 7 this volume), and so there can be many successful players in the competitive game, each to some extent learning from and interacting with the somewhat alternative paths to capability creation being taken by others. Put in these terms few could object that the pursuit of competitiveness through innovation is a laudable objective of national policy, and indeed an increasingly important objective as the role of innovation has risen in the modern knowledge-driven economy, even for (actually especially for) countries that start behind and wish to catch up (Fagerberg and Godinho, Ch. 19 in this volume).

To be meaningful, competitiveness must be thought of as entailing a relative comparison of growth rates or benchmarking of performance to assess how well each participant has done in developing the capabilities for innovation and growth, and not be about the mutual potential for damaging one another (a misleading interpretation of competitiveness criticized by Krugman 1994a, 1996). It is reasonable to expect that, at least on average, the spillover benefits for others of a good performance in one location or by one agent tend to outweigh the costs of that good performance for others. This argument is largely applicable whether the unit of analysis is countries in the world economy or firms in an industry. At a country level the efforts of each national system of innovation to promote the competitiveness of businesses sited locally are increasingly complementary as scientific and engineering communities become more international, and cross-border knowledge flows are more common (as discussed by Narula and Zanfei, Ch. 12 in this volume). Likewise, much of the growth achieved by the leading corporations in an industry reflects the wider growth of that industry. The competitive race between firms stimulates innovation, and this innovation lowers costs and improves product quality in the industry, and thereby increases industry demand. All firms benefit that contribute successfully to what is often a combined and interactive process of innovation.

20.2 COMPETITIVENESS AT THE NATIONAL LEVEL

When looking at the country level, competitiveness is about the way in which the pattern of international trade evolves over time to reflect changing patterns of capabilities and hence competitive advantage (what might be thought of as the evolution in the comparative advantage of countries), rather than about the established pattern of comparative advantage which is the usual focus of trade theory. While the earliest theories of trade and growth can be traced back to the classical economists, such dynamic accounts of the paths of international trade and investment were revived in recent times by the technology gap approach (Posner 1961) and the product cycle model (Vernon 1966). However, a major shortcoming of the product cycle model was its reliance upon an overly simplistic demand-driven theory of innovation (which reflected the spirit of the 1960s, when it was devised), through which the firm was assimilated to the product, and innovation was supposed to be concentrated in just one leading country—the US (see Cantwell 1989, for a further discussion of the model). Sadly, when the product cycle model broke down in the 1970s, in large part owing to the reemergence of multiple centers for innovation in a number of international industries, the amended versions of the model (Vernon 1974, 1979) focused upon considerations of oligopolistic strategy rather than revisiting the underlying theory of innovation and competitiveness. It was only in the 1980s that scholars based at Sussex once again wedded an analysis of structural shifts over time in the pattern of international trade to a more realistic approach to innovation: see Soete 1981; Dosi and Soete 1988; Dosi, Pavitt, and Soete 1990; and Fagerberg's 1987 paper on structural changes in international trade (repr. as ch. 7 in Fagerberg 2002).

Part of the inspiration for Fagerberg's research had been that economists sometimes use the term "competitiveness" in various different ways, and especially in macroeconomic policy discussions not always in the way that has been defined here.

This chapter is concerned with innovation and competitiveness, and this is some-times distinguished as being about longer-term technological competitiveness, as opposed to shorter-term price competitiveness. There are two different ways of discussing competitiveness in the latter sense of shorter-term price competitiveness. In the context of conventional demand management policy discussions, if (say) lower government borrowing means a fall in interest rates and so a rise in net outward investment, and if this leads to a decline in the value of the domestic currency, then the price "competitiveness" of domestically produced goods and services can be said to have increased, as export prices fall in foreign currency terms while import prices rise in domestic currency terms. However, this type of competi-tiveness is unlikely to be sustainable, especially if (for example) the rise in import prices sparks off domestic inflation, or if lower net inward investment has adverse consequences for domestic productivity growth. The second and for our purposes more substantive context is the conventional cost-based account of competitiveness, in which a fall in relative unit labor costs means lower prices (or a lower rate of inflation), which in turn leads to a rise in exports and fall in imports, and so an increase in the value of the domestic currency.

Longer-term technological competitiveness is more akin to the second of these two versions of price competitiveness, in supposing that a faster growth of (output and) exports drives up the domestic currency, rather than it being a falling currency that promotes net exports. In the context of what is sometimes termed "non-price" competitiveness to distinguish it more clearly from the kind of cost-based competi-tiveness just referred to, innovation and new lines of value creation may mean higher average prices as an indicator of higher quality, but in any event they lead to a faster growth of productivity and trade, and thus an upward trend in the value of the domestic currency. Given what has been said already, it is worth stressing here that in this perspective the rise in the value of the currency is simply the reflection of competitiveness, defined as a relatively rapid growth in productivity and the value of (output and) exports. The rise in the value of the domestic currency is not itself the achievement of competitiveness (an improvement in the terms of trade that is essentially a potential side effect resulting from competitiveness). It is also worth making explicit that the departures from comparative advantage associated with trade imbalances are merely a temporary result of competitiveness in this framework, and again not in themselves the objective of competitiveness. What is implicitly supposed here is that faster productivity growth is associated with a rising share of world trade, and that in this process the growth of exports leads the growth of imports. So net exports rise until imports catch up, and this catching up is facilitated by the conse-quent rise in the value of the domestic currency and in domestic wage rates.

Now neo-Schumpeterian approaches to international competitiveness focus on this kind of process of forging technological competitiveness, which for those whose innovative efforts are most successful implies a sustainable increase in the share of world trade (or at the firm level, a sustainable increase in the share of the relevant world market). However, as has been discussed at length already, in the Schumpe-terian perspective competition entails the positive sum game of establishing new spheres of value creation, so innovations expand the overall magnitude of world trade and the world market. Those that contribute most to this process of expansion see their shares rise as they are responsible for more of the new element of value creation, and not because of a substitution effect within some fixed total level of world trade or some fixed and given world market (or even within some steadily exogenously growing world market). In this context, the neo-Schumpeterian analy-sis of innovation and competitiveness is unlike equilibrium growth accounts, even when those accounts incorporate an acknowledgement of research activity, if invest-ment in innovation is treated as being inherently like investment in any other economic activity, and if the only difference between activities is treated as lying in their wider impact on other activities through externalities. Instead, in the neo-Schumpeterian story the very nature and purpose of innovative activity is to disturb and add to the existing circular flow of income generation, in an experimental and non-equilibrium fashion.

Such neo-Schumpeterian models of innovation and growth might be specified in at least two alternative ways. The first of these leans heavily on the distinction just drawn between shorter-term price competitiveness and longer-term non-price technological competitiveness. In Fagerberg's (1987, 1988) technology gap formula-tion of international competitiveness across countries, the impact on growth of national rates of innovation and distance behind the technology leader are treated primarily as additive elements, to be added on to the more traditional determinants of economic growth in the form of capital accumulation (the share of investment in national output) and relative unit labor costs. The origins of viewing cross-country growth in this kind of additive framework can be traced to Abramowitz (1956), Solow (1957), and Denison (1967), for whom technological improvements (and the productivity growth to which they led) were an obvious means of accounting for the substantial "residual" in variations in growth that remained after allowing for the effect of the increase in factor inputs. So in this context capital accumulation proxies for the extension of the scale of established activities, relative unit labor costs capture cost-based "price" competitiveness, while the contribution of corporate research and the capacity to catch up through imitating the achievements of a leader represent "non-price" technological competitiveness.

Setting the problem up in this way is convenient, as the empirical evidence then generally suggests that technological competitiveness is more important than the more commonly considered traditional influences upon competitiveness. Techno-logical competitiveness is judged to be more significant than relative unit labor costs; although Krugman (1994b) and Young (1995) point to the continuing importance of capital accumulation within this kind of framework. The evidence for three coun-tries—Japan, the UK, and the US—over the period 1960–73 is illustrated in Table 20.1. Based on the estimation of his empirical model of international

Table 20.1 The decomposition of the predicted growth in national market shares from an estimated empirical model of cross–country competitiveness, for 1961–73 (%)

	Japan	UK	USA
Growth in technological capabilities	66.9	6.9	−0.6
Rise in relative unit labour costs	−0.9	0.8	1.6
Initial technological capabilities (catch–up)	20.9	15.9	7.3
Investment as share of GDP, and growth of world demand	16.5	−39.8	−38.2
Total growth in market share (predicted by model)	103.3	−16.2	−29.8

Source: Fagerberg (1988).

competitiveness, Fagerberg (1988) was able to decompose the model's predicted change in each country's share of world trade (which were reasonably good predictions of the actual changes in market shares) into four elements, as shown. What emerges is that the traditional consideration of relative wage costs contributed rather little to overall competitiveness in any of these countries (although it was statistically significant in all the equations of the model in which it appeared). In contrast, the growth in indigenous technological capabilities in Japan, and the diffusion of foreign frontier technologies, account for a good deal of the Japanese competitive success of that period. The loss of world trade shares by the UK and the US over the same period can be attributed mainly to weak capital accumulation, and Fagerberg explained this mainly by the drain placed on national resources by the high shares of military spending in these two countries.

However, when capital accumulation contributes positively to a favorable growth rate, at least some element of it reflects the establishment of new fields of activity, and is a response to the creation of new innovative opportunities. Therefore, it is not clear that the contribution of the growth of traditional factor inputs can really be cleanly distinguished from that of innovation, unlike in the logic of a standard production function approach. So to exclude capital accumulation from the contribution of technological competitiveness provides only a conservative lower bound estimate of the significance of the latter, and perhaps concedes too much to orthodox skeptics of the role of innovation in growth and competitiveness. Fagerberg (1988) was aware of this issue, and so he included a separate equation in his simultaneous system for capital accumulation as a function of the growth of output, which in turn depended as we have seen upon the increase in technological capacity, so that he acknowledged indirectly the influence of technological competitiveness upon capital investment. This need to revise the traditional production function logic becomes especially relevant when trying to compare innovative or technological "assimilationist" explanations of (East Asian) competitive success with those of "accumulationists," if using aggregate measurements in the context of substantial structural change (Nelson and Pack 1999). As we have already noted, neo-Schumpeterian economists have particularly emphasized the connection between structural change and growth through innovation.

Some evidence on what distinguished the East Asian growth experience from that of other countries that sustained similarly high rates of capital accumulation over the 1960–89 period is set out in Table 20.2. The table shows eleven countries that enjoyed very high shares of investment in GDP, of over 20 per cent, as indicated in the first column. The right-hand column shows the residuals of a regression across 101 countries of GDP per capita on the investment share as a proxy measure of the rate of capital accumulation, and on three other control variables (a catching-up effect proxied by the intial level of GDP per capita in 1960, the growth of population to capture the availability of labor supplies, and the proportion of the relevant cohort of the population educated to at least secondary school standards). This was part of the study of Nelson and Pack (1999). What emerges is that among high investment countries, the East Asian tigers—Hong Kong, Korea, Singapore, and Taiwan—stand

Table 20.2 Actual growth rates achieved by countries, 1960–89, over and above that predicted by (*inter alia*) their rates of capital accumulation

	Investment/GDP (%)	Actual minus predicted growth rate of GDP per capita
Hong Kong	27.3	0.031
Korea	24.9	0.032
Singapore	34.3	0.017
Taiwan	25.0	0.047
Gabon	40.0	−0.030
Algeria	35.0	−0.026
Greece	24.2	0.008
Panama	24.0	0.002
Portugal	23.7	−0.002
Jamaica	25.0	−0.037
Ireland	22.2	0.011

Source: Nelson and Pack (1999).

out as managing to achieve growth rates well in excess of what might have been predicted from their favorable rates of capital accumulation alone. What was different in these economies was their greater ability to innovate, to upgrade and restructure their indigenous industries, and to learn and absorb more effectively from foreign technologies. Capital accumulation can embody innovation to the extent that it is linked to the transformation of the productive activities being conducted.

So an alternative approach also in the Schumpeterian tradition is to treat technological accumulation and capital accumulation as simply aspects of a common process, rather than as independent (even if complementary) contributions to growth. In this case innovation can be seen as driving up profitability and hence lowering the share of wages in output (even though wages are rising faster, and so may be unit labor costs), which leads to a higher share of investment in output, and so higher capital accumulation and growth as a result of higher technological accumulation (Cantwell 1989, 1992). The basic idea here is that in fast-growing countries just as an increase in imports tends to follow an increase in exports with a lag, so wages tend to follow productivity increases with a lag, enabling innovation to create a fresh source of profitability and growth. Yet this also suggests that technological competitiveness is in part cost-based. It should be noted, though, that labor productivity is defined here simply as the value of output per worker employed, which implies that productivity growth is as much attributable to product quality improvements (that raise the value or unit price of output, as stressed in Fagerberg's approach), as it is to the cost reductions associated with process improvements. In this alternative neo-Schumpeterian formulation we need worry less about the distinction between embodied and disembodied technological change, or the distinction between improvements in product quality and delivery times as opposed to improvements in processes that are reflected in costs and prices.

The renewed interest in international competitiveness and variations in growth rates has spawned a substantial new literature on cross-country convergence or catching-up versus divergence or falling-behind (see e.g. Baumol, Nelson, and Wolff 1994). The evidence suggests that whether one observes convergence or divergence depends upon the period studied and the countries selected. In any case, the overall trend in cross-country variance at a world level may not be the most important issue. Rather than trying to work out whether East Asian convergence statistically outweighed the effect of African divergence in aggregate, the issue is more why and how firms in East Asia had the capabilities to catch up in the period since 1960, while those in Africa did not. The concepts of a techno-socio-economic paradigm (Freeman and Perez 1988), or of an evolution in the institutional characteristics of capitalism (Lazonick 1991, 1992), can be useful in this respect as a means of explaining occasional shifts in technological leadership or longer term competitiveness, and in the direction of those shifts. Emphasizing again the role of structural change in growth, and in particular during periods of paradigm change, when the prevailing characteristics of innovation undergo transformation, helps to explain the existence of windows of opportunity in which the catching-up of selected countries may be especially dramatic. At these times leaders can have special difficulty in adjusting to the new conditions since they have become most locked in to the types of innovation favored under the earlier paradigm, while others that lie behind initially may find that their rather different institutions and methods of social organization are in fact quite well suited to adapting so as to promote just the kinds of structural change in which lie now the greatest opportunities for fresh innovation.

20.3 COMPETITIVENESS AT THE INDUSTRY LEVEL: THE NEXUS OF RELATIONSHIPS BETWEEN FIRMS AND THEIR ENVIRONMENT

When speaking of shifts in competitiveness between firms or between different national groups of firms that constitute the major players in an international industry, Mowery and Nelson (1999) prefer the term industrial leadership, so as to emphasize that such leadership may be due as much to the national or regional environment in which firms operate, or to institutions that are specific to an industry, as to factors that are purely internal to the firms in question. From detailed historical case studies of the evolution of national industries, they conclude that competitiveness derives from the contributions of each of, and the interactions between, firms, regions, and countries, and the sectoral support systems that connect these different levels of analysis. Their account provides a clear justification for a section that covers competitiveness at the international industry level, rather than attempting to move directly from the country level to the firm level.

In this framework the factors that are thought to influence competitiveness may be grouped under the headings of resources or capabilities, institutions (notably for higher education and science, and in financial systems), markets or demand conditions, and inter-company networks. Models that are based at the level of the evolution of particular products or technologies, or which are predicated on a notion of competence-destroying innovation (when moving from one type of product or technology to another that represents a radical departure from the past), may each have some relevance. Yet Mowery and Nelson argue that this applicability is limited when dealing with the performance of large multiproduct firms over very long periods of time. Likewise, the policy debate surrounding the relationship between government policy interventions and trends in competitiveness is overly polarized, when each side of the debate has in mind a rather simple model of that relationship (or lack of a relationship) which may apply at some places and at certain times, but cannot be universally applied in the way that some advocates seem to imagine.

The relationships that exist between the development of the technological capabilities in firms that are responsible for competitiveness and the institutions of the wider society vary from one country to another, but in particular they tend to be different in countries that belong to an already leading industrialized group and those that are attempting to catch up with them (see also Fagerberg and Godinho, Ch. 19 in this volume). It is noticeable that there have been a greater number of cases in which governments in catching-up economies, partly through measures of domestic protection, have contributed more actively to the fostering of capabilities in local infant industries and in indigenous companies. This was true of the US and Germany when they were catching up with Britain in the nineteenth century (Landes 1969), it was true of Japan when it was catching up with the West during the twentieth century (Ozawa 1974), and it was true of Korea when it was catching up after 1960 (Enos and Park 1988). It is true that there are occasionally other cases of catching-up economies, like those of Singapore or Mexico in recent years, that have taken advantage instead of various aspects of trade liberalization. However, what is most noticeable in all these instances of successful catching-up is that the trade policies of governments were merely part of a much wider package of support for the longer term nourishment of capabilities in indigenous firms. Since the emergence of science-based industries towards the end of the nineteenth century this has meant especially investing in science and higher education, in the training of engineers, and in the learning of skills more widely (Freeman and Louçã 2001). Equally important, where there were measures of trade protection local firms accepted their part of the bargain to invest very substantially in capability creation in an outward-looking and export-oriented fashion, rather than simply remaining an inefficient enclave as in so many other cases of protectionism or so-called import-substituting industrialization.

Of course, the institutional structures of catching-up economies changed markedly (and any protectionist measures were largely reversed) as their firms caught up and themselves sometimes forged ahead and became innovative leaders in their own right. This is perhaps the most vivid illustration of the general observation that the development of technological capabilities in firms and the character of the institutions that support these competitive efforts in the wider society tend to coevolve with one another (Nelson 1995), through a process of continual interaction. Another perspective on these interrelated national systems for the construction of competitiveness is offered by Porter (1990), as represented through the four corners of a diamond of factor conditions; demand conditions; related and supporting industries; and firm strategy, structure, and rivalry. In Porter's view the capacity of firms to innovate depends critically on having sufficient domestic rivalry in their own home country of origin, but also on the presence of spillovers between firms associated with localized clusters (to which issue we return below). In other words, innovation requires an appropriate mix of interfirm rivalry and cooperation or exchange (Richardson 1972). Lazonick (1993) has argued that when confronted with a major new competitive challenge from some new source of innovation from outside, domestic industries may need to shift this balance away from rivalry and towards cooperation in order to respond effectively. To express this another way in the light of the trend towards globalization mentioned earlier, it may be that firms in some national industry may need to increasingly collectively focus their efforts in what they do locally (as opposed to activities they may locate abroad) to be better mutually aligned with whatever may be the fields of particular local excellence or of specialization in innovation. This would have the effect of tending to reinforce national patterns of comparative advantage in innovation.

As has been mentioned already, with the emergence of science-based industries over a century ago, the need for an infrastructure that suitably supports relevant education, skill formation, and training became critical to the competitiveness of industries, and is widely believed to have become more important still in the modern techno-socio-economic paradigm associated with computerization and information processing. For firms to be able to create capabilities requires costly and difficult internal learning processes, but these in their turn depend upon having suitable organizational and technical skills in the management and workforce on which they rely. The composition of skills in the workforce of the home base of firms is therefore critical to the success or failure of countries that are trying to catch up, but it also becomes a central influence upon the fields in which any national group of firms has its specific pattern of comparative advantage in innovation and capability creation. Of course, this is not just a one-way street, since the types of investments and commitments to training that are made by firms themselves in the course of learning, the professional associations they help to form, and the pressures they place upon governments and others imply again a process of coevolution between firms and their environment in this respect too.

Table 20.3 helps to illustrate the significance of education and skills in the catching-up of the four East Asian tiger economies (see also Fagerberg and Godinho, Ch. 19 in this volume). Korea stands out as having surpassed even the commitment of the traditional industrialized countries to higher education in the natural sciences and mathematics. Yet a key to the success of these countries as a group lies more in the investments they have undertaken in support of engineering graduates—while Hong Kong lies behind the industrialized group (and this may help to account for why its local learning and upgrading has been more limited than in the other three, as discussed by Lall 2001), Singapore is above the industrialized country average, and Korea and Taiwan are way ahead of that average for tertiary level engineering enrolments as a proportion of the population. Other developing countries are generally well behind the achievements in engineering education of the tiger economies, although the Philippines, Argentina, and Mexico have at least matched the

Table 20.3 Educational enrollments in technical subjects at tertiary level as a % of
the total population in selected countries, in 1995 or closest year
available

	Natural science, maths and computing	Engineering
Japan	0.07	0.39
France	0.53	0.09
Germany	0.39	0.49
UK	0.31	0.38
USA	0.39	0.31
Hong Kong	0.20	0.25
Singapore	0.10	0.47
Korea	0.56	0.98
Taiwan	0.24	0.86
Indonesia	0.02	0.11
Malaysia	0.07	0.07
Philippines	0.22	0.33
Thailand	0.14	0.19
China	0.03	0.10
India	0.10	0.02
Argentina	0.21	0.29
Brazil	0.09	0.10
Mexico	0.06	0.27

Source: Lall (2001).

position of Hong Kong. Considering the enormous size of its population, it is also
clear that China has been catching up fast in this area.

The Japanese and German systems are known to focus on a broad and deep skill
base by emphasizing the acquisition of general engineering skills and the good
standard of basic education of the population as a whole, while the US and UK
systems tend to be more elitist and focus on the development of higher-end skills
over a narrow range of people (Prais 1995; Lazonick and O'Sullivan 1997; Lazonick
1998). This helps to explain why the Japanese and German fields of comparative
advantage in innovation (and thus, of competitive advantage) include motor
vehicles and engineering that rely (and have increasingly come to rely) on a broad

skill base, while the US and UK advantages include aerospace, software, pharmaceuticals, biotechnology, and medical equipment that rely on very highly skilled individuals and an intensive R&D effort.

When examining the extent of path-dependency in the specific technological traditions of national groups of large firms, and in their patterns of technological specialization as a measure of their relative contributions to each of the major international industries (the cross-sectoral pattern of their technological competitiveness), some cross-border interdependencies appear to emerge. This raises again the issue of whether or not there have been any elements of convergence across countries, but in this context in the specific mix of strengths and weaknesses in international industries, rather than in aggregate productivity or performance. Examining patterns of technological specialization among national groups of the largest firms from six countries (the US, Germany, the UK, France, Switzerland, and Sweden) based on their patterns of corporate patenting, it has been observed that these profiles are path-dependent and tend to persist to some extent even over periods of sixty years, from the interwar period to the present day (Cantwell 2000). This may be taken to imply that the positive effect on the continuity of collective technological trajectories of intercompany technological cooperation and spillovers within national groups has tended to outweigh the negative effect of mobility in cross-company distributions of activity. There is some evidence that through the evolution in these patterns of technological competence that has occurred, certain national groups have come somewhat closer to one another than they were in the past, or they have changed in similar ways. Indeed, it might be argued that the six national groups examined can now be divided into three clusters of two countries each.

The first cluster comprises the largest US and UK firms, in which the profile of technological competence can be characterized as being resource-based, oil-related, and defence-related. It is increasingly also health-related. As shown in Table 20.4, in the US case since the interwar period a continuing comparative advantage in innovative activity in the largest industrial firms has been sustained in the oil, food products, rubber products, aerospace (defence and larger-scale transport systems) and building materials industries. The greatest continuing strengths for the largest British companies over the same historical period has been in textiles, other transport (defense) and oil since the 1930s. Thus, it can be argued that there has been some convergence in the profiles of the US and UK innovation systems (Vertova 1998). UK firms have also seen a post-war shift into technological competence in the pharmaceutical industry, although it can be claimed that this too represents the revival of a much earlier nineteenth-century tradition in biological and medical technologies. In any event, consistent with the overall UK or US pattern of technological development, the British pharmaceutical industry had links with the food industry, unlike in Germany where it derived purely from the chemicals industry (Cantwell and Bachmann 1998). In the US there has been a related post-war continuation and strengthening of the medical instrument industry (grouped under professional and scientific instruments in Table 20.4), and a more recent move into biotechnology, although this has not yet been reflected in a comparative advantage in the pharmaceutical industry as a whole.

The second cluster is that of the German-and Swiss-owned corporate groups, in which technological development since the end of the nineteenth century has been largely science-based, and revolved around the dominance of the chemicals industry. In the post-war period this has been increasingly complemented by engineering excellence, although some recent commentators have seen this direction of change (as opposed to a move into the other science-based area of electronics) as a weakness of the modern German innovation system (Albach 1996; Audretsch 1996). The leading German firms have held a consistent focus on development in the chemicals and metal product industries, with some recent shift towards industries more reliant on engineering-based technologies, linked in part to the emergence of a wider range of innovative smaller specialist supplier companies. The Swiss concentration histor-

Table 20.4 The industries in which the largest nationally owned firms have persistently held comparative advantage in innovation, 1920–39 and 1978–95

US-owned	UK-owned
Food and drink	Textiles
Office equipment and computing	Other transport equipment
Other transport equipment (other than motor vehicles)	Coal and petroleum products
Rubber and plastic products	
Non-metallic mineral products	
Coal and petroleum products (oil)	
Professional and scientific instruments	

German–owned	Swiss–owned
Chemicals	Chemicals
Pharmaceuticals	Pharmaceuticals
Metal products	Mechanical engineering
Motor vehicles	

French–owned	Swedish–owned
Metal products	Mechanical engineering
Rubber and plastic products	
Non-metallic mineral products	

Source: Cantwell (2000).

ically on chemicals and pharmaceuticals makes it a microcosm of (part of) the German innovation system, which has also been shifting in the direction of engineering excellence.

The third cluster may be more a matter of coincidence than due to any historical, geographical, or cultural ties, involving as it does the French and Swedish national groups of companies. This grouping has emphasized infrastructural types of technology, spanning engineering, construction, transport and communications systems, and some recent moves into health care. In the French case comparative advantage in large firm innovation has been sustained since the interwar years in metal products, rubber products, and building materials, while some earlier strengths in electrical communications technologies have been subsequently consolidated. This infrastructural orientation is less reliant upon large-scale private corporate R&D than the German system has been, but is not as resource-oriented as the US or UK company systems of technological development. Swedish technological excellence has also been engineering-based (and has become increasingly so) around the metals and vehicles industries, but it has been shifting more closely towards the French pattern with the recent rise of development in the areas of telecommunications and pharmaceuticals.

The apparent convergence of certain national systems of large-firm innovation with continuing differentiation between these clusters of groupings may be an aspect of the rise in technological interrelatedness and interlinked systems of technologies, which have eroded the more highly specialized national systems of the past. Hence, the significance of technological lock-in and path-dependency in each respective system has still been accompanied by some selected convergences between particular national groups.

20.4 COMPETITIVENESS AT THE REGIONAL AND FIRM LEVELS

The significance of the "regional dimension" of an innovation system has emerged as another aspect of an interactive model (Kline and Rosenberg 1986) that emphasizes the relationships of local companies with knowledge sources external to the firm—see also Asheim and Gertler (Ch. 11 in this volume). Such relationships—between firms and the science infrastructure, between producers and users of innovations at an interfirm level, between firms and the wider institutional environment—are strongly influenced by spatial proximity mechanisms that favor processes of polarization and cumulativeness (Lundvall 1988; von Hippel 1989). Furthermore, the employment of informal channels for knowledge diffusion (of so-called tacit or uncodified knowledge) provides another argument for the tendency of innovation to be geographically confined (Hägerstrand 1967; Lundvall 1992). The lack of existent capabilities in weaker regions hampers the potential for inter-regional technology diffusion (Fagerberg, Verspagen, and Caniëls 1997).

Some evidence on the extent of the locational concentration of the corporate capabilities for innovation that underpin competitiveness in Europe is set out in Table 20.5. It shows that except in the case of Germany in which there are four major regions that are each responsible for 13 per cent or more of the innovative capacity of the largest companies, of which the leading region accounts for 27 per cent, in all the other European economies the equivalent share of the biggest region is around 50 per cent or higher. This represents a quite remarkable geographical concentration of innovative capacity, far more than the extent of concentration of population or the total value of output. Given therefore that the interactions between the development of technological capabilities by firms and the supporting institutions found in their environment takes place mainly in such regionally bounded areas, this section begins by discussing the relationship between regional concentrations of activity or "industrial districts" and the competitiveness of individual firms (Malmberg, Sölvell, and Zander 1996; Porter and Sölvell 1998; Enright 1998; Scott 1998).

One particular aspect of regional systems that is underlined here is their interplay with the international dispersion of the creation of new technology and the new innovatory strategies of multinational corporations (MNCs), which have been associated with a restructuring of MNC technological operations at a subnational

Table 20.5 The shares of patenting of the largest industrial firms attributable to research facilities located in the biggest single region of selected European countries, in 1969–95

Country/Region	Percentage
Belgium (Flanders-Brussels)	78.6
France (Île de France)	57.9
Germany (Nordrhein Westfalen)	27.0
Italy (Lombardia)	52.3
Netherlands (South Netherlands)	62.7
Sweden (Stockholm-Östra Mellansverige)	49.5
Switzerland (Basel)	57.5
UK (South East England)	46.9

Source: Cantwell and Iammarino (2001), and (for Germany) Cantwell, Iammarino, and Noonan (2001).

level. On the one hand, as seen above, there are general external economies and spillover effects which attract all kinds of economic activities in certain regions and determine, in the case of corporate integration, the localization of new research units. These centripetal forces strengthen the interborder intrafirm integration and the feedback of knowledge, expertise, and information which occurs within networks of affiliates. On the other hand, sector-specific localization economies intensify intra-border sectoral integration, implying local external networks between affiliates, indigenous firms and local non-market institutions. By tapping into local knowledge and expertise, foreign affiliates gain a competitive advantage which can not only be exploited locally but may also be transferred back to the parent company, enhancing its global technological competence. Thus, Narula and Zanfei (Ch. 12 in this volume) refer to the recent shift away from asset-exploiting and towards asset-augmenting investments by MNCs, which is typically associated with a greater dispersion of innovative activity in the international network of an MNC. However, the nature of the relationship between MNC international innovation systems and local systems varies across regions (Cantwell and Iammarino 2000, 2001). This entails different types of regional strategy for technological competitiveness.

Evidence has now emerged that the choice of foreign location for technological development in support of what is done in the home base of the MNC depends upon whether host regions within countries are either major centers for innovation or not (termed "higher-order" or "lower-order" regions by Cantwell and Iammarino 2000). Whereas most regions are not major centers and tend to be highly specialized in their profile of technological development, and hence attract foreign-owned activity in the same narrow range of fields, in the major centers much of the locally sited innovation of foreign-owned MNCs does not match very well the specific fields of local specialization, but is rather geared towards the development of general purpose technologies (GPTs) that are core to cross-industry innovation today (notably information and communication technologies, ICT) or in the past (notably mechanical technologies). The need to develop such GPTs is shared by the firms of all industries, and the knowledge spillovers between MNCs and local firms in this case may be inter-industry in character. Thus, ICT development in centres of excellence is not the prerequisite of firms of the ICT industries, but instead involves the efforts of the MNCs of other industries in these common locations.

Turning to competitiveness at the level of an individual firm, the determinants of cross-company growth summarized in Table 20.6 derive from a cross-sectional regression analysis of 143 of the world's largest firms between 1972 and 1982 (Cantwell and Sanna-Randaccio 1993). As has been remarked earlier, a key aspect of innovation and growth in the firm has to do with the largely industry-specific environment that firms have in common and which regulates their individual behavior and partially reflects their mutual interactions (Levin, Cohen, and Mowery 1985). Thus, the growth of demand and of technological opportunities in their own industry are key influences on corporate performance.

Table 20.6 The statistically significant determinants of comparative growth among the world's largest industrial firms, 1972–82

Regressor	Sign of coefficient
Growth of own-industry demand	+
Growth of own-industry technological opportunities	+
Firm size	−
Firm-specific technological competitiveness	+
Degree of market power	+
Relative multinationality within own-industry	+
Increase in multinationality over period	+

Source: Cantwell and Sanna-Randaccio (1993).

It is curious that although the work of Penrose and Downie mentioned at the start of this chapter emphasized issues of the creation of firm-specific capabilities and intra-industry competitive rivalry, until quite recently even that minority of economists that did work on firm growth paid relatively little attention to these issues. However, now the notion of corporate competence has moved center stage in the strategic management literature (see Lam, Ch. 5 in this volume). Penrose had argued that the competitive advantages of a firm derive essentially from the cumulative and incremental learning experience of its management team, which experience differentiates it from other firms. The distinctiveness of the firm's accumulated experience and knowledge determines the set of opportunities for growth which it perceives ahead of its rivals when screening the external environment (the growth of demand and technological opportunities in its industry). Corporate technological competitiveness is the principal advantage of this kind associated with differentiated learning (Cantwell 1989; Teece, Pisano, and Shuen 1997). A higher technological capability lowers the unit costs and raises the demand curve of the firm at a given rate of growth, and it facilitates new entry into related product lines. A more technologically competent firm is able to utilize its existing experience to lower the costs of expanding its managerial and technical team in related areas. Technological competitiveness comes out as one of the most statistically significant influences upon firm growth of the variables listed in Table 20.6. In that large firm study, technological competitiveness was measured by the firm's share of patenting in its industry relative to its market share (of industry output), in terms of the relevant world industry.

20.5 CONCLUSIONS

To return to where this chapter began, competitiveness derives from the creation of the locally differentiated capabilities needed to sustain growth in an internationally competitive selection environment. Such capabilities are created through innovation, and because capabilities are varied and differentiated, and since the creative learning processes for generating capabilities are open-ended and generally allow for multiple potential avenues to success, a range of different actors may improve their competitiveness together. Innovation is a positive sum game that consists of the efforts often of many to develop new fields of value creation, in which on average the complementarities or spillovers between innovators tend to outweigh negative feedback or substitution effects, even if there are generally at least some actors that lose ground or fail. The basic conclusion is that efforts to promote competitiveness through innovation can rarely be understood in isolation from what others are achieving at the same time. This applies whether we are speaking of countries, of national groups of firms in an industry, of subnational regions, or of individual companies. Indeed, it is worth emphasizing that the degree of interaction between innovators in search of competitiveness has tended to rise substantially historically, and has attained new heights in recent years.

Firms are less independent than they were, and they now all float in a much deeper sea of background knowledge, which Nelson (1992) refers to as the "public" element of technology. There are at least four aspects to this: intercompany knowledge flows have increased, there is a growing role for governments and other non-corporate institutions in knowledge development and transfer, the importance of science for technology has risen and diversified in its impact, and there has been a tendency towards more rapid codification and the formation and spreading of professional and scientific communities. We can now think of firms and the individuals aboard them like ships floating in a sea of public knowledge which connects them, or more accurately potentially public knowledge since the extent that they can draw upon it depends upon their own absorptive capacity and on their membership of the appropriate clubs (whether intercompany alliances or professional associations and the like). Over time, especially since 1945, firms have been designed to float deeper down in the water, but they still always leave a critical part comprising their own tacit capabilities above the surface, which does not sink down or fall into the general mass. Indeed, holding stronger capabilities above the surface is positively related to the depth to which one can reach below the surface, both for the absorptive capacity to extract complementary knowledge and for the extent to which one contributes oneself to the public knowledge pool. Universities and governments have increasingly contributed to the sea of public knowledge as well. Additionally, among firms that deliberately cooperate through technology-based alliances personnel can be exchanged so as to coordinate learning efforts.

The sharing of knowledge between firms implies not just that technology must be developed through an interactive social and cultural evolution rather than through a biological evolutionary process involving competition between genetically independent entities, but also that followers and innovative adapters may stand to make greater gains than the original leaders in some new field of technological endeavor. For example, knowledge developed in one context may ultimately prove to have a bigger impact in another, which was not foreseen by the originator or even perhaps initially by the most innovatively successful recipients. Firms also now devote much greater effort to attempts to understand their own technological practice and that of others. Codification of knowledge is the outcome of a conscious effort, shifting back the dividing line between what is potentially public and what is tacit (Cohendet and Steinmueller 2000; Cowan, David, and Foray 2000). So firms that become especially adept at codification may find that this is a source of competitive advantage since they can then more readily draw on the public pool.

To engage in this intercompany interaction fruitfully, firms must maintain an adequate diversification of their in-house technological efforts, since the closer that knowledge is to the proprietary interests of a firm the more likely that it will only be shared in return for something else that is probably technologically complementary, which is what each firm needs to join the relevant corporate club (Cantwell and Barrera 1998). The entrepreneurial function has not been eliminated but it is more institutionally embedded in an ability to network and make new connections (see Powell and Grodal, Ch. 3 in this volume).

The interaction effects between innovators has been further compounded by the role of ICT as a means of combining fields of knowledge creation that were previously kept largely apart (or what Kodama 1992, terms technology fusion, and has led to the creation of new fields such as bioinformatics). ICT thus broadens the field for potential innovation by linking formerly separate areas of innovative activity. ICT potentially combines the variety of technological fields themselves and so increases the scope for wider innovation.

So, in the light of these recent changes in emphasis in the context for innovation and competitiveness, how can we reevaluate some of the earlier literature in the field that has been summarized here? Which lines of research have perhaps run out of steam for the time being, and which offer the most promising new avenues or opportunities for future research? The revival of the classical issues of competitiveness at a national level (Section 20.2) has been useful in that it represents a return of interest into the major questions of the wealth of nations, which are of primary importance from a social viewpoint. However, this new literature has also exposed the limitations of trying to tackle what are really issues of structural change at an aggregate level. It might perhaps be judged that we have gone about as far as we can for now with purely cross-country models. It may well be for this reason that the focus of the latest research has tended to shift to the industry and firm levels, allowing for (indeed now emphasizing more) the scope for technological interactions between firms and industries, so as not to lose sight of the wider context or the aggregate effects.

When recasting the analysis at an industry level (Section 20.3), we now know a good deal about historical shifts in patterns of industrial leadership between countries, about the role of education and skills in catching up economies, and how particular kinds of skills help to account for inter-industry discrepancies in innovative potential. Arguably we still know too little about the interaction between governments, non-business institutions, and firms (especially large firms) in the process of establishing competitiveness. In particular, we would like to know more about university–industry (science–technology) interaction over a wider range of countries, beyond the relatively clear picture that we have for the US (see Mowery and Sampat, Ch. 8 in this volume). The context here is what seems to be the growing significance of a local science base for the construction of corporate capabilities and hence competitiveness, including and perhaps especially in latecomer economies. Note that this newly emerging view reverses the "traditional" perspective that developing countries should concentrate on (organizational innovation in) lower-skill activities, and leave science to the largest most developed economies.

Coming to competitiveness at the firm or cluster level (Section 20.4), the latest research has also had a renewed focus on the role of intercompany interaction in knowledge creation and innovation, especially in subnational regional areas, and through alliances or cooperative agreements. Here we now know more of the details of the localized character of innovation, and of the steady growth in technology-based alliances as a means of facilititating competitiveness through knowledge exchange and spillovers. Work on firm size and innovation or growth seems to have rather run out of steam for now, at least insofar as it had regarded individual firms as quite independent entities. We need to know more about the specificities of knowledge flows between regions and between firms, of how and where technological knowledge is sourced by firms, and then how such knowledge is effectively combined in networks of interrelated innovation within and between firms. This is surely an exciting agenda for further research.

REFERENCES

ABRAMOWITZ, M. (1956), "Resources and Output Trends in the United States since 1870," *American Economic Review* 46(1): 5–23.

ALBACH, H. (1996), "Global Competitive Strategies for Scienceware Products," in G. Koopmann and H.-E. Scharrer (eds.), *The Economics of High-Technology Competition and Cooperation in Global Markets*, Baden-Baden: Momos Verlagsgesellschaft, 203–17.

AUDRETSCH, D. B. (1996), "International Diffusion of Technological Knowledge," in G. Koopmann and H.-E. Scharrer (eds.), *The Economics of High-Technology Competition and Cooperation in Global Markets*, Baden-Baden: Momos Verlagsgesellschaft, 107–35.

BAUMOL, W. J., NELSON, R. R., and WOLFF, E. N. (eds.) (1994), *Convergence of Productivity: Cross-National Studies and Historical Evidence*, Oxford and New York: Oxford University Press.

*CANTWELL, J. A. (1989), *Technological Innovation and Multinational Corporations*, Oxford: Basil Blackwell.

—— (1992), "Japan's Industrial Competitiveness and the Technological Capabilities of the leading Japanese Firms," in T. S. Arrison, C. F. Bergsten, E. M. Graham and M. C. Harris (eds.), *Japan's Growing Technological Capability: Implications for the US Economy*, Washington, DC: National Academy Press, 165–88.

—— (2000), "Technological Lock-in of Large Firms since the Interwar Period," *European Review of Economic History* 4(2): 147–74.

—— and BACHMANN, A. (1998), "Changing Patterns of Technological Leadership: Evidence from the Pharmaceutical Industry," *International Journal of Innovation Management* 2(1): 45–77.

—— and BARRERA, M. P. (1998), "The Localisation of Corporate Technological Trajectories in the Interwar Cartels: Cooperative Learning versus an Exchange of Knowledge," *Economics of Innovation and New Technology* 6(2–3): 257–90.

—— and IAMMARINO, S. (2000), "Multinational Corporations and the Location of Technological Innovation in the UK Regions," *Regional Studies* 34(4): 317–22.

—— —— (2001), "EU Regions and Multinational Corporations: Change, Stability and Strengthening of Technological Comparative Advantages," *Industrial and Corporate Change* 10(4): 1007–37.

—— —— and NOONAN, C. A. (2001), "Sticky Places in Slippery Space—the Location of Innovation by MNCs in the European Regions," in N. Pain (ed.), *Inward Investment, Technological Change and Growth: The Impact of MNCs on the UK Economy*, Oxford: Pergamon, 210–39.

*—— and SANNA-RANDACCIO, F. (1993), "Multinationality and firm growth," *Weltwirtschaftliches Archiv* 129(2): 275–99.

COHENDET, P., and STEINMUELLER, W. M. (2000), "The Codification of Knowledge: A Conceptual and Empirical Exploration," *Industrial and Corporate Change* 9(2): 195–209.

COWAN, R., DAVID, P. A., and FORAY, D. (2000), "The Explicit Economics of Knowledge Codification and Tacitness," *Industrial and Corporate Change* 9(2): 211–53.

DENISON, E. F. (1967), *Why Growth Rates Differ: Post-War Experience in Nine Western Countries*, Washington, DC: Brookings Institute.

DOSI, G., PAVITT, K. L. R., and SOETE, L. L. G. (1990), *The Economics of Technical Change and International Trade*, London: Harvester Wheatsheaf.

—— and SOETE, L. L. G. (1988), "Technical Change and International Trade," in G. Dosi, C. Freeman, R. R. Nelson, G. Silverberg, and L. L. G. Soete (eds.), *Technical Change and Economic Theory*, London: Pinter.

DOWNIE, J. (1958), *The Competitive Process*, London: Duckworth.

ENOS, J. L., and PARK, W. H. (1988), *The Adoption and Diffusion of Imported Technology: The Case of Korea*, London: Croom Helm.

ENRIGHT, M. J. (1998), "Regional Clusters and Firm Strategy," in A. D. Chandler, P. Hagström, and Ö. Sölvell (eds.), *The Dynamic Firm: The Role of Technology, Strategy, Organization and Regions*, Oxford and New York: Oxford University Press, 315–42.

* Asterisked items are suggestions for further reading.

FAGERBERG, J. (1987), "A Technology Gap Approach to Why Growth Rates Differ," *Research Policy* 16(1): 87–99.

*(1988), "International Competitiveness," *Economic Journal* 98: 355–74.

*(2002), *Technology, Growth and Competitiveness: Selected Essays*, Cheltenham: Edward Elgar.

—— VERSPAGEN, B., and CANIËLS, M. (1997), "Technology, Growth and Unemployment across European Regions," *Regional Studies* 31(5): 457–66.

FREEMAN, C., and LOUÇÃ, F. (2001), *As Time Goes By: From the Industrial Revolutions to the Information Revolution*, Oxford and New York: Oxford University Press.

—— and PEREZ, C. (1988), "Structural Crises of Adjustment, Business Cycles and Investment Behaviour," in G. Dosi, C. Freeman, R. R. Nelson, G. Silverberg, and L. L. G. Soete (eds.), *Technical Change and Economic Theory*, London: Pinter 38–66.

HÄGERSTRAND, T. (1967), *Innovation Diffusion as a Spatial Process*, Chicago: University of Chicago Press.

KLINE, G. J., and ROSENBERG, N. (1986), "An Overview of Innovation," in R. Landau and N. Rosenberg (eds.), *The Positive Sum Strategy: Harnessing Technology for Economic Growth*, Washington, DC: National Academy Press, 275–306.

KODAMA, F. (1992), "Technology Fusion and the New R&D," *Harvard Business Review* (July–August): 70–8.

*KRUGMAN, P. R. (1994a), "Competitiveness: A Dangerous Obsession," *Foreign Affairs* 73(2): 28–44.

—— (1994b), "The Myth of Asia's Miracle," *Foreign Affairs* 73(6): 62–78.

—— (1996), "Making Sense of the Competitiveness Debate," *Oxford Review of Economic Policy* 12(3): 17–25.

*LALL, S. (2001), *Competitiveness, Technology and Skills*, Cheltenham: Edward Elgar.

LANDES, D. S. (1969), *The Unbound Prometheus: Technological and Industrial Development in Western Europe from 1750 to the Present*, Cambridge and New York: Cambridge University Press.

LAZONICK, W. (1991), *Business Organization and the Myth of the Market Economy*, Cambridge and New York: Cambridge University Press.

*(1992), "Business Organization and Competitive Advantage: Capitalist Tranformations in the Twentieth Century," in G. Dosi, R. Giannetti, and P. A. Toninelli (eds.), *Technology and Enterprise in a Historical Perspective*, Oxford and New York: Oxford University Press, 119–63.

—— (1993), "Industry Clusters versus Global Webs: Organizational Capabilities in the American Economy," *Industrial and Corporate Change* 2(1): 1–24.

—— (1998), "Organizational Learning and International Competition," in J. Michie and J. Grieve-Smith (eds.), *Globalization, Growth and Governance*, Oxford and New York: Oxford University Press, 204–38.

—— and O'SULLIVAN, M. (1997), "Big Business and Skill Formation in the Wealthiest Nations: The Organizational Revolution in the Twentieth Century," in A. D. Chandler, F. Amatori, and T. Hikino (eds.), *Big Business and the Wealth of Nations*, Cambridge and New York: Cambridge University Press, 497–521.

LEVIN, R. C., COHEN, W. M., and MOWERY, D. C. (1985), "R&D Appropriability, Opportunity and Market Structure: New Evidence on Some Schumpeterian Hypotheses," *American Economic Review* 75: 20–4.

LUNDVALL, B. Å. (1988), "Innovation as an Interactive Process: From User–Producer Inter-action to the National System of Innovation," in G. Dosi, C. Freeman, R. R. Nelson, G. Silverberg, and L. L. G. Soete (eds.), *Technical Change and Economic Theory*, London: Pinter, 349–69.

—— (ed.), *National Systems of Innovation*, London: Pinter.

MALMBERG, A., SÖLVELL, Ö., and ZANDER, I. (1996), "Spatial Clustering, Local Accumulation of Knowledge and Firm Competitiveness," *Geografiska Annaler* 78B (2): 85–97.

MARRIS, R. (1964), *The Economic Theory of Managerial Capitalism*, London: Macmillan.

MOWERY, D. C., and NELSON, R. R. (1999), *Sources of Industrial Leadership: Studies of Seven Industries*, Cambridge: Cambridge University Press.

NELSON, R. R. (1992), "What is 'Commercial' and What is 'Public' about Technology, and What Should Be?" in N. Rosenberg, R. Landau, and D. C. Mowery (eds.), *Technology and the Wealth of Nations*, Stanford: Stanford University Press, 57–71.

—— (1995), "Co-evolution of Industry Structure, Technology and Supporting Institutions, and the Making of Comparative Advantage," *International Journal of the Economics of Business* 2: 171–84.

*—— and PACK, H. (1999), "The Asian Miracle and Modern Growth Theory," *Economic Journal* 109: 416–36.

OZAWA, T. (1974), *Japan's Technological Challenge to the West, 1950–1974: Motivation and Accomplishment*, Cambridge, Mass.: MIT Press.

PENROSE, E. T. (1959), *The Theory of the Growth of the Firm*, Oxford: Basil Blackwell.

*PORTER, M. E. (1990), *The Competitive Advantage of Nations*, New York: Free Press.

—— and SÖLVELL, Ö. (1998), "The Role of Geography in the Process of Innovation and the Sustainable Competitive Advantage of Firms," in A. D. Chandler, P. Hagström, and Ö. Sölvell (eds.), *The Dynamic Firm: The Role of Technology, Strategy, Organization and Regions*, Oxford and New York: Oxford University Press, 440–57.

POSNER, M. V. (1961), "International Trade and Technical Change," *Oxford Economic Papers* 13(3): 323–41.

PRAIS, S. J. (1995), *Productivity, Education and Training: Facts and Policies in International Perspective*, Cambridge and New York: Cambridge University Press.

RICHARDSON, G. B. (1972), "The Organization of Industry," *Economic Journal* 82: 883–96.

SCOTT, A. J. (1998), "The Geographic Foundations of Industrial Performance," in A. D. Chandler, P. Hagström, and Ö. Sölvell (eds.), *The Dynamic Firm: The Role of Technology, Strategy, Organization and Regions*, Oxford and New York: Oxford University Press, 384–401.

SOETE, L. L. G. (1981), "A General Test of Technological Gap Trade Theory," *Weltwirtschaf-tliches Archiv* 117(4): 638–60.

SOLOW, R. (1957), "Technical Change and the Aggregate Production Function," *Review of Economics and Statistics* 39: 312–20.

*TEECE, D. J., PISANO, G., and SHUEN, A. (1997), "Dynamic Capabilities and Strategic Management," *Strategic Management Journal* 18(7): 537–56.

TYSON, L. D'A. (1992), *Who's Bashing Whom: Trade Conflict in High Technology Industries*, Washington, DC: Institute for International Economics.

VERNON, R. (1966), "International Investment and International Trade in the Product Cycle," *Quarterly Journal of Economics* 80(2): 190–207.

—— (1974), "The Location of Economic Activity," in J. H. Dunning (ed.), *Economic Analysis and the Multinational Enterprise*, London: Allen and Unwin, 89–114.

—— (1979), "The Product Cycle Hypothesis in the New international Environment," *Oxford Bulletin of Economics and Statistics* 41(4): 255–67.

VERTOVA, G. (1998), *Historical Evolution of National Systems of Innovation and National Technological Specialisation.* Ph.D. thesis, University of Reading.

VON HIPPEL, E. (1989), *The Sources of Innovation.* Oxford and New York: Oxford University Press.

YOUNG, A. (1995), "The Tyranny of Numbers: Confronting the Statistical Realities of the East Asian Growth Experience," *Quarterly Journal of Economics* 110(3): 641–80.

CHAPTER 21

INNOVATION AND EMPLOYMENT

MARIO PIANTA

21.1 INTRODUCTION

THE relationship between innovation and employment is a complex one and has long been a topical issue in economic theory.[1] Moving from the classical question "does technology create or destroy jobs?" recent research has investigated the impact of different types of innovation and the structural and institutional factors affecting the *quantity* of employment change. *Quality* aspects have received increasing attention, with questions of "what type of jobs are created or destroyed by innovation?" This line of research has asked, "how does the composition of skills change" and "how does the wage structure change," leading to a large literature on skill biased technical change and on wage polarization.

This chapter examines the enormous body of scholarly research on this topic within the advanced economies. First, the perspectives, scope, and types of innovations are considered, identifying the different employment effects they may have. Second, the effects on the *quantity* of employment are reviewed at the firm, industry, and macroeconomic level. Third, changes in the *quality* of employment are examined, considering the effects on skills and wages, and the impact of organizational innovation, again at different levels of analysis. A summary of stylized facts concludes the chapter with a discussion of future research issues.

21.2 PERSPECTIVES AND SCOPE
OF INNOVATIONS

The literature on innovation and employment has addressed different research questions rooted in several streams of research. Table 21.1 summarizes the main perspectives in terms of assumptions as to the economic system, methodologies, and levels of empirical analysis (for a review of theoretical approaches, see Petit 1995).

The mainstream approach looks at innovation as a change in technology across the economy, leading to economic growth and employment. The lack of an explicit conceptualization of technology, either in the decisions of private agents, or in the tools of public policy, and its exogenous nature remain, however, a limitation of Keynesian growth theories, as well as of Solow's growth model (see Ch. 18 by Verspagen in this volume). Modern theories have approached the question of the employment impact of technology from two perspectives. On the one hand, growth theories have moved from an exogenous view of technological change to efforts, in the *new growth theory*, to conceptualize innovation—proxied by technology, learning, and educational variables—as an engine of endogenous growth. Labor economists, on the other hand, have explained changes in employment (and wages) with reference to the demography of jobs, macroeconomic factors, wage costs, bargaining modes, and the flexibility of labor markets, moving later to consider competitiveness and technology factors (see Table 21.1).

However, the fundamentally *disequilibrating* nature of technological change is usually treated in a context that still assumes a general (or partial) equilibrium of markets, that is, all output finds its demand, and all workers ready to accept the current wage find employment. Technological change is often reduced to new production processes (and new production functions), with the models rarely envisaging the emergence of product innovation. When employment losses appear in such studies, they rarely lead to permanent (or structural) unemployment; rather, they lead to downward adjustments in wages so that the additional jobless are returned to work. If this state of affairs cannot be found in the real world, then the responsibility is attached to the lack of flexibility of labor markets, with excessive union power or institutional rigidities, such as the minimum wage.

A more convincing approach to the study of innovation and its consequences is one that addresses from the start the *disequilibrium* nature of economic change. This view has been developed by neo-Schumpeterian perspectives, by Kaldorian, Structural, and Evolutionary approaches (see Table 21.1).

Neo-Schumpeterians have argued that advanced economies are witnessing the emergence of a new techno-economic paradigm based on Information and Communication Technologies (ICTs) (see Ch. 1 by Fagerberg in this volume). Such powerful technological changes *do create and destroy a large amount of jobs*. The

Table 21.1 A summary of approaches to innovation and employment

Main research questions *Does technology create or destroy jobs?*	General approach	Major streams of literature and key findings	Key assumptions and methodology	Main level of analysis
Equilibrium of product and labor markets				
What is the amount of jobs created/lost	Labour economics	Job demography and flexibility of labor markets	Product and labor markets are in equilibrium The absolute level of job lost/gained is irrelevant	Firms, industries, macroeconomy
What is the skill composition What is the structure of wages		Technical change favors more skilled workers, replaces the unskilled, exacerbates inequality Supply of educated workers shapes technical change Technological unemployment is irrelevant	Complementarity between ICTs and high skills	
What are the returns from innovation	Growth	Technology, productivity, growth, employment: innovation may raise the natural rate of unemployment	Standard production function, focus on process innovation	Industries, macro
What is the innovation input to growth	New growth theory	Endogenous innovation, growth and employment: unemployment may happen	Innovating and non innovating firms, spillovers, focus on process innovation	Macroeconomy
Disequilibrium perspectives				
What is the type of innovation What is the amount/nature of unempl.	Evolutionary	Technological opportunities, variety, regimes: firms' strategies and industries' outcomes are different	Innovation brings disequilibrium in markets	Firms, industries
	Neo-Schumpeterian	Techno-economic paradigms and long waves: mismatches can lead to unemployment	New product markets emerge Radical innovations, pervasiveness, diffusion of new technology systems and ICTs	Industries, macro
What are the structural factors What are the demand factors	Structural	Sectoral composition of economies: specificity of innovation and demand, different job results	Innovation is differentiated: constrasting effects of new products and processes	Industries
What are the distribution effects What are the institutions	Regulationist	Macro models for testing indirect effects of innovation: compensation mechanisms may not work	Industries are different, demand is important Countries are different, institutions are important	Macroeconomy

question as to where and to what extent jobs are created or destroyed depends upon the highly dynamic process that shapes the content of specific technological innovations and the speed of their adoption, with an often blurry distinction between the two (see below). Moreover, jobs lost and new jobs offered may take place in different areas or require different skills, leading to mismatches. The speed of adjustment is therefore crucial and makes the difference between frictional unemployment (easily absorbed by well-functioning labor markets) and technological unemployment.

Sustained and sustainable growth can be expected only once the mismatches between the new technologies and the old economic and social structures and institutions are overcome, with a two-way adjustment. Innovation has to be adapted to social needs and economic demands; economic and social structures evolve under pressure from new technologies. New technologies need to be matched by organizational changes, new institutions and rules, learning processes, the emergence of new industries and markets, and the expansion of new demand. Several studies on the emergence of technological paradigms and key technologies in the past have pointed to the long time required before the impact of these elements (positive or negative) on economic growth and employment become evident (Freeman, Clark, and Soete 1982; Freeman and Soete 1987, 1994; Freeman and Louçã 2001). While extremely powerful in its explanation of long term economic changes and historical evolution, this approach has yet to be "operationalized" with more specific questions on the type of innovation and on the interaction with economic and employment variables.

Box 21.1 Technology and unemployment: a classical debate

Since the emergence of the Industrial Revolution, the extensive substitution of labor by machinery incorporating the new technology of the time has led economists and policy makers to debate the economic and social consequences. At the end of the eighteenth century, James Steuart drew attention to the difficulty of reabsorbing the unemployment caused by sudden mechanization, in spite of the positive effects from the construction of new machines and price reductions, and already envisioned a role for the government. Adam Smith linked the invention of machines to the division of labor and emphasized its labor-saving effects. Jean-Baptiste Say had less doubts about the ability of markets to adjust, while Thomas Malthus emphasized the positive effects resulting from the strong demand dynamics experienced by England at the time. The optimism of classical economists in the early nineteenth century contrasted with the dramatic impoverishment of the English working classes—industrial workers, small artisans, and displaced peasants—who had started to organize trade unions and to launch Luddite struggles against the job losses and deskilling brought about by mechanization. David Ricardo was convinced that the economy could compensate the negative employment effects, but in a famous passage in the chapter "On machinery," added in the third edition of his *Principles of Political Economy and Taxation,* argued that "The opinion, entertained by the labouring class, that the employment of machinery is frequently detrimental to their interests, is not founded on prejudice and error, but is conformable to the correct principles of political economy" (Ricardo 1951:392).

The most articulate criticism of compensation theory was developed by Karl Marx, who emphasized the losses for workers in terms of jobs, skills, wages, and control over their work resulting from the way mechanization was proceeding at the time. Arguing that unemployment grows as technical change displaces labor more rapidly that the accumulation of capital demands new workers, Marx developed important insights on the functioning of capitalism. The drive to capital accumulation leads to a constant search for new production techniques and new products (a key starting point in Joseph Schumpeter's theory of innovation). High unemployment assures lower wages and greater control over workers, but capital accumulation ultimately encounters the problems of finding adequate markets and demand, and making adequate profits (for a reconstruction of the debate see Heertje 1973; Vivarelli 1995).

21.2.1 New Products, Processes and Organizations

Schumpeter (1934) defined *product innovation* as "the introduction of a new good... or a new quality of a good," and *process innovation* as "the introduction of a new method of production ... or a new way of handling a commodity commercially."[2] The development (or the adoption) of process innovations leads to greater efficiency of production, with savings in labor and/or capital, and with a potential for price reductions. The usual outcome is higher productivity and loss of employment; to the extent that process innovations increase product quality or reduce prices, a rise in demand (when elasticity is high) may result in more jobs.

New products (or services) can be radical innovations (new to the world), incremental improvements on previous innovations, or imitation of goods already produced in other countries or firms. Generally, product innovations increase the quality and variety of goods and may open up new markets, leading (when elasticity is high) to greater production and employment. But new products can simply replace old ones, with limited economic effects, or be designed in order to reduce costs, with an impact similar to process innovations (Pianta 2001).

New goods enter the economy as consumption goods, intermediate goods, or investment goods, following the demands of consumers, firms, and investors. Innovative investment goods have a dual nature: they start as new products in the industries producing them, but become process innovations in the industries acquiring them. Their employment consequences are likely to be positive in the machinery producing sectors, and negative (when demand offsets are insufficient, as noted above) in the industries making new investments (Edquist, Hommen, and McKelvey 2001).

The distinction between process and product innovations should not be carried too far. Most innovative firms introduce both at the same time, but in most firms and industries (see Ch. 14 by Malerba in this volume) it is possible to identify the dominant orientation of innovative efforts, associated with strategies of either price competitiveness (and mainly process innovations) or technological competitiveness (and mainly product innovations). In addition to *product and process* innovation, *organizational* innovation also can affect the quantity and quality of employment, and is usually closely linked to the introduction of new technologies (Caroli 2001). Sections 21.22 and 21.3 below will consider the impact on job creation and loss of each of these types of innovation.

21.2.2 Innovation, Imitation, Adoption, and Use

The employment consequences of new technologies obviously require the application of these technologies in production activities. Firms *innovate* when they first market a new product or introduce a new process; followers in the same industry may *imitate* them (perhaps with incremental improvements). Firms may *adopt* new processes or *use* new products generated in other industries, leading to the diffusion of innovations—and to their employment impact—throughout the economy (see Ch. 17 by Hall in this volume).

For empirical analyses, this distinction can be captured either by studies on the emergence of particular technologies, following their evolution through all the previous steps, or by studies on firms (or industries) based on surveys that identify product innovations novel to the market (original innovators) and those new to the firm only (imitators), and the introduction of new processes (adopters). Traditional indicators such as R&D and patents fail to capture a large part of the latter two modes of technological development.

Box 21.2 Evidence on innovation and employment

Conceptual and empirical difficulties have long led economists to adopt a rather homogeneous view of innovation, described either by R&D expenditure (one of its inputs), or proxied by patenting activity (one of its output). In the last decade, the spread of innovation surveys in Europe and of surveys on panels of firms in the US has provided important new evidence on the variety of innovative activities.

The key results for Europe in the period 1994–6 are that innovation is present in 51 per cent of manufacturing firms and in 40 per cent of services firms (see Chapters 6 and 16 by Smith and Miles in this volume). Close to 40 per cent of manufacturing firms change their production processes, while a fifth develop minor product improvements

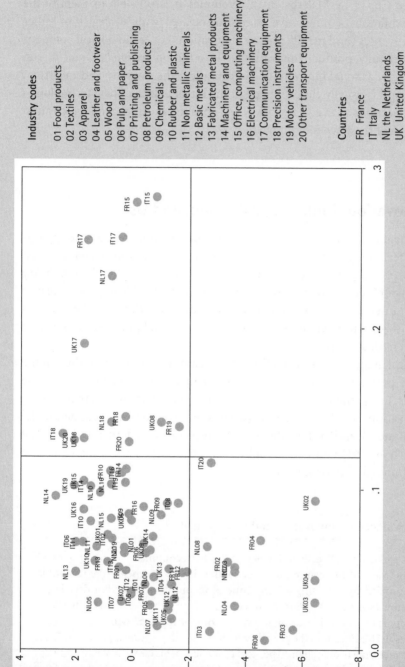

Industry codes

01 Food products
02 Textiles
03 Apparel
04 Leather and footwear
05 Wood
06 Pulp and paper
07 Printing and publishing
08 Petroleum products
09 Chemicals
10 Rubber and plastic
11 Non metallic minerals
12 Basic metals
13 Fabricated metal products
14 Machinery and equipment
15 Office, computing machinery
16 Electrical machinery
17 Communication equipment
18 Precision instruments
19 Motor vehicles
20 Other transport equipment

Countries

FR France
IT Italy
NL the Netherlands
UK United Kingdom

Fig. 21.1 Share of new products in sales and employment change

Source: CIS2–SIEPI Innovation Database, University of Urbino and OECD STAN data.

and another fifth introduce products new to their market, whose impact on sales is limited to 12 per cent. In all indicators, important differences are found across countries, associated with the nature of their industrial structure and national innovation system (European Commission-Eurostat 2001).

The latter variable offers the most accurate description of the economic relevance of innovations, and is related, in Figure 21.1, to the employment performance (average annual rate of change) of 20 manufacturing industries in four EU countries (France, Italy, the Netherlands, and the UK). Looking at industry data for the share of new products in sales (drawn from the CIS 2-SIEPI database, which provides data at the two-digit industry level for major countries), a very strong variability is evident, from more than 20 per cent in office computing and telecommunications, to close to zero in more traditional industries (again, with important cross-country differences due to national specializations). Looking at overall employment change, we first see that, between 1994 and 2000, the majority of industries lost jobs. The relationship between product innovation and employment experienced in European industries looks like a positively oriented curve, but a closer look at the distribution of cases is important.

ICT industries (computing, telecommunications, precision instruments, and other transport, including aerospace) are generally in the top right quadrant, where new products in sales have a share higher than the average value for the EU and where no dramatic job losses are found. As expected, industries characterized by the new technology (and by a high concentration of product innovations) show the highest impact of new products in their turnover and better employment performances. In a few cases, however, moderate job losses are found even in this group, as strong international competition may lead to the decline of some ICT industries in some countries.

Traditional industries (textiles, wearing apparel, leather, and a few others) tend to concentrate in the bottom left quadrant, where a below average innovativeness (and a strong dominance of process innovations) is matched by dramatic job losses.

The remaining sectors, in the top left quadrant, combine a low or intermediate impact of innovation with modest job losses or substantial gains, showing again a positive association. While several factors alongside innovation affect employment change (macroecomic conditions, competitiveness, etc., see Pianta 2000), the distribution of Figure 21.1 highlights, on the one hand, the generally positive link we expect between product innovation and jobs; on the other hand, it shows the presence of winners and losers in all industries, reflecting the importance of national specializations, of economic structures, and of the intensity of international competition in open economies.

21.3 THE EFFECTS ON THE QUANTITY OF EMPLOYMENT

The relationship between innovation and jobs is investigated in this section, which looks at the impact on the *quantity* of employment, defined in terms of the number of existing jobs, or, more accurately, in terms of the total hours of work. This link can be examined at different levels of analysis: firms, industries, and the aggregate economy. Table 21.2 summarizes the most relevant empirical evidence emerging from the literature.

21.3.1 Direct Effects at the Firm Level

Firms are where innovations are introduced and where they show their immediate direct effects on employment. A growing literature has explored the issue with a variety of models, national studies, and panels of firms (for reviews, see Petit 1995; Chennells and Van Reenen 1999; Spiezia and Vivarelli 2002). Empirical work in this field has generally used annual surveys of firms in panel data; however, panels are usually not representative of the whole manufacturing industry, and in most cases leave out services altogether; and therefore it is difficult to generalize their conclusions.

The evidence on the overall employment impact of innovation at the level of firms tends to be positive: firms that innovate in products, and also in processes, grow faster and are more likely to expand their employment than non-innovative ones, regardless of industry, size, or other characteristics.[3] The variety of innovative strategies, job creation, and destruction patterns have been highlighted in such studies, together with the firms' characteristics (structural factors, flexibility, competences, etc.) that tend to be associated with better performances.

A reverse relationship can also be considered. In the long-run development of firms, phases of rapid growth of employment may be seen as determinants of innovations as firms innovate in order to cope with the rigidity of production processes and increasing wages, while trying to capture, through greater productivity and quality, the opportunities of expanding markets.[4]

Some studies also suggest that the positive employment effects of technological innovation are linked to organizational changes. A study on a large and representative sample of French firms found that firms that adopted advanced manufacturing systems in the period 1988–93 and introduced in parallel organizational change had greater employment growth than others, regardless of size or sector, and that this positive effect was greater than in firms that introduced organizational innovation only (Greenan 2003; see Section 21.3.3 below).

However, firm-level studies on the innovation–employment link are unable to point out whether the output and job gains of innovating firms are achieved at the expense of competitors, or whether there is a net effect on aggregate industry employment.[5] It is often difficult to generalize beyond the groups of firms investigated and to compare results across countries. When panels are used, a large part of

Table 21.2 Effects of innovation on the quantity of employment: Selected empirical studies

Study	Countries	Years	Level of analysis	Innovation data sources	Results on employment
Firm-level studies					
Machin and Wadhwani 1991	UK	1984	Cross firm, manuf.	British workplace industrial relations survey	Positive
Brouwer, Kleinknecht, and Reijnen 1993	Netherlands	1983–8	Cross firm, manuf.	Dutch survey	Negative Positive with product innovation
Meghir, Ryan, and Van Reenen 1996	UK	1976–82	Panel of firms, manuf.	SPRU innovation database and patents	Positive more flexibility
Van Reenen 1997	UK	1976–82	Panel of manuf. firms	Survey on UK firms	Positive
Smolny 1998	Germany	1980–92	Panel of manuf. firms	Survey on German firms	Positive
Greenan and Guellec 2000	France	1986–90	Cross firm, manuf. Cross sector	Innovation survey	Positive at the firm level Negative at the industry level for process innovations
Industry-level studies					
Meyer-Kramer 1992	Germany	1980s	Input output model all economy	Industry data	Negative, differentiated by sector
Vivarelli, Evangelista, and Pianta 1996	Italy	1985	Cross sector 30 manuf. industries	Innovation survey	Negative of process innovation Positive of product innovation
Pianta 2000, 2001	5 EU countries	1989–93	Cross sector 21 manuf. industries	Innovation survey	Overall negative Positive of product innovation
Antonucci and Pianta 2002	8 EU countries	1994–9	Cross sector 10 manuf. industries	Innovation survey	Overall negative Positive of product innovation
Evangelista and Savona 2002, 2003	Italy	1993–5	Cross sector service industries	Innovation survey	Overall negative, differentiated by service industries and size

Table 21.2 (cont.)

Study	Countries	Years	Level of analysis	Innovation data sources	Results on employment
Macroeconomic–level studies					
Layard and Nickell 1985	UK	1954–83	Macro model	Labor productivity	Neutral
Vivarelli 1995	US and Italy	1966–86	Macro model	R&D linked to product and process innovations	Differentiated by compensation mechanism and country
Simonetti, Taylor, and Vivarelli 2000	US, Italy, France, Japan	1965–93	Macro model	R&D linked to product and process innovations	Differentiated by compensation mechanism
Simonetti and Tancioni 2002	UK and Italy	1970–98	Macro model quarterly data	R&D linked to product and process innovations	Differentiated by compensation mechanism
Simulation studies					
Leontief and Duchin 1986	US	1980–2000	Input output model all economy	Assumptions on performance	Negative
Kalmbach and Kurz 1990	Germany	2000	Input output model all economy	Assumptions	Negative
IPTS–ESTO 2001	Europe	2000–2020	Gen. equil. model all economy	Assumptions on productivity growth	Positive, differentiated by innovation policy

the jobs created or lost may be accounted for by the entry or exit of firms left outside the panel. In order to address these issues, we need to turn to industry-level studies.

21.3.2 The Effects at the Industry Level

Analyses at the industry level address not only the direct employment effects of innovation within firms, but also any indirect effects that operate within the industry. These indirect effects include the competitive redistribution of output and jobs from low to high innovation-intensive firms, and the evolution of demand (and therefore output and jobs) resulting from the lower prices due to innovation, given the price elasticities of the industry's goods. The industry level may be the most satisfactory level of analysis, as it is able, on the one hand, to differentiate between the variety of technological regimes and strategies and, on the other hand, to bring in the demand dynamics of specific sectors, taking into account country differences in economic structures.

Studies on industries (see Table 21.2) show that the employment impact of innovation is positive in industries (both in manufacturing and services) characterized by high demand growth and an orientation toward product (or service) innovation, while process innovation leads to job losses. The overall effect of innovative effort depends on the countries and periods considered, but, in general, the higher the demand growth, the greater the importance of innovative industries (both in manufacturing and services) and the greater the orientation of firms within an industry toward product innovation, the more positive are the employment effects of innovation.

Demand factors are important because an industry's demand is constrained by the composition and dynamics of domestic and foreign demand. High demand growth leaves room for a variety of firm strategies and for better employment outcomes, while stagnant demand deepens the selection process among firms and emphasizes the role of technological competition (Pianta 2001).[6]

The empirical investigations of the employment impact of innovation at the industry level include studies using R&D or patenting as innovation proxies,[7] input–output models,[8] and more recent works based on innovation surveys (unfortunately, available only for Europe). Set in the context of the European debate on "jobless growth" in the 1990s, the evidence points to an extensive process of restructuring in many manufacturing sectors where the growth of value added is not matched by increases in jobs (see Pianta, Evangelista, and Perani 1996; Fagerberg, Guerrieri, and Verspagen 1999).

Studies using innovation survey data[9] show that, in Europe, employment change (in most cases a decrease) is positively affected by the dynamics of demand and by the relevance of product innovations, while a higher intensity of innovative expenditure *per se* has a negative impact on jobs, suggesting a prevailing pattern of labor-replacing technological change.

Similar findings come from a model in which employment is affected by demand dynamics, labor costs, and innovation variables on product or process innovations; in the context of the modest aggregate growth of the 1990s, European industries (in eight countries) were dominated by process innovations, with generally negative effects on jobs (Antonucci and Pianta 2002).

The findings for service industries do not differ substantially from those of manufacturing. Studies on Italy have found an overall negative effect, concentrated among the largest firms, on low-skilled workers, on capital-intensive and finance-related sectors, and where the impact of ICTs has been most widespread. Smaller firms and technology-oriented activities show, on the other hand, net employment gains (Evangelista and Savona 2002, 2003; see also Ch. 16 by Miles in this volume).

The evidence at the sectoral level, especially for European countries, suggests a less optimistic view on the employment impact of innovation. The slow aggregate growth of the 1990s has constrained demand, limiting the potential employment benefits of technological change. While product innovation has had positive effects on output and jobs, increased international competition has pushed firms toward restructuring, and process innovations have dominated several industries in many European countries, leading to a prevalence of labor-saving effects. This contrasts with the experience of the US in the 1990s, marked by high demand and employment growth.

21.3.3 Direct and Indirect Compensation Effects at the Macroeconomic Level

The most complete view of the employment impact of innovation is provided by a macroeconomic perspective that can integrate all the indirect effects through which technological change affects employment. This is the approach typical of the debate on "compensation mechanisms" since the times of Ricardo and Marx, pitting those who argue that the economic system has built-in mechanisms which assure the recovery of jobs lost due to innovation, against critics who point out the limitations of their effects and the possibility of technological unemployment. Most of this debate conceptualizes technological change as the introduction of new capital goods, typical of nineteenth-century mechanization, thus conceptualizing "innovation" as "process innovation." Second, the debate typically assumes equilibrium in product markets, i.e. no demand constraint is considered, following Say's law. Recent research in this direction has relied on Kaldorian approaches, on the work by Pasinetti (1981) and Boyer (1988a and b). A detailed treatment of this issue is in

Vivarelli (1995), who has summarized the compensation mechanisms and the way they may (or may not) operate in the economy.

The compensation mechanism *via decrease in prices* is one of the most important ones: new technologies may make lower prices possible, increasing international competitiveness and output, offsetting job losses due to the original innovation. This outcome, however, is contingent on the lack of demand constraints, on the decision of firms to transfer in lower prices the productivity gains due to the innovation, and on the lack of oligopolistic power in the relevant markets (Sylos Labini 1969).

The compensation mechanism *via new machines* may create jobs in the industries in which the new means of production are made, responding to the increased demand for equipment by users. However, at the aggregate as well as at the firm level, the rationale for mechanization is by definition saving on the overall use of labor, putting a limit on the relevance of this mechanism.

The compensation mechanism *via new investment* argues that the temporary extra profits available to the innovator may be turned into new investment if profit expectations are favorable (and assuming that Say's law operates); this, however, may expand production capacity and jobs, or may introduce additional labor saving effects.

The compensation mechanism *via decrease in wages* is typical of the neoclassical view of the labor market. As technological unemployment appears, wages should fall and firms should hire more workers. This mechanism, however, is based on strong assumptions as to the feasibility of any combination of labor and capital, competitive markets, flexibility of wages, and labor markets.

The compensation mechanism *via increase in incomes* operates in the opposite way, through the increased demand associated with the distribution of part of the gains from innovation through higher wages, as has happened in large, oligopolistic firms in mass production industries. However, any wage increases can hardly be large enough to sustain additional aggregate demand.

Finally *new products* may lead—as discussed above—to new economic activities and new markets (*welfare effects*) or, on the other hand, they may simply replace existing goods (*substitution effect*, see Katsoulacos 1986).

Aggregate empirical patterns for the US are examined in a descriptive way by Baumol and Wolff (1998). Considering five innovation indicators for the whole economy and their link to the structure and changes in unemployment in the US in the 1950–95 period, they conclude that faster innovation leads to a higher "natural rate of unemployment" and to longer spells of frictional unemployment.[10] Layard and Nickell (1985), on the other hand, have argued that the working of various compensation mechanisms ruled out the possibility of technological unemployment in the UK.

Building on Boyer (1988*a* and *b*) and on the Regulationist approach, Vivarelli (1995) has developed a simultaneous equations model for testing the compensation mechanisms in the US and Italy. He has found that the mechanism *via decrease in prices* is the most effective one, and that the positive effects of new products and labor markets operate in the US (where new jobs are created), but not in Italy, where net job losses have been found. This approach has been further developed by Simonetti, Taylor, and Vivarelli (2000), who have considered four countries, and by Simonetti and Tancioni (2002), who have developed a model for an open economy in the case of the UK and Italy. All have found a differentiated impact of compensation mechanisms in different national economies.

While this approach is the most comprehensive and satisfactory for explaining the overall impact of technological change on employment, the complexity of the construction of the model, the problems in specifying all relevant relationships, and the lack of adequate data limit the feasibility of this approach. The overall findings of these studies point to a differentiated impact of innovation depending on countries' macroeconomic conditions and institutional factors. The employment impacts of innovation generally are more positive in economies in which new-product generation and investment in new economic activities are higher, and in which the demand-increasing effects of price reductions are greater.

21.3.4 Simulation Studies

The employment impact of innovation has also been studied through the use of a simulation approach. Leontief and Duchin (1986) have estimated that the diffusion of computer technology and automation in the US economy would have negative employment effects using an input–output model incorporating strong assumptions on the productivity-enhancing effects of process innovation, but no demand dynamics.[11] Likewise, a study on the impact of microelectronics in forty sectors of the UK economy in the 1980s suggested that either net job gains or losses were possible, depending on the assumptions on the speed of diffusion and users' demand of microelectronics (Whitley and Wilson 1982). A different approach—a general equilibrium model with a sectoral structure, which assumes full employment—has been used for simulating the employment impact of different scenarios of technology-based productivity growth and of the composition of consumption, in a recent study by IPTS-ESTO (2001) on the European Union. The results show an overall positive impact on jobs, differentiated according to the alternative sectoral distributions of R&D and innovation efforts; the best outcomes result from the concentration of efforts in high technology industries. While they are interesting as explorations of alternative futures, the results of such simulations are weakened by the models' inability to identify either technological unemployment (when general equilibrium is used) or most compensation effects (when input–output models are used), and on the arbitrariness of the assumptions on the diffusion and productivity of new technologies.

Summarizing the results of this section, both sectoral and aggregate studies generally point out the possibility of technological unemployment, which emerges when industries or countries see the prevalence of process innovations in contexts of weak demand. Firms innovating in both products and processes may be successful in expanding output and jobs regardless of the economic context, but often do so at the expense of non-innovating firms. The specificities of industries, countries, and macroeconomic conditions are crucial determinants of the results obtained in empirical studies.

All the analyses of this section refer to national economies, sometimes even assuming a closed economy. When we consider an open economy, the picture becomes more complex, as, on the one hand, innovation may lead to competitiveness and exports, weakening the demand constraint while, on the other hand, domestic demand may increase imports when foreign competitors are more innovative in terms of price or quality. No empirical analysis has so far addressed the innovation-employment question in a truly global dimension. In manufacturing sectors that are highly internationalized, the introduction and diffusion of innovation leads to job creation and losses in a large number of different countries, making the distribution of the benefits and costs of technological change very complex. The case of developing countries in this context is particularly interesting, as new technologies are at the center of the structural changes and dynamic learning economies typical of catching-up countries which nevertheless face greater difficulties in capturing the employment benefits of technological change (see Karaomerlioglu and Ansal 2000 and Ch. 19 by Fagerberg and Godinho in this volume).

21.4 THE EFFECTS ON THE QUALITY OF EMPLOYMENT

Most approaches that assume equilibrium in labor markets (and therefore no technological unemployment, see Table 21.1) have disregarded the effects of innovation on the quantity of jobs, focusing instead on change in the skill mix of employment and on wage polarization. A large (mainly) US literature on skill-biased technical change (reviewed in Acemoglu 2002: 7) argues that technical change is biased towards skilled workers as it replaces unskilled labor and increases wage inequality. In fact, a strong complementarity between technology and skills has characterized most of the twentieth century, when innovation has probably always been skill biased, in contrast to the unskilled bias typical of the nineteenth century, when mechanization led to the deskilling of artisans (see Braverman 1974).[12]

This section considers the streams of research that have found evidence on the complementarity between ICTs and high skills, as well as those highlighting the deskilling process associated with the greater control over production made possible by ICTs. A related finding in much of this research emphasizes the tendency for ICT adoption to increase polarization in the wage structure as a result of changes in the skill structure of employment. Another stream of this research examines the effects of organizational innovation—alone or combined with technological innovation—on skills, wages, and employment (for reviews see Chennells and Van Reenen 1999; Sanders and ter Weel 2000). Table 21.3 summarizes the most relevant literature.

21.4.1 Skill-biased Technical Change

Many studies—mainly on US firms and industries—argue that, in the last two decades, a long-established trend of increase in the skill-intensity of employment has been accelerated by the introduction of information technology and computers. The issue has generally been investigated using a factor substitution framework, showing that direct or indirect measures of technology are important explanatory factors for the relative increase of skilled labor (see Berman, Bound, and Griliches 1994; Autor, Katz, and Krueger 1998). One stream of work compares the effects of technology with those of increased international trade, finding that the former accounted for most of the fall in demand for less-skilled workers (see Berman, Bound, and Machin 1998). Other studies have found that new technologies are adopted more extensively in plants with more skilled workers, but do not increase the demand for skills (Doms, Dunne, and Trotske 1997. See the review in Chennells and Van Reenen 1999).

When more refined measures of skill are used, however, the evidence on skills bias is less clear. Moving from the simple measures of blue or white collar jobs, or of years of schooling, to more refined indicators of skills, including cognitive (typical of technical staff), interactive (typical of supervisory staff), and motor competences (typical of manual workers) in US industries between 1970 and 1985, Howell and Wolff (1992) found that expenditures on computers and new investment were associated with raising demand for high cognitive skill workers, although with differences across occupations and industries.

Howell (1996) rejects the idea of a link between computerization, upskilling, and wage inequality, and finds that major shifts in skill structure occurred between 1973 and 1983, with little change taking place afterwards, when the diffusion of ICTs accelerated and computer-related investment per employee increased dramatically.[13] In recent years, the employment shares of high-skilled blue collars and low

Table 21.3 Effects of innovation on the quality of employment: skills, wages, and organizations: Selected empirical studies

Study	Countries	Years	Level of analysis	Innovation data sources	Results
Skill biased technical change					
Firm-level studies					
Machin 1996	UK	1984–90	Cross firm, manuf. cross sectors	British workpl. ind. rel. surv. R&D, innov., computer use	Positive effect on high skill jobs Negative on least skilled ones
Doms, Dunne, and Troske 1997	US	1988, 1993	Panel and cross firm	Use of 5 manuf. technologies	Higher skill where technologies are used; in panel: only for computers Higher wages where technologies are used; in panel no effects
Bresnahan, Brynjolfsson, and Hitt 2002	US	1987–94	Panel and cross firm	IT stock and use	Positive effect on skill demand combined with organ. change
Industry-level studies					
Howell and Wolff 1992	US	1970–85	Cross sector all private economy	Exp. on computers, new investment	Positive on cognitive skill jobs Negative on interactive and motor skills
Berman, Bound, and Griliches 1994	US	1979–89	Cross sector 4 digit manuf. industr.	R&D and computer invest.	Higher skill where technology is higher
Wolff 1996	US	1970–90	Cross sector all private economy	Exp. on computers, R&D	Positive effect on complexity of tasks Negative on motor skills
Autor, Katz, and Krueger 1998	US	1960–95	Cross sector all economy	Computer use, R&D, TFP	Faster upskilling in high tech industries, and after 1970
Machin and Van Reenen 1998	7 OECD countr.	1970–89	Cross sector 2 digit manuf. industr.	R&D intensity	Higher skill where R&D is higher in all countries

Table 21.3 (cont.)

Study	Countries	Years	Level of analysis	Innovation data sources	Results
Technology and wage dispersion					
Firm–level studies					
Casavola, Gavosto, and Sestito 1996	Italy	1986–90	Cross firms	% of intangible capital INPS wage data	Positive effect on skills Limited wage dispersion
Van Reenen, 1996	UK	1976–82	Panel and cross firm	Innov. counts and patents	Positive effects on wages
Black and Lynch 2000	US	1993–96	Panel and cross firm manuf.	Computer use, organ. changes	Higher wages and productiv. with tech. use and organ. change involving employees
Industry level studies					
Bartel and Lichtenberg 1991	US	1960, 1970 1980	Cross sector manuf.	Computer invest., R&D	Wage premia with newer technologies
Bartel and Sicherman 1999	US	1979–1993	Cross sector manuf.	Computer invest, TFP, etc Nat'l. Long. Survey of Youth	Wage premia in high tech industries
Organizational innovation					
Firm–level studies					
Caroli and Van Reenen 2001	UK and France	1984–90 1992–96	Panel and cross firm	% of workers affected Computer use	Positive effects on skills in the UK Weak in France Organ. innov. has stronger negative effects on low skill
Piva and Vivarelli 2002	Italy	1991–97	Panel and cross firm	R&D, organ. change	Positive effects of organ. inn. on skills, no effect of technology
Greenan 2003	France	1988–93	Panel and cross firm	Tech. invest, organ. change	Positive effects of organ. inn. on skills, less effect of technology

skilled white collars have declined more rapidly, and little change in the skill structure of services employment is apparent. ICT investment since the mid-1980s therefore appears to have negatively affected the low skilled white collars (mainly female) much more than the lowest blue collar skills.

21.4.2 Wage Polarization and the Labor Market

Although many studies do not separate the effects of technological change on skills and wages, some have investigated the evolution of wage differentials in three contexts: across industries with varying R&D and capital intensities; among workers and firms with different levels of use of computer technology and ICTs; and among workers or social groups with different educational levels (see Sanders and ter Weel 2000; Acemoglu 2002). These studies find weak evidence of polarization, although the relationship of polarization to technological change is confounded by the tendency for computers to be used by more competent workers who already earn higher wages (Chennells and Van Reenen 1999). On the other hand, the technology–wage polarization link has been questioned by studies pointing out the lack of an acceleration in these effects in recent years, and suggesting instead that sectoral shifts in employment and growing international trade have exercised more influence on the structure of wages (Addison and Teixeira 2001).

In comparing US and European results, a paradox emerges. Empirical patterns show that, in the last two decades, low skill and low wage jobs have grown more rapidly in the US than in Europe, a trend associated with the faster growth of the supply of labor, and greater polarization in the wage structure. Skill-biased technical change thus may be more a European than an American phenomenon. In this view, a stagnant labor supply and more educated workers, as well as slower aggregate growth and greater competitive pressure, have led European firms and industries to adopt technological and organizational innovations during the past two decades that have favored relatively skilled workers. The effects of this skills bias in European employment on wage polarization have been attenuated by European labor market institutions (e.g. stronger labor unions). But in the United States, faster growth of new jobs at the top and bottom end of the skill structure, combined with weaker labor unions, have produced considerable polarization in wages.[14]

21.4.3 The Effects of Organizational Innovation on Employment and Skills

In an extensive survey of organizational change in US manufacturing and service firms, Appelbaum and Batt (1994) find widespread adoption of new management practices in conjunction with the introduction of new technologies, but no information is provided on any change in employment levels. Both technology adoption (such as the number of non-managers using computers) and new workplace practices are closely associated with higher productivity and wages (Black and Lynch 2000). Improved performances and upskilling of the workforce are also found in US firms when ICT introduction is matched by new organizations delegating authority to workers and teams (Brynjolfsson and Hitt 2000; Bresnahan et al. 2002).

Several European studies (Caroli and Van Reenen 2001 on France and Britain; Greenan 2003 on France; Piva and Vivarelli 2002 on Italy) have shown that organizational innovation is more important than technological innovation in shaping changes in occupational structure and skills. This is generally not associated with an increase in the number of employees, with the exception of management occupations.

Organizational and technological changes in services, on the other hand, have reflected the opportunities offered by ICT to overcome time and space constraints in the provision of services, leading to major flows of job creation and destruction, and to rapidly changing skill requirements. A variety of strategies of restructuring, emergence of networks, subcontracting, and outsourcing has resulted, leading to polarization effects in skills and wages (Petit and Soete 2001b; Frey 1997).

The rather fragmented evidence so far available on organizational innovation suggests that it plays a crucial role alongside technological innovation in shaping productivity and employment outcomes. The two can have a complementary relationship (especially when a virtuous circle of growth is in place) leading to a combined effect on performance and upskilling that can be greater than their sum. On the other hand, changes in organizations or in technologies may be pursued as alternative roads in contexts of restructuring and job losses.

21.4.4 A Broader View

Although there is little doubt that, in the long term, technological change is associated with improvements in the skills and wages of some workers, the specific effects of innovation on employment in particular countries and periods reflect the operation of many other factors, including economic structures, the strategies of firms, the operation of labor markets rules and institutions, and national economic policies. The macroeconomic dimension is missing from much of the current literature on technological innovation and employment. Little research has addressed the broader distribution effects of technological change among profits, rents, wages, shorter work hours, and lower prices.

This literature also has restricted its focus to the demand for labor, despite abundant evidence that new technologies interact with broader social changes that affect the supply of labor (see Acemoglu 2002). Slower population growth, higher immigration flows, an ageing population and the forms of women's presence in the labor force all influence labor supply. In addition, the quality of labor supply—and the potential for innovation—are affected by education and training, learning processes, and the accumulation of competences by workers and firms.

21.5 Conclusion: Stylized Facts and Directions for Future Research

This chapter has examined the relationship between innovation and employment, reviewing a large empirical literature. The complexity of the issues is such that no single approach can account for the direct and indirect consequences of technological change, or for its effects on the quantity and quality of labor employed in the economy.[15] Theories and empirical research have to proceed in parallel, with a close interaction between concepts and measures, hypotheses and tests, building links to related research areas. This final section summarizes some key stylized facts emerging from the empirical evidence and highlights possible directions for future research.

21.5.1 A Few Stylized Facts

The never-ending race of innovation and employment. The evolution of (most) economies shows that, when growth, structural change, and demand dynamics take place together, in the long run the jobs lost to technological change are found elsewhere in the economy. If no innovation took place, economic activities facing competitive pressure would cut costs, wages, and eventually, jobs. The key question is the rate at which technological innovation and diffusion eliminate jobs, versus the pace at which new economic activities create new jobs.

Technological unemployment can happen. On the basis of the available evidence, current technological change may be a cause of unemployment. There is no automatic mechanism ensuring that national or regional economies fully compensate for innovation-related job losses. Indeed, Europe in the 1990s appears to have been a region in which technological innovation and diffusion, in combination with a long list of supporting influences, eroded employment.

The type of innovation is important. The evidence shows that it is essential to discriminate between *product* innovation (novel or imitative) that has a generally positive employment impact, and *process* innovation (adoption and use of new technologies) usually with negative effects; these findings emerge regardless of the theoretical approach used.

Organizational innovation is closely linked to technological change. Organizational innovation is frequently an indispensable complement to the adoption of new technologies, especially ICT (see Ch. 5 by Lam in this volume). These complementary organizational changes critically affect the productivity and employment consequences of technological innovation and diffusion, especially in ICT.

One important consequence of technological innovation and adoption is change in the "skill bias" of employment. Unskilled jobs have long been declining in absolute terms (in Europe) or growing slowly (in the US), while jobs for more educated workers have been created at a faster pace in most countries and are associated with greater innovative efforts. Nevertheless, considerable uncertainty remains over the specific effects of ICT-related innovation since the 1980s on the skill requirements of new jobs, as opposed to coincidental changes in the supply of more educated labor.

Wage polarization has been significant, although the specific effects of new technologies on this trend are difficult to identify. Since the 1980s, most countries have experienced a growing divide within wages (and more generally in incomes). This is the result of changes in economic structures, in firms' strategies, and in government policies. Technological change has almost certainly contributed to this trend, but the specific link between innovation and wage differentials is unclear. Trends in wage dispersion and inequality reflect a number of broader, coincidental trends in the evolution of labor markets, employment forms, social relations, and national policies.

Aggregate demand and macroeconomic conditions are important. Although the role of demand has generally been downplayed in the innovation literature, it may help explain the negative employment impact of innovation found in Europe in the low-growth decade of the 1990s (see Ch. 18 by Verspagen in this volume).

Innovation interacts with trade. In open economies, trade is an important factor—alongside innovation—affecting employment and wages. The evidence available (mainly for the US) suggests that technological change has affected job losses, particularly among unskilled workers, and lower wages more substantially than has increased international trade. But the interactions between innovation and trade shape competitiveness (see Ch. 20 by Cantwell in this volume), the direction of technological change, the evolving division of labor, and the employment outcomes.

The national innovation system is a critical mediating influence in the effects of technological innovation and diffusion on employment. A country's technological opportunities and ability to develop long-term learning and innovating capabilities are rooted in the nature and characteristics of its national innovation system. Its strengths, orientation, and priorities are likely to be reflected in the employment effects of a country's innovative efforts (see Ch. 7 by Edquist in this volume).

Labor market conditions and institutions matter. The employment outcomes of technological change depend on the operation of job markets, the operation of institutions that set wages, and the institutions influencing worker learning, flexibility, and welfare protection. Labor market institutions also influence the supply of labor, which can affect the match between labor supply and the skill and competence requirements, as well as the employment effects, of new technologies.

21.5.2 Directions for Future Research

The following issues emerge from this chapter as promising directions for future studies.

Both theories and empirical studies should strive to develop explanations and predictions that are consistent at the various levels of analysis of the innovation–employment relationship—firms, industries, and the aggregate economy. All too often, the findings at one level of analysis contradict the patterns at a more aggregated level. Achieving this goal requires that firm-level data be representative of whole sectors, and further requires that industry coverage includes much if not all of the economy (including services). Innovation survey data, when accessible, could make this effort possible.

Studies on *firms* should address the innovation–employment question in the context of the evolution of firms and market structures through the processes of diversity generation (through innovation) and selection (in the marketplace, Nelson and Winter 1982). This avenue for research recognizes that the potential for job creation depends on mechanisms of firm growth and new firm generation.

An important challenge to innovation studies is to extend empirical research to cover the *service sector*, rather than confining such work to manufacturing industries. From a theoretical perspective, an analysis linking structural and technological (and organizational) change is especially important for understanding the expansion of services in advanced economies. The increasing availability of industry level data also for services (such as in European innovation surveys) makes this extension possible. Since most new jobs are created in services, this extension also is essential for policy.

The need to devote greater attention to *organizational innovations* and to link them to technological ones has already been pointed out; this appears as a major direction for future studies.

Bringing the analysis of the innovation–employment link into the *macroeconomic context of an open economy* is also important, in order to highlight the interactions between technological developments on the supply side, and growth potential on the demand side, as well as the interactions between innovation and trade in shaping growth and employment outcomes.

As *international production*, especially in manufacturing industries, becomes a key aspect of industrial structures, an additional level of analysis, looking at individual industries at the global level, may emerge. Data constraints are serious, but substantial evidence is now available on the international production networks of multinational corporations, and the link to employment variables across the relevant countries may become possible.

The appropriate *labor market* arrangements for favoring a virtuous circle between innovation and a high quantity and quality of jobs and wages, through greater learning, competence building and improved working conditions, is an additional area for future research.

Finally, as already suggested, research could address the *distributional effects* of innovation throughout the economy. So far, innovation has mainly benefited firms and consumers, in the form of higher profits and lower prices. When we consider also its effect on wage polarization, the result has been an increasingly uneven distribution of incomes, an issue with major policy relevance.

NOTES

1. For their comments and suggestions, I wish to thank the editors, Charles Edquist, Bengt-Åke Lundvall, Marco Vivarelli, and several contributors to this volume.
2. The emergence of new forms of organization, the opening of new markets and of new sources of materials were other types of innovation considered by Schumpeter.
3. Generally positive effects of indicators of innovation on the number of jobs (once the characteristics of firms have been controlled for) have been found in the studies of UK firms by Van Reenen (1997), who related a large panel of manufacturing firms to the SPRU database of British innovations, and by Machin and Wadhwani (1991), using the British workplace industrial relations survey on the adoption of ICTs. In a panel of German firms, Smolny (1998) has found a positive effect of product innovation and no effect of process innovations. Dutch firms investigated by Brouwer, Kleinknecht and Reijnen (1993) showed an overall negative link between innovation and jobs, but where product innovations were dominant, better employment outcomes have been found.
4. This view is proposed by Antonelli (2001: 173) in a study on the Italian car maker Fiat over the time period 1900–1970, which shows that employment growth Granger-caused patent growth with different time lags. In turn, patent growth led to productivity growth.
5. In France, Greenan and Guellec (2000) found a positive relationship between both product and process innovation and employment at the firm level, but at the industry level, only the former played a positive role. Using data from the French household survey, Entorf, Gollac, and Kramarz (1999) found that computer use reduces the risk of unemployment in the short term, but not in the long term.
6. High demand is a necessary, but not sufficient, condition for employment growth. In order to expand jobs, demand and output have to grow faster than productivity, a condition that has been often found in the US, but not so much in Europe.

7. Typical of the former approach is the OECD Jobs Study (OECD 1994), which investigated the high and persisting unemployment in advanced countries and downplayed the role of technological change. The report emphasized the positive role of new technologies associated to structural change and showed employment decline in low-technology sectors and also in some R&D intensive industries.

8. Meyer Kramer (1992) has used a model of the whole German economy to assess the impact of ICTs, proxied by direct and indirect R&D in 51 sectors. The findings suggest a generally negative employment impact, with some positive effects in higher technology industries.

9. An early use of these data for Italian manufacturing industry is in Vivarelli, Evangelista, and Pianta (1996), who found a generally negative employment impact of technological change and the expected contrasting consequences of product and process innovations. Pianta (2000, 2001) investigated five European countries (Denmark, Germany, Italy, the Netherlands, and Norway) in 1989–93 across twenty-one manufacturing industries, finding a positive employment effect of changes in demand and product innovations, and an overall negative impact of innovation intensity.

10. They argue that "The evidence supports the conclusion that an increase in the pace of innovation (all else equal) will raise both the natural rate of unemployment and the average length of time during which an unemployed worker is 'between jobs'" (p.10).

11. Their assumptions on productivity growth were based on the improvements in the engineering performance of robots. The long time-lags in their adoption and their poor initial performance have meant that their actual economic effect on productivity has been much lower.

12. Petit (1995) suggests that techniques and human labor had a complementary relationship in pre-industrial times, following strong social norms. Industrialization was characterized by the substitution of machines for labor on the basis of profit seeking; and the post-war rapid growth was associated with technologies designed to overcome labor shortages and increase productivity, while the current emergence of a new technological system based on ICTs is marked by uncertainties.

13. Computer-related investment per employee rose from 150 $US in 1982 to about 1000 $US in 1992 for both manufacturing and services (Howell 1996: 292). The share of low-skilled blue collar US workers in manufacturing declined from 45.1 per cent in 1978 to 39.7 per cent in 1982 and has since then remained stable, ending in 1990 with a 41 per cent share. In the late 1980s, when ICTs became important, the most serious reduction in the employment shares concerned the high-skilled blue collars (which since 1978, had a stable share around 21.7 per cent, a share which declined after 1985 to 19.7 per cent in 1990) and the low-skilled white collars (a share which stayed stable at around 12 per cent until 1985 and then declined to 10.6 per cent in 1990). At the opposite end, the share of high-skilled white collars increased in the 1978–83 period (from 19.5 to 24.1 per cent), then was stable until a rise in 1989–1990 to 25.8 per cent (Howell 1996: 299, tables 1 and 2).

14. This may amount to a shift in "wage norms": "in the face of mounting competition, employers reduced unit labour costs and increased flexibility in the production process by following the 'low road'—lower wages, little training, and fewer permanent employees" (Howell 1996: 301). Combined with a large use of part-time and temporary workers, anti-union practices, relocation to low-wage sites and inflows of low-wage foreign workers, these developments have led to an increase in the supply of labor competing

for low-skill jobs, leading to a major fall in their wages, accompanied by a 25 per cent reduction in the real minimum wage over the 1980s (ibid. 292).

15. There is also a "national specificity" to most of the different approaches. The "skills" research is largely US-focused, while most industry-level studies focus on Europe. There is little English-language empirical work on the employment–innovation relationship in Japan or other Asian high-growth economies.

REFERENCES

*ACEMOGLU, D. (2002), "Technical Change, Inequality and the Labor Market," *Journal of Economic Literature*, 40(1): 7–72.

*ADDISON, J., and TEIXEIRA, P. (2001), "Technology, Employment and Wages," *Labour* 15(2): 191–219.

ADLER, P. (ed.) (1992), *Technology and the Future of Work*, New York: Oxford University Press.

ANTONELLI, C. (2001), *The Microeconomics of Technological Systems*, Oxford: Oxford University Press.

ANTONUCCI, T., and PIANTA, M. (2002), "The Employment Effects of Product and Process Innovations in Europe," *International Review of Applied Economics* 16(3): 295–308.

APPELBAUM, E., and BATT, R. (1994), *The New American Workplace*, Ithaca: ILR Press.

AUTOR, D., KATZ, L., and KRUEGER, A. (1998), "Computing Inequality: Have Computers Changed the Labor Market?" *Quarterly Journal of Economics* 113: 1169–214.

BARTEL, A. P., and LICHTENBERG, F. R. (1991), "The Age of Technology and its Impact on Employee Wages," *Economics of Innovation and New Technology* 1: 215–31.

—— and SICHERMAN, N. (1999), "Technological Change and Wages: An Interindustry Analysis," *Journal of Political Economy* 107: 285–325.

BAUMOL, W., and WOLFF, E. (1998), *Side Effects of Progress: How Technological Change Increases the Duration of Unemployment*, Jerome Levy Economics Institute of Bard College, Public Policy Brief 41.

*BERMAN E., BOUND, J., and GRILICHES, Z. (1994), "Changes in the Demand for Skilled Labor Within US Manufacturing Industries: Evidence from the Annual Survey of Manufactures," *Quarterly Journal of Economics* 109: 367–98.

———— and MACHIN, S. (1998), "Implications of Skill Biased Technological Change: International Evidence," *Quarterly Journal of Economics* 113: 1245–79.

BLACK, S., and LYNCH, L. (2000), "What's Driving the New Economy: The Benefits of Workplace Innovation," NBER working paper 7479.

BOYER, R. (1988a), "Technical Change and the Theory of Régulation," in Dosi et al. 1988: 67–94.

—— (1988b), "Formalizing Growth Regimes," in Dosi et al. 1988: 608–35.

BRAVERMAN, H. (1974), *Labour and Monopoly Capital*, New York: Monthly Review Press.

* Asterisked items are suggestions for further reading.

BRESNAHAN, T. F., BRYNJOLFSSON, E., and HITT, L. M. (2002), "Information Technology, Workplace Organization and the Demand for Skilled Labor: Firm-Level Evidence," *Quarterly Journal of Economics* 117: 339–76.

BROUWER, E., KLEINKNECHT, A., and REIJNEN, J. O. N. (1993), "Employment Growth and Innovation at the Firm Level: An Empirical Study," *Journal of Evolutionary Economics* 3: 153–9.

BRYNJOLFSSON, E., and HITT, L. M. (2000), "Beyond Computation: Information Technology, Organisational Transformation and Business Performance," *Journal of Economic Perspectives* 14: 23–48.

CAROLI, E. (2001), "New Technologies, Organizational Change and the Skill Bias: What do we Know?" in Petit and Soete 2001: 259–92.

—— and VAN REENEN, J. (2001), "Skill Biased Organizational Change? Evidence From a Panel of British and French Establishments," *Quarterly Journal of Economics* 116(4): 1149–92.

CASAVOLA, P., GAVOSTO, A., and SESTITO, P. (1996), "Technical Progress and Wage Dispersion in Italy: Evidence from Firms' Data," *Annales d'Economie et de Statistique* (Jan.–June): 387–412.

CHENNELLS, L., and VAN REENEN, J. (1999), "Has Technology Hurt Less Skilled Workers? An Econometric Survey of the Effects of Technical Change on the Structure of Pay and Jobs," London: Institute for Fiscal Studies working paper 27.

CYERT, R. M., and MOWERY, D. C. (eds) (1988), "The Impact of Technological Change on Employment and Economic Growth," Cambridge, Mass.: Ballinger.

DOMS, M., DUNNE, T., and TROTSKE, K. (1997), "Workers, Wages, and Technology," *Quarterly Journal of Economics* 112: 253–89.

DOSI, G., FREEMAN, C., NELSON, R., SILVERBERG, G., and SOETE, L. (eds), (1988), *Technical Change and Economic Theory*, London: Pinter.

EDQUIST, C., HOMMEN, L., and McKELVEY, M. (2001), *Innovation and Employment: Product versus Process Innovation*, Cheltenham: Edward Elgar.

ENTORF, H., GOLLAC, M., and KRAMARZ, F. (1999), "New Technologies, Wages and Worker Selection," *Journal of Labor Economics* 17(3): 464–91.

EUROPEAN COMMISSION-EUROSTAT (2001), *Statistics on Innovation in Europe. Data 1996–1997*, Luxembourg: European Commission.

EUROPEAN FOUNDATION FOR THE IMPROVEMENT OF LIVING AND WORKING CONDITIONS (2001), *Third European Survey on Working Conditions 2000*. Dublin: European Foundation for the Improvement of Living and Working Conditions.

EVANGELISTA, R., and SAVONA, M. (2002), "The Impact of Innovation on Employment and Skill in Services. Evidence from Italy," *International Review of Applied Economics*.

————(2003), "Innovation, Employment and Skills in Services. Firm and Sectoral Evidence," *Structural Change and Economic Dynamics* 14: 449–74.

FAGERBERG, J., GUERRIERI, P., and VERSPAGEN, B. (eds) (1999), *The Economic Challenge for Europe: Adapting to Innovation Based Growth*, Northampton: Elgar.

*FREEMAN, C., and LOUÇÃ, F. (2001), *As Time Goes By. From the Industrial Revolution to the Information Revolution*, Oxford: Oxford University Press.

—— and SOETE, L. (1994), *Work for All or Mass Unemployment?* London: Pinter

————(eds.) (1987), *Technical Change and Full Employment*, Oxford: Basil Blackwell.

——CLARK, J., and SOETE, L. (1982), *Unemployment and Technical Innovation*, London: Pinter.

FREY, L. (1997), "Il lavoro nei servizi verso il secolo XXI," *Quaderni di Economia del Lavoro* 57, Milan: Angeli.

GREENAN, N. (2003), "Organisational Change, Technology, Employment and Skills: An Empirical Study of French Manufacturing," *Cambridge Journal of Economics* 27: 287–316.

—— and GUELLEC, D. (2000), "Technological Innovation and Employment Reallocation," *Labour* 14(4): 547–90.

*GREENAN, N., L'HORTY, Y., and MAIRESSE, J. (eds.) (2002), *Productivity, Inequality and the Digital Economy*, Cambridge: Mass.: MIT Press.

HANSEN, A. H. (1964), *Business Cycles and National Income*, London: Allen and Unwin (expanded edn.).

HEERTJE, A. (1973), *Economics and Technical Change*, London: Weidenfeld and Nicolson.

HOWELL, D. (1996), "Information Technology, Skill Mismatch and the Wage Collapse: A Perspective on the US Experience," in OECD 1996: 291–306.

—— and WOLFF, E. (1992), "Technical Change and the Demand for Skills by US Industries," *Cambridge Journal of Economics* 16: 128–46.

IPTS-ESTO (Institute for Prospective Technology Studies, European Science and Technology Observatory) (2001), *Impact of Technological and Structural Change on Employment. Prospective Analysis 2020, Synthesis Report and Analytical Report*, Seville: European Commission Joint Research Centre.

KALMBACH, P., and KURZ, H. D. (1990), "Microelectronics and Employment: A Dynamic Input–Output Study of the West German Economy," *Structural Change and Economic Dynamics* 1: 317–86.

KARAOMERLIOGLU, A., and ANSAL, T. (2000), "Compensation Mechanisms in Developing Countries," in Vivarelli and Pianta 2000: 165–81.

KATSOULACOS, Y. S. (1986), *The Employment Effect of Technical Change*, Brighton: Wheatsheaf.

*KRUEGER, A. (1993), "How Computers have Changed the Wage Structure: Evidence from Micro Data 1984–1989," *Quarterly Journal of Economics* 108: 33–60.

LAYARD, R., and NICKELL, S. (1985), "The Causes of British Unemployment," *National Institute Economic Review* 111: 62–85.

LEONTIEF, W., and DUCHIN, F. (1986), *The Future Impact of Automation on Workers*, Oxford: Oxford University Press.

LUNDVALL, B. Å. (1987), "Technological Unemployment in a Small Open Economy," in R. Lund, P. Pedersen, and J. Schmidt-Sorensen (eds), *Studies in Unemployment*, Copenhagen: New Social Science Monographs, Institute of Organisation and Industrial Sociology, 21–48.

MACHIN, S. (1996), "Changes in the Relative Demand for Skills," in A. Booth, and D. Snower (eds), *Acquiring Skills*, Cambridge: Cambridge University Press, 129–46.

—— and VAN REENEN, J. (1998), "Technology and Changes in Skill Structure: Evidence from Seven OECD Countries," *Quarterly Journal of Economics* 113: 1215–44.

—— and WADHWANI, S. (1991), "The Effects of Unions on Organisational Change and Employment: Evidence from WIRS," *Economic Journal* 101: 324–30.

MEGHIR, C., RYAN, A., and VAN REENEN, J. (1996), "Job Creation, Technological Innovation and Adjustment Costs: Evidence from a Panel of British Firms," *Annales d'economie et statistique* 41–2: 256–73.

MEYER-KRAMER, F. (1992), "The Effects of New Technologies on Employment," *Economics of Innovation and New Technology* 2: 131–49.

NELSON, R., and PHELPS, E. (1966), "Investment in Humans, Technological Diffusion and Economic Growth," *AEA Papers and Proceedings* 56: 69–75.

—— and WINTER, S. (1982), *An Evolutionary Theory of Economic Change*, Cambridge, Mass.: Belknap Press.

NOBLE, D. (1984), *Forces of Production: A Social History of Industrial Automation*, New York: Knopf.

OECD (1994), *The OECD Job Study. Evidence and Explanations. Part I, Labour Market Trends and Underlying Forces of Change*, Paris: OECD.

—— (1996), *Employment and Growth in the Knowledge-Based Economy*, Paris: OECD.

PADALINO, S., and VIVARELLI, M. (1997), "The Employment Intensity of Economic Growth in the G-7 Countries," *International Labour Review* 136: 191–213.

PASINETTI, L. (1981), *Structural Change and Economic Growth*, Cambridge: Cambridge University Press.

PEREZ, C. (1983), "Structural Change and the Assimilation of New Technologies in the Economic and Social Systems," *Futures* 15(5): 357–75.

*PETIT, P. (1995), "Employment and Technological Change," in P. Stoneman (ed), *Handbook of the Economics of Innovation and Technological Change*, Amsterdam: North Holland, 366–408.

*—— and SOETE, L. (eds) (2001*a*), *Technology and the Future of European Employment*, Cheltenham: Edward Elgar.

—— —— (2001*b*), "Technical Change and Employment Growth in Services: Analytical and Policy Challenges," in Petit and Soete 2001*a*: 166–203.

PIANTA, M. (2000), "The Employment Impact of Product and Process Innovation," in Vivarelli and Pianta 2000: 77–95.

—— (2001), "Innovation, Demand and Employment," in Petit and Soete 2001*a*: 142–65.

—— EVANGELISTA, R., and PERANI, G. (1996), "The Dynamics of Innovation and Employment: An International Comparison," *Science, Technology Industry Review* 18: 67–93.

PINI, P. (1995), "Economic Growth, Technological Change and Employment: Empirical Evidence for a Cumulative Growth Model with External Causation for Nine OECD Countries, 1960–1990," *Structural Change and Economic Dynamics* 6: 185–213.

—— (1996), "An Integrated Cumulative Growth Model: Empirical Evidence for Nine OECD Countries, 1960–1990," *Labour* 10: 93–150.

PIVA, M., and VIVARELLI, M. (2002), "The Skill Bias: Comparative Evidence and an Econometric Test," *International Review of Applied Economics* 16(3): 347–58.

RICARDO, D. (1951), *Principles of Political Economy and Taxation*, in P. Sraffa (ed.), *The Works and Correspondence of David Ricardo*, vol. 1, Cambridge: Cambridge University Press (3rd edn. 1821).

SANDERS, M., and TER WEEL, B. (2000), "Skill Biased Technical Change: Theoretical Concepts, Empirical Problems and a Survey of the Evidence," Druid working paper, Copenhagen Business School and Aalborg University.

SCHUMPETER, J. A. (1934), *Theory of Economic Development*, Cambridge, Mass.: Harvard University Press (1st edn. 1911).

SHAIKEN, H. (1984), *Work Transformed: Automation and Labor in the Computer Age*, New York: Holt, Rinehart and Winston.

SIMONETTI, R., TAYLOR, K., and VIVARELLI, M. (2000), "Modelling the Employment Impact of Innovation: Do Compensation Mechanisms Work?" in Vivarelli and Pianta 2000.

SIMONETTI, R. and TANCIONI, M. (2002), "A Macroeconometric Model for the Analysis of the Impact of Technological Change and Trade on Employment," *Journal of Interdisciplinary Economics* 13: 185–221.

SMOLNY, W. (1998), "Innovation, Prices and Employment: A Theoretical Model and an Application for West German Manufacturing Firms," *Journal of Industrial Economics* 46: 359–81.

SPIEZIA, V., and VIVARELLI, M. (2002), "Innovation and Employment: A Critical Survey," in Greenan, L'Horty, and Mairesse 2002: 101–31.

SYLOS LABINI, P. (1969), *Oligopoly and Technical Progress*, Cambridge, Mass.: Harvard University Press (1st edn. 1956).

VAN REENEN, J. (1996), "The Creation and Capture of Economic Rents: Wages and Innovation in a Panel of UK Companies," *Quarterly Journal of Economics* 111(1): 195–226.

—— (1997), "Employment and Technological Innovation: Evidence from U.K. Manufacturing Firms," *Journal of Labor Economics* 15: 255–84.

—— EVANGELISTA, R., and PIANTA, M. (1996), "Innovation and Employment in the Italian Manufacturing Industry," *Research Policy* 25: 1013–26.

*VIVARELLI, M. (1995), *The Economics of Technology and Employment: Theory and Empirical Evidence*, Aldershot: Edward Elgar.

*and PIANTA, M. (eds.) (2000), *The Employment Impact of Innovation: Evidence and Policy*, London: Routledge.

WHITLEY, J. D., and WILSON, R. A. (1982), "Quantifying the Employment Effects of Micro-Electronics," *Futures* 14(6): 486–95.

WOLFF, E. (1996), "Technology and the Demand for Skills," *Science Technology Industry* 18: 95–124.

CHAPTER 22

....

SCIENCE, TECHNOLOGY, AND INNOVATION POLICY

....

BENGT-ÅKE LUNDVALL
SUSANA BORRÁS

22.1 INTRODUCTION

....

THIS chapter is about what governments have done and could do to promote the production, diffusion, and use of scientific and technical knowledge in order to realize national objectives.

We begin the chapter with "story-telling" based on sketchy historical facts. The aim of the two stories is to illustrate that innovation policy covers a wide set of issues that have been on the agenda far back in history while still remaining important today. We move on to sketch the history of innovation policy, splitting it up into the three ideal types: science, technology, and innovation policy. We use OECD documents and other sources to do so. Finally we point to future challenges, and highlight research opportunities.

22.1.1 From Guns...

One important early example of successful technology policy was the initiative by Henry VIII to develop the competitive production of cannons made out of iron in England in the first half of the sixteenth century. The reason for establishing this technology policy program was that England desperately needed more cannons to win the ongoing war with France, and that the bronze cannons at use were too scarce and too expensive (Yakushiji 1986).

One factor making the success of the program possible was that England had access to iron ore as well as forests for fuel. Another important factor was that skilled people who were expert forgers had moved to England from the Continent, especially from France. Some came because they had been expelled because of their religious faith, while others were brought to England because of their unique skills. The program was realized through establishing a consortium under the leadership of William Leavitt, head of the royal iron works in Newbridge, and bringing together expertise coming from different countries, especially founders with their roots in France.

Many of the instruments of technology and innovation policy that were used at that time are still in use today. Technological diffusion through immigration was important. Competition policy and property right regimes were manipulated, and public procurement used. After a period when several competing companies were allowed to export cannons (pirates were among the more advanced users!) monopolist rights were guaranteed to one producer (Ralph Hogge). Barriers to the movement of skilled people to the Continent and export licensing were established to make sure that only Protestant countries got access to the technology. Public procurement was certainly a policy instrument that promoted the refinement of the technology, in this case with a national security objective.

22.1.2   ...to Butter

In the middle of the nineteenth century the Danish economy was highly dependent on the export of corn to England. Danish rye served as the "fuel" driving the English horse-based transport system. In the second half of the century and especially in the 1870s a dramatic reduction in overseas transport costs gave Russia and the US easier access to this and other markets. Prices were brought down to a level where Danish farming could not compete anymore. A major agricultural crisis broke out.

At that time there was already some export of animal products including butter especially from the manor houses. But the most important barrier was the uneven quality of the products. Standardization was a key element in the transformation of Danish production and exports away from corn toward butter and other animal products. This transformation involved a combination of technical and social innovation and a combination of state and "non-state intervention." The public policies involved were "diffusion oriented" rather than "mission oriented." The episode made deep imprints on the Danish innovation system that are still possible to identify.

The most important technology brought into use was related to dairy processing. To separate the fat from the milk the "separator" was developed in Sweden, however it was first produced in Denmark and first brought into use in Danish dairies. But the single most important innovation driving the transformation was social rather than technical and it came neither from the state nor from the market. Grundtvig— priest, philosopher, and nationalist—was a key person who founded a social movement that changed the mode of production in Danish agriculture. In his writings and in the lectures he held all over the country, he emphasized the need for farmers to get educated and take on responsibility for their own fate. Over a short timespan local "folk high schools" spread quickly in the countryside.

Inspired by this ideology and forced by economic necessity farmers got engaged in cooperative ownership around new local dairy plants. In the 1880s, the number of cooperatively owned dairies grew from three to 700. Between 1850 and 1900 the share of butter in Danish exports to the UK went from 0 per cent to 60 per cent. The cooperative form of ownership created a framework supportive for rapid diffusion of dairy techniques and for standardization.

State policies supported these developments. The Agricultural University in Copenhagen was established as early as 1856 (Wagner 1998). Dairy consultants played an important role in the process of diffusing good practice and on this basis the dairy oriented Agricultural Research Laboratory was established in 1883.

This example might be of special importance for developing countries with "soft states." In these countries ordinary state-led science, technology and innovation policy might not be sufficient to overcome the obstacles for economic development. A broader social mobilization might be necessary in order to overcome barriers to socio-economic development.

22.1.3 Modern Innovation Policy between Guns and Butter

If we go to the leading economy in the world in terms of technology—the US—we find elements in the innovation system that remind us of both these stories. The successful development of the atomic bomb at Los Alamos has many characteristics in common with the "gun-story." It was a crash-mission project aimed at winning a war and it brought together skilled people from different parts of the world. It also set the agenda for much of what followed in terms of technology policy not only in the US but all over the world.

But the US innovation system does not have characteristics in common solely with the gun-story. One of the most successful examples of innovation policy in the US has been the upgrading of agricultural activities in terms of products and productivity. The establishment of land universities and extension services were crucial for training, and also for research and development of new technologies and products. It was especially important for the rapid diffusion of new ideas among farmers. In spite of this success there have been few attempts to introduce similar diffusion oriented policies in relation to manufacturing and services. The US now has a "Manufacturing Extension Partnership," created in the late 1980s, which seems likely to survive politically although its effectiveness is uncertain. When attempts were made to apply the model to construction industries in the US, the attempts failed (Nelson 1982).

22.2 SCIENCE POLICY, TECHNOLOGY POLICY, AND INNOVATION POLICY

As can be seen from the other chapters in this book—and from the two stories told above—innovation policy covers a wide range of initiatives and it is necessary to give some structure to the complex reality. One way to do so is to introduce "ideal types" that bring out in a more distinct form classes of phenomena that are muddled and mixed in the real world.

The distinction to be used here is between science policy, technology policy, and innovation policy. For each we will discuss how it became an explicit policy field, what the major issues at stake are, what part of the innovation system is in focus, which are the major actors involved and what are the instruments used. We will point to the critical sets of data supporting the policies. And since the OECD has played a unique role among international organizations in the diffusion of ideas about innovation policy we will use OECD documents to illustrate some of the more important shifts in the post-war debate.

It would be misleading to argue that we pass from science policy to technology policy and then to innovation policy as we pass from one historical stage to another. For instance some of the classical issues in science policy are very high on the current policy agenda (Pavitt 1995; Martin and Salter 1996). Current developments in biotechnology and pharmaceuticals have made the distance from basic research to commercial application much shorter and therefore the organization of universities has become a major issue (see Ch. 8 by Mowery and Sampat). Genetic engineering

Box 22.1 OECD and the evolving discourse around science, technology and innovation policy

The OECD has played a key role in the evolution of the understanding of the policy fields discussed in this chapter. It is certainly one of the best sources for internationally comparable data on science technology and innovation. Data are accessible through regular publications in the form of periodical policy reviews and through data bases that are regularly updated. But it is also interesting to follow the policy discourse organized at the OECD secretariat. What has been said at OECD meetings and recommended by its expert groups might not always be transformed into practical use in member countries but it reflects the new ideas. Therefore a brief summary of the most important reports (in fact they all tend to arrive at the beginning of a new decade) will be made here.

OECD 1963: Rationalizing science policy and linking it to economic growth. In an OECD document from the beginning of the sixties (OECD 1963a) a shift in attention toward economic objectives was signaled for science policy. This document with Christopher Freeman, Raymond Poignant, and Ingvar Svennilsson as major contributors is quite remarkable in the emphasis it gives to national and rational planning. It came the same year as the Frascati meeting on a new manual for gathering R&D statistics took place and it argues for a strong link between better data on R&D and more systematic policy. This report obviously aimed at giving science policy legitimacy outside the narrow circles of ministries of education and science.

OECD 1970: Bringing in human and social considerations on technology policy. This report (the "Brooks Report") introduced a broader social and ecological perspective to science and technology policy. It also gave strong emphasis to the need to involve citizens in assessing the consequences of developing and using new technologies. The new focus reflected a combination of growth satiation and dissatisfaction with the social consequences of technical change. Therefore it was also assumed that room should be given for wider concerns and the uncritical optimism was challenged.

OECD 1980: Innovation policy as an aspect of economic policy. The OECD expert report "Technical Change and Economic Policy" (OECD 1980) redefined the agenda for innovation policy in the light "of a new economic and social context." Its message was that the slow-down of growth and the increase in unemployment could not be seen as something that macroeconomic expansionary policies could solve. Neither did the report restrict itself to recommending an increase in investments in science and R&D. As compared to earlier reports on science and technology, the focus was moved toward the capacity of society to absorb new technology. The experts included, among others, Freeman, Nelson, and Pavitt. It is interesting to note that this report was published simultaneously with a more traditional document on innovation policy (1980). This was obviously a period when innovation policy began to be regarded as a legitimate policy field.

OECD 1990: Innovation defined as an interactive process. The TEP project, edited and given its flavor by Francois Chesnais and with Luc Soete as one of the leading external experts brought together in one coherent framework a broad set of new research results on innovation and used those to point in new policy directions

(OECD 1990). The report took as its starting point "innovation as an interactive process" and gave a prominent role to national innovation systems as an organizing concept. There was a stronger emphasis on network formation, new forms of organization and industrial dynamics than in earlier OECD contributions to innovation policy. The demand side and strengthening the absorptive capacity of firms as well as the feedback from users to the supply side were given strong emphasis. This report gave the systemic version of innovation policy an analytical foundation.

OECD 2001: **The new economy beyond the hype.** In the middle of the 1990s the idea of a new economy based upon ICT and vibrant entrepreneurship began to be diffused in the US with Federal Reserve President Alan Greenspan as an important source. At the end of the millennium OECD began to analyze the phenomenon and its first reports came in 2001. These reports were interesting because they were coordinated by the Economics Department at OECD that so far had treated innovation as a secondary phenomenon and had been a spokesman for a pure market innovation policy. The catalog of recommendations with regard to innovation policy is more extensive than what OECD economists had ascribed to so far—see for instance the Jobs Study (OECD 1995). The new economy episode is interesting because it is the first time that innovation becomes widely accepted among economists as a fundamental factor that needs to be analyzed and understood. At the same time, it is clear that the basic hypothesis was too simple and the policy recommendations remained colored by the traditional pro-market and anti-state philosophy of the OECD economists.

brings in fundamental issues about ethics that we normally would connect with science policy. In the real world the forms overlap and mix.

22.2.1 From Science Policy...

The two historical examples given above belong to the realm of technology policy rather than science policy. Science policy is a concept that belongs to the post-war era. Before the war, regional and federal governments were funding university research and the training of scientists. But they did so primarily for historical and cultural reasons and, before World War II, the idea of science as a productive force was taken up mainly in the planned economies.

According to Christopher Freeman science policy was recognized as a policy area through the pioneering work by Bernal (1939). Bernal was a pioneer in measuring the R&D effort at the national level in England and he strongly recommended a dramatic increase in the effort since he was convinced that it would stimulate economic growth and welfare. In the US, the Vannevar Bush report from 1945, "Science: The Endless Frontier," has a specific status in defining an agenda for the US post-war science (and technology) policy. It defined the task for science policy as contributing to national security, health and economic growth. Like Bernal, the Bush report gave strong emphasis to the potential economic impact of investments in science.

The real reason for the breakthrough of science policy and the increased public investment in research was probably the way World War II ended and the Cold War started. The success of the Los Alamos project made plausible the idea that a massive investment in science (especially physics but also chemistry and biology), applied science and technological development could produce solutions to almost any difficult problem, and underscored the importance of science and technology for national security. This pressure to invest in the promotion of science was reinforced in the arms and the space races between the US and Soviet Union. In 1957 the launch of Sputnik put extra pressure on the West, and especially the US, to invest in defense and space-related research.

The major issues in science policy are about allocating sufficient resources to science, to distribute them wisely between activities, to make sure that resources are used efficiently and contribute to social welfare. Therefore, the quantity and quality of students and researchers receives special attention. The objectives for science policy that are actually pursued by governments are mixed and include national prestige and cultural values besides social, national security, and economic objectives.

The elements of the innovation system that are focused upon are universities, research institutions, technological institutes, and R&D laboratories. Science policy is both about the internal regulation of these parts of the innovation system and about how they link up to the environment—not least to government and industry. However, strengthening this linkage becomes even more crucial in technology and innovation policy.

There are two more or less standing debates within the science policy community. The first debate has to do with how far scientific progress is identical with progress in general. Critical scholars would point to how science is abused in the control of people and nature—including genetic manipulation and undermining ecological sustainability. Those more positive to science would respond that none of this can be seen as emanating from science: rather, it should be seen as the result of unwise use of science. Scholars from humanities and sociology might more frequently join the first camp while scientists and technologists have a tendency to join the second.

The second debate is about to what degree science should be made the obedient servant of the state and/or capital and to what degree it should be autonomous. Sociologists of knowledge have devoted considerable attention to this question, also in relation to different national research policy styles—with more or less governmental steering and with different levels of aggregation—large/small research activities (Rip and van der Meulen 1997; Jasanoff 1997). Others have argued that there is a shift in the mode of research and knowledge production, with different effects upon the autonomy of science (Cozzens et al. 1990; Gibbons et al. 1994; Jasanoff 2002).

University scholars tend to argue that "freedom" and "autonomy" of academic research is important for at least two reasons. One is the long-term value of serendipity—only when basic research is allowed to move along its own trajectories will it produce the unsuspected that can open up new avenues for applied research and technical solutions. The second reason is that critical science is an important element in modern democracy, because scientific knowledge from independent sources is an important input for open, transparent and representative political decision making. Most scholars of innovation would agree that this is an area where there is a trade-off. While the idea of basic research as "free" science—signaling a complete absence of direction and use—is an illusion, the massive subordination of science under political and economic interests would certainly undermine its long-term contribution to society and economy.

The main policy actors in the public sector are ministries of education and research and research councils. But sector ministries in charge of health, defense, energy, transport, and environment may also play a role since they organize their own research communities, and in some industrial economies account for the majority of public spending on R&D. Ministries of finance play a role when it comes to decide the total budget for research. Civil organizations representing consumers and citizens may be invoked as corrections to a bias in favor of commercial interests.

The instruments used are budgetary decisions on allocating funds to public research organizations, such as universities, and subsidies or tax relief for private firms. Finding institutional mechanisms that link universities and public laboratories to the users of research is of course a fundamental issue. But it becomes even more so when we turn to technology and innovation policy and therefore we will save the discussion of such instruments until later. Designing intellectual property rights for universities has recently become a major issue (see Chapters 10 by Granstrand and 8 by Mowery and Sampat in this volume).

The evaluation of research is an important policy tool and it can be seen both as creating incentives for scholars and institutions to become more effective and as a means to allocate public money. Academic life has built evaluation procedures into it. To make an academic career exams must be passed and books and articles will be peer reviewed. At well-functioning departments weekly seminars expose ongoing work to criticism from colleagues. These days such processes increasingly involve international expertise.

The criteria used for these ongoing evaluations are mainly internal to scientific communities and they might be regarded as misdirected by external authorities. Evaluation by peers organized according to disciplines may promote "middle of the road" work rather than new ideas coming from crossing disciplinary borders. Where this is the case, it might be helpful to establish alternative sources of regarding interdisciplinary efforts more kindly. The European framework programs for R&D have represented an opening in this direction for European researchers.

Policy makers may also see internal evaluation as being too slack. This seems to have been the case in the UK where a very ambitious and detailed reporting system has been imposed on scholars doing research and teaching. So far the effect seems to be a lot of time used to report on research and that in England university scholars have become the professional group most unsatisfied with working conditions. While this kind of draconian reform linking quantitative indicators of performance to access to public money might be useful in shaking up conservative institutions for a brief period they tend to become a nuisance for the innovation system as a whole if not loosened up again.

One of the most fundamental questions in science policy is whether it is true that "good research is always useful research" or, more demanding, "the higher the scientific quality the more useful is the research." If it were true it would be a strong argument for leaving at least some of the allocation to the academic communities. However, the evidence on this question is contradictory. While Zucker and Darby (1998) demonstrate that star scientists are important for the success of biotechnology firms, Gittelman and Kogut (2003) show that at least within biotech there is no clear relationship between prestigious publishing and high impact innovations—if anything the relationship is negative (Mowery and Sampat 2001).

It might be expected that the requirements between doing outstanding research and creating well-designed technology are even more distinct in other fields where the distance from scientific discovery to innovation is longer. The truth may be that also science has a lot to learn from interacting with its users and that the best system is one with several layers and with career shifts—some going only for "excellence" within a scientific discipline, others focusing only on user-needs and some operating in a mode between the two.

Since the early 1990s, basic science has been constantly pushed by politicians to demonstrate its social and economic usefulness. This has been termed the "new social contract of basic science" in the age of budgetary retrenchments, particularly in Europe and the US (Martin and Salter 1996). Several authors argue that this "value for money" attitude disregards some essential aspects of science policy, such as the training of scientists and technicians, and the development of knowledge capabilities in areas where uncertainty about exploitation is so high that private investors lack incentives (Sharp 2003) and that lessons from the US experience should be examined critically, and along with for instance the Scandinavian and Swiss experiences (Pavitt 2001).

22.2.2 ... to Technology Policy

Technology policy refers to policies that focus on technologies and sectors. The era of technology policy is one where especially science-based technologies such as nuclear power, space technology, computers, drugs and genetic engineering are seen as being at the very core of economic growth. These technologies get into focus for several reasons. On the one hand they stimulate imagination because they make it possible to do surprising things—they combine science with fiction. On the other hand they open up new commercial opportunities. They are characterized by a high rate of innovation and they address rapidly growing markets.

Technology policy means different things for catching-up countries than it does for high-income countries and it might also mean different things for small and big countries. In big high-income countries the focus will be on establishing a capacity in *producing* the most recent science-based technologies, as well as applying these innovations. In smaller countries it might be a question about being able to absorb and use these technologies as they come on the market. Catching-up countries may make efforts to enter into specific promising established industries using new technologies in the process of doing so.

Common for these strategies is that they tend to define "strategic technologies" and sometimes the sectors producing them are also defined as strategic sectors. The idea of strategic sectors may be related to Perroux and to Hirschman, both students of Schumpeter. Perroux used concepts such as "industrializing industries" and growth poles while Hirschman introduced unbalanced growth as a possible strategy for less-developed countries (Perroux 1969; Hirschman 1969).

In the lead countries, government initiatives of the technology policy kind were triggered when national political or economic interests were threatened and the threats could be linked to the command of specific technologies. Sputnik gave extra impetus to a focus on space technology and the Cold War motivated the most ambitious technology policy effort ever in the US. In Europe, Servan Schreiber's book *Le Défi américain* (Servan Schreiber 1967) gave a picture of a growing dominance of the US multinational firms especially in high technology sectors. It gave the big European countries such as France, the UK, and Germany incentives to develop a policy of promoting national champions in specific sectors. A specific important event triggering French and, later, European efforts was the export embargo of computer technology that was seen in France as blocking its progress in the development of nuclear technology.

The motivation behind the technology policy in Japan—and later on in countries such as Taiwan and Korea—is different. It is driven by a national strategy aiming at catching up and in the Japanese case it has roots back to the Meiji revolution when the first ideas of modernization based on imitating the technology of the West were formed.

At this point we need to be aware of a number of fundamental questions regarding technology policies.

- Is it at all legitimate and effective for the state to intervene for commercial reasons in promoting specific sectors or technologies? Or is the only legitimate technology policy one where national societal issues are at stake, including establishing national military power? It is a paradox that in the country having the most massive public intervention in terms of technology policy (the US), most of the policy has been motivated by non-commercial arguments and the discourse has been anti-state. Japan is the country with the most explicitly commercially driven technology policy with a recognition of a role for the state, and here the intervention has been much more modest, at least in terms of the amount of public money involved.

- What technologies should be supported? Is it always the case that high-tech and science-based sectors should be given first priority? Again the Japanese government as well as governments in smaller countries has been more apt to think about the modernization of old industries than the US and the big European countries.

- At what stage should the support be given? Should it be given only to "precompetitive" stages or should it also be helpful in bringing the new products to the market? In the second case there might be a combination of government support of new technology and more or less open protectionism.

- What limits should be set for public sector competence? Technology policy may be pursued with competence where government operates as a major user but when it comes to developing new technologies for the market, the role of governments must be more modest. To be more specific, there are several historical examples of how government ambitions to make technological choices that reduce diversity have ended in failure, for example, the "minitel" experience in France, and the high definition TV policy in the EU, both in the early 1990s.

- How can promoting a technology or a sector best be combined with competition? The period in the 1980s of promoting single firms as national champions in the bigger European countries was not a great success while the Japanese public strategy to promote "controlled competition" among a handful of firms was more successful.

The objectives of technology policy are not very different from those of science policy but—at least to begin with—it represented a shift from broader philosophical considerations to a more instrumental focus on national prestige and economic objectives. Technology policies were developed in an era of technology optimism. But later on—in the wake of the 1968 student revolt—more critical and broader concerns relating to technology assessment and citizen participation came onto the agenda (OECD 1971).

The elements of the innovation system in focus remain universities, research institutions, technological institutes, and R&D laboratories. But the attention moves from universities toward engineering and from the internal organization of universities toward how they link to industry. Technology policy may go even further and include the commercialization of technologies, but then we approach what we will call innovation policy.

In some countries such as the US, the main technology policy actors in the public sector are sector ministries promoting and sometimes procuring technology for purposes of telecommunications, defense, health, transport, energy etc. while in others, such as Japan, they are ministries in charge of industry and trade. Ministries of education and research are important since they organize the education and training of scientists and engineers. Authorities in charge of regulating competition as well as other regulating authorities may have a major impact on technology policy and on technological development. Public authorities may, as elements of technology policy, organize technology assessment and other ways of involving citizens.

There are many possible instruments to be used in promoting specific technologies and sectors. Most efficient may be combinations of instruments in fields where public procurement is involved. When the government has the leading user competence, it is in a better position to judge what kind of instruments will work (Edquist et al. 2000). Besides public procurement direct economic incentives in terms of subsidies and tax reductions may be offered to firms. Supporting research at universities in the science fields in which the new technologies are rooted may be an important part of a public mission policy. The danger of these kinds of policies is that "industrial complexes" combining the vested interests of a segment of public users with those of a segment of industry emerge and that a lack of transparency is exploited by vested interests. A more subtle problem is the kind of convergence and agreement on the direction of technological trajectories that might develop in such complexes, excluding new and more promising venues (Lundvall 1985).

In areas where the main application of the new technologies is commercial, the set of instruments used may be a combination of sector or technology specific economic incentives with more or less protectionist trade policy. An example might be the high definition TV policy of the EU in the early 1990s, where the attempt to define a compulsory analogical standard would have been a technical trade barrier to emerging digital standards, combined with specific economic incentives for European producers. Such packages may create a sheltered atmosphere for the firms involved. More promising may be project-organized support bringing different firms and knowledge institutions together in order to focus on generic and common and *new* technological problems while making sure that the use of the new knowledge takes place in a global competitive climate. Experience also shows that making the projects well defined both in terms of content and time but open in terms of what specific type of technical solutions should be aimed at, limits the negative impact on competition.

While the evaluation of research is important in science policy there are similar general policy tools that are useful when designing and redesigning technology policy. Technology forecasting is a way of capturing new technological trends. Asking leading experts among scientists and among the most advanced producers and users about what technologies are rising on the horizon helps to scout the next generation of "strategic technologies." In order to limit the capturing of public interest by private firms, independent policy evaluation of specific initiatives may be useful. Many evaluations end up addressing users of the programs with questions about the efficacy of the program. Not surprisingly, such studies often end up reporting that the program was very good and that more of the same would be welcome. In this situation, as in many other situations, where too much agreement among partners threatens to become a lock-in, it should be considered whether to give "outsiders" a strong role as evaluators. It is as important for public policy as it is for science-based firms to promote "job rotation" and "interfunctional teams."

As pointed out, science and technology policy are ideal types, which serve our broad analytical purposes. In the real world of advanced capitalist economies, however, the policy focus, instruments and actors involved in science and technology policy-making are not always easily grouped in one or the other of these categories. As we will examine now, innovation policy takes a step further by bringing in an even broader set of policy issues.[1]

22.2.3 . . . and to Innovation Policy

Innovation policy appears in two different versions. One—the laissez-faire version—puts the emphasis on non-interventionism and signals that the focus should be on "framework conditions" rather than specific sectors or technologies. This often goes with a vocabulary where any kind of specific measure gets grouped under the negative heading "picking the winners." The extreme version of this type of innovation policy is one where basic research and general education are seen as the only legitimate public activities and intellectual property right protection as the only legitimate field for government regulation. In more moderate versions public initiatives aiming at fostering "entrepreneurship" and promoting a positive attitude to science and technology in the population may be endorsed.

The other version may be presented as the "systemic" version and by referring to the concept of "innovation system." This perspective implies that most major policy fields need to be considered in the light of how they contribute to innovation. A fundamental aspect of innovation policy becomes the reviewing and redesigning of the linkages between the parts of the system. The first approach is built upon the standard assumption made in economics that firms always know what is best for them and that they normally (in the absence of market failure) act accordingly. The second perspective takes into account that competence is unequally distributed among firms and that good practice in terms of developing, absorbing and using new technology is not immediately diffused among firms; and that "failures" may extend beyond neoclassical "market failure" to subsume "failures" of institutions to coordinate, link, or address various systemic needs, etc. (see Ch. 7 by Edquist in this volume).

Both of these approaches cover all aspects of the innovation process—including diffusion, use and marketing of new technologies—and in a sense they may be seen as an important form of "economic policy" where the focus is more on innovation than on allocation. Both tend to put stronger emphasis on "institutions" and "organizations" than do science and technology policy. In the laissez-faire version, the predominance of the market and of competition becomes the most important prerequisite for innovation—there is in principle one single recommendation for institutional design valid for all countries.

In the systemic approach the importance of competition is recognized but so is the need for closer cooperation vertically between users and producers and sometimes even horizontally among competitors when it comes to develop generic technologies. In the systems approach it is recognized that the institutional set-up differs across national economies and that this has implications for what types of technologies and sectors thrive in the national context. To design a suitable innovation policy requires specific insights in the institutional characteristics of the national system.

Innovation policy does not imply any a priori preference for high versus low technology. The systems approach introduces a vertical perspective on the industrial system, seeing it as a network and as value chains where certain stages might be more suitable for firms in a specific country.

The respective theoretical foundations of the two different versions of innovation policy are (1) an application of standard neoclassical economics on innovation, and (2) a long-term outcome of research on innovation and economic evolution (Metcalfe 1995; Metcalfe and Georghiou 1998). The innovation system approach may be seen as bringing together the most important stylized facts of innovation. It makes use of empirical material and analytical models developed in innovation research, as well as in institutional and evolutionary economics.

The major reason for innovation policy becoming more broadly used as a concept was the slow-down in economic growth around 1970 and the persistence of sluggish growth as compared to the first post-war decades. The reasons for the slow-down in the growth in "total factor productivity" were, and still are, not well understood but there was a feeling that it had to do with a lack of capability to exploit technological opportunities. At the same time, the restrictions imposed on general economic policy by fear of inflation made it important to understand the possibilities to promote growth from the supply side.

This implies that the major objectives of innovation policy are economic growth and international competitiveness. In the European Union discourse these objectives are combined with "social cohesion" and equality. Innovation might also be seen as a way to solve important problems relating to pollution, energy, urbanism, and poverty. But the main focus is on the creation of economic wealth.

Among the instruments to be used are the regulation of intellectual property rights and access to venture capital. One fundamental distinction in innovation

Box 22.2 The neoclassical economics of innovation policy

According to neoclassical economics a necessary condition for public policy intervention is market failure. If markets can do the job there is no need to intervene. Market failure may have different causes but the ones most often raised in the context of innovation policy are lack of incentives to invest in knowledge production. Knowledge tends to be seen as a public good from which it is difficult to exclude others and also as being non-rival since its user value may not suffer from the fact that others use it. When knowledge is rival but non-excludable, intellectual property rights can be guaranteed and enforced by governments. When knowledge is non-excludable and non-rival governments should subsidize knowledge production addressed for public use or take charge of producing the knowledge by itself.

The problem with this analysis is not the conclusions reached. There are certainly good reasons for governments to support knowledge production and innovation in the ways referred to—actually we have seen that most of these instruments were taken into use long before neo-classical economics was established. The problem is that the argument for support comes from a theory based upon assumptions that are incompatible with a dynamic economy where innovation is a widely spread and ongoing process.

On rationality, markets, and competition
Innovation research has demonstrated that innovation is a ubiquitous phenomenon in the modern economy. In such an economy the idea that the "representative firm" can operate on precise calculations and choose among well-defined alternatives is a dubious abstraction. The point, made by Kenneth Arrow and others, is that innovation by definition involves fundamental uncertainty. Or as Rosenberg puts it "it is not possible to establish the knowledge production function" since the output is unknown (Rosenberg 1972: 172).

To assume that agents know what there is to know and to disregard competence building tends to miss what is at the very heart of competition in a learning economy. The assumption that markets are "pure" with arm's-length and anonymous relationships between producers and users is logically incompatible with the fact that a major part of innovative activities aims at product innovations. The only solution to the paradox is that real markets are organized and constitute frameworks for interactive learning between users and producers (Lundvall 1985).

policy lies between initiatives aiming at promoting innovation within the institutional context and those aiming at changing the institutional context in order to promote innovation. The first category overlaps with instruments used in science and technology policy. The second may include reforms of universities, education, labor markets, capital markets, regulated industries, and competition laws.

One fundamental question is how far these other policy fields can and should be adapted to the needs to promote innovation. Universities and schools obviously have other missions than to support innovation and economic growth. And the ministries and authorities in charge of them are normally acting according to a different logic than the one implied by a focus on innovation. One interesting example is labor market policy where it has been assumed that what matters is flexible markets where flexibility refers both to workers' mobility and to flexible wages. To think about how to design labor market institutions so that they promote competence building and innovation might lead to different conclusions regarding what is a "well-functioning labor market."

A second issue is about the limits for public sector intervention. Assume, for instance, that it can be demonstrated that the way firms organize themselves internally has a major impact on innovation performance and that economic growth can be explained by differences in this respect. Is there a role for governments in promoting the diffusion of good practices in this respect or should it still be left to management and owners to cope with such problems? It is not self-evident that government should help to diffuse technical solutions while keeping away from supporting the diffusion of more efficient organizational solutions.

As Figure 22.1 indicates, the elements of the innovation system still include universities, research institutions, technological institutes, and R&D laboratories. However, the focus of policy moves from universities and technological sectors, as in science and technology policies, toward all parts of the economy that have an impact on innovation processes. This is the reason why, in Figure 22.1, the instruments of innovation policy are also those of science and technology policy. Innovation policy pays special attention to the institutional and organizational dimension of innovation systems, including competence building and organizational performance. Innovation policy calls for "opening the black box" of the innovation process, understanding it as a social and complex process.

Ministries of economic affairs or ministries of industry may be the ones playing a coordinating role in relation to innovation policy but in principle most ministries could be involved in efforts to redesign the national innovation system. This is actually the case for some specific countries, which have experienced a truly "innovation policy turn" since the late 1990s, like for example Finland, The Netherlands, and Denmark (Biegelbauer and Borrás 2003). Developing an interaction and dialogue on policy design between government authorities on the one hand and the business community, trade unions and knowledge institutions on the other is a necessary condition for developing socially relevant and clear policy programs that can be implemented successfully.

The analytical basis of innovation policy could be a combination of general insights about what constitutes good practice, given the global context in terms of technology and competition, with specific insights in the characteristics of the national innovation system. The system can be analyzed in terms of specialization, institutional set-up, and insertion in the global economy. Attempts may be made to locate strengths, weaknesses, threats, and opportunities through intelligent benchmarking. Missing links and links constituting lock-ins may be especially important

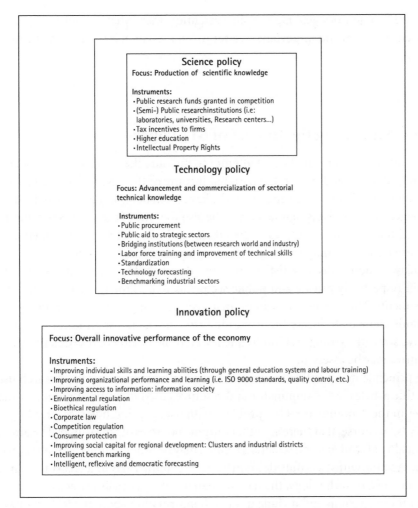

Figure 22.1 Relationship between science, technology, and innovation policy

to locate. Human resource development and use is another important dimension. Finally, innovation policy analysis must increasingly give attention to the international dimension.

22.3 STI POLICY EVALUATION AND IMPACT MEASUREMENT

It is characteristic for the evolution of new policy fields that measurement and quantitative guidance gives more legitimacy to the field. Within the field what can be measured sometimes gets more policy attention than what cannot be measured. A special problem and important area for further research is to measure the impact of policy.

22.3.1 Measuring the Impact of STI Policy

In the 1930s, Bernal made the first attempt to measure the effort made in science by relating R&D expenditure to the national income of the UK. In the late 1950s and early 1960s, Christopher Freeman played a key role in developing the analytical basis of science policy and it is significant that he also was one of the architects behind the Frascati manual that in 1963 gave the OECD and national authorities methods to measure R&D and compare the effort across countries (OECD 1963b).

Today, national R&D statistics are quite detailed. They show the effort made within respectively private and public sectors as well as the financial source of the investments. The expenditure can be analyzed according to purpose. Bibliographic methods allow us to locate the scientific fields in which a specific country has its relative strength—using citation frequency even the quality of research in different countries may be assessed.

In principle it is possible to construct productivity measures for research using scientific articles in the nominator and resources used in terms of money or manpower in the denominator. One problem with using such crude measures to guide policy is, of course, that there are other outputs not so easily measured. The amount and quality of students and scientists trained may be brought into the analysis, while the interaction with users outside knowledge institutions may be less easy to quantify.

In the field of technology, the data on patents are especially attractive since they exist for long periods and include quite rich information about the technology and agents taking out the patent. Patent statistics can be used to compare national systems in terms of technological specialization—as revealed technological advantage—and it might even be possible to distinguish between more or less important patents using citation patterns.

However, it has to be taken into account that patents play very different roles in different sectors and in some (e.g. pharmaceuticals and biotechnology) they might be more relevant when it comes to judging performance than they are in others (e.g. software and service production). The major use of patent statistics may therefore be to help mapping the evolution of national innovation systems rather than as performance indicators to judge the efficiency of technology policy.

The systemic view of the innovation policy means that these previous measurements are necessary but far from sufficient to investigate the innovative performance of an economy. The Oslo manual for gathering information and data on innovation, which was agreed upon in 1990, is an important step in this direction. In Europe, the

Community innovation surveys have been collected several times in most of the member states (see Ch. 6 by Smith in this volume).

Among the more interesting information that can be obtained through these surveys is the share of new products in total sales in firms in different sectors and countries. This is a measure of diffusion of product innovations in the economy and may be seen as an important intermediate performance variable. Another field where the diffusion of technology has been mapped quite thoroughly is in relation to information and communication technologies. These indicators are important since for innovation policy one major performance indicator should be the diffusion and effective use of new technologies.

The most important remaining tasks for building indicators to support innovation policy relate to the diffusion of process innovations, innovation in services, organizational innovations—and their diffusion—and, finally, to experience-based learning. Even with better indicators in these fields, we cannot expect to get very simple and clear conclusions from quantitative evaluation exercises.

Therefore case studies bringing together qualitative and quantitative information and dialogue with policy practitioners will remain important sources of insight when designing policy. Richard Nelson has more than any other scholar developed this approach (Nelson, Peck, and Kalachek 1967; Nelson 1982, 1984, 1988, 1993).

Box 22.3 Innovation systems and innovation policy

Innovation systems is not an economic theory in the same sense as neoclassical or evolutionary economics, but the concept integrates theoretical perspectives and empirical insights based on several decades of research.

Innovation is seen as a cumulative process that is path-dependent and context-dependent. This is why innovation policy needs to build upon insight in a specific context and why "best-practice" cannot be transplanted from one innovation system to another. Innovation is also seen as an interactive process. The competence of single innovating firms is important but so is the competence of suppliers, users, knowledge institutions, and policy makers. The linkages and the quality of interaction is important for outcomes. This is why innovation policies that focus on subsidizing and protecting suppliers of knowledge at best are incomplete—at worst they increase the gap between technological opportunities and absorptive capacity. At least the same attention needs to be given to users and to linkages.

Innovation systems may be seen as frameworks both for innovation and for competence building. Competence building involves learning and renews the skills and insights necessary to innovate. Innovation processes are processes of joint production where innovations and enhanced competence are the two major outputs. Learning takes place in an interaction between people and organizations. The "social climate" including trust, power, and loyalty contributes to the outcome of learning processes. This is why innovation policy needs to take into account the broader social framework even when the objective is to promote economic wealth creation.

Not least, in the era of innovation policy, where institutions matter more than ever, it is difficult to see how quantitative analysis could stand alone as the basis for policy.

22.3.2 Evaluating STI Programs and Policies

With the growth of STI policies and programs, public authorities have been increasingly interested in evaluating the effects and impacts of public expenditure in these areas. Evaluation is the systematic assessment of programs or related public expenditures in terms of how far they have attained their goals. Evaluation should be considered as an element in a political process, namely, when public administrations try to elaborate conclusions and lessons from past performance in order to become better in the future, or to decide upon the fate of the activity in question. The evaluative exercise is typically conducted by external and independent actors, who use a range of methodologies, including the self-assessment of those persons involved in the implementation of the program. There are as many evaluation methodologies as evaluators, and as many policy styles as public administrations.

Several authors emphasize that evaluating STI policies and programs is particularly difficult given the wide effects throughout the system. It has been argued, for example, that micro-level evaluations (program-specific) are more reliable than macro-level evaluations where issues such as whether a specific program or policy enhanced the competitiveness of an economy are almost impossible to determine (Luukkonen 1998). Similarly, it has been pointed out that most programs have important effects beyond their strict initial goals. This is the case when STI policies

---

Box 22.4 Normative principles for design of STI policies

Robustness: Decisions and social structures should withstand the occurrence of different future scenarios.

Flexibility: In the occurrence of sudden socio-economic change institutions must be able to change direction rapidly.

Internal diversity: Structurally dissimilar characteristics must be built in to allow survival if the selection environment changes.

External diversity: Variety of links to different kinds of agents will help adaptation when change in the environment arises.

Window of opportunity: Attention to timing and sequence in face of path-dependent systemic context.

Incremental approach: The whole can be changed only through the cumulative impact of small steps.

Experimentation and prudence: new policy ideas should be submitted to trial in localized contexts before full deployment.

Source: Sandro Mendonça

have helped the creation of standards, have induced more risk-taking attitudes of innovators, have fostered long-term rather than short-term strategies in firms' research, and have enhanced the acquisition of new skills and knowledge (Peterson and Sharp 1998). This has also been framed as the problem of attribution. Closely related is the problem of synchronizing the evaluators' time horizon with the political time horizon of the political consumers of their work. Big programs' effects may not be realized for years if not decades.

22.4 STI Policy in the US, Japan, and Europe

We have used OECD documents to organize a stylized presentation of the evolution of science, technology and innovation policy. But while what is discussed at OECD is one thing, what national governments actually do to affect science, technology and innovation is quite another. No country has focused on just one of the kind of policies described above. All countries have combined elements of science, technology and innovation policy. But the mix and the policy design has been quite different between countries. Here we will try to capture the most basic characteristics for respectively the US, Japan, and Europe—understood as the big countries in Europe and as the European Union. Finally, the future challenges for each of them will be discussed.

22.4.1 Public Mission Technology Policy in the US

We have already mentioned the crucial Vannevar Bush report published in 1945 entitled "Science: The Endless Frontier." The neglect of one of its major recommendations had a very important effect upon the evolution of technology policy. Bush recommended the establishment of a coordinating authority at the national level, "the National Research Foundation." But it took another five years before "the National Science Foundation" was set up. In the meantime, different sector authorities in charge of contracting out research on nuclear power, defense, space, and health had already established ambitious research programs of their own and the total resources of NSF never came close to the budgets of these sector-oriented activities (Mowery 1994).

Technology policy in the US may therefore be seen as being organized in parallel *industrial complexes*—vertically organized complex networks crossing disciplines, technologies as well as industrial sectors, but with user interests in the public and private sector driving and defining the policy. These complexes were operated with little coordination between them and they went all the way from procurement of specific technical systems and components to the support of basic research and research training. In particular, the defense budget was used for promoting activities with little direct connection to short-term military needs. Research in computer science and software got substantial support from this budget and produced quite generic knowledge, in addition to producing short-term solutions to military needs (Langlois and Mowery 1996).

One major issue in relation to the public mission technology policy is the lack of coordination. One negative aspect is the resulting bias in the direction of military and space expenditure. It seems as if it is much easier to mobilize taxpayers' money and development efforts to go to the moon than it is to solve the problem of the ghetto (Nelson 1977). In part, this reflects the presence of well-organized lobbies and interest groups among private sector beneficiaries; it also reflects the sheer difficulty, noted by Nelson, of "solving" such intractable problems as urban poverty or primary educational achievement. Another problem is that the calculation of expected costs and benefits of specific projects may either be completely neglected because national pride is at stake or be systematically biased downwards to make a project look more attractive. A third weakness may be a one-sided focus on the development of science-based industries and technologies and a general assumption that science and technology is a quick fix for all kinds of problems. More mundane industries aiming at consumers, and problems where the solutions are less glamorous in technological terms, may be neglected.

But the US-type national innovation system certainly has its strong points beyond its sheer scale advantages. The fact that several of the agencies are prepared to finance more or less generic research and research training while still having the use of research in mind tends to overcome the bias toward the supply side. It tends to establish a "chain-linked model of innovation" with strong feedback elements (Kline and Rosenberg 1986). One strong element is diversity. The fact that different agencies "compete" in funding good research may be beneficial in supporting the diversity of research efforts. The coexistence of big private foundations that commit resources to research teams on the basis of their track record also contributes to diversity. In these respects current European centralistic initiatives in research policy in connection with the sixth framework program and the European Research Era could learn from the US, which benefits not only from scale *but perhaps even more from* diversity.

22.4.2 Sector Technology Policy in Japan

While the international competitiveness of US firms has been a concern in US technology policy, it has, with few exceptions, been an implicit rather than explicit objective. This is different when we look at technology policy in Japan. More than perhaps any other market economy Japan has made use of an explicit national policy to promote specific sectors and industries—with the ultimate aim of stimulating economic growth and competitiveness. According to Freeman (1987), the radical character of the policy had to do with the fact that since the mid-1950s policy ended up being designed by experts with an engineering background in the Ministry of Technology (MITI) who were much less concerned with "comparative advantage" than the economists in the Bank of Japan.

But the MITI has not acted alone. In the area of telecommunications, NTT—a public company in monopolistic control of telecommunication services—has played an important role in coordinating technology development efforts in major electronic firms such as Hitachi and NEC. The policies pursued have not involved massive subsidies and, actually, the public sector has been much less involved in financing R&D in the private sector than has been the case in the US (Nelson 1984).

The strategic promotion of the car industry, consumer electronics, and "mega-tronics" typically combined different policy instruments such as subsidies to research and development of generic technologies with elements of "infant-industry" protection. Bringing together competing firms in consortia aiming at solving common problems has been an important role for MITI. This has been done on the basis of attempts to map new trends in technology and markets through, for instance, technology forecasting.

One interesting aspect of MITI's technology policy has been that it did not focus exclusively on high technology sectors. For instance, consortia initiated by MITI aiming at promoting the modernization of textile and clothing industries brought together firms producing textiles and textile machinery with electronics firms.

22.4.3 European STI Policies: From the Promotion of National Champions to EU Framework Programs

Europe is certainly a much more diverse region than the US and Japan. University systems are different in the UK, France, and Germany. The role of engineering in industry and corporate governance differ as well. Bringing less rich countries such as Portugal and Greece into the picture makes it even less homogeneous in terms of R&D efforts and innovation styles. Therefore, to treat Europe as one region and compare it with the US and Japan in terms of, for instance, R&D effort or patenting, without being explicit about the dispersion of the variable, is not helpful.

The formation of a common European approach to science and technology is still evolving. Figure 22.2 indicates the elements of the European scientific and technological architecture, within and outside the functional–administrative borders of the EU. As can be seen, some international research organizations were established in the 1950s, whereas the main thrust came in the 1970s. This figure does not include the European standardization bodies (ETSI, CEN, and CENELEC[2]) and the European agencies granting intellectual property rights (like EPO, OHIM, and Community Plant Variety Office[3]), which have also played important roles in promoting scientific, technological, and innovative synergies.

CERN is the oldest and one of the most successful organizations, and is dedicated to nuclear research. Its undeniable scientific success contrasts sharply with the troubles of its "twin," the JRC (Joint Research Center), under the European Communities and also dedicated to nuclear energy research. Disputes between France and Germany about nuclear reactor designs in the 1960s weakened this institution, until the launch of the Joint European Torus (JET) concerned with nuclear fusion (Guzzetti 1995). Other European but non-EU scientific and technological organizations emerged in the 1960s and 1970s, among them the European Southern Laboratory (Astronomy), the European Molecular Biology Laboratory (EMBL), the European Space Agency (ESA), the European Science Foundation (ESF), gathering national research councils, and Airbus, a public–private enterprise with heavy investment in aviation technology. All these bodies are internationally constituted and directed toward one specific scientific–technological field.

In 1971, COST (Cooperation in Science and Technology) was established as an intergovernmental program. The novelty with COST was that it covered several

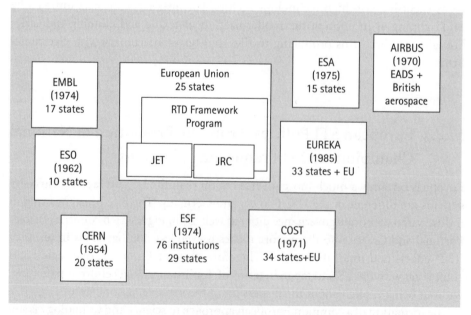

Figure 22.2 The scientific and technological architecture of Europe, 2001

Source: Borrás (2003)

scientific areas, and envisaged a very flexible form of cooperation. Later on, EUREKA developed this concept further, and became a successful tool of European collaboration outside the EU, mainly due to the high degree of participation by firms and to its market orientation.

In the 1980s attempts to build strength in ICT technology were still national in France, the UK and Germany and the basic strategy was to promote national champions. The success was limited and this was one important reason why the ESPRIT program was developed in the early 1980s, under the realm of the EU. ESPRIT was inspired by Japanese technology policy style and its actual design came out of intense consultation of the EU commission with the top leaders of the fifteen biggest European ICT firms (Peterson 1991).

The first EU framework program was established in 1984 as a multiannual, multisectoral (covers several scientific fields) and multinational program (grants funding to projects submitted by researchers of at least three EU countries). The most recent framework program (the sixth) is very ambitious in terms of promoting European-wide networks of excellence, as a means of creating a European Research Area, reducing national barriers. Furthermore, in 2002 European ministers of science set as their ambitious goal that the share of R&D should reach 3 per cent of GNP in member states (2 per cent private and 1 per cent public as the rule of thumb). This should be seen in the light of declarations made by prime ministers at the Lisbon Summit held in 2000, that Europe should become the world's most competitive knowledge economy by 2010, with social cohesion as a twin goal.

In general, it is a problem that the European construction calls for dramatic declarations to build support for European STI policy. It contrasts with Japan and the US where less is said and more is done. The share of total R&D expenditures that the commission distributes to member states is still quite small. Another problem is that the Brussels administration tends to take on more than it can master in terms of defining research programs, evaluating applications and administering projects; the result being that it is quite demanding to be a project coordinator for EU projects. Last, but not least, another major problem is that the general idea behind the European Research Area is scale, rationalization, coordination, and concentration of effort. Much less weight is given to the dimensions of diversity and competition, two key elements for successful innovation systems (Lundvall and Borrás 1998; Borrás 2003).

The framework programs have been used as instruments to promote European integration and there is no doubt that the programs have had an enormous effect in terms of building research collaboration of a lasting kind across Europe. And in spite of administrative problems the money is still regarded as attractive. This is especially the case in countries that have very little alternative free funding such as France and in countries where the European efforts have been used to reduce national efforts such as the UK.

22.4.4 The Challenges for the US, Japan, and Europe

There is a tendency to glorify the innovation system and the innovation policy of the country doing especially well in international competition. In the beginning of the 1990s Japan was seen as a model and Europe as a second-best alternative, while the US was seen as a threatened but big power in the field of technology. Today the roles have been reversed. The truth of the matter is that each system has its own strengths and weaknesses and that these do not go away in periods of rapid economic growth.

The US economy could probably do even better if it invested more broadly and more equally in human resources. For instance it is remarkable that the unwillingness of citizens in California to pay taxes has resulted in students in schools in Silicon Valley having more limited access to computers than students have in the Nordic countries. The work organization in industry with a strong division between trained and unskilled workers—and the resulting income gaps—probably makes the diffusion and use of new technologies less efficient than would a more even distribution of competence. We believe that for the US these are some of the major challenges in the broad field of innovation policy.

One major weakness of the Japanese system of innovation has to do with science policy. The universities do not have the same incentives and traditions for promoting high quality research. In Europe, the major challenge might be calling for a combination of science and technology policy. The so-called European paradox—that Europe is doing well in science while being weak in technology—might be somewhat off the mark due to its high intra-EU disparities (Pavitt 1998). Rather, Europe seems to be weak both in science and technology in some of the most rapidly growing fields and markets—not least within biotechnology and pharmaceuticals.

22.4.5 The Triad Game

As long as the Cold War was a reality, the existence of a common enemy supported scientific collaboration between US, Europe, and Japan. Given the current predominant unilateral approach of the US and the attempts to build a coherent European research area there are growing risks for conflicts within the triad regarding STI policy. There is especially a risk that access to knowledge is used as a political instrument as it was used by England for 500 years against the Catholic states in Europe. Since more and more technologies can be argued to have military relevance this could become a major problem for the global knowledge society.

The growing emphasis on the protection of intellectual property rights in relation to trade and extension to software and services may establish great obstacles for the catching-up countries and reduce their possibilities of pursuing the successful strategies of Japan and NICs. An interesting question, in this context, is whether

Europe and Japan are prepared to compete with the US in becoming centers for global interaction with "the rest." The more positive scenario would be one where the three poles used their capacities to promote the diffusion and use of knowledge to less privileged parts of the world. In any case, the extent to which there is a true globalization of knowledge production and diffusion, and the different forms it assumes, is a topic that has started to be analyzed (Archibugi and Michie 1997; Georghiou 1998), but deserves more attention.

22.5 CONCLUSIONS

It should be clear from what has been said that the most pressing issues on the policy agenda are specific to each national system. Even so, there are some issues that are common for all countries, and these represent new research challenges for scholars in this field. We have presented a stylized sequence, beginning with science and ending with innovation policy. In the most recent debates about the learning economy and the knowledge-based society we can see the contours of a new policy that we might call "knowledge policy."[4] It recognizes that innovation and competence building involve many different sources of knowledge and that innovation itself is a learning process. This raises the need for new analytical efforts and for rethinking the organization and implementation of policy in several respects.

It is increasingly important to understand better the connection between science and technology on the one hand and economic performance on the other. The rise and fall of the new economy demonstrates that assumptions about simple and direct connections are problematic. Between the new technologies and the performance of the economy the organizational characteristics of innovation systems and firms including "slippery" elements such as "social capital" affect the impact. This is an issue that remains understudied. In terms of public policy there is a need for innovative thinking about how governments can support the diffusion of good and sustainable practices in cooperation with management and employees. In terms of research opportunities, this links with the importance and need to devote more analytical efforts examining how technical innovation interacts with organizational change. The academic traditions of business organization and of innovation systems' research have to come closer to each other in order to answer questions regarding how organizational change affects innovation processes in the economy.

A second issue is about aggregate demand. In a period of growing fears for deflation and with little room left for expansionary monetary policy it might be useful to reconsider what Keynesian policy could mean in "a knowledge-based economy." Establishing large-scale technological–military programs (like the Star Wars programs under the Reagan and Bush administrations) may perhaps be seen as the modern version of building "pyramids"; however, other more socially oriented and low-scale options could also be considered. Establishing firm- or sector-specific funds through tax exemptions for upgrading the skills of all categories of employees and for making extra development efforts in periods of low economic activity could be one option.

A third common concern in the era of innovation policy is how to coordinate policies affecting innovation. The prevailing institutional set-up means that ministries of finance are the only agencies taking on a responsibility for coordinating the many specialized area policies. Area-specific ministries, on the other hand, tend to identify the interests of their own "customers" and take less interest in global objectives of society. It could be decided to establish new types of institutions such as cross-sector and interdisciplinary *Councils on Innovation and Competence Building* at the subnational and national level (in Finland the prime minister is chairman of a National Council of Science and Technology). This should be complemented by the much-needed research efforts towards developing sophisticated measuring methods about the innovation system trends and about the impact of STI policy on it. More advanced innovation indicators would be a crucial input for such holistic perspective of public authorities.

In 1961, the OECD expert group—Freeman, Svennilsson, and others—presented a kind of manual for how to design science policy in such a way that it became integrated with economic policy and provided a real impact on economic growth. Perhaps a suitable closing of this chapter might be to produce a similar set of recommendations for designing national innovation policy? But the message here is that there is no way to design an effective innovation policy without analyzing the domestic innovation system, including the way it produces and reproduces knowledge and competence, and comparing it with others. The stage of development and the size of the respective economy will affect the resulting plan of action. In small countries and developing countries the structures and institutions that affect absorption and efficient use of technology are more important to understand and act upon than those promoting the production of the technologies at the front. Big countries will necessarily be more focused on the production of the new technologies, but they too would have much to gain from taking into account the absorption and efficient use of innovations and new knowledge.

Notes

1. Ergas has suggested the distinction between "mission-oriented" and "diffusion-oriented" policy designs, based respectively on the massive support to a small number of science–technology fields, and on supporting the scientific–technological infrastructure (Ergas

1987). This analytical distinction has also been used when comparing policy styles in science and technology policies across countries. (It has been argued that the Japanese and French policies are mission-oriented, whereas the German policy has been depicted as an example of a diffusion-oriented policy style.) Again it is important to note that what is referred to is ideal types and that simple groupings of this kind might miss some of the most important complexities in the set-up of national systems of innovation.

2. ETSI: European Telecommunications Standards Institute; CEN: Comité Européen de normalisation; and CENELEC: Comité Européen de Normalisation Electrotechnique. These three bodies have divided their standardization activities by technological sectors.

3. EPO: European Patent Office; OHIM: Office for the Harmonization of the Internal Market; dedicated respectively to patents, and to trademarks and other intellectual property rights.

4. The concept of knowledge policy has evolved in connection with European policy making and as a follow-up to the Lisbon ministerial meeting 2000. One of the architects behind the Lisbon-strategy, Maria Rodrigues, defines knowledge policies as "policies aimed at fostering and shaping the transition to a knowledge-based society" (Conceçãio, Heitor, and Lundvall 2003: p. xx).

REFERENCES

ANDREASEN, L., CORIAT, B., DEN HARTOG, F., and KAPLINSKY, R. (eds.) (1995), *Europe's Next Step—Organisational Innovation, Competition and Employment*, London: Frank Cass.

ARCHIBUGI, D., and MICHIE, J. (1997), *Technology, Globalisation and Economic Performance*, Cambridge: Cambridge University Press.

—— and LUNDVALL, B.-Å. (eds.) (2000), *The Globalizing Learning Economy: Major Socio-economic Trends and European Innovation Policy*, Oxford: Oxford University Press.

AROCENA, R., and SUTZ, J. (2000), "Interactive Learning Flaces and Development Policies in Latin America," DRUID Working Paper no. 13.

ARROW, K. J. (1971), "Political and Economic Evaluation of Social Effects and Externalities," in M. Intrilligator (ed.), *Frontiers of Quantitative Economics*, North Holland, 3–23.

ARUNDEL, A., and SOETE, L. (eds.) (1993), *An Integrated Approach to European Innovation and Technology Diffusion Policy: A Maastricht Memorandum*, Brussels. CEC: Sprint.

BERNAL, J. D. (1939), *The Social Function of Science*, London: Routledge & Kegan Paul.

*BIEGELBAUER, P., and BORRÁS, S. (eds.) (2003), *Innovation Policies in Europe and the US. The New Agenda*, Aldershot: Ashgate.

BORRÁS, S. (2003), *The Innovation Policy of the EU*, Cheltenham: Edward Elgar.

CANTNER, U., and PYKA, A. (2001), "Classifying Technology Policy from an Evolutionary Perspective," *Research Policy* 30: 759–75.

COLEMAN, J. (1988) "Social Capital in the Creation of Human Capital," *American Journal of Sociology* 94 (Suppl.): 95–120.

CONCEIÇÃO, P., HEITOR, M., and LUNDVALL, B.-Å. (eds.) (2003), *Innovation, Competence Building and Social Cohesion in Europe: Towards a Learning Society*, Cheltenham: Edward Elgar.

* Asterisked items are suggestions for further reading.

COZZENS, S., HEALEY, P., RIP, A., and ZIMAN, J. (eds.) (1990), *The Research System in Transition*, Dordrecht: Kluwer Academic.

DALUM, B., JOHNSON, B., and LUNDVALL, B.-Å. (1992), "Public Policy in the Learning Society," in Lundvall 1992: 296–317.

*DODGSON, M., and BESSANT, J. (1996), *Effective Innovation Policy: A New Approach*, London: International Thomson.

DERTOUTZOS, M. L., LESTER R. K., and SOLOW, R. M. (1989), *Made in America*, Cambridge, Mass.: The MIT-Press.

EDQUIST, C. (ed.) (1997), *Systems of Innovation: Technologies, Institutions and Organizations*, London: Pinter.

——and HOMMEN, L. (1998), "Government Technology Procurement and Innovation Theory," *TSER-publication*, Linköping: University of Linköping, Department of Social and Technological Change.

————and TSIPOURI, L. (eds.) (2000), *Public Technology Procurement and Innovation*. London: Kluwer Academic.

ERGAS, H. (1987), "The Importance of Technology Policy," in P. Dasgupta and P. Stoneman (eds.), *Economic Policy and Technological Performance*, Cambridge: Cambridge University Press.

FAGERBERG, J. (1995), "Is there a Large Country Advantage in High-Tech?" NUPI Working Paper no. 526, Oslo: NUPI.

——(2002), *Technology, Growth and Competitiveness*, Cheltenham: Edward Elgar.

——and VERSPAGEN, B. (1996), "Heading for Divergence? Regional Growth in Europe Reconsidered," *Journal of Common Market Studies*, No. 34.

————and MARJOLEIN, C. (1997), "Technology, Growth and Unemployment across European Regions," *Regional Studies* 31(5): 457–66.

FRANSMAN, M. (1995), *Japan's Computer and Communications Industry*, Oxford: Oxford University Press.

FREEMAN, C. (1982), "Technological Infrastructure and International Competitiveness," Draft paper submitted to the OECD Ad hoc-group on Science, Technology and Competitiveness, August 1982, mimeo.

——(1987), *Technology and Economic Performance: Lessons from Japan*, London: Pinter.

*——(1992), *The Economics of Hope*, London: Pinter.

——(1995), "The National Innovation Systems in Historical Perspective," *Cambridge Journal of Economics* 19(1): 5–24.

——and LUNDVALL, B.-Å. (eds.) (1988), *Small Countries Facing the Technological Revolution*, London: Pinter.

——and PEREZ, C. (1988), "Structural Crises of Adjustment: Business Cycles and Investment behaviour," in G. Dosi et al., *Technology and Economic Theory*, London: Pinter, 38–66.

GEORGHIOU, L. (1998), "Global Cooperation in Research," *Research Policy* 27: 611–26.

GIBBONS, M., et al. (1994), *The New Production of Knowledge*, London: Sage.

GITTELMAN, M., and KOGUT, B. (2003), "Does Good Science Lead to Valuable Knowledge? Biotechnology Firms and the Evolutionary Logic of Citation Patterns," *Management Science* 49(4): 366–82.

GUZZETTI, L. (1995), *A Brief History of European Union Research Policy*, Luxembourg: Office for Official Publications.

HIRSCHMAN, A. O (1969), *The Strategy of Economic Development*, New Heaven: Yale University Press.

JASANOFF, S. (ed.) (1997), *Comparative Science and Technology Policy*, Cheltenham: Edward Elgar.

*—— (ed.) (2002), *Handbook of Science and Technology Studies*, 2nd edn., London: Sage.

KATZENSTEIN, P. J. (1985), *Small States in World Markets. Industrial Policy in Europe*, New York: Cornell University Press.

KLINE, S. J., and ROSENBERG, N. (1986), "An Overview of Innovation," in R. Landau and N. Rosenberg (eds.), *The Positive Sum Game*, Washington, DC: National Academy Press, 275–305.

KRISTENSEN, P. H., and STANKIEWICZ, R. (eds.) (1982), *Technology Policy and Industrial Development in Scandinavia*, Lund: Research Policy Unit.

KRULL, W., and MEYER-KRAMER, F. (eds.) (1996), *Science and Technology in Germany*, London: Cartemills.

KUENH, T. J., and PORTER, A. L. (eds.) (1981), *Science, Technology and National Policy*, London: Cornell University Press.

KUTZNETS, S. (1960), "Economic Growth of Small Nations," in E.A.G. Robinson (ed.), *Economic Consequences of the Size of Nations*, Proceedings of a Conference held by the International Economic Association, London: Macmillan.

LANGLOIS, R., and MOWERY, D. C. (1996), "The Federal Role in the Development of American Computer Software Industry: An Assessment," in D. C. Mowery (ed.), *The International Computer Software Industry*, New York: Oxford University Press.

LIST, F. (1841), *Das Nationale System der Politischen Ökonomie*, Basel: Kyklos (trans. and pub. as *The National System of Political Economy* by Longmans, Green and Co., London, 1841).

LUNDVALL, B.-Å. (1985), *Product Innovation and User–Producer Interaction*, Aalborg: Aalborg University Press.

—— (1988), "Innovation as an Interactive Process: From User–Producer Interaction to the National Innovation Systems," in G. Dosi, C. Freeman, R. R. Nelson, G. Silverberg, and L. Soete (eds.), *Technology and Economic Theory*, London: Pinter, 349–69.

—— (ed.) (1992), *National Innovation Systems: Towards a Theory of Innovation and Interactive Learning*, London: Pinter.

—— (1996), "The Social Dimension of the Learning Economy," DRUID Working Paper, No. 1.

—— (2002), *Innovation, Growth and Social Cohesion: The Danish Model*, Cheltenham: Edward Elgar.

—— and JOHNSON, B. (1994), "The Learning Economy," *Journal of Industry Studies* 1(2): 23–42.

*—— and BORRÁS, S. (1998), *The Globalising Learning Economy. Implications for Innovation Policy*, Brussels: European Commission.

—— and TOMLINSON, M. (2001), "Policy Learning by Benchmarking National Systems of Competence Building and Innovation," in G. P. Sweeney (ed.) *Innovation, Economic Progress and Quality of Life*, Cheltenham: Edward Elgar.

—— MENDONÇA, S. (2003), 'Política de Ciência, Tecnologia e Inovação: Princípios, tendências e questões,' mimeo, ISCTE University.

LUUKKONEN, T. (1998), "The Difficulties in Assessing the Impact of EU Framework Programmes," *Research Policy* 27(6): 599–610.

MARTIN, B., and SALTER, A. (1996), *The Relationship Between Publicly Funded Basic Research and Economic Performance. A SPRU Review* (July 1996). Report prepared for HM Treasury.

*METCALFE, S. (1995), "The Economic Foundations of Technology Policy: Equilibrium and Evolutionary Perspectives," in P. Stoneman (ed), *Handbook of the Economics of Innovation and Technological Change*, Oxford: Oxford University Press, 409–512.

—— and GEORGHIOU, L. (1998), "Equilibrium and Evolutionary Foundations of Technology Policy," *STI Review* 22: 75–100.

*MOWERY, D. C. (1994), "US Post-war Technology Policy and the Creation of new Industries," in OECD, *Creativity, Innovation and Job-Creation*, Paris: OECD.

—— and SAMPAT, B. N. (2001), "University Patents and Patent Policy Debates in the USA, 1925–1980," *Industrial and Corporate Change* 10: 781–814.

NATIONAL ACADEMY OF ENGINEERING (1993), *Prospering in a Global Economy*, Washington, DC: National Academy Press.

*NELSON, R. (1959), "The Simple Economics of Basic Economic Research," *Journal of Political Economy* 67: 323–48.

—— (1977), *The Moon and the Ghetto: An Essay on Public Policy Analysis*, New York: Norton.

—— (ed.) (1982), *Government and Technical Progress; A Cross Industry Analysis*, New York: Pergamon Press.

—— (1984), *High-Technology Policies: A Five-Nation Comparison*, Washington DC: American Enterprise Institute.

—— (1988), "Institutions Supporting Technical Change in the United States," in G. Dosi, C. Freeman, R. R. Nelson, G. Silverberg, and L. Soete (eds.), *Technology and Economic Theory*, London: Pinter, 312–29.

—— (ed.) (1993), *National Innovation Systems: A Comparative Analysis*, Oxford: Oxford University Press.

—— PECK, M. J., and KALACHEK, E. (1967), *Technology, Economic Growth and Public Policy*, Washington, DC: The Brookings Institution.

OECD (1963a), *Science, Economic Growth and Government Policy*, Paris: OECD.

—— (1963b), "Proposed Standard Practice for Surveys of Research and Development," Directorate for Scientific Affairs, DAS/PD/62.47, Paris: OECD.

—— (1971), *Science, Growth and Society*, Paris: OECD.

—— (1980), *Technical Change and Economic Policy*, Paris: OECD.

—— (1981), *Science and Technology Policy for the 1980s*, Paris: OECD.

—— (1991), *Technology and Productivity*, Paris: OECD.

—— (1992), *Technology and the Economy*, Paris: OECD.

—— (1994), *The OECD Jobs Study: Facts, Analysis, Strategies*, Paris: OECD.

—— (1996), *Transitions to Learning Economies and Societies*, Paris: OECD.

*—— (2000), *Knowledge Management in the Learning Society*, Paris: OECD.

—— (2001), *The New Economy: Beyond the Hype*, Paris, OECD.

OSTRY, S. (1996), "Policy Approaches to System Friction: Convergence Plus," in S. Berger and R. Dore (eds.), *National Diversity and Global Capitalism*, Ithaca and London: Cornell University Press.

PAVITT, K. (1995), "Academic Research, Technical Chance and Government Policy," in J. Krige and D. Pestre (eds.), *Science in the 20th Century*, Harwood Academic Publishers, 143–58.

—— (1998), "The Inevitable Limits of EU R&D Funding," *Research Policy* 27(6): 559–68.

*—— (2001), "Public Policies to Support Basic Research: What Can the Rest of the World Learn from the US Theory and Practice (And What they Should not Learn)," *Industrial and Corporate Change* 10(3): 761–79.

PEREZ, C. (2002), *Technological Revolutions and Financial Capital*, Cheltenham: Edward Elgar.

PERROUX, F. (1969), *L'Economie du 20eme siècle*, Paris: Presses universitaires de France.

PETERSON, J. (1991), "Technology Policy in Europe: Explaining the Framework Programme and Eureka in Theory and Practice," *Journal of Common Market Studies* 29(3): 269–90.

—— and SHARP, M. (1998), *Technology Policy in the European Union*, London: Macmillan.

PUTNAM, R. D. (1993), *Making Democracy Work: Civic Traditions in Modern Italy*, Princeton: Princeton University Press.

RIP, A., and VAN DER MEULEN, B. J. R. (1997), "The Post-Modern Research System," in R. Barré, M. Gibbons, J. Maddox, B. Martin, and P. Papon (eds), *Science in Tomorrow's Europe*, Paris: Economica International, 51–67.

ROMER, P. M. (1990), "Endogenous Technological Change," *Journal of Political Economy* 98: 71–102.

ROSENBERG, N. (1972), *Technology and American Growth*, New York: M. E. Sharpe.

ROTHWELL, R., and ZEGVELD, W. (1980), *Innovation and Technology Policy*, London: Pinter.

SHARP, M. (2003), "The UK-Experiment: Science, Technology and Industrial Policy in Britain 1979–2000," in P. Biegelbauer and S. Borrás (eds), *Innovation Policies in Europe and the US. The New Agenda*, Aldershot: Ashgate, 17–41.

SVENNILSON, I. (1960), "The Concept of the Nation and its Relevance to Economic," in E. A. G. Robinson (ed.) *Economic Consequences of the Size of Nations*, Proceedings of a Conference held by the International Economic Association, London: Macmillan.

WAGNER, M. F. (1998), *Det polytekniske gennembrud*, Aarhus: Aarhus Universitetsforlag.

WESTLING, H. (1991), *Technology Procurement: For Innovation in Swedish Construction*, Stockholm: Council for Building Research.

—— (1996), *Cooperative Procurement. Market Acceptance for Innovative Energy-Efficient Technologies*, Stockholm: NUTEK.

WOOLCOCK, M. (1998), "Social Capital and Economic Development: Toward a Theoretical Synthesis and Policy Framework," *Theory and Society* 27(2): 151–207.

YAKUSHIJI, T. (1986), "Technological Emulation and Industrial Development," Paper presented at The Conference on Innovation Diffusion in Venice, mimeo.

ZUCKER, L., and DARBY, M. (1998), "Intellectual Human Capital and the Birth of US Biotechnology Enterprises," *American Economic Review* 88(1): 290–306.

Index

Note: Bold entries refer to boxed features.